P-Stereogenic Ligands in Enantioselective Catalysis

RSC Catalysis Series

Series Editor:
Professor James J Spivey, *Louisiana State University, Baton Rouge, USA*

Advisory Board:
Krijn P de Jong, *University of Utrecht, The Netherlands*, James A Dumesic, *University of Wisconsin-Madison, USA*, Chris Hardacre, *Queen's University Belfast, Northern Ireland*, Enrique Iglesia, *University of California at Berkeley, USA*, Zinfer Ismagilov, *Boreskov Institute of Catalysis, Novosibirsk, Russia*, Johannes Lercher, *TU München, Germany*, Umit Ozkan, *Ohio State University, USA*, Chunshan Song, *Penn State University, USA*

Titles in the Series:
1: Carbons and Carbon Supported Catalysts in Hydroprocessing
2: Chiral Sulfur Ligands: Asymmetric Catalysis
3: Recent Developments in Asymmetric Organocatalysis
4: Catalysis in the Refining of Fischer–Tropsch Syncrude
5: Organocatalytic Enantioselective Conjugate Addition Reactions: A Powerful Tool for the Stereocontrolled Synthesis of Complex Molecules
6: *N*-Heterocyclic Carbenes: From Laboratories Curiosities to Efficient Synthetic Tools
7: *P*-Stereogenic Ligands in Enantioselective Catalysis

How to obtain future titles on publication:
A standing order plan is available for this series. A standing order will bring delivery of each new volume immediately on publication.

For further information please contact:
Book Sales Department, Royal Society of Chemistry, Thomas Graham House, Science Park, Milton Road, Cambridge, CB4 0WF, UK
Telephone: +44 (0)1223 420066, Fax: +44 (0)1223 420247,
Email: books@rsc.org
Visit our website at http://www.rsc.org/Shop/Books/

P-Stereogenic Ligands in Enantioselective Catalysis

Arnald Grabulosa
Department of Inorganic Chemistry, University of Barcelona, Barcelona, Spain

RSCPublishing

RSC Catalysis Series No. 7

ISBN: 978-1-84973-123-2
ISSN: 1757-6725

A catalogue record for this book is available from the British Library

Published by The Royal Society of Chemistry,
Thomas Graham House, Science Park, Milton Road,
Cambridge CB4 0WF, UK

Registered Charity Number 207890

For further information see our web site at www.rsc.org

Printed and bound in Great Britain by CPI Antony Rowe, Chippenham and Eastbourne

Preface

We felt strongly that, if one wanted to get high ee values, the asymmetry would have to be directly on the phosphorus. That is where the action is.

William S. Knowles, Nobel Lecture (December 8th, 2001)

There appears to be no end to the thirst for new phosphorus ligands for homogeneous catalysis. In enantioselective catalysis, in which chiral ligands are required, the first family to be explored was that of *P*-stereogenic phosphines (*i.e.* those that contain phosphorus atoms bound to three different substituents), with spectacular results. Despite that, these ligands were soon outnumbered and supplanted by others possessing other stereogenic elements further away from the phosphorus atom. Nowadays, the number of reported chiral non-racemic ligands is counted in thousands, and hundreds of them are commercially available. Often these ligands possess sophisticated scaffolds as a result of a careful optimisation process to adapt the ligand for a specific reaction and substrate. In contrast, even simple *P*-stereogenic ligands are still infrequent despite their long history. For a long time, lack of general preparative methods as well as misconceptions about their configurational stability have slowed down the development of the family of *P*-stereogenic ligands. In spite of that, the last 15 years have witnessed a renaissance of the field. This renewed interest has provided synthetic strategies with certain generality and even methods based on enantioselective catalysis. All that effort has given birth to dozens of new ligands, some of them displaying an exceptional performance in several reactions. At this point, it is clear that the ever-increasing need for enantiopure ligands can no longer forget phosphorus as a stereogenic element.

RSC Catalysis Series No. 7
P-Stereogenic Ligands in Enantioselective Catalysis
By Arnald Grabulosa
© Arnald Grabulosa 2011
Published by the Royal Society of Chemistry, www.rsc.org

The present book attempts to provide a comprehensive account of the preparation methods for enantiopure *P*-stereogenic ligands and give an overview of their performance in enantioselective homogeneous catalysis with transition metal complexes.

The book begins with a brief chapter on the generalities of *P*-stereogenic compounds: history, configurational stability and interconversions among them. The next five chapters describe the main preparative methods from resolution of racemates to enantioselective catalysis. The last two chapters are devoted to catalytic applications of *P*-stereogenic ligands. Chapter 7 describes the use of the ligands in catalytic hydrogenation and related reactions whereas chapter 8 deals with other reactions, mainly C–C bond forming reactions. The aim of those two chapters is to give an outline of the usefulness of the ligands in homogeneous catalysis. It is evident that the outcome of a catalytic run depends not only on the ligand but also on the substrate and reaction conditions. As only the best results are usually listed, comparisons should be done with utmost care.

Textbooks on enantioselective homogeneous catalysis focus on the application of ligands in catalysis, but not on their synthesis. This is understandable for space restrictions, but leaves the reader unaware of the rich chemistry involved in ligand preparation, especially of chiral non-racemic ligands. Furthermore, it compels the interested reader to go to primary literature to look for the preparation of a particular ligand, even when only a rough idea is sought. This book aims to overcome this difficulty for *P*-stereogenic ligands.

I have tried to make a modest contribution to the increasing interest in *P*-stereogenic ligands in catalysis. In spite of the inestimable efforts of the Royal Society of Chemistry, I am sure that this book contains errors and probably some missing references. These inaccuracies are of course mine alone. For this reason, I will greatly appreciate comments and suggestions to improve the manuscript.

For me, writing this book has been as demanding as it is rewarding now that I write these lines. Fortunately, many people have helped me in one or another way. I am firstly grateful to Prof. Guillermo Muller for introducing me to the research with *P*-stereogenic ligands during my PhD in Barcelona and to Dr Antonio Mezzetti for a very pleasant stay in Zürich. I am also indebted to Profs Philippe Gros and Marc Beley in Nancy and to Prof. Paul Kamer and Dr Matthew Clarke in St Andrews for trusting me as a postdoctoral researcher. Still in St Andrews, Ms Jacorien Coetzee, Dr Deborah Dodds, Ms Tanja Eichelsheim, Mr Jason Gillespie, Dr Gregorio Guisado, Dr Wouter Laan and Mr Michiel Samuels were very helpful by reading several parts of the manuscript and suggesting many changes that doubtlessly improved the quality of the book. Back to Barcelona, Ms Charo Aznar, Dr Zoraida Freixa and Dr Susanna Jansat also read with interest the whole manuscript whereas Edgar Segués, Edurne Arrondo and Joaquín Montón helped me with the cover of the book. My gratitude goes also to Dr Laia Haurie and Ms Gabriela Haurie, who allowed me not only to stay at their flat while I was finishing this book, but also to convert their flat into an office. Dr Merlin Fox, from the Royal Society of

Chemistry, has always been very helpful and supportive in the production of the book. Last, but not least, I owe my deepest gratitude to my family and to my parents in particular, who have always encouraged me to work hard to achieve my goals.

Arnald Grabulosa
Departament de Química Inorgànica, Universitat de Barcelona

Contents

RSC Catalysis Series No. 7
P-Stereogenic Ligands in Enantioselective Catalysis
By Arnald Grabulosa
© Arnald Grabulosa 2011
Published by the Royal Society of Chemistry, www.rsc.org

Abbreviations

Ac	acetyl
Ad	adamantyl
AIBN	azoisobutyronitrile
An	anisyl
aq.	aqueous
Ar	aryl group
Asc	ascorbate
BARF	*tetrakis*[3,5-*bis*(trifluoromethyl)phenyl]borate
BINAP	2,2′-*bis*(diphenylphosphino)-1,1′-binaphthyl
Biph	biphenylyl
Bn	benzyl
Boc	*tert*-butoxycarbonyl
BSA	*bis*(trimethylsilyl)acetamide
Bu	butyl
Bz	benzoyl
cat.	catalyst or catalytic
Chiraphos	2,3-*bis*(diphenylphosphino)butane
COD	1,5-cyclooctadiene
Cy	cyclohexyl
DABCO	1,4-diazabicyclo[2.2.2]octane
de	diastereomeric excess
dcpe	1,2-*bis*(dicyclohexylphosphino)ethane
DFT	density functional theory
DIOP	2,3-(isopropylidenedioxy)-2,3-dihydroxy-1,4-*bis*(diphenylphos-phanyl)butane
DME	dimethoxyethane
DMPP	3,4-dimethyl-1-phenylphospholane
dppe	1,2-*bis*(diphenylphosphino)ethane
dppp	1,3-*bis*(diphenylphosphino)propane
E	electrophile
EDC	1-ethyl-3-(3-dimethylaminopropyl)carbodiimide hydrochloride
ee	enantiomeric excess
equiv.	equivalent

Et	ethyl
Fc	ferrocenyl
HMDS	hexamethyldisilazane
HOBt	*N*-hydroxybenzotriazole
HPLC	high-performance liquid chromatography
i	iso
L	ligand
L*	chiral ligand
LDA	lithium *N*,*N*-diisopropilamide
LDBB	lithium 4,4'-di-*tert*-butylbiphenylide
Ln	lanthanide
M	metal
MCPBA	*m*-chloroperbenzoic acid
Me	methyl
Men	menthyl
Mes	mesityl (2,4,6-trimethylphenyl)
MPLC	medium-performance liquid chromatography
Ms	methanesulfonyl (mesyl)
n	normal
Naphth	naphthyl
NBD	norbornadiene
NMM	*N*-methylmorpholine
NMR	nuclear magnetic resonance
[O]	oxidative treatment
o.p.	optically pure
P*	chiral phosphine
Ph	phenyl
Phen	phenanthryl
Piv	pivalate
PTSA	*para*-toluenesulfonic acid
purif.	purified
py	pyridine
quant.	quantitative
R	general organic group
R*	chiral organic group
recr.	recrystallised
[Red]	reductive treatment
rt	room temperature
s	secondary
S	solvent
sc	*supercritical*
SPOs	secondary phosphine oxides
t	tertiary
TBAF	tetra-*n*-butylammonium fluoride
TBDMS	*tert*-butyldimethylsilyl
tert	tertiary

Tf	trifluoromethanesulfonyl
TFA	trifluoroacetic acid
THF	tetrahydrofuran
TMEDA	N, N, N', N'-tetramethylethylenediamine
TMS	trimethylsilyl
Tol	tolyl
Ts	4-methylbenzenesulfonyl (tosyl)
X	halide
Y	anion

CHAPTER 1

Introduction

1.1 Historical Overview

A century has passed since the discovery that compounds bearing tetrahedral sp^3 phosphorus atoms bound to different substituents (*P*-stereogenic compounds) are in general configurationally stable and can be separated into a pair of enantiomeric forms.[1] During the first 50 years, very few resolved *P*-stereogenic compounds were prepared (Figure 1.1).

In these early days, only quaternary phosphorus compounds were resolved. Ethylmethylphenylphosphine oxide (**1**) constitutes the first ever racemic *P*-stereogenic compound separated into individual enantiomers in 1911 by Meisenheimer and Lichtenstadt.[1] It took 15 years for the same laboratory to repeat that feat and apply the same method to resolve benzylmethylphenyl-phosphine oxide (**2**).[2] In 1944 sulfide **3** was resolved by Davies and Mann[3] and 15 years later McEwen and co-workers[4] resolved the first acyclic phosphonium salt (**4**).

In 1961 Horner and co-workers[5] were the first to isolate optically pure *trivalent* phosphines, proving that, unlike amines, they do not racemise at room temperature.[5,6] Soon afterwards, a few dialkylaryl- and triarylphosphines (**5–7**) were prepared in optically pure form (Figure 1.2).[5–11]

Those compounds were synthesised by alkaline hydrolysis or cathodic reduction of phosphonium salts or by reduction of phosphine oxides with silanes. In both cases, the optically pure precursor had to be prepared by chemical resolution (see Chapter 2, Section 2.2), a step that is time-consuming, inefficient and very dependent on the structure of the compound being resolved.

Two key breakthroughs occurred in 1967–1968. The first one was the development of a new, relatively flexible route for the preparation of phosphine oxides, based on the separation of asymmetrically substituted menthylpho-sphinates.[12,13] This method, described in Chapter 2, allowed the preparation of

RSC Catalysis Series No. 7

P-Stereogenic Ligands in Enantioselective Catalysis

By Arnald Grabulosa

© Arnald Grabulosa 2011

Published by the Royal Society of Chemistry, www.rsc.org

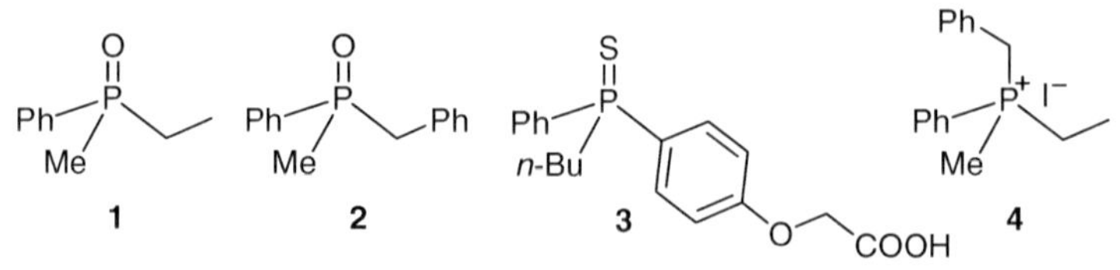

Figure 1.1 Examples of *P*-stereogenic compounds resolved between 1911 and 1960.

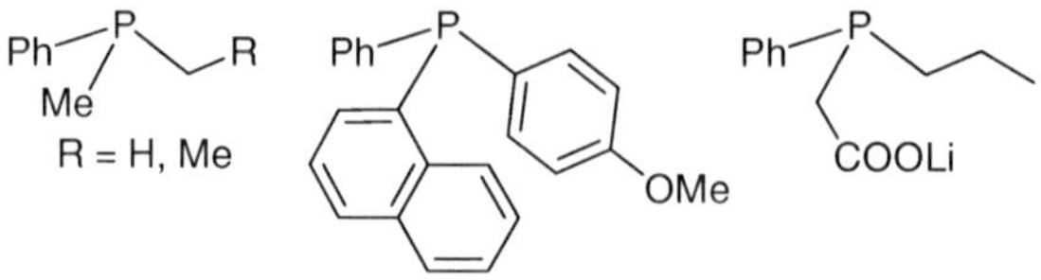

Figure 1.2 Examples of *P*-stereogenic phosphines resolved between 1961 and 1966.

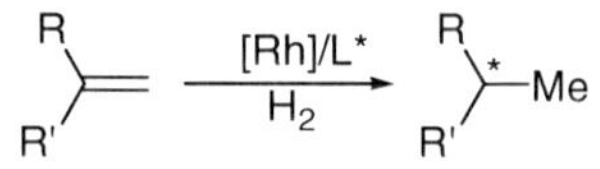

Scheme 1.1 First examples of enantioselective hydrogenation reactions (Table 1.1).

many optically pure phosphine oxides, which were reduced to the parent phosphines by known methods (see Section 1.3).

The second breakthrough was the discovery of enantioselective hydrogenation, which has had an enormous impact in homogeneous catalysis. In 1965 Wilkinson and co-workers[14,15] reported that a soluble Rh(I) complex, [RhCl(PPh$_3$)$_3$], was an excellent precatalyst for alkene hydrogenation under mild conditions. The research groups led by Knowles[16] and Horner[17] independently replaced the achiral triphenylphosphine ligands in the Wilkinson catalyst by optically enriched *P*-stereogenic phosphines. The chiral precatalysts were tested in the hydrogenation of prochiral alkenes (Scheme 1.1 and Table 1.1).

The optical yields of **9** were low, but proved that enantioselective catalytic homogeneous hydrogenation was feasible. The extremely low enantioselectivity with the phosphine of entry 5 (Table 1.1) was justified by the location of the stereogenic centres further away from the metal in comparison to *P*-stereogenic phosphines.

At the same time, it was discovered that (*S*)-amino-3-(3,4-dihydroxyphenyl) propanoic acid (L-DOPA) was very efficient in the treatment of Parkinson's disease, thereby creating a sudden demand for this rare amino acid. Knowles and co-workers, who were working at Monsanto, produced vanillin that was used in the synthesis of an L-DOPA precursor. They decided to use the newly

Table 1.1 First results in enantioselective hydrogenation (Scheme 1.1).

Entry	L*	R	R'	ee of 9 (%)	References
1[a]	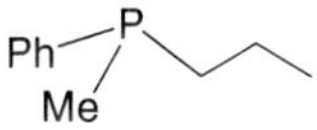 Ph–P(Me)–propyl	Ph	–COOH	15	16
2[a]	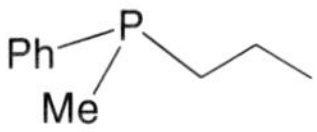 Ph–P(Me)–propyl	–CH$_2$COOH	–COOH	3	16
3	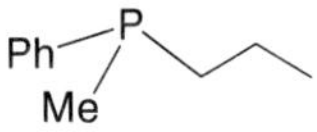 Ph–P(Me)–propyl	Ph	Et	7–8	17
4	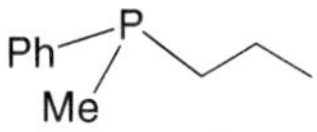 Ph–P(Me)–propyl	Ph	OMe	3–4	17
5	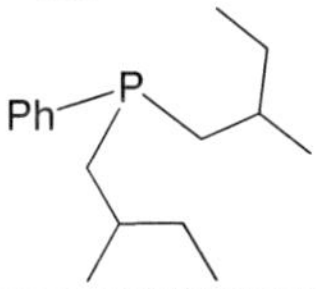 Ph–P(branched alkyl)	Ph	–COOH	~1	16

[a]The phosphine was not enantiomerically pure (69% ee).

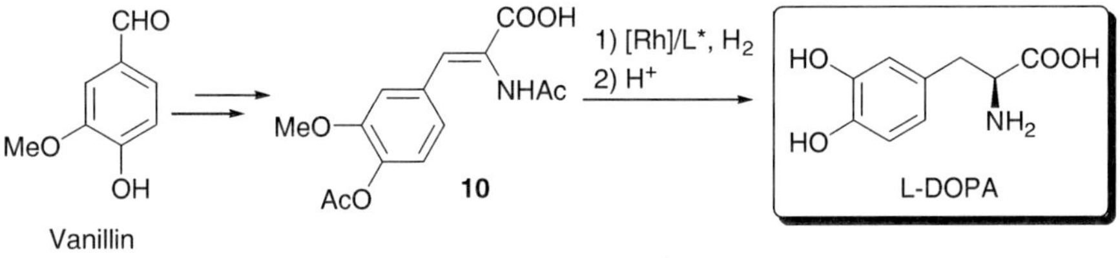

Scheme 1.2 Monsanto's L-DOPA synthesis.

developed enantioselective hydrogenation for the efficient synthesis of L-DOPA (Scheme 1.2).[18]

To optimise the hydrogenation of **10**, a model reaction with a simpler substrate, 2-α-acetamidocinnamic acid (**11**), was developed and a range of phosphines was screened (Scheme 1.3).

After a quick screening with phosphines giving low enantioselectivities, it was found that the introduction of an *o*-anisyl group was very beneficial and finally led to the phosphine CAMP, which hydrogenated **11** (and **10**) in high enantioselectivity.[19] This was the first time that enzyme-like selectivity was achieved with an artificial, 'man-made' catalyst. Indeed, the Rh/CAMP system was so good that the process was scaled up for industrial production of L-DOPA. The first ever industrial catalytic enantioselective process was born. It seems common sense that to achieve high enantioselectivities the stereogenic elements of the ligands have to be as close as possible to the rhodium centre, as Knowles himself stated in his Nobel Lecture.[18] Therefore, at that time a bright future was anticipated for *P*-stereogenic ligands.

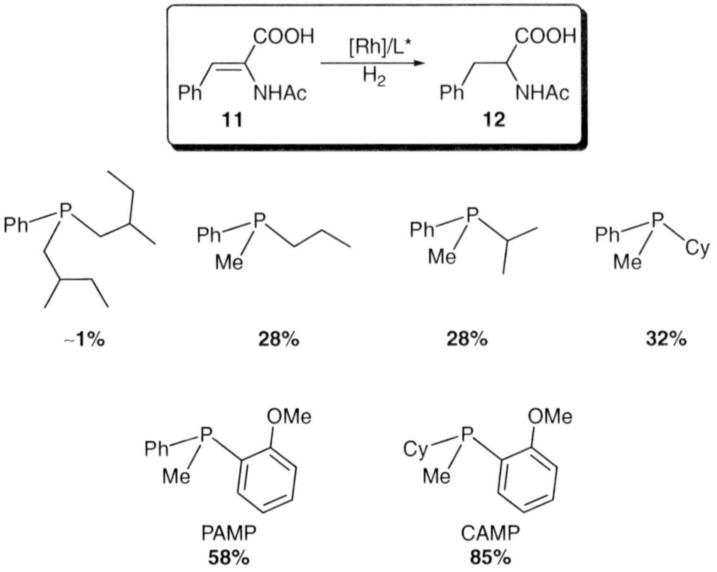

Scheme 1.3 Hydrogenation of 2-α-acetamidocinnamic acid. The numbers refer to the ee of **12**.

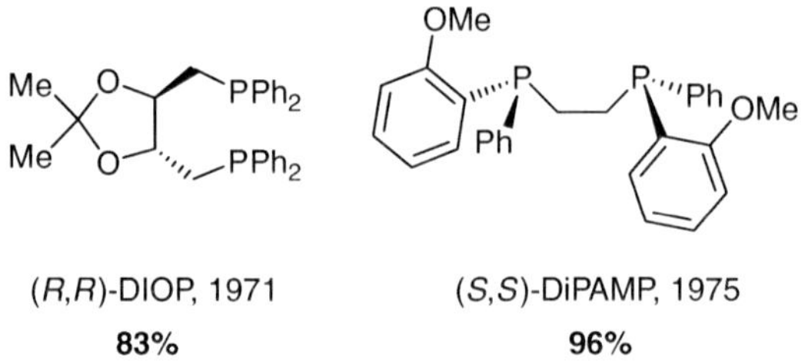

Figure 1.3 Chelating C_2-diphosphines and their ee values in Rh-catalysed hydrogenation of **11**.

At the same time, however, Dang and Kagan[20,21] prepared the chelating C_2-diphosphine DIOP (Figure 1.3), which was only marginally worse than CAMP in the hydrogenation of **11** in spite of having the stereogenic elements on the backbone and not at the phosphorus atoms.

Knowles and co-workers[22] reported that DiPAMP, the most well-known *P*-stereogenic phosphine, gave **12** with an outstanding (at that time) 96% ee. With this phosphine many other enamides, enol esters and similar compounds could be hydrogenated in high ee.[23,24] This work can be considered a milestone in the field of enantioselective homogeneous catalysis and have been profusely cited.[25] William S. Knowles was awarded (along with Ryoji Noyori and Barry Sharpless) the Nobel Prize in Chemistry for the year 2001 for his contributions to enantioselective catalysis.[18]

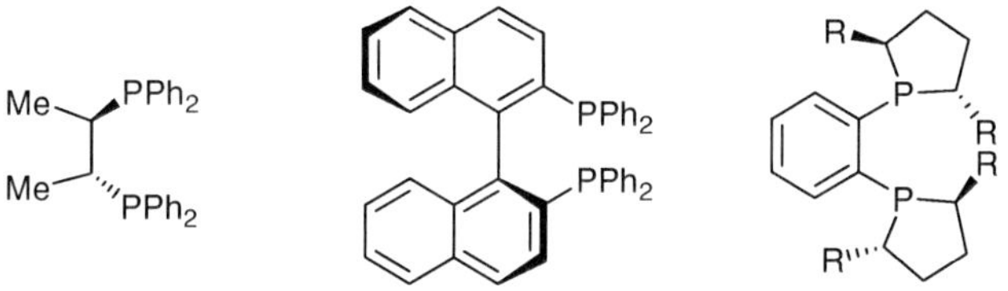

Figure 1.4 Chelating C_2-diphosphines appeared after DiPAMP, their ee values in Rh-catalysed hydrogenation of **11** are given here.

In spite of the success of DiPAMP, the good results with DIOP as well as those of Bosnich[26] with Chiraphos, Noyori[27] with BINAP, Burk[28] with DuPHOS and many others showed that stereogenic phosphorus atoms were not required to obtain high enantioselectivities. This result prompted the preparation of many more phosphines bearing stereogenic carbon atoms or axes (Figure 1.4).

These ligands were configurationally stable and often more easily prepared than those bearing *P*-stereogenic atoms. This is the reason than soon after the discovery of DiPAMP, the attention was shifted away from *P*-stereogenic systems, leaving them in the shadow for decades.

In spite of this apparent oblivion,[29] the development of methods to prepare *P*-stereogenic compounds kept improving slowly. An important event was the introduction of borane as a versatile protecting group in *P*-stereogenic chemistry (see Section 1.3),[30] displacing phosphine oxides. Phosphine boranes are key intermediates in the majority of the modern methods of synthesis of *P*-stereogenic phosphines, described in Chapters 3 to 5, which were developed in the decade of 1990. Since 2000, those methods have reached a considerable maturity and nowadays are widely employed. In addition, novel exciting routes, such as the preparation of ligands by enantioselective catalysis, are also being explored (see Chapter 6).

In 1994 an excellent review on *P*-stereogenic phosphines and derivatives was published by Pietrusiewicz and Zablocka.[31] It summarised much of the classic synthetic work but also described some of the emerging new methods. The review ended with the sentence 'the time has come for the *P*-chiral ligands to merge the stream and to bring in the *P*-chirality factor into play again'. More than 15 years later, this prediction is becoming progressively more accurate as the methods of synthesis develop and the use of *P*-stereogenic ligands in homogeneous catalysis spreads.

1.2 Configurational Stability of *P*-Stereogenic Compounds

Any atom bearing three different substituents and an electron pair in a trigonal pyramid structure is stereogenic. These compounds can be configurationally

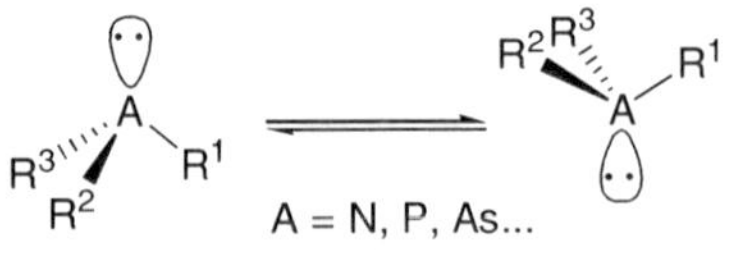

A = N, P, As...

Scheme 1.4   Pyramidal inversion of a tricoordinated, pyramidal compound.

unstable and invert following a mechanism where the molecule turns inside out, like an umbrella. This mechanism is usually referred as pyramidal inversion (Scheme 1.4).[32,33]

For ammonia and amines the energy barrier for pyramidal inversion is around $25 \, kJ \, mol^{-1}$.[34] This precludes, for instance, the isolation of optically active *N*-stereogenic amines, except in especial cases.

In contrast, the energy barrier for phosphine pyramidal inversion is high enough (around $155 \, kJ \, mol^{-1}$)[34] to prevent inversion occurring appreciably at room temperature. The same happens, in general, for substituted phosphines. In spite of that, in some compounds the phosphorus atom can racemise upon heating at moderate temperatures or under certain reaction conditions. As a result, the risk of racemisation is always a concern for the *P*-stereogenic chemist. The value of the pyramidal inversion energy depends on steric hindrance, electronic effects (electronegativity of the substituents and conjugative effects) and other parameters.

The evaluation of the value of inversion barriers has been done experimentally, often by variable temperature NMR methods, or by calculation.[34] Munro and Horner[6,35] showed that the pyramidal inversion follows first-order kinetics with little influence of the solvent polarity in the rate of racemisation. There are several classical studies that give reliable estimates of the inversion barrier values of many phosphines. Some examples are collected in Table 1.2.

Entries 1–6 show that typical acyclic tertiary phosphines have inversion energies in the range 120 to $150 \, kJ \, mol^{-1}$. In general, trialkyl phosphines have higher inversion barriers compared to diaryl and triaryl counterparts (compare entry 1 with entries 2–5). In spite of that, it seems that structural parameters do not profoundly affect the barriers for pyramidal inversion (compare entries 2–4). Entry 5 shows a small but interesting correlation between the inversion barrier and the electronegativity of the aryl group in phosphines of the type PPhArMe. It was observed that electron-withdrawing substituents in *para* position of the aryl group lower the energy of pyramidal inversion.[36]

Phosphines of entries 8 and 9 are examples of compounds whose configurational stability is inferior due to other effects. Acylphosphines (entry 8) have inferior inversion barriers due to conjugation in the planar transition state.[40] Comparison between entries 7 and 9 shows a dramatic lowering of inversion energies in phospholes compared to phospholanes, probably caused by an increase in the delocalisation of the planar transition state in the phosphole inversion.[39,41] In spite of that, more conjugated fused phospholes (entries 10 and 11) are more configurationally stable.

Table 1.2 Barriers for pyramidal inversion of *P*-stereogenic phosphines.

Entry	Compound	Barrier $(kJ\,mol^{-1})^a$	References
1	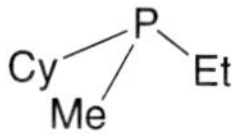 Cy–P(Me)–Et	149	36
2	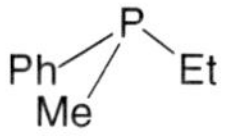 Ph–P(Me)–Et	126–134	6, 35, 36
3	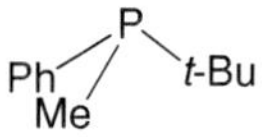 Ph–P(Me)–*t*-Bu	137	36
4	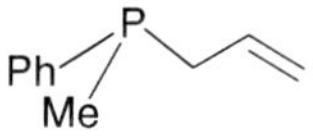 Ph–P(Me)–allyl	135–137	36, 37
5	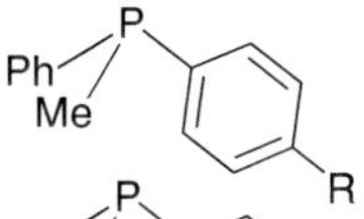 Ph–P(Me)–C₆H₄–R	129 (R = Me) 127 (R = OMe) 122 (R = CF$_3$)	36
6	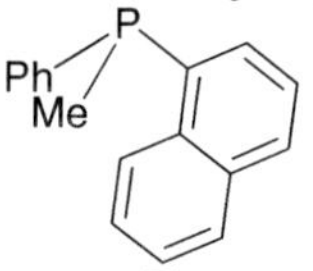 Ph–P(Me)–(1-naphthyl)	121	38
7	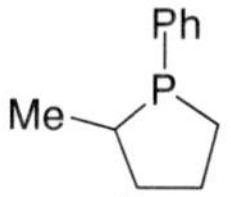 2-Me-1-Ph-phospholane	151	39
8	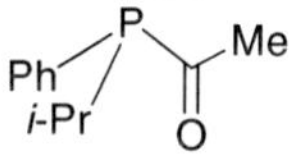 Ph–P(*i*-Pr)–C(=O)Me	81	40
9	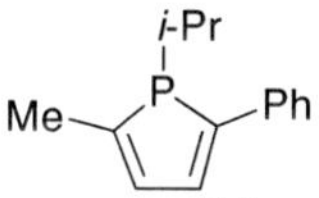 2-Me-1-(*i*-Pr)-5-Ph-phosphole	67	39, 41
10	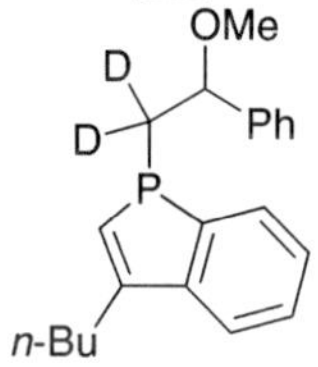 3-(*n*-Bu)-phosphindole, P–CD$_2$–CH(OMe)Ph	98	41
11	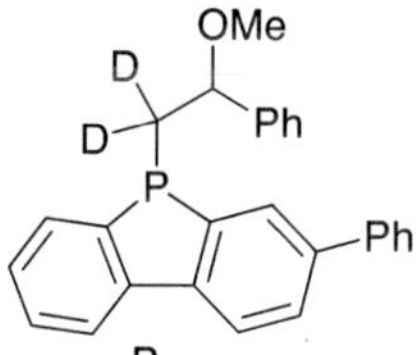 2-Ph-dibenzophosphole, P–CD$_2$–CH(OMe)Ph	110	41
12	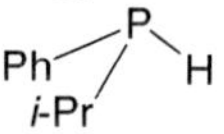 Ph–P(*i*-Pr)–H	98	42

^{a}Experimental or calculated values of *pyramidal inversion barrier energy*. The reader is directed to the original references for the exact meaning of this expression on each case.

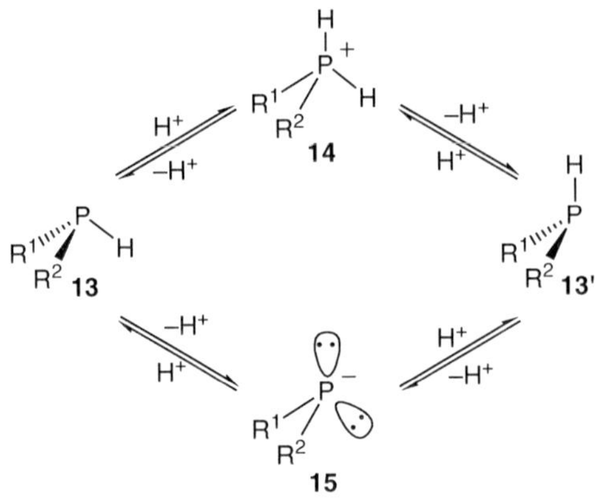

Scheme 1.5   Acid- and base-catalysed racemisation of secondary phosphines.

Although intramolecular pyramidal inversion is the most important and general pathway, some compounds can racemise following other mechanisms. A typical example is secondary phosphines, which are relatively stable towards pyramidal inversion (Table 1.2, entry 12) but can racemise following acid- and base-catalysed mechanisms (Scheme 1.5).[42,43]

The formation of achiral phosphonium salts (**14**) or phosphide anions (**15**) explains the easy racemisation of **13** in the presence of traces of acids or bases. It has been found that **13** can be stabilised by the presence of a mild base, capable of neutralising any trace of protons but unable to form **15**.[44] A different way to stabilise secondary phosphines is the quenching of the electron pair by the formation of phosphine boranes (see Section 1.3).

Another family of configurationally unstable compounds are chlorophosphines, which racemise extremely fast through the formation of an achiral pentacoordinate intermediate (see Chapter 4, Section 4.3.2.2).

Inversion at the phosphorus atom is usually regarded as a serious drawback of *P*-stereogenic ligands but it can also be an advantage from the synthetic point of view in molecules containing additional stereogenic elements. In the ideal case, even optically pure compounds can be obtained if the phosphorus atoms are left to epimerise until the most thermodynamically stable isomer is formed. The best part is that only heating is required. This strategy, however, has been seldom used in *P*-stereogenic chemistry.[45,46]

The conclusion of this section is that, except in special cases, tertiary phosphines have racemisation barrier energies higher than $120 \, kJ \, mol^{-1}$, corresponding to equilibration temperatures exceeding $100 \,°C$. For example, solutions of methyl(1-naphthyl)phenylphosphine (Table 1.2, entry 6) have an inversion half-life of only 7 hours at $95 \,°C$ but of 10 years at $20 \,°C$.[38] When coordinated to transition metal centres, even better configurational stabilities are in general expected. Therefore, it can be concluded that *P*-stereogenic phosphines are stable enough to be used in most transition metal-catalysed reactions.

1.3 Interconversions between *P*-Stereogenic Compounds

Optically pure, trivalent *P*-stereogenic compounds are prepared by several methods, discussed in Chapters 2–6. Most of the syntheses are multi-step processes that involve tetra- or pentacoordinated intermediates. Two families of compounds, phosphine oxides and phosphine boranes are by far the most important direct precursors of free phosphines. The interconversions between phosphines and phosphine boranes and oxides are briefly discussed in this section.

1.3.1 Phosphines, Phosphine Oxides and Sulfides

Phosphine oxides were the classic precursors to *P*-stereogenic phosphines until they were (partially) supplanted by phosphine boranes. Here some of the methods of oxidation and reduction of phosphines (Scheme 1.6) are briefly considered.

Many synthetic routes lead to phosphine oxides although phosphines themselves are rarely oxidised on purpose, except for purification or analysis. If required, however, there are several reagents that deliver the phosphine oxide quantitatively and stereoselectively, usually with retention of configuration at the phosphorus atom.[31] The most typical reagents are peroxo compounds such as H_2O_2, *t*-BuOOH and MCPBA.[6,47–49] The most available oxidant, air, can be also used but often leads to impure products.[31] Treatment of *P*-stereogenic phosphines with halogens or pseudohalogens in water leads to inverted phosphine oxides.[50,51]

Due to the stability of the phosphoryl group, reduction of phosphine oxides to free phosphines requires strongly oxophilic reagents under harsh conditions. In spite of that, is such an important transformation in *P*-stereogenic chemistry that dependable methods with either retention or inversion of configuration have been developed during the last 50 years.[31] The reductants of choice have traditionally been silane reagents, sometimes mixed with amines. Some selected examples of reductions with those reagents are listed in Table 1.3.

Trichlorosilane is a very powerful reducing agent. On its own or mixed with mildly basic amines or phosphines it leads to products with retention of configuration in the reduction of acyclic phosphines (entry 1) whereas with more basic amines it affords inverted phosphines (entries 2 and 3) with the exception of phosphetane oxides (entry 4). Entry 3 shows that the famous ligand DiPAMP was prepared for the first time using this reagent.[23] Hexachlorodisilane has

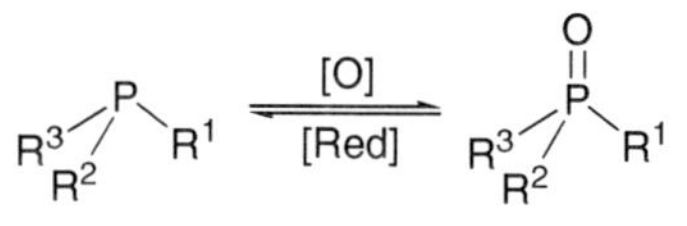

Scheme 1.6 Interconversion between phosphines and phosphine oxides.

Table 1.3 Selected examples of successful reduction of phosphine oxides with silanes.

Entry	Reductant	Substrate	Stereochemistry	References
1	HSiCl$_3$ HSiCl$_3$/py	Ph, Me, R on P=O; R = Bn, *o*-An	Retention	10, 52
2	HSiCl$_3$/PPh$_3$ HSiCl$_3$/NEt$_3$	Ph, Me, R on P=O; R = Bn, Fc	Inversion	10, 53
3	HSiCl$_3$/N(*n*-Bu)$_3$	**DiPAMPO**	Inversion	22, 23
4	HSiCl$_3$/NEt$_3$	Men on P=O	Retention	54
5	Si$_2$Cl$_6$ Si$_3$Cl$_8$	Ph, Me, R on P=O; R = *n*-Pr, *n*-Bu, Bn, *p*-An	Inversion	55–57
6	Si$_2$Cl$_6$		Retention	58
7	PhSiH$_3$	Ph, Me, *n*-Pr on P=O	Retention	59
8	PhSiH$_3$	Me, Ph, Ph, Me on diphosphine dioxide	Retention	60
9	PhSiH$_3$	Ph on P=O	Retention	61

Table 1.3 (*Continued*).

Entry	Reductant	Substrate	Stereochemistry	References
10[a]	HSi(OEt)$_3$/Ti(O*i*-Pr)$_4$ or PMHS/Ti(O*i*-Pr)$_4$	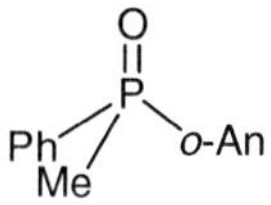 **PAMPO**	Retention	62
11[a]	PMHS/Ti(O*i*-Pr)$_4$	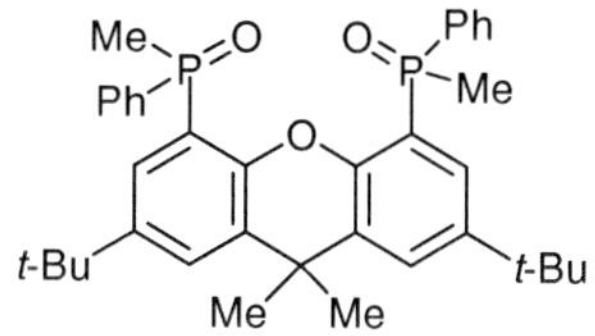	Retention	48
12[a]	PMHS/Ti(O*i*-Pr)$_4$	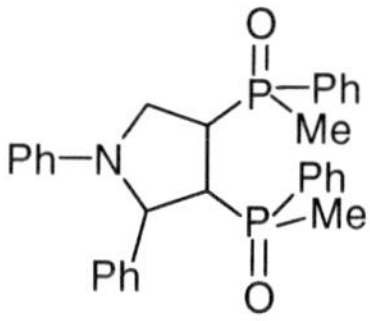	Retention	63

[a]PMHS = polymethylhydrosiloxane = Me$_3$SiO[(CH$_3$)HSiO]$_n$SiMe$_3$.

comparable reducing capacity often under milder conditions. With this reagent, acyclic phosphine oxides invert upon reduction (entry 5) but with phosphetane oxides retention is observed (entry 6). Entries 7–11 show that non-halogenated silanes have also been used. Phenylsilane (entries 7–9) and triethoxysilane or polymethylhydrosiloxane with a catalytic amount of Ti(IV) isopropoxide (entries 10–12) are milder reductants than chlorosilanes and provide phosphines with retention of configuration. In the systems of entries 10–12 it is thought that titanium hydride species are the active reductant.[62]

Reduction reactions are typically performed in refluxing benzene (bp 80 °C), acetonitrile (bp 82 °C) or toluene (bp 111 °C). Depending on the substrate and conditions, the reaction time can be anything from minutes to days. Although the reductions themselves are frequently very stereoselective, exposure of the free phosphines to the reducing agents and their oxidation products degrades the optical purity of the phosphines.[31] Therefore, the general rule is to keep the temperature and reaction time as low as possible to maximise the optical purity of the free phosphines. In spite of that, variable amounts of racemic products are often unavoidable.[56,64,65]

Other reductive systems have been explored in much lower intensity (Table 1.4).

Lithium aluminium hydride is a very powerful reducing agent for phosphine oxides but unfortunately produces racemic products (entries 1 and 2) due to pseudorotation of the pentacoordinated species formed upon addition of LiAlH$_4$ to phosphine oxides.[68] Imamoto and co-workers[38] (entry 3) had the idea to introduce an alkylating reagent, methyl triflate, to stabilise the phosphorus centre prior to reduction with LiAlH$_4$. With this system, phosphines with clean inversion of configuration were obtained (entries 3 and 4). It is likely

Table 1.4 Examples of reduction of phosphine oxides with non-silane based reagents.

Entry	Reductant	Substrate	Stereochemistry	References
1	LiAlH$_4$	Ph–P(=O)(Me)–n-Pr	Racemisation	66
2	LiAlH$_4$ CeCl$_3$	Ph–P(=O)(Me)–o-An **PAMPO**	Racemisation	67
3	1) MeOTf 2) LiAlH$_4$	Ph–P(=O)(Me)–o-An **PAMPO**	Inversion	38
4	1) MeOTf 2) LiAlH$_4$	t-Bu–P(=O)(Me)–CH$_2$CH$_2$–P(=O)(Me)–t-Bu **t-Bu-BisP*(O)$_2$**	Inversion	38

that methyl triflate forms a phosphonium salt that is attacked by a hydride from the backside of the methoxy group to give the reduced product.[38]

Many phosphine sulfides are known but their use in *P*-stereogenic chemistry has been rather limited until recently. They are easily prepared by reaction of phosphines with elemental sulfur, a step which proceeds with retention of configuration.[69] Hexachlorodisilane and Raney nickel have been found to be stereoselective reagents for desulfurisation of phosphine sulfides. Some examples of reductions are collected in Table 1.5.

Hexachlorodisilane is an appropriate reagent to reduce phosphine sulfides, with retention of configuration even for the acyclic substrates. This is in contrast to the reduction of phosphine oxides (compare entries 1 and 2 of Table 1.5 with entry 5 of Table 1.3). Raney nickel has also been used successfully (entries 5 and 6) with the advantage of being heterogeneous, which greatly simplifies the work-up after the reaction. It has to be mentioned that the use of phosphine sulfides is gaining relevance in *P*-stereogenic chemistry, where they are comparable or even outperform phosphine boranes.[76,77]

The stringent conditions required for the reduction of phosphine oxides are incompatible with many functional groups and often cause severe racemisation of the phosphine. Those problems have been largely circumvented by switching to phosphine boranes, as discussed in the next section.

Table 1.5 Examples of successful reductions of phosphine sulfides.

Entry	Reductant	Substrate	Stereochemistry	References
1	Si_2Cl_6	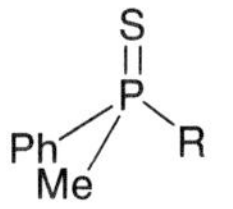 Ph, Me, R (R = *n*-Pr, allyl) phosphine sulfide	Retention	69
2	Si_2Cl_6	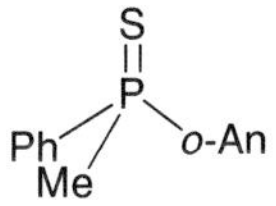 Ph, Me, *o*-An phosphine sulfide — **PAMPS**	Retention	70
3	Si_2Cl_6	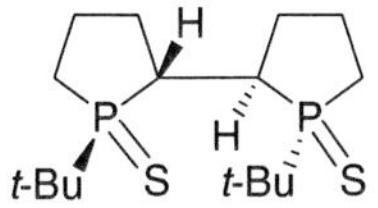 **TangPhos(S)$_2$**	Retention	71
4	Si_2Cl_6	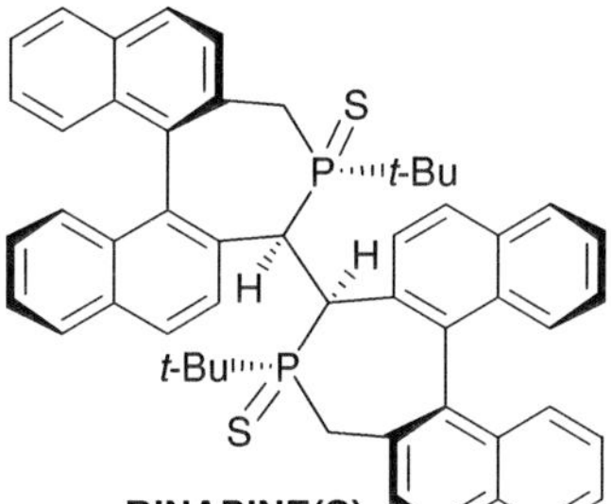 **BINAPINE(S)$_2$**	Retention	72
5	Raney Ni	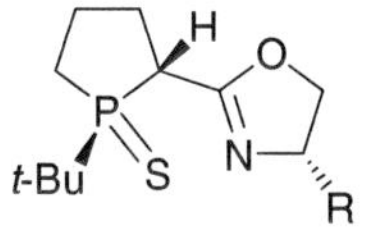 *t*-Bu oxazoline phosphine sulfide (R = *i*-Pr, *i*-Bu, *t*-Bu, Bn, Ph)	Retention	73
6	Raney Ni	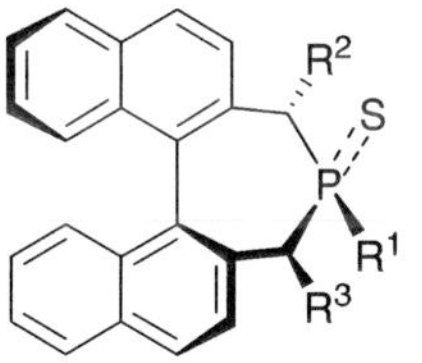 R^1, R^2, R^3 binaphthyl phosphine sulfide		
	Si_2Cl_6		Retention	74, 75

1.3.2 Phosphines and Phosphine Boranes

Phosphine boranes, adducts between phosphines and a BH_3 unit, have been known for over 70 years[78] but it was not until the 1990s that they became essential intermediates in the synthesis of *P*-stereogenic compounds, thanks to the work of Imamoto and co-workers.[30] Since then their use has only increased year after year.[79]

The borane group quenches the electron pair at the phosphorus atom, stabilising its configuration and protecting it from oxidation. The low polarity and polarisability of the P–B and B–H bonds makes them rather unreactive. For the same reasons, the hydrogen atoms of the borane do not show any hydridic character. This makes phosphine boranes remarkably stable, with the exception of $PH_3 \cdot BH_3$ and $PF_3 \cdot BH_3$. For example, $PPh_3 \cdot BH_3$ can be heated at 150 °C in concentrated HCl for hours without any alteration.[80] This stability has simplified enormously the handling and purification of key intermediates of *P*-stereogenic phosphines. At the same time, the borane group modifies the reactivity of the phosphine coordinated to it, activating hydrogen atoms or methyl groups adjacent to the P atom, a fact that has been used to prepare many phosphines as will be seen throughout this book.

There are excellent reviews on phosphine boranes with different degrees of focus on *P*-stereogenic compounds.[79–82] Here only a brief discussion about the formation and cleavage of phosphine boranes is given (Scheme 1.7).

Phosphine boranes are usually prepared by direct reaction between phosphines and commercially available solutions of borane complexed to labile Lewis bases (method **A**). The most frequent ones are $BH_3 \cdot THF$ and $BH_3 \cdot SMe_2$, the latter being somewhat more stable and with a higher *boronating* capacity than the former. The higher Lewis basicity of phosphines compared to THF of SMe_2 causes the migration of the borane group to form the phosphine borane quantitatively. This reaction always occurs with strict retention of configuration at the phosphorus atom.

Imamoto and co-workers[67,83,84] developed procedure **B**, which is a one-pot transformation of phosphine oxides into phosphine boranes. The cerium cation plays two roles: it coordinates to the phosphoryl group to make the deoxygenation with $LiAlH_4$ possible and it also facilitates the reaction between

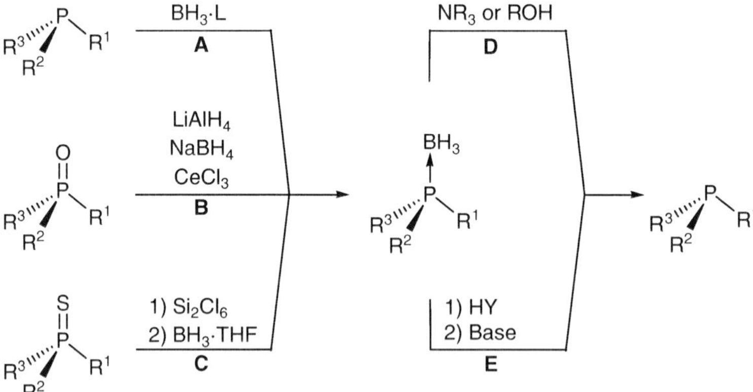

Scheme 1.7   Boronation and deboronation of phosphines and derivatives.

$NaBH_4$ and the intermediate free phosphine. Unfortunately, starting from optically pure phosphine oxides yielded only racemic phosphine boranes. Finally, Corey and co-workers[70] prepared optically pure phosphine boranes by *in situ* desulfurisation–complexation (method **C**), with retention of configuration.

Phosphine boranes would not be so popular if the borane group could not be easily removed. Two reliable methods have been developed. The first one is the treatment of phosphine boranes with amines (**D**) or alcohols, which act as borane acceptors. Although phosphine boranes are in general more stable than amine boranes, the presence of a large excess of a basic amine frequently suffices to displace the equilibrium and deprotect completely the phosphine. Typical amines used for this transformation are diethylamine,[85] TMEDA,[86] pyrrolidines,[87,88] morpholine[89] and DABCO.[90,91] The reactions are usually carried out at room temperature or with moderate heating in neat amine with the exception of DABCO, which as it is a solid is typically used in toluene.

Jugé and co-workers[92] found that stirring some phosphine boranes in ethanol was sufficient to cleave the borane group, while classical decomplexation with DABCO produced many unidentified products. This method has been extended, with slight modifications, to other alcohols and phosphine boranes.[93,94] Not surprisingly, basic trialkylphosphine boranes are not deprotected following this procedure.[94] Although no detailed studies on the mechanism has been reported, it is thought to involve the formation of tertiary borates and hydrogen gas.[93–95]

Trialkylphosphine boranes are electron rich and possess very inert P–B bonds. These compounds are incompletely deprotected by amines[96,97] but are efficiently deboronated by method **E**, developed by McKinstry, Livinghouse and co-workers.[96–98] Treatment of phosphine boranes with an excess of certain strong acids such as methanesulfonic,[99] trifluoromethanesulfonic[100] or more often tetrafluoroboric[96,98] in dichloromethane at low temperature ($-5\,°C$ to room temperature) cleaves the borane group producing phosphonium salts, which are extracted with an aqueous base ($NaOH$ or $NaHCO_3$) to afford tertiary phosphines in good yields.[101,102] The mechanism of the decomplexation is not clear, but it is thought to involve nucleophilic substitutions at the borane moiety,[97–99] eventually leading to decoordination.

Both deprotection methods are very stereoselective, with complete retention of configuration at the phosphorus atom. Losses of optical purity have been reported very rarely.[103] The reader must be aware that deboronation is not always required since phosphine boranes can be used directly in transition metal homogeneous catalysis. Jugé and co-workers[104] combined the reducing properties of the borane moiety to prepare, *in situ*, Cu(I), Pd(0) and Rh(I) metal complexes by mixing Cu(II), Pd(II) and Rh(III) salts and phosphine boranes.

There are also procedures to transform phosphine boranes into other derivatives by one-pot stereoselective reactions (Scheme 1.8).

Phosphine oxides with retention of configuration were prepared by Imamoto and co-workers[105] using MCPBA and by Jugé and co-workers[106] using *t*-BuOOH. The oxidation with iodine in the presence of water was found to occur

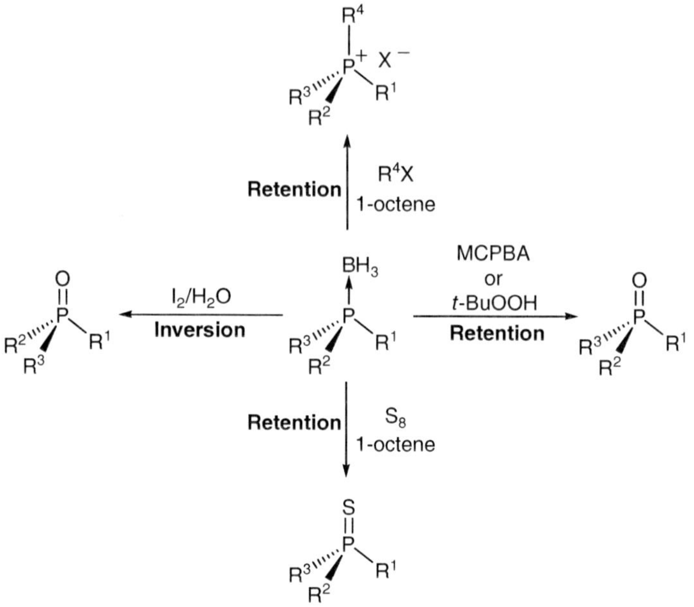

Scheme 1.8 One-pot transformations of phosphine boranes into phosphonium salts and phosphine chalcogenides.

with inversion of configuration (75–85% ee), a fact that was justified by a mechanism involving phosphine–iodoborane complexes.[105] Phosphine sulfides were prepared by the same authors[105,106] in very high optical purity whereas Jugé and co-workers[107] reported a one-pot deprotection–quaternisation. Both processes take place with retention of configuration.

References

1. J. Meisenheimer and L. Lichtenstadt, *Chem. Ber.*, 1911, **44**, 356.
2. J. Meisenheimer, J. Casper, M. Horing, W. Lauter, L. Lichtenstadt and W. Samuel, *Liebigs Ann. Chem.*, 1926, 213.
3. W. C. Davies and F. G. Mann, *J. Chem. Soc.*, 1944, 276.
4. K. F. Kumli, W. E. McEwen and C. A. VanderWerf, *J. Am. Chem. Soc.*, 1959, **81**, 248.
5. L. Horner, H. Winkler, A. Rapp, A. Mentrup, H. Hoffmann and P. Beck, *Tetrahedron Lett.*, 1961, **2**, 161.
6. L. Horner, *Pure Appl. Chem.*, 1964, **9**, 225.
7. L. Horner, F. Schedlbauer and P. Beck, *Tetrahedron Lett.*, 1964, **5**, 1421.
8. L. Horner and H. Winkler, *Tetrahedron Lett.*, 1964, **5**, 175.
9. L. Horner and H. Winkler, *Tetrahedron Lett.*, 1964, **5**, 3265.
10. L. Horner and W. D. Balzer, *Tetrahedron Lett.*, 1965, **6**, 1157.

11. L. Horner, W. D. Balzer and D. J. Peterson, *Tetrahedron Lett.*, 1966, **7**, 3315.
12. O. Korpiun and K. Mislow, *J. Am. Chem. Soc.*, 1967, **89**, 4784.
13. O. Korpiun, R. A. Lewis, J. Chickos and K. Mislow, *J. Am. Chem. Soc.*, 1968, **90**, 4842.
14. J. F. Young, J. A. Osborn, F. H. Jardine and G. Wilkinson, *Chem. Commun.*, 1965, 131.
15. J. A. Osborn, F. H. Jardine, J. F. Young and G. J. Wilkinson, *J. Chem. Soc. (A)*, 1966, 1711.
16. W. S. Knowles and M. J. Sabacky, *Chem. Commun.*, 1968, 1445.
17. L. Horner, H. Siegel and H. Büthe, *Angew. Chem. Int. Ed.*, 1968, **7**, 942.
18. W. S. Knowles, *Adv. Synth. Catal.*, 2003, **345**, 3.
19. W. S. Knowles, M. J. Sabacky and B. D. Vineyard, *J. Chem. Soc., Chem. Commun.*, 1972, 10.
20. T. Dang and H. B. Kagan, *J. Chem. Soc., Chem. Commun.*, 1971, 481.
21. H. B. Kagan and T. Dang, *J. Am. Chem. Soc.*, 1972, **94**, 6429.
22. W. S. Knowles, M. J. Sabacky, B. D. Vineyard and D. J. Weinkauff, *J. Am. Chem. Soc.*, 1975, **97**, 2567.
23. B. D. Vineyard, W. S. Knowles, M. J. Sabacky, G. L. Bachman and D. J. Weinkauff, *J. Am. Chem. Soc.*, 1977, **99**, 5946.
24. W. S. Knowles, *J. Chem. Educ.*, 1986, **63**, 222.
25. W. S. Knowles, *Acc. Chem. Res.*, 1983, **16**, 106.
26. M. D. Fryzuk and B. Bosnich, *J. Am. Chem. Soc.*, 1977, **99**, 6262.
27. A. Miyashita, A. Yasuda, H. Takaya, K. Toriumi, T. Ito, T. Souchi and R. Noyori, *J. Am. Chem. Soc.*, 1980, **102**, 7932.
28. M. J. Burk, J. E. Feaster, W. A. Nugent and R. L. Harlow, *J. Am. Chem. Soc.*, 1993, **115**, 10125.
29. F. Maienza, M. Wörle, P. Steffanut, A. Mezzetti and F. Spindler, *Organometallics*, 1999, **18**, 1041.
30. T. Imamoto, T. Hoshiki, T. Onozawa, T. Kusumoto and K. Sato, *J. Am. Chem. Soc.*, 1990, **112**, 5244.
31. K. M. Pietrusiewicz and M. Zablocka, *Chem. Rev.*, 1994, **94**, 1375.
32. A. Rauk, L. C. Allen and K. Mislow, *Angew. Chem. Int. Ed.*, 1970, **9**, 400.
33. R. D. Baechler, J. D. Andose, J. Stackhouse and K. Mislow, *J. Am. Chem. Soc.*, 1972, **94**, 8060.
34. A. Rauk, J. D. Andose, W. G. Frick, R. Tang and K. Mislow, *J. Am. Chem. Soc.*, 1971, **93**, 6507.
35. H. D. Munro and L. Horner, *Tetrahedron*, 1970, **26**, 4621.
36. R. D. Baechler and K. Mislow, *J. Am. Chem. Soc.*, 1970, **92**, 3090.
37. R. D. Baechler, W. B. Farnham and K. Mislow, *J. Am. Chem. Soc.*, 1969, **91**, 5686.
38. T. Imamoto, S. Kikuchi, T. Miura and Y. Wada, *Org. Lett.*, 2001, **3**, 87.
39. W. Egan, R. Tang, G. Zon and K. Mislow, *J. Am. Chem. Soc.*, 1970, **92**, 1442.
40. W. Egan and K. Mislow, *J. Am. Chem. Soc.*, 1971, **93**, 1805.

41. W. Egan, R. Tang, G. Zon and K. Mislow, *J. Am. Chem. Soc.*, 1971, **93**, 6205.

42. A. Bader, T. Nullmeyers, M. Pabel, G. Salem, A. C. Willis and S. B. Wild, *Inorg. Chem.*, 1995, **34**, 384.

43. J. Albert, J. Magali Cadena, J. R. Granell, G. Muller, D. Panyella and C. Sañudo, *Eur. J. Inorg. Chem.*, 2000, 1283.

44. A. Bader, M. Pabel, A. C. Willis and S. B. Wild, *Inorg. Chem.*, 1996, **35**, 3874.

45. S. Kudis and G. Helmchen, *Angew. Chem. Int. Ed.*, 1998, **37**, 3047.

46. G. Hoge, *J. Am. Chem. Soc.*, 2004, **126**, 9920.

47. D. B. Denney and J. W. Hanifin, *Tetrahedron Lett.*, 1963, **4**, 2177.

48. Y. Hamada, F. Matsuruta, M. Oku, K. Hatano and T. Shiori, *Tetrahedron Lett.*, 1997, **38**, 8961.

49. S. Matsukawa, H. Sugama and T. Imamoto, *Tetrahedron Lett.*, 2000, **41**, 6461.

50. L. Horner and H. Winkler, *Tetrahedron Lett.*, 1964, **5**, 455.

51. J. A. MacKay and E. Vedejs, *J. Org. Chem.*, 2006, **71**, 498.

52. H. Wu, J. Yu and J. B. Spencer, *Org. Lett.*, 2004, **6**, 4675.

53. L. L. Troitskaya, Z. A. Starikova, T. V. Demeshchik and V. I. Sokolov, *Russ. Chem. Bull.*, 1999, **48**, 1738.

54. A. Marinetti and L. Ricard, *Tetrahedron*, 1993, **49**, 10291.

55. K. Naumann, G. Zon and K. Mislow, *J. Am. Chem. Soc.*, 1969, **91**, 2788.

56. K. Naumann, G. Zon and K. Mislow, *J. Am. Chem. Soc.*, 1969, **91**, 7012.

57. E. P. Kyba, *J. Am. Chem. Soc.*, 1975, **97**, 2554.

58. T. Imamoto, K. V. L. Crépy and K. Katagiri, *Tetrahedron: Asymmetry*, 2004, **15**, 2213.

59. K. L. Marsi, *J. Org. Chem.*, 1974, **39**, 265.

60. T. Miura and T. Imamoto, *Tetrahedron Lett.*, 1999, **40**, 4833.

61. Z. Pakulski, O. M. Demchuk, J. Frelek, R. Luboradzki and K. M. Pietrusiewicz, *Eur. J. Org. Chem.*, 2004, 3913.

62. T. Coumbe, N. J. Lawrence and F. Muhammad, *Tetrahedron Lett.*, 1994, **35**, 625.

63. N. Vinokurov, K. M. Pietrusiewicz, S. Frynas, M. Wiebcke and H. Butenschön, *Chem. Commun.*, 2008, 5408.

64. U. Nettekoven, M. Widhalm, P. C. J. Kamer, P. W. N. M. van Leeuwen, K. Mereiter, M. Lutz and A. L. Spek, *Organometallics*, 2000, **19**, 2299.

65. L. J. Higham, E. F. Clarke, H. Müller-Bunz and D. G. Gilheany, *J. Organomet. Chem.*, 2005, **690**, 211.

66. P. D. Henson, K. Naumann and K. Mislow, *J. Am. Chem. Soc.*, 1969, **91**, 5645.

67. T. Imamoto, T. Takeyama and T. Kusumoto, *Chem. Lett.*, 1985, 1491.

68. W. B. Farnham, R. A. Lewis, R. K. Murray and K. Mislow, *J. Am. Chem. Soc.*, 1970, **92**, 5808.

69. G. Zon, K. E. DeBruin, K. Naumann and K. Mislow, *J. Am. Chem. Soc.*, 1969, **91**, 7023.

70. E. J. Corey, Z. Chen and G. J. Tanoury, *J. Am. Chem. Soc.*, 1993, **115**, 11000.
71. W. Tang and X. Zhang, *Angew. Chem. Int. Ed.*, 2002, **41**, 1612.
72. W. Tang, W. Wang, Y. Chi and X. Zhang, *Angew. Chem. Int. Ed.*, 2003, **42**, 3509.
73. W. Tang, W. Wang and X. Zhang, *Angew. Chem. Int. Ed.*, 2003, **42**, 943.
74. P. Kasák, K. Mereiter and M. Widhalm, *Tetrahedron: Asymmetry*, 2005, **16**, 3416.
75. Q. Dai, W. Li and X. Zhang, *Tetrahedron*, 2008, **64**, 6943.
76. J. J. Gammon, S. J. Canipa, P. O'Brien, B. Kelly and S. Taylor, *Chem. Commun.*, 2008, 3750.
77. J. J. Gammon, P. O'Brien and B. Kelly, *Org. Lett.*, 2009, **11**, 5022.
78. E. L. Gamble and P. Gilmont, *J. Am. Chem. Soc.*, 1940, **62**, 717.
79. N. Andrushko and A. Börner, in *Phosphorus Ligands in Asymmetric Catalysis: Synthesis and Applications*, A. Börner, (ed.), Wiley-VCH, Weinheim, 2008, p. 1275.
80. J. M. Brunel, B. Faure and M. Maffei, *Coord. Chem. Rev.*, 1998, **178–180**, 665.
81. M. Ohff, J. Holz, M. Quirmbach and A. Börner, *Synthesis*, 1998, 1391.
82. B. Carboni and L. Monnier, *Tetrahedron*, 1999, **55**, 1197.
83. T. Imamoto, T. Kusumoto, N. Suzuki and K. Sato, *J. Am. Chem. Soc.*, 1985, **107**, 5301.
84. T. Imamoto, *Pure Appl. Chem.*, 1993, 655.
85. S. Jugé, M. Stephan, J. A. Laffitte and J. P. Genêt, *Tetrahedron Lett.*, 1990, **31**, 6357.
86. T. Imamoto, K. Sugito and K. Yoshida, *J. Am. Chem. Soc.*, 2005, **127**, 11934.
87. H. Heath, B. Wolfe, T. Livinghouse and S. K. Bae, *Synthesis*, 2001, 2341.
88. K. Katagiri, H. Danjo, K. Yamaguchi and T. Imamoto, *Tetrahedron*, 2005, **61**, 4701.
89. A. Grabulosa, G. Muller, J. I. Ordinas, A. Mezzetti, M. A. Maestro, M. Font-Bardia and X. Solans, *Organometallics*, 2005, **24**, 4961.
90. S. Yamago, M. Yanagawa, H. Mukai and E. Nakamura, *Tetrahedron*, 1996, **52**, 5091.
91. Y. Morisaki, H. Imoto, K. Tsurui and Y. Chujo, *Org. Lett.*, 2009, **11**, 2241.
92. D. Moulin, S. Bago, C. Bauduin, C. Darcel and S. Jugé, *Tetrahedron: Asymmetry*, 2000, **11**, 3939.
93. M. Schröder, K. Nozaki and T. Hiyama, *Bull. Chem. Soc. Jpn.*, 2004, **77**, 1931.
94. M. van Overschelde, E. Vervecken, S. G. Modha, S. Cogen, E. van der Eycken and J. van der Eycken, *Tetrahedron*, 2009, **65**, 6410.
95. J. M. Camus, J. Andrieu, P. Richard, R. Poli, C. Darcel and S. Jugé, *Tetrahedron: Asymmetry*, 2004, **15**, 2061.
96. L. McKinstry and T. Livinghouse, *Tetrahedron Lett.*, 1994, **35**, 9319.
97. L. McKinstry and T. Livinghouse, *Tetrahedron*, 1995, **51**, 7655.

98. L. McKinstry, J. J. Overberg, C. Soubra-Ghaoui, D. S. Walsh and K. A. Robins, *J. Org. Chem.*, 2000, **65**, 2261.
99. T. Imamoto and T. Oshiki, *Tetrahedron Lett.*, 1989, **30**, 383.
100. T. Hamada and S. L. Buchwald, *Org. Lett.*, 2002, **4**, 999.
101. T. Imamoto, J. Watanabe, Y. Wada, H. Masuda, H. Yamada, H. Tsuruta, S. Matsukawa and K. Yamaguchi, *J. Am. Chem. Soc.*, 1998, **120**, 1635.
102. A. Ohashi and T. Imamoto, *Org. Lett.*, 2001, **3**, 373.
103. U. Nettekoven, P. C. J. Kamer and P. W. N. M. van Leeuwen, *Organometallics*, 2000, **19**, 4596.
104. C. Darcel, E. B. Kaloun, R. Merdès, D. Moulin, N. Riegel, S. Thorimbert, J. P. Genêt and S. Jugé, *J. Organomet. Chem.*, 2001, **624**, 333.
105. T. Imamoto, K. Hirose, H. Amano and H. Seki, *Main Group Chem.*, 1996, **1**, 331.
106. J. Uziel, C. Darcel, D. Moulin, C. Bauduin and S. Jugé, *Tetrahedron: Asymmetry*, 2001, **12**, 1441.
107. J. Uziel, N. Riegel, B. Aka, P. Figuiere and S. Jugé, *Tetrahedron Lett.*, 1997, **38**, 3405.

Resolution of Racemic and Diastereomeric Mixtures

2.1 Introduction

This chapter describes the preparation of optically pure phosphines and derivatives by resolution of racemic/diastereomeric mixtures or by taking advantage of additional stereogenic elements in the scaffold of the phosphine.

- Section 2.2 deals with the classical resolution of racemic mixtures *via* formation of diastereomeric adducts with chiral auxiliaries or by chromatographic methods.
- Section 2.3 discusses the covalent introduction of enantiopure chiral auxiliaries to form separable diastereomers from which the optically pure compound can be obtained after stereoselective transformations. In this part, mainly the methods where chiral auxiliaries are introduced in order to perform the resolution, but not present in the final structure are discussed.
- Section 2.4 is devoted to those ligands prepared taking advantage of the chirality of particular scaffolds like *o*-substituted ferrocenes, certain diols or 1,1′-biaryls which are intended to be part of the final *P*-stereogenic ligand.
- Finally, Section 2.5 describes the use of stoichiometric Pd-templated synthesis of certain *P*-stereogenic phosphines.

Some of the chemistry discussed in this chapter has been used for over a century and indeed it constituted the only way to prepare enantiopure *P*-stereogenic compounds until new methods appeared in the 1990s, discussed in Chapters 4 to 6. In spite of the considerable versatility of those new methodologies, they allow

RSC Catalysis Series No. 7
P-Stereogenic Ligands in Enantioselective Catalysis
By Arnald Grabulosa
© Arnald Grabulosa 2011
Published by the Royal Society of Chemistry, www.rsc.org

the preparation of only certain types of phosphines hence some of the methods discussed in this chapter are still quite frequently used nowadays.

2.2 Resolution of Racemic Mixtures

2.2.1 Direct Resolutions with Non-metallic Auxiliaries

The first methods of preparation of optically pure *P*-stereogenic compounds consisted of the separation of diastereomeric adducts of phosphine oxides with chiral auxiliaries.[1] Several quaternary phosphonium bromides or iodides were also resolved after exchanging the halide for a chiral anion. Some of the resolving agents employed in these studies are represented in Figure 2.1 and typical examples of compounds successfully resolved are listed in Table 2.1.

In the case of phosphonium salts (entries 3–6) once the individual epimers had been separated, stereoselective alkaline hydrolysis[32–34] or electroreductive cleavage of one of the groups[5,6] afforded the phosphine oxide or the free phosphine respectively.

Resolution of phosphine oxides involves protonation of the weakly basic phosphoryl group by the chiral auxiliary. This method showed limited success for simple monophosphine oxides (entries 1, 2 and 10) but is better suited to resolve compounds bearing carboxylic acid groups (entries 11 and 12) or diphosphine oxides (entries 16–20). It is still sporadically used nowadays, as for the compounds of entries 19 and 20.

The secondary phosphine oxide shown in entries 7 and 8 is the most important representative of the family of *P*-stereogenic secondary phosphine oxides (SPOs),[35] which can be used directly as P-donor ligands to transition metals. Furthermore, this particular compound has been also used to prepare many mono- and diphosphine oxides by stereoselective alkylation[16,36] and other reactions.[37–39]

To resolve phosphine boranes, an additional protic group is required (entries 13–15). Enantiomerically pure phosphinous acid boranes of entries 14 and 15 are versatile compounds that were transformed, often stereoselectively, to many other *P*-stereogenic compounds.[23]

Finally, the resolutions of the phosphines in entries 21–23 were carried out whilst the ligand was coordinated to Rh or Pd.

All the resolutions mentioned above involved non-covalent interactions. There are a few cases of classical resolutions involving covalent binding of the chiral auxiliary to the racemic phosphine and cleavage after the separation of diastereomers.

A recent example has been reported by Tang and co-workers,[40,41] who prepared series of bidentate ligands named POPs and BIBOPs (Scheme 2.1). The resolution of **rac-1** is performed through menthyl carbonate **2**, from which a single diastereomer can be obtained by recrystallisation. A simple hydrolysis step affords **(R)-1** quantitatively. This reaction has been performed on a kilogram scale.[40] From this enantiopure intermediate, several steps lead to the desired ligands.[40,41]

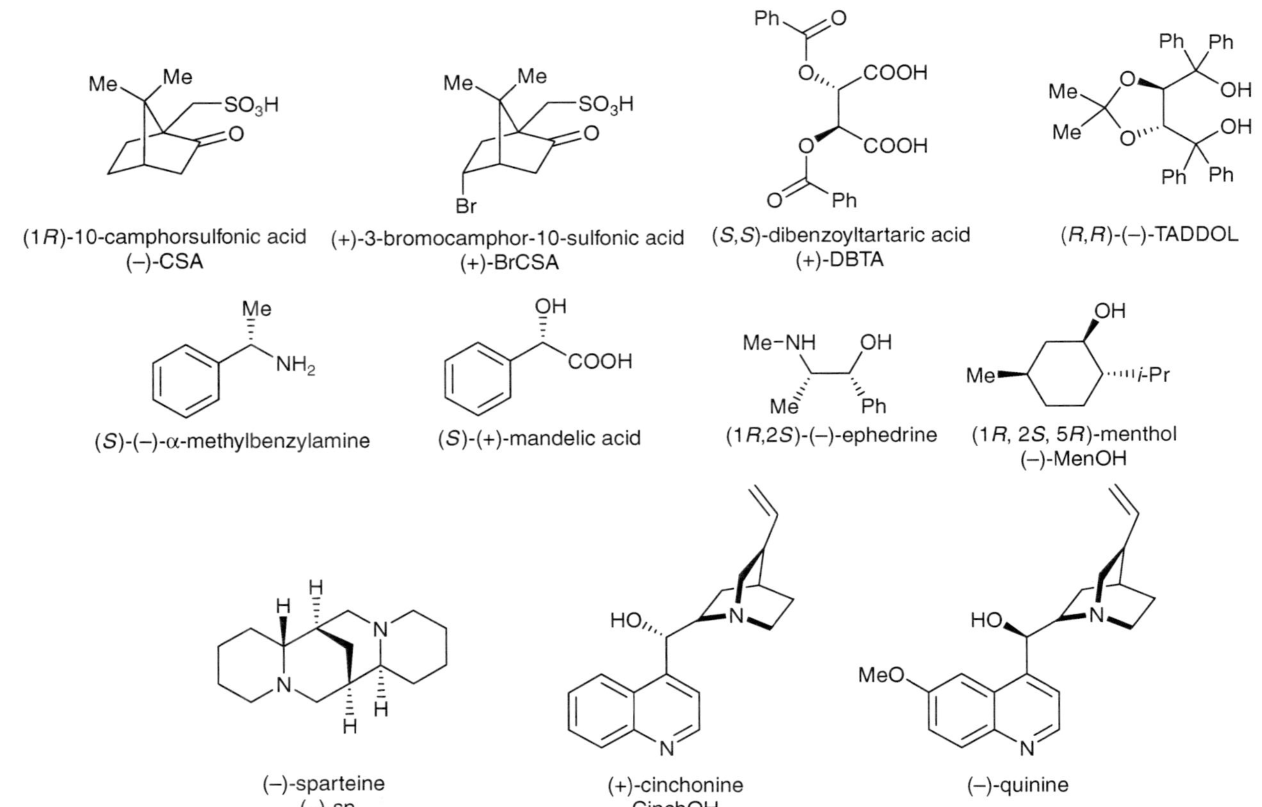

Figure 2.1 Chiral auxiliaries used in resolutions of phosphine oxides and phosphonium salts.

Table 2.1 Classical resolutions of phosphine oxides and phosphonium salts.

Entry	Compound	Auxiliary[a]	References
1	Ph–P(=O)(Me)Et	BrCSA	2,3
2	Ph–P(=O)(Me)Bn	CSA	3
3	Ph–P⁺(Bn)(Me)(R) X⁻ R = Et, *n*-Pr, *i*-Pr, allyl...	DBTA	4–7
4	phospholanium (Me-substituted) Br⁻, Ph, Bn	CSA	8
5	Ph–P⁺(Bn)(Me)–CH₂CH₂–P⁺(Bn)(Me)Ph 2Br⁻	DBTA	9
6	bis-phosphonium, 2PF₆⁻	DBTA	10
7	Ph–P(=O)(H)*t*-Bu	Mandelic acid	11, 12
8	Ph–P(=O)(H)*t*-Bu	DBTA	13
9	Ph–P(=S)(OH)*t*-Bu	α-methylbenzylamine	14–16

Table 2.1 (*Continued*).

Entry	Compound	Auxiliary[a]	References
10	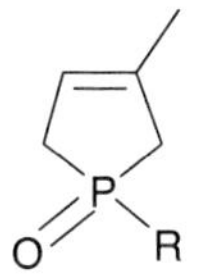 R = Me, Et, Ph, *o*-Tol, OEt...	TADDOL DBTA	17–19
11	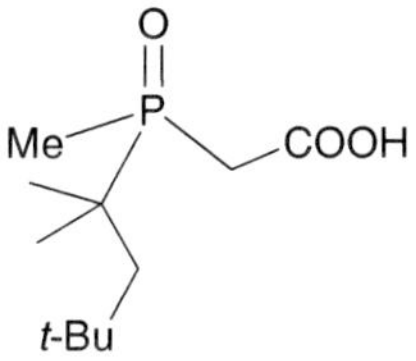	α-methylbenzylamine	20
12	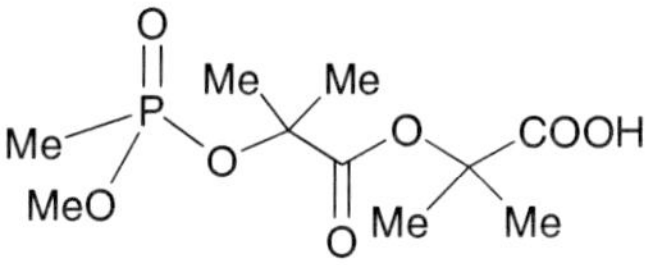	Quinine or ChinhOH	21
13[b]	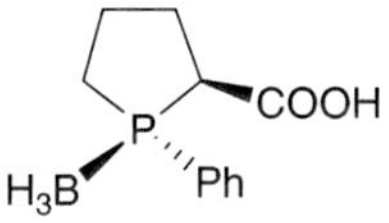	Quinine	22
14	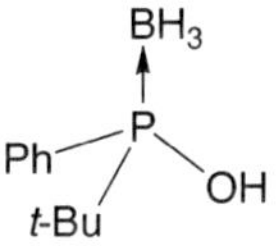	Ephedrine/CinchOH	23
15	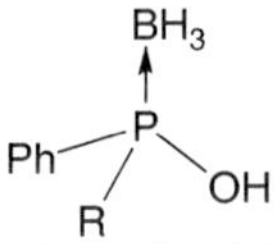 R = Me, Bn, *o*-An, 1-Naphth	CinchOH	24
16[c]	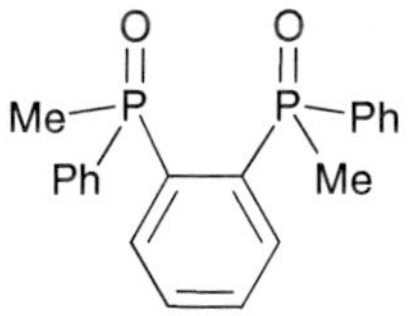	DBTA	25
17[c, d]	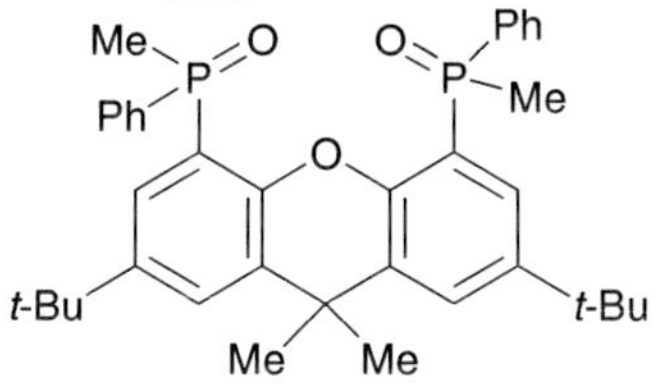	DBTA	26

Table 2.1 (*Continued*).

Entry	Compound	Auxiliary[a]	References
18	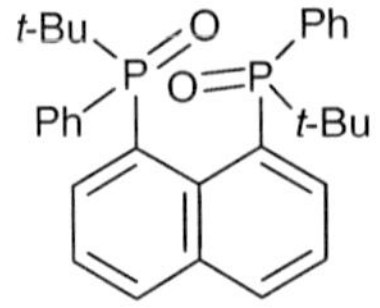	DBTA	27
19[e]	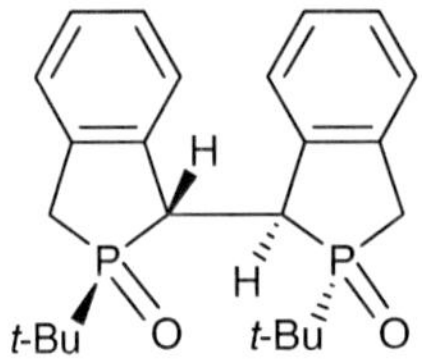 DuanPhos(O)$_2$	DBTA	28
20[e]	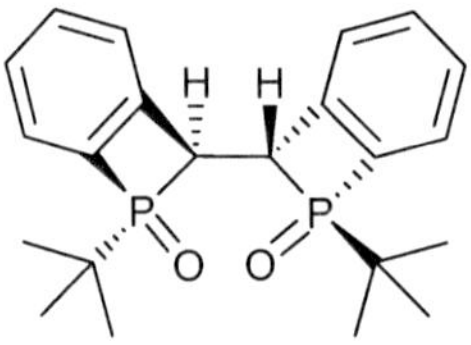	DBTA	29
21[f]	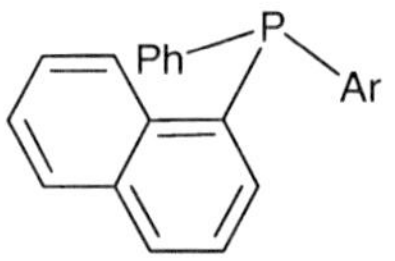 Ar = o-, m-An NPTP	BrCSA	30
22[f]	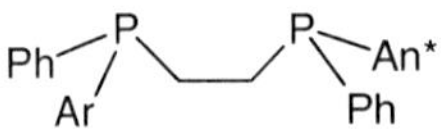 Ar = o-, m-An, 1-Napth	BrCSA	30
23[g]	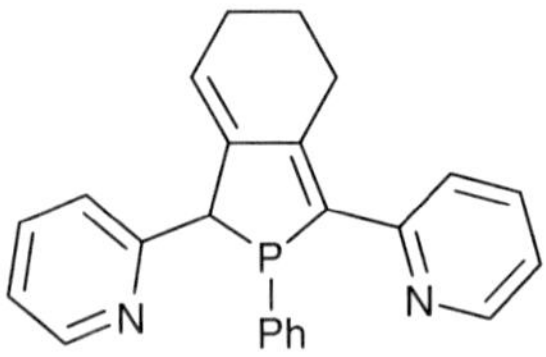	α-methylbenzylamine	31

[a]See Figure 2.1.
[b]The *cis* diastereomer was removed by recrystallisation, only one enantiomer of the *trans* isomers is shown.
[c]The racemic mixture was obtained by recrystallisation and resolved.
[d]The *meso* isomer was epimerised to the *rac* and recycled.
[e]Only one enantiomer shown.
[f]The resolution was carried out with the ligand complexed to Rh.
[g]The resolution was carried out with the ligand complexed to Pd.

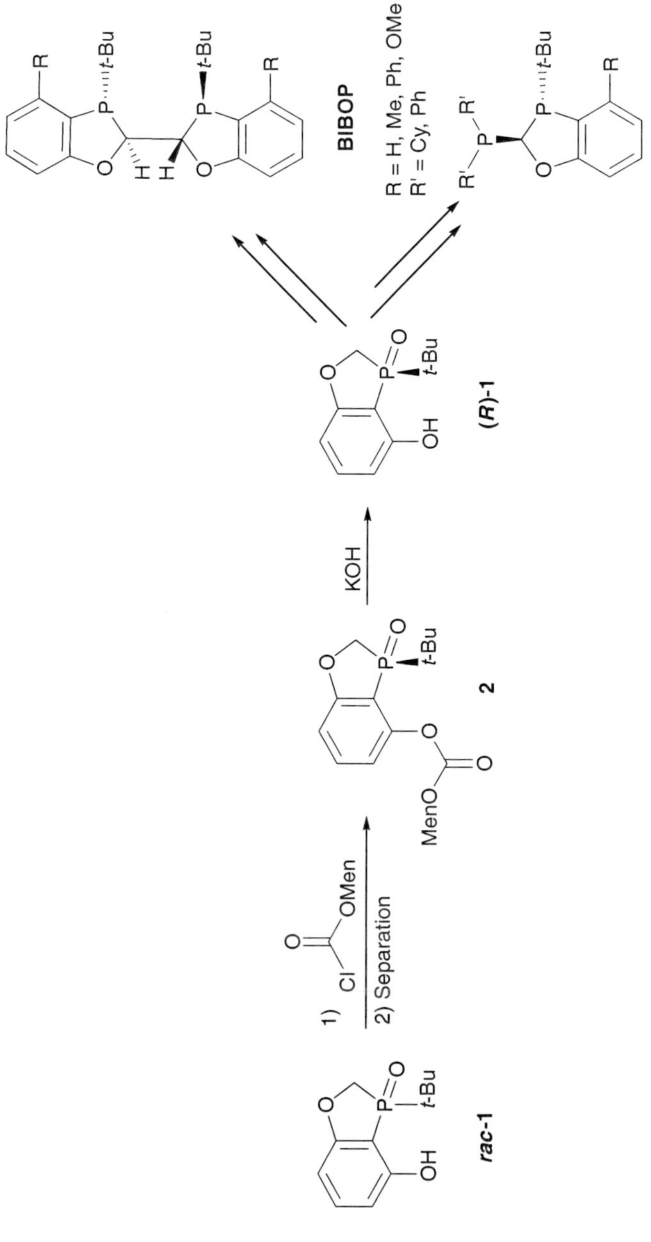

Scheme 2.1 Preparation of **POP** and **BIBOP** ligands.

Enantiopure 2-formyl-1-phosphanorbornadiene **3** was prepared by Mathey and co-workers (Scheme 2.2).[42] The resolution was carried out by acetylisation with chiral diol **4**, separation of diastereomers and deprotection back to the aldehyde.

Once separated, optically pure **3** was transformed into other monophosphines such as brominated phosphine oxide **5**[42] and iminophosphines **6**.[43] The P_2N_2 ligand **7** was prepared by condensation with ethylenediamine or (*S,S*)-1,2-diaminocyclohexane.[44] The imine double bond in **6** and **7** could be easily reduced with $NaBH_4$. The McMurry coupling of **3** produced the corresponding

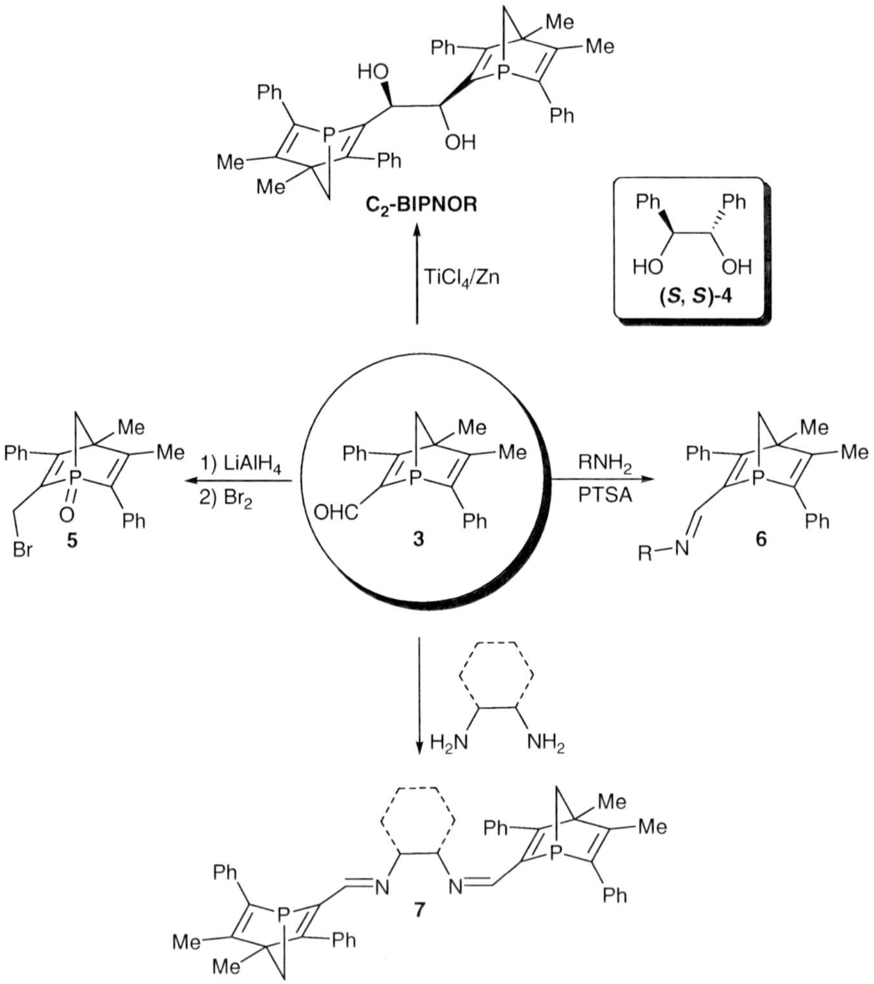

Scheme 2.2 Preparation of 1-phosphanorbornadienes from **3**. Only one enantiomer of each is shown.

optically pure diol in 70% yield. This diol was dubbed C_2-BIPNOR, because is similar to BIPNOR (see Section 2.2.2) but with an ethyl bridge.[45]

A special variant of resolution is called *oxidative resolution*, because it yields optically pure phosphine oxides from racemic phosphines. To this end, Keay and co-workers[46] used the Staudinger reaction between racemic phosphines and an optically pure azide (Scheme 2.3).

The azide **9** was obtained from (+)-CSA[47] and reacted with several racemic phosphines affording phosphinimines **10** in very good yields. They could be separated by conventional methods and each diastereomer was easily hydrolysed under acidic conditions giving the phosphine oxides **11** in excellent yields and with clean inversion of configuration at the phosphorus atom.[46]

Pietrusiewicz and co-workers[48] resolved a bicyclic phosphole by an oxidative resolution protocol *via* a phosphonium salt (Scheme 2.4).

Quaternisation of ***rac*-12** with (–)-menthyl bromoacetate and separation of the salts by recrystallisation afforded optically pure **13**, which was hydrolysed to **(R_P)-14** in 25% yield from ***rac*-12**. The other enantiomer could also be obtained from the mother liquor that had furnished the crystals of **13**. Compound **(S_P)-14** was hydrogenated to the saturated phosphole oxide, which was reduced with phenylsilane with retention of configuration at the P atom.

Finally, there are some examples of miscellaneous resolutions such as kinetic resolutions,[1] or dynamic thermodynamic resolutions.[49] Some of them are listed in Table 2.2.

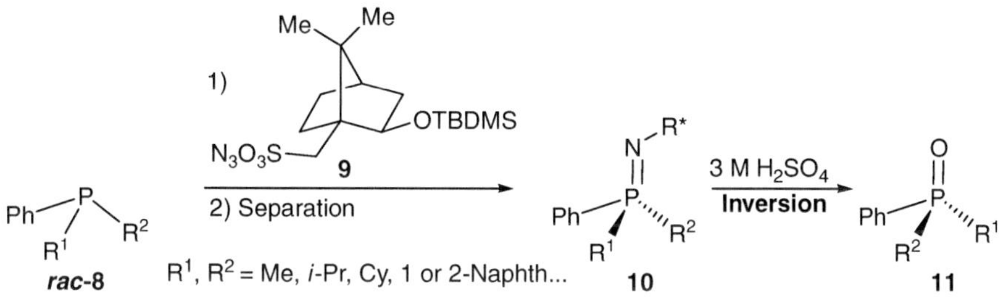

Scheme 2.3 Oxidative resolution of phosphines using intermediate phosphinimines.

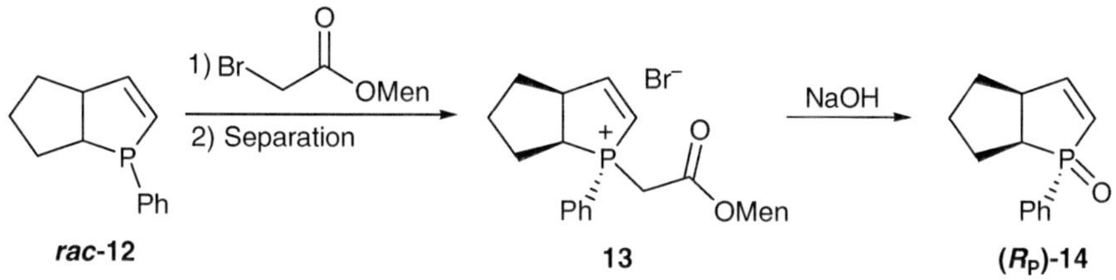

Scheme 2.4 Oxidative resolution of ***rac*-12** using an intermediate phosphonium salt.

Table 2.2 Miscellaneous resolutions of phosphine oxides and phosphonium salts.

Entry	Compound	Auxiliary	References
1	Ph–P(1-Napth)(*p*-Biph)	CSA	50
2	Ph–P^{+}(Et)(Bn)(Et$_2$N) Br^{-}	DBTA	51
3	Ph–P(Me)(R); R = *o*-An, *o*-Tol, 1-Naphth	See text	52,53
4	Ph–P(Me)(R); R = *t*-Bu, *o*-An, *o*-Tol....	MenOH and related alcohols	54
5	*t*-Bu–P(Me)(R)→BH$_3$; R = Et, Ph	(–)-sp	55

The first (partially) successful resolution of a phosphine (entry 1) is due to Wittig and co-workers[50] and was based on the formation of a diastereomeric phosphonium salt. Several years later, Horner and Jordan[51] resolved the aminophosphonium salt of entry 2. Much more recently, Mikolajczyk, Daran and co-workers[52,53] have resolved several phosphines (entry 3) using *P*-stereogenic *bis*(phosphinoyl) and *bis*(phosphinothioyl) disulfides, but the optical purity of the resulting oxides or sulfides remained relatively low. Gilheany and co-workers[54] have studied the oxidative dynamic resolutions of phosphines of entry 4 under Appel conditions with hexachloroacetone as chlorinating agent. Under these conditions, the phosphine oxide and an alkyl chloride are obtained. Quantitative yields and ee of up to 80% were obtained, the highest ever reported enantioselective oxidation of a phosphine.[54]

The first kinetic resolutions of phosphine boranes have been recently reported by O'Brien and co-workers[55] (entry 5), *via* deprotonation with *s*-BuLi/(–)-sp. With this procedure, products with up to 82% ee were obtained.

2.2.2 Direct Resolutions with Cyclometallated Pd Complexes

Given the high tendency of trivalent phosphorus compounds towards coordination to transition metal ions, is not surprising that enantiopure complexes

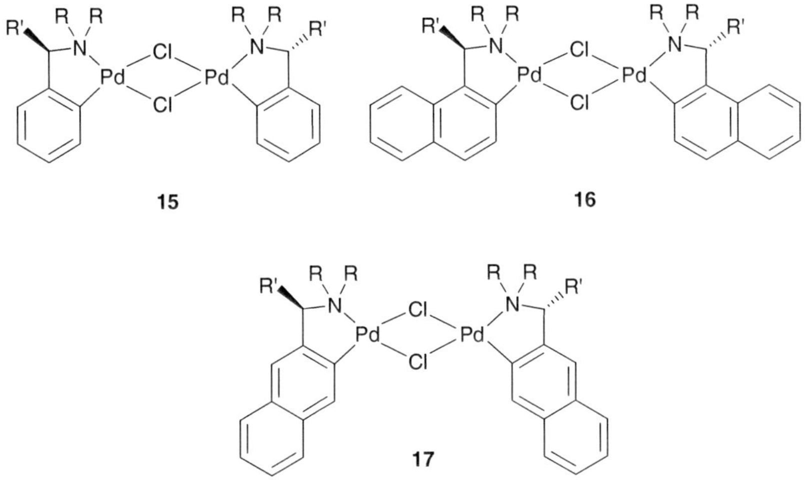

15   **16**

17

Figure 2.2 Cyclometallated Pd(II) dimers used to resolve racemic phosphines. Only one enantiomer of the chiral amine and only the *cis* isomer for each complex are represented. Usually, R = R′ = Me.

have been used to resolve racemic mixtures of phosphines.[1] As early as 1968 Chan[56] used a chiral platinum complex to resolve *tert*-butylmethylphenylphosphine. Up to now, however, the most successful systems are certain chlorobridged dimeric palladacycles containing *ortho* cyclometallated optically pure amines (**15–17**, Figure 2.2).[1,57,58]

These dimers can be easily prepared from the chiral amines and exist as mixtures of *cis* and *trans* isomers in solution, but high yields of the *cis* isomers can be obtained upon concentration of solutions of **16**.[59] The rearrangement probably occurs through a highly reactive mononuclear intermediate generated after splitting of the chloro bridge.[58] This bridge can also be easily cleaved by racemic *P*-stereogenic phosphines, with the formation of diastereomeric mononuclear complexes **18** amenable to separation by conventional methods (Scheme 2.5).

Complexes **18** always present a relative *cis* disposition between the phosphine and the metallated carbon atom. The separation of **18** and **18′** by column chromatography or fractional crystallisation has been successful for a number of mono- and polydentate phosphines. The desired optically pure phosphines can then be easily liberated by the addition of a strongly coordinating ligand L such as cyanide or dppe, forming insoluble complexes and leaving the free *P*-stereogenic phosphine in solution. The exact experimental procedure for the resolution depends on the phosphine. Sometimes 4 equivalents of racemic phosphine are added to the dimer to quantitatively complexate the phosphine while sometimes only 2 equivalents of phosphine are used with the aim of coordinating only one – ideally – of the enantiomers of the phosphine and leaving the other in the mother liquor after precipitation and filtration of the

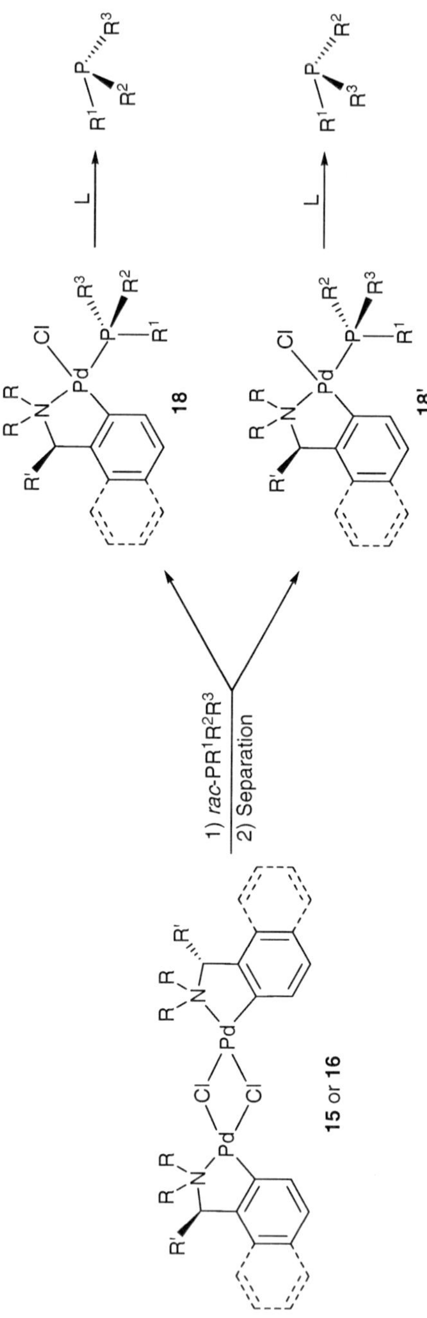

Scheme 2.5   Resolution of monophosphines by means of chiral palladacycles.

complex **18**. In theory both enantiomers of the phosphine should be accessible by a single resolution,[60] but in practice that is not always the case.[61] The driving force of these resolutions seems to be the difference in solubility between the diastereomeric complexes **18** and **18′** and not differences in complexation rates.[1] Single crystal X-ray analysis of **18** has been used to unambiguously determine the absolute configuration of the coordinated phosphine[60–63] although that is not always required because careful NMR studies have also been used with satisfactory results.[60,62–65] Some work has also been done for the recovery of the resolving agent after the resolution has been carried out.[60,66]

This types of resolution were originally described in 1971 by Otsuka and co-workers[67] with **15** (R = R′ = Me) and since then they have been applied to numerous racemic *P*-stereogenic phosphines. In spite of an early use of complex **17** (R = R′ = Me),[68] **15** or **16** are more typically used. A limitation of the method is that the optimum complex for the resolution depends on the exact phosphine being resolved.[60,63,69] Naphthyl complex **16** is frequently a better resolving agent than **15**, but also more expensive.[58,70,71]

Most of the monodentate phosphines resolved to date by means of cyclo-metallated palladacycles are listed in Table 2.3.

Table 2.3 shows that the procedure has been successfully employed to resolve triaryl (entries 1 and 2), diaryl (entry 3) or monoaryl phosphines (entries 4–6, 10 and 11), all of them bearing a phenyl group.

Several of the phosphines are either relatively bulky (entries 1, 6, 8 and 9) or bear two very similar substituents (entry 2). Those phosphines would be difficult to prepare by other methods in optically pure form, proving the usefulness of the method described here. The phosphine of entry 9 contains two stereogenic centres and was reacted to **16** forming **18** as a 1:1:1:1 mixture of all the possible diastereomers. Three of them could be obtained in optically pure form after careful chromatographic separation and recrystallisation.[75] Entries 10 and 11 show that racemic secondary phosphines can also be resolved. Once decomplexed, however, those phosphines were found to racemise in minutes (entry 10) or hours (entry 11) probably by an acid-catalysed process.[70] Entry 12 constitutes the first resolution of a *P*-stereogenic fluorophosphine, which could be isolated in optically pure form (*S*) although it was found to racemise in benzene solution after only 6 h.[58,77] Chlorophosphine of entry 13 racemised immediately after decoordination.[79] The methylphosphinite of entry 14 was also prepared by the same resolution procedure. More interestingly, it can also be prepared by reaction of the complex **18** containing the coordinated fluoro- or chlorophosphines of entries 12 and 13 respectively with NaOMe or MeOH/NEt$_3$. These reactions occur with inversion of configuration at phosphorus.[78,79] The methylphosphinite was isolated in 92% ee after liberation and was found to be configurationally stable.

The palladacycles discussed here are very effective for the resolution of bidentate mono- and diphosphines. For C_2-symmetric diphosphines, a single pair of cationic complexes is formed but for C_1-bidentates two isomers can be formed.[80] For the resolution of these complexes the chloride counterion is usually exchanged by hexafluorophosphate for solubility reasons.[58]

Table 2.3   Monodentate phosphines resolved by means of chiral palladacycles.

Entry	Phosphine	References
1	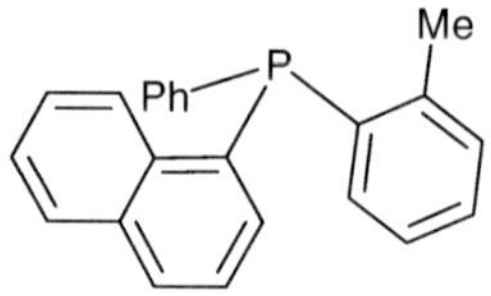	67
2	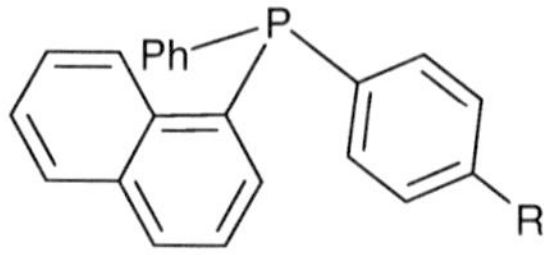 R = Ph, OMe, OEt	67, 68
3	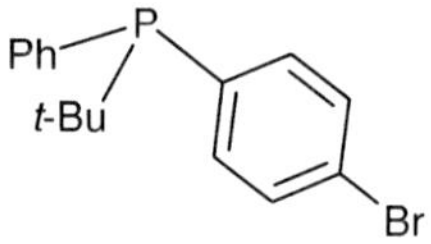	60
4	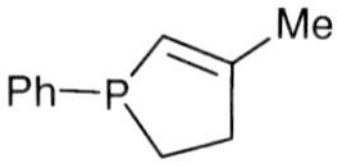	68
5	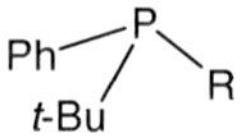 R = Me, *i*-Pr	66, 68, 69
6	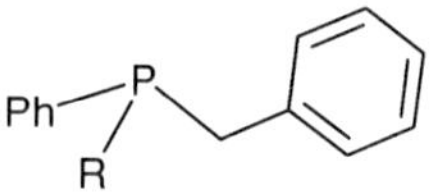 R = *i*-Pr, Cy, Mes	62–64, 72
7	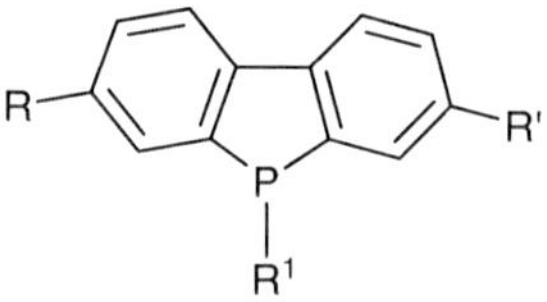	73
8	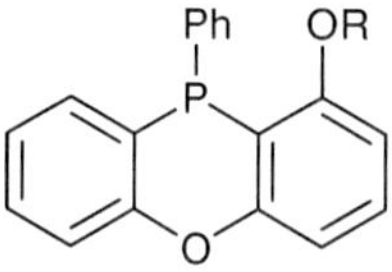 R = H, OC(O)OMen	74
9	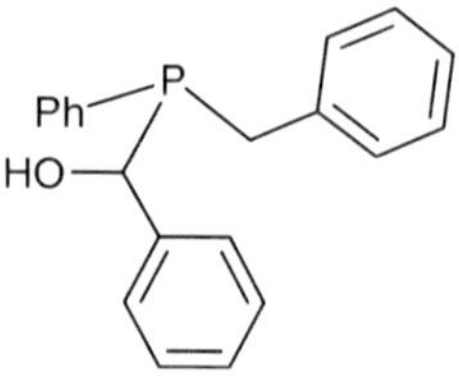	75

Table 2.3 (*Continued*).

Entry	Phosphine	References
10	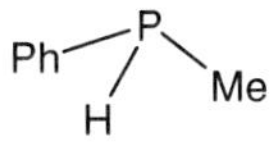 Ph–PH–Me	70, 76
11	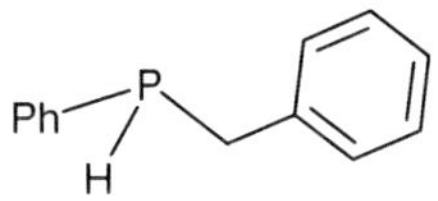 Ph–PH–CH₂Ph	63, 70
12	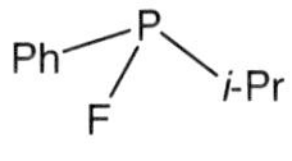 Ph–PF–*i*-Pr	77, 78
13	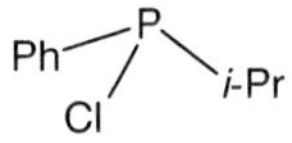 Ph–PCl–*i*-Pr	79
14	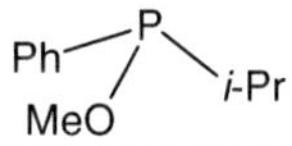 Ph–P(MeO)–*i*-Pr	78, 79

Most of the bi- and polydentate phosphines resolved to date by means of palladacycles are listed in Table 2.4.

Both enantiomers of the rigid C_2-symmetric phosphine of entry 1, named diph, were obtained in very good yields (85 and 90%) using complex **15** (R = R′ = Me) in multi-gram scale and used in Rh-catalysed hydrogenation.[82] This diphosphine racemises readily upon heating.[81] All four isomers of the interesting P- and As-stereogenic bidentate ligand of entry 4 could be obtained in enantiomerically pure form.[84] Upon heating at 140 °C for 2 h, the phosphorus atom was found to epimerise while the configuration of the As centre remained intact. The coordination chemistry of the ligands in entries 1 and 4 towards groups 11 and 12 cations has been explored by Wild and co-workers.[96–100] The tetra- and pentadentate ligands of entries 7 and 8 were obtained from the resolved (*R*)-phosphine of entry 6 and separated by complexation to Co(III).[61]

The diphosphine BIPNOR of entry 10 is particular because it bears non-racemisable bridgehead stereogenic phosphorus atoms. The resolution of the racemate with **15** is very efficient, giving over 90% yield of each enantiomer.[87] An identical procedure was followed to prepare the 1,1′-ferrocenylene-bridged analogue of entry 11.[89]

Bidentate P,X ligands of entries 13–15 were also efficiently resolved. For the liberation of those phosphines from the complex, racemic diph (entry 1) had to be used. Interestingly, enantiomerically pure (*R*)-β-thiophosphine of entry 15 rearranges in light by a radical mechanism giving optically pure (*S*)-ethyl-methylphenylphosphine sulfide with complete retention of configuration.[94]

The unsymmetrical epmpe phosphine of entry 18 was obtained from dpmpe (entry 16) by selective cleavage of a phenyl group of the diphenylphosphino moiety.

Table 2.4 Bi- and polydentate phosphines resolved by means of chiral palladacycles.

Entry	Phosphine	References
1[a]	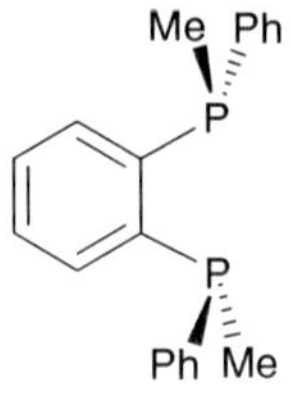 Me Ph P P Ph Me **diph**	81, 82
2	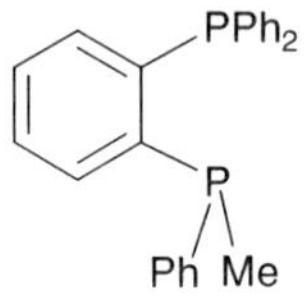 PPh_2 P Ph Me	71
3	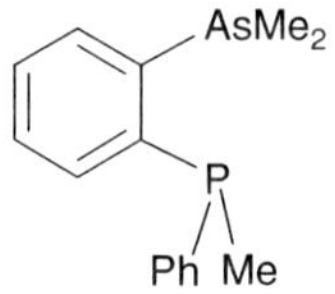 $AsMe_2$ P Ph Me	83
4[a]	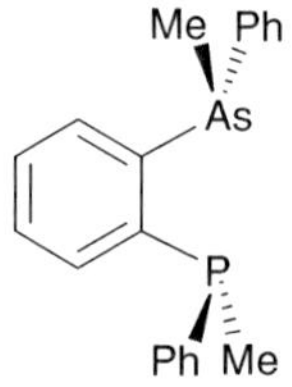 Me Ph As P Ph Me **phas**	84
5	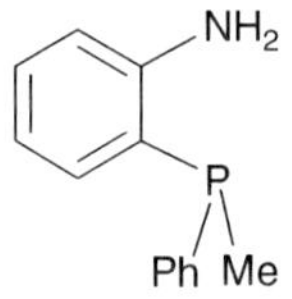 NH_2 P Ph Me	85
6	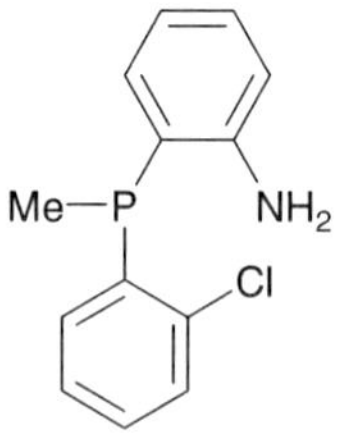 Me—P NH_2 Cl	61

Table 2.4 (*Continued*).

Entry	Phosphine	References
7[b]	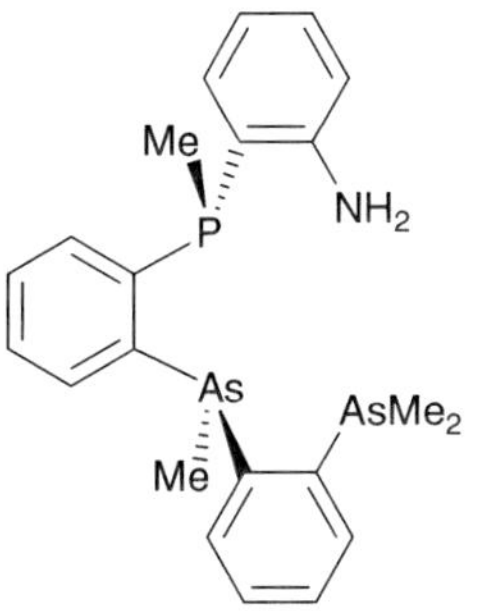	61
8[c]	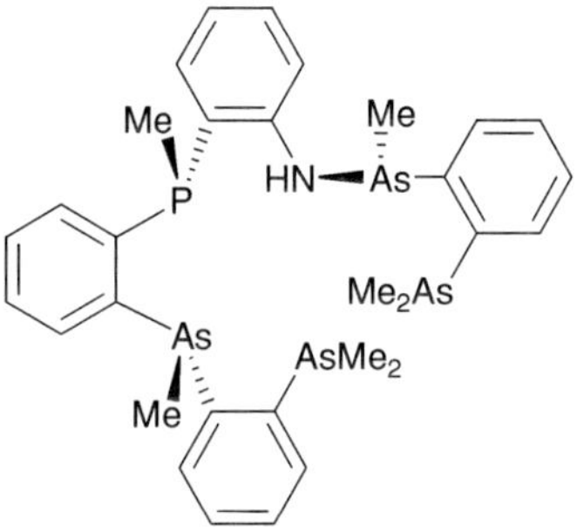	61
9	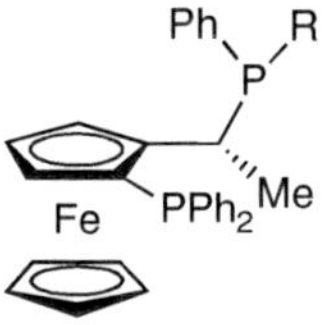 R = Cy, *o*-An	86
10[a]	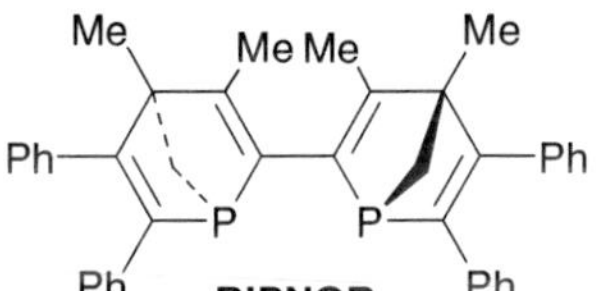 **BIPNOR**	87, 88
11	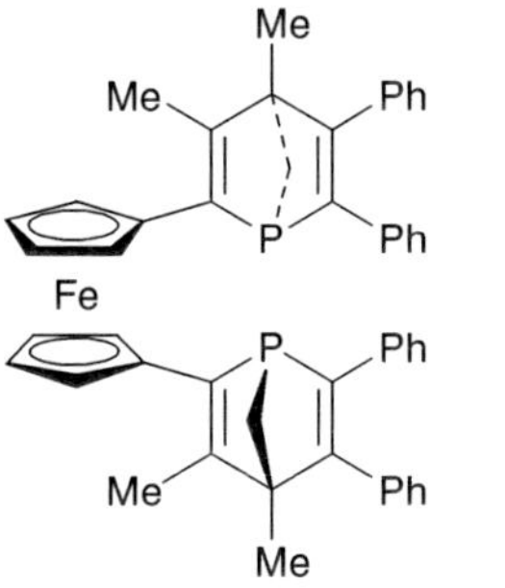	89

Table 2.4 (*Continued*).

Entry	Phosphine	References
12	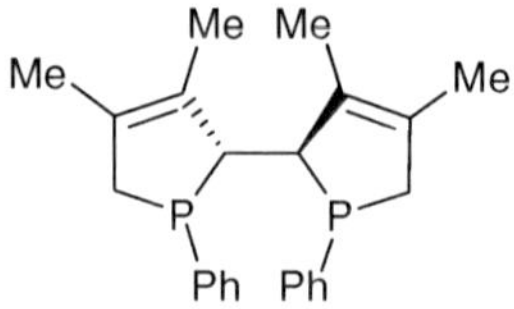	90, 91
13	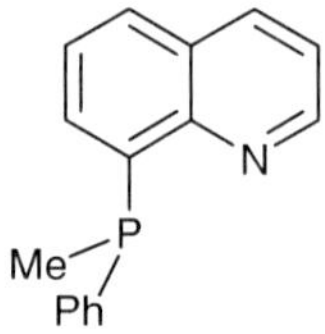	92
14	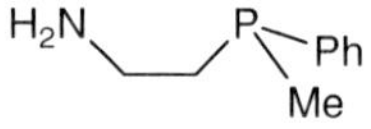	93
15	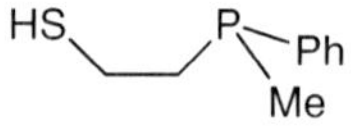	94
16	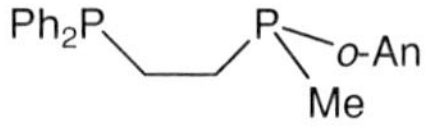	95
17	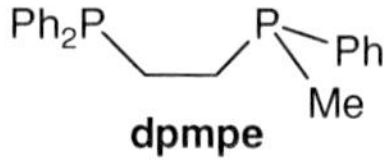 **dpmpe**	80
18[a,d]	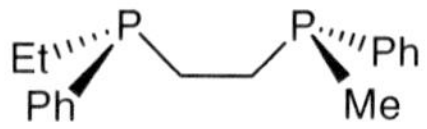 **epmpe**	80

[a]The (*R*,*R*) isomer is represented.
[b]The (S_{As},R_P) isomer is represented.
[c]The (R_{As},S_P,R_P) isomer is represented.
[d]The enantiomers of this phosphine were not separated.

A few linear polyphosphines have been resolved into the homochiral individual enantiomers by cyclometallated complexes (Figure 2.3).

Tetraphosphine tetraphos was the first tetra(tertiary) phosphine to be resolved,[101,102] by using complex **15** (R = R′ = Me), once the *meso*-isomer had been removed by recrystallisation. A similar method was employed to prepare the tetradentate ligand diphars.[103] From enantiomerically pure (*S*,*S*)-tetraphos, penta- and hexaphosphines **19** and **20** were prepared by reductive cleavage of one (or two) of the phenyl groups at the terminal phosphorus atoms to generate the mono- or *bis*(secondary phosphine), which were reacted with diphenylvinylphosphine giving a statistical, isomeric mixture of **19** and **20**, but with the

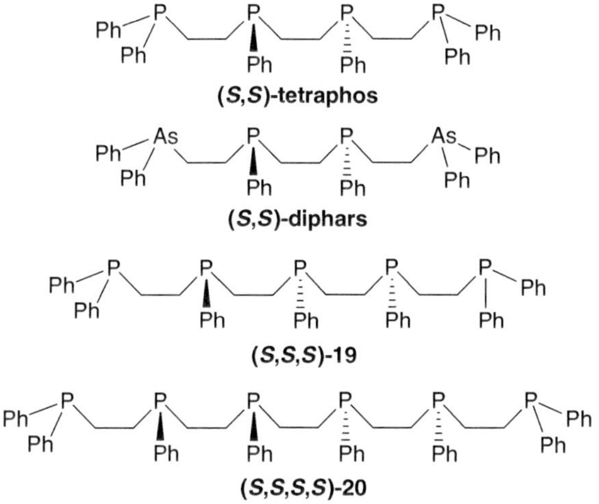

(S,S)-tetraphos

(S,S)-diphars

(S,S,S)-19

(S,S,S,S)-20

Figure 2.3 Polyphosphines and arsines resolved by palladacycles. Only the *all-S* enantiomers are represented.

inner phosphorus stereocentres configuration untouched. The separation of the isomers was achieved by HPLC purification of the phosphine borane adducts.[104] The specific optical rotation of dichloromethane solutions of enantiopure tetraphos and **20** present a pronounced dependence on temperature, indicating that single-handed helical conformations are present in solution.[105]

Even more interesting is the fascinating coordination chemistry of these fully flexible polyphosphines with many transition metals, which has been studied extensively by Wild and co-workers.[102–104,106–108] For example, it has been found that optically pure tetraphosphine (*S,S*)-tetraphos spontaneously self-assembles into double-stranded helicates in the presence of Ag(I) or Au(I) salts.[102,106] The phosphines presented here are exceptional in the sense that there are only a few other examples of resolved *P*-stereogenic polyphosphines, most of them discussed in Chapter 5, Section 5.2.1.5.

2.2.3 Direct Resolutions by Chromatographic Methods

Determination of the optical purity of compounds by HPLC systems equipped with chiral stationary phases has become an indispensable method in enantioselective synthesis and *P*-stereogenic compounds are not an exception. The optical purity of those compounds was traditionally determined by optical rotation or by NMR integration of diastereomeric adducts with several chiral shift reagents,[65,109–111] but nowadays the most common way is to analyse the phosphine borane, oxide or sulfide by chiral HPLC. Given the high sensitivity of phosphines to oxygen, analyses are carried out in protected derivatives. In theory, the same methods could be used for the separation of racemates in gram or even larger scale in an extremely time-efficient way. In practice, however, that requires special apparatus, preparative columns and large volumes of high quality solvents, which are beyond the economic possibilities of many research laboratories. In spite of that, there are examples of successful separations and some of them are presented in this section. When there are no enantioselective synthetic routes to a particular phosphine and the chemical resolution methods fail, a chromatographic separation of the racemate is worth considering, as may be the only way to obtain the enantiopure compound.

Initially, a relatively long list of phosphine oxides were resolved with several degrees of efficiency.[1] Since then the performance of the chromatographic equipment has improved dramatically and nowadays many more phosphine oxides, along with an increasing number of phosphine boranes, have been resolved. Some representative examples of both types of compounds are listed in Table 2.5, which is by no means comprehensive.

In the early studies, the phosphine oxides resolved contained substituents with aromatic condensed rings (entries 2 and 3) but later other oxides without that substitution were also resolved (entries 1 and 5) and even secondary phosphine oxides (entry 7). Entries 8–15 show some resolutions of the phosphine boranes including examples with several stereogenic centres in the backbone (entries 11–14). Compound of entry 11 can be regarded as a hybrid between BPE and DiPAMP. A few diastereomeric pairs have also been resolved by HPLC (entries 16 and 17), because conventional methods such as column chromatography or recrystallisation failed.[127,128]

An important practical parameter of these resolutions is the scale at what they are practicable, but unfortunately this detail is not always available even in primary literature. Ideally, for laboratory applications, multi-gram separations are sought, a feature that has been achieved in a number of occasions (entries 5, 8 and 16). The separation of entry 8 is remarkable for the relative large scale although the dimensions of the custom-designed MPLC column are also equally remarkable: 1 m high, 20 cm of diameter fitted with 17 kg of stationary phase with a flux of $5\,L\,h^{-1}$ of ethanol![119] Although the performance of this separation could be probably improved with modern equipment, it gives an idea of the costs involved in such a large-scale separations.

In spite of the economic costs mentioned, it is appropriate to cite here the statement of Hoge and co-workers[120] related to the separation of the two enantiomers of Trichickenfootphos borane (entry 9): 'despite the stigma associated with chiral preparatory HPLC separations, this type of separation is becoming increasingly important to the pharmaceutical industry'. Therefore, if the use of *P*-stereogenic ligands increases as expected, then the chromatographic methods for their resolution will become important, particularly in industry.

2.2.4 Dynamic Resolution of Secondary Phosphine Boranes

Wolfe and Livinghouse[129] reported a powerful method for resolution of the racemic *tert*-butylphenylphosphine borane (**21**) *via* deprotonation in the

Table 2.5 Some phosphine oxides and boranes resolved by HPLC.

Entry	Compound	Scale	References
1	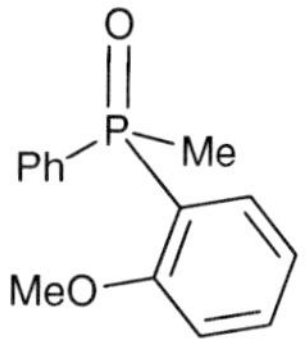 PAMPO	–	112
2	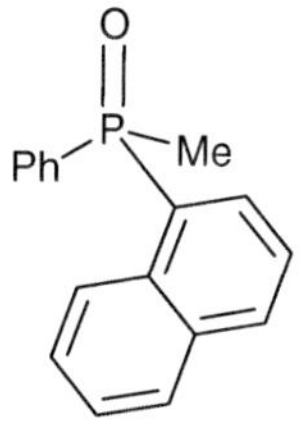	150 mg racemate	113,114
3	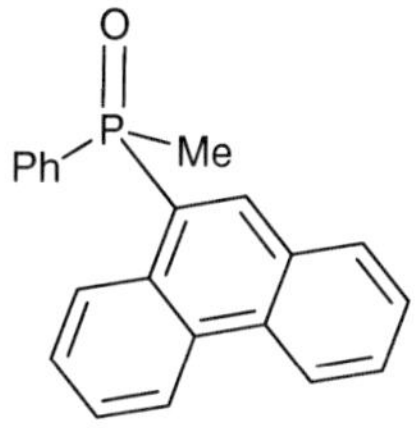	–	113,114
4	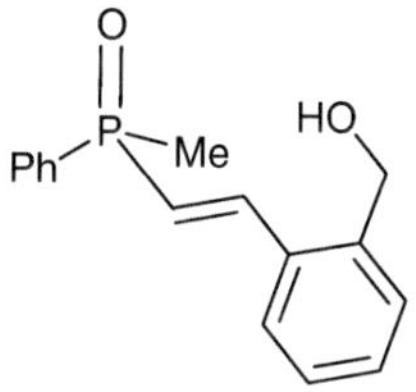	30 mg racemate 14 mg (R), 13 mg (S)	112
5	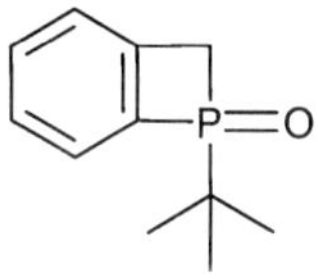	3 g racemate 1.3 g (R), 1.3 g (S)	29
6	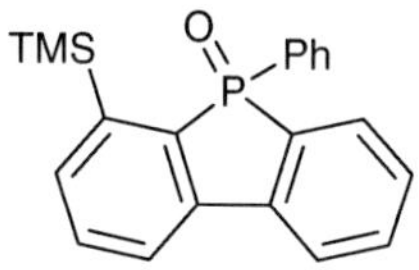	300 mg racemate 95% mass recovery	115
7	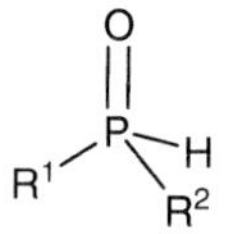 R^1, R^2 = i-Pr, t-Bu, Ph, 1-Naphth...	100 mg racemate	116–118

Table 2.5   (*Continued*).

Entry	Compound	Scale	References
8[a]	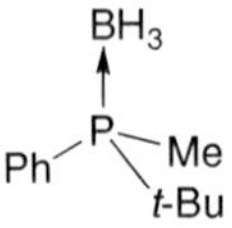	25 g racemate 11.4 g (*R*), 10.7 g (*S*)	119
9	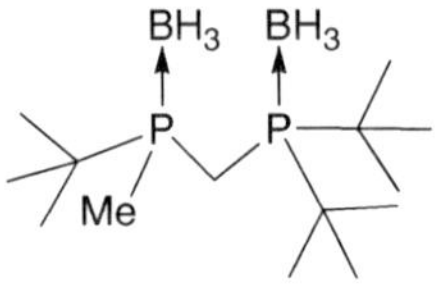 **Trichickenfootphos·(BH$_3$)$_2$**	90% mass recovery	120
10	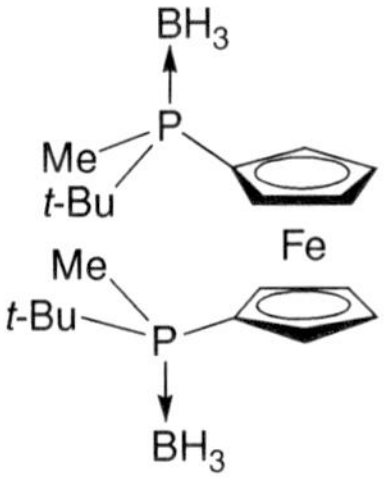	200 mg racemate	121
11	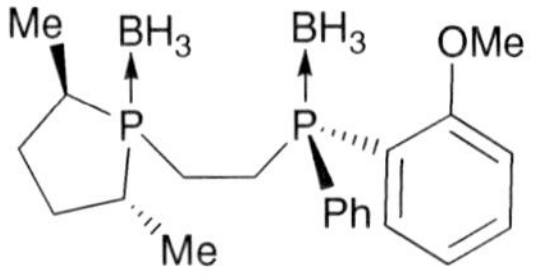	–	122
12[b]	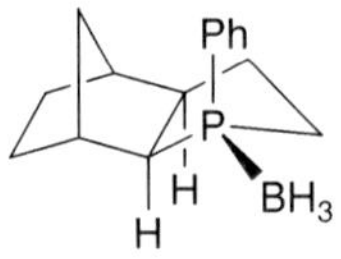	–	123
13[b]	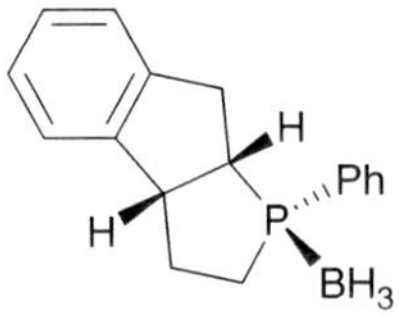	–	123
14	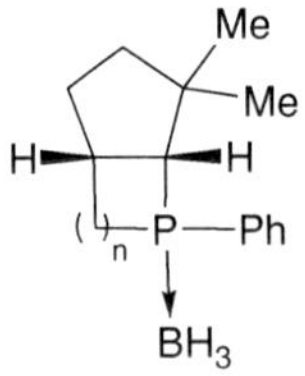 n = 1, 2	–	124
15[c]	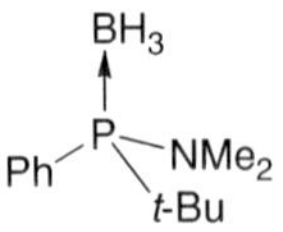	30 mg (*R*), 30 mg (*S*)	125

Table 2.5　(*Continued*).

Entry	Compound	Scale	References
16[d]	BH₃ ... P, Me, SBor (adamantyl structure)	2 g diast. mixture 0.64 g (R_P), 0.67 g (S_P)	126
17	BH₃ ... R, P, OMen, R¹ (aryl structure) R = Me, Et, Ph R¹ = *t*-Bu, Ph	–	127,128

[a]The resolution was performed in a custom-designed MPLC system.
[b]The racemate was separated, only one of the enantiomers is shown
[c]Amounts obtained after 12 injections.
[d]Bor = (–)-borneol, see Section 2.3.2.

presence of a chiral amine (Scheme 2.6), a method which is an example of dynamic thermodynamic resolution.[49]

Rac-**21**, which can be easily prepared from dichlorophenylphosphine, was deprotonated in the presence of (–)-sparteine (a natural product with high affinity to complexate the lithium cation, as will be seen in Chapter 5 in more detail), to generate a mixture of the lithium phosphides **22** and **22′**. Initially, both isomers are probably formed in similar amounts but the chiral environment generated by the coordinated sparteine makes one of them (**22**) more stable and upon equilibration it becomes predominant. Electrophilic quenching of this mixture with alkyl halides therefore forms enantioenriched tertiary phosphine boranes. The degree of optical purity of the products depends on the time and temperature of the reaction. To achieve high enantioselectivities is very important to ensure that thermodynamic equilibration of **22** and **22′** takes place efficiently. Both the initial metallation and the final electrophilic quenching take place easily at − 78 °C but the equilibration step in between needs higher temperatures: warming the reaction at 0 °C for 30 min affords a product with 35% ee, whereas at 25 °C for 1 h the product presents 95% ee.[129] The solvent is also important. Good results are obtained in diethyl ether, but changing to THF rends the reaction completely nonstereoselective, probably due to the higher coordinating ability of THF to the Li⁺ cation, disturbing its coordination to sparteine.[130] An interesting experimental observation is the formation of a 'voluminous precipitate'[129] during the equilibration **22** and **22′**, suggesting that the driving force of the dynamic resolution is related to solubility.

The sense of the asymmetric induction was determined by analysing the known[131] monophosphine **23** with R = Me and was independently confirmed by the determination of the crystal structure of the adduct **22**,[132] which was

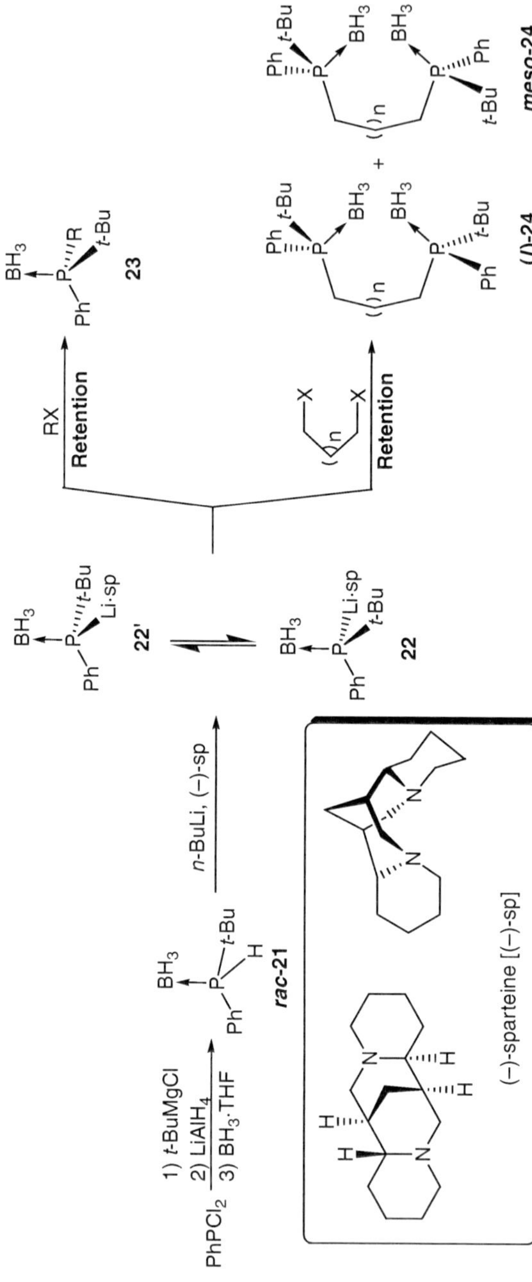

Scheme 2.6 Dynamic thermodynamic resolution of **22** and electrophilic quenching, (*l*) = (*R*,*R*) + (*S*,*S*).

found to contain the enantiomerically pure [P(BH$_3$)(t-Bu)Ph] anion arranged into dimers with C_2-symmetry. The crystal structure also shows the optimum fitting of anion with the rings of sparteine. This method has been employed with several electrophiles. Most of the mono- and diphosphine boranes prepared to date are listed in Table 2.6.

Table 2.6 shows that mono- and bidentate ligands were efficiently prepared in good yields and excellent optical purities, usually after a single recrystallisation. As expected, the optical purities of the products are all similar because they arise from the same diastereomeric mixture **22** and **22'**, without any noticeable racemisation in the alkylation step. In the case of the phosphine borane of entry 4 the synthesis was carried out in relatively large scale (38 g) without loss of optical purity of the product. The alcohol group of the β-hydroxyphosphine borane of entry 8 was reacted with a bulky phosphochloridite to afford phosphine phosphite ligands containing *C*- and *P*-stereogenic atoms.[135] Phosphine boranes of entries 11 (X=C) and 12 were direct precursors of the first *P*-stereogenic PCP pincer ligands and were prepared to study their metallation reactions towards Ir, Pd and Pt.[130,137]

A limitation of the method is that all the phosphine boranes of Table 2.6 bear the –P(BH$_3$)(t-Bu)Ph group. The dynamic resolution procedure could become much more versatile method if it could be applied to wide variety of secondary phosphine boranes. Surprisingly, this subject has been scarcely studied and only a few other aryldialkylphosphine boranes have been prepared (Scheme 2.7 and Table 2.7).

Table 2.7 shows that the extension to other substrates is challenging at least under the optimised conditions for **rac-21**. Only the closely related substrate of entry 5 gave an enantioselectivity comparable to the parent system. Interestingly, in the other cases no precipitate was observed during the equilibration of the two enantiomers of the phosphide borane, suggesting that the precipitation is a key event in the success of the reaction and inducing it (with other solvents, for example) may be a way to improve the results of Table 2.7.[130]

Another limitation of the original method is that only one enantiomer of the phosphines can be prepared, because (+)-sparteine is difficult to obtain. O'Brien and co-workers[139] used some chiral amines as (+)-sparteine surrogates, in other words, molecules with '(+)-sparteine-like' properties, as described in more detail in Chapter 5, Section 5.4.1. The best of those surrogates, under the same conditions used with (−)-sparteine, gave only racemic products, in contrast to the good results of the surrogate in other reactions. Not surprisingly, no precipitation was observed during the equilibration of the species **22** and **22'**.

2.3 Resolution of Diastereomeric Mixtures

2.3.1 Menthol as a Chiral Auxiliary

The inexpensive natural alcohol (−)-menthol is obtained from the natural source, *Mentha arvensis*, or from myrcene *via* a homogenously catalysed route (Takasago process) in multi-tonne quantities (Scheme 2.8).

Table 2.6 Mono- and diphosphine boranes prepared by the dynamic resolution of ***rac*-21** (Scheme 2.6).

Entry	Product	Yield (%)	de (%)	ee (%)[a]	References
1	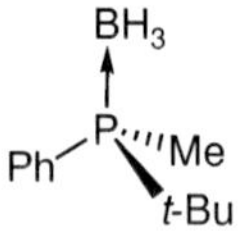	83–88	–	93–94	129,133,134
2	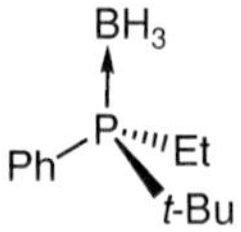	85	–	ND	134
3	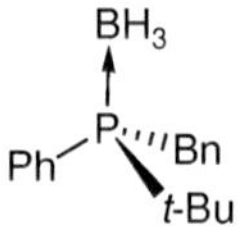	70	–	92	134
4	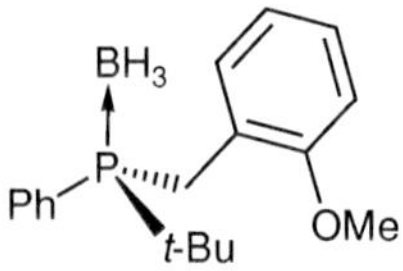	63–94	–	95–o.p.	129, 132, 133
5	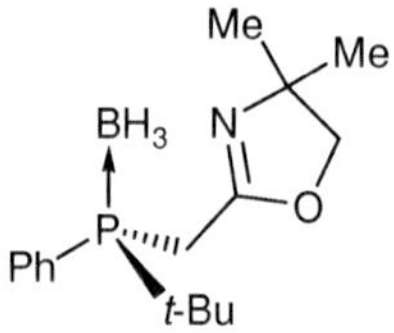	90	–	>82	129, 133
6	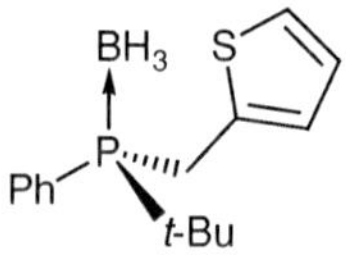	85	–	95	129, 133
7	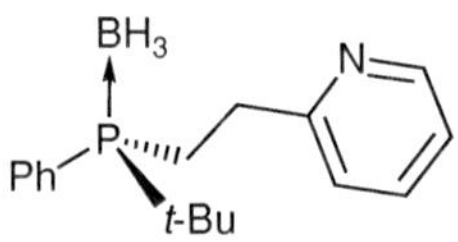	90	–	92	129, 133
8[b]	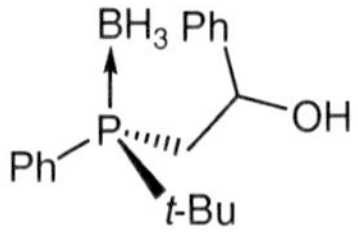	65–68	–	o.p.	135
9	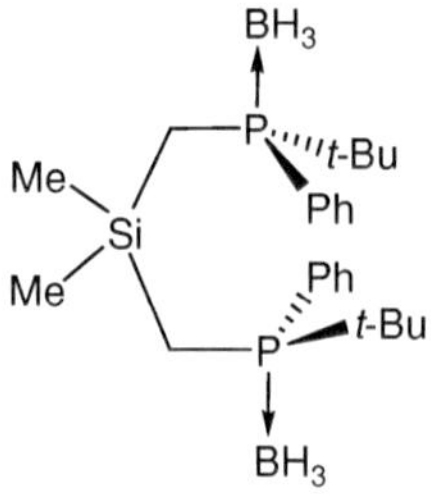	67	87	o.p.	129, 133

Table 2.6 (*Continued*).

Entry	Product	Yield (%)	de (%)	ee (%)[a]	References
10	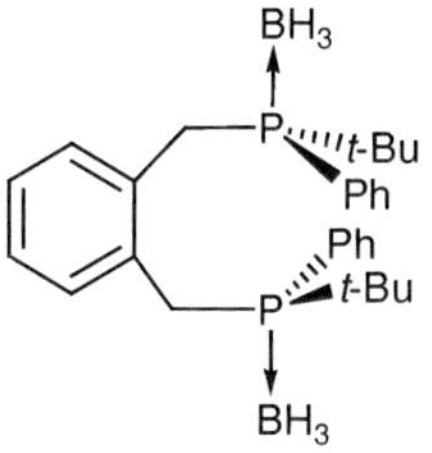	68	92	o.p.	129, 133
11	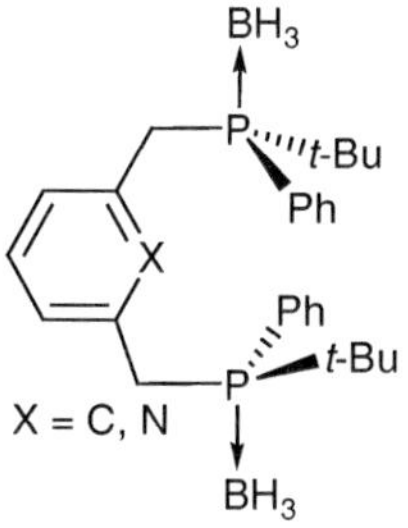 X = C, N	42–95 (C) 71 (N)	90 (C) 84 (N)	98.5 (C) o.p. (N)	129, 130, 133, 136, 137
12	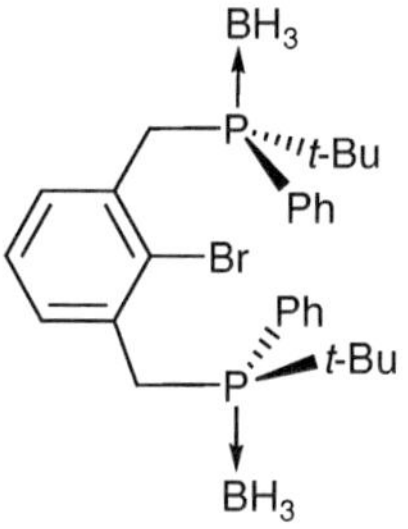	99	–	o.p.	130
13	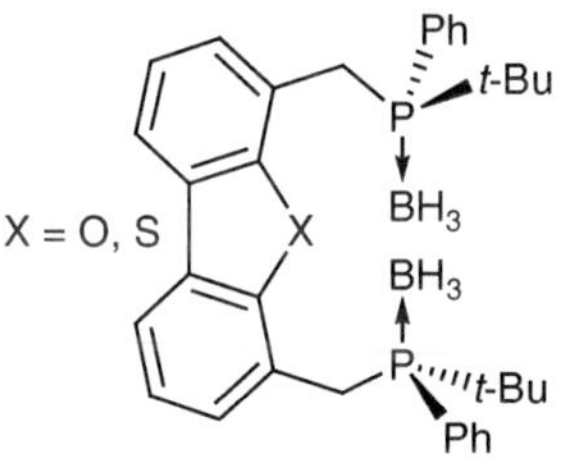 X = O, S	75 (O) 76 (S)	83 (O) 90 (S)	o.p.	129, 133

[a]Value for the (*l*)-isomers.
[b]This compound was prepared from both enantiomers of styrene oxide.

Although the other enantiomer is also commercially available, the natural isomer is usually employed in phosphine synthesis and is the only one considered here.

The development of synthetic methods based on diastereomers containing this alcohol supposed a big step forward in the preparation of enantiopure *P*-stereogenic compounds because it allowed a much greater flexibility compared to previous methods.

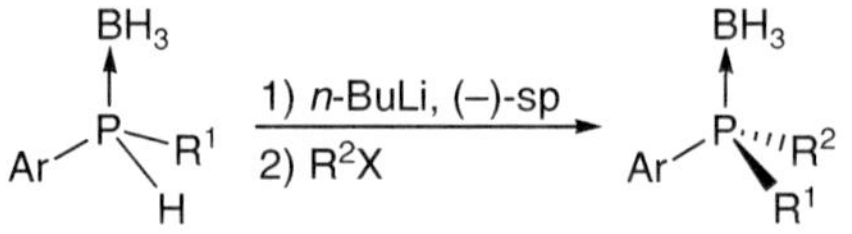

Scheme 2.7 Synthesis of dialkylarylphosphine boranes by the dynamic resolution protocol.

Table 2.7 Dialkylarylphosphine boranes **26** obtained by the dynamic resolution protocol.

Entry	Ar	R^1	R^2	Yield (%)	ee (%)[a]	References
1	Ph	Me	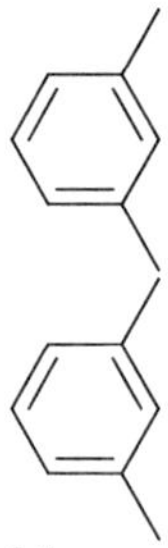	–	0	130
2	Ph	*i*-Pr		31	77	138
3	Ph	Cy	Me	41	27	134
4	Ph	2-Ad	Me	99	21	134
5	*o*-Tol	*t*-Bu	Me	83	92	134
6	1-Naphth	*t*-Bu	Me	65	61	134
7	*o*-An	*t*-Bu	Me	80	0	134
8	*o*-Biph	*t*-Bu	Me	83	0	134

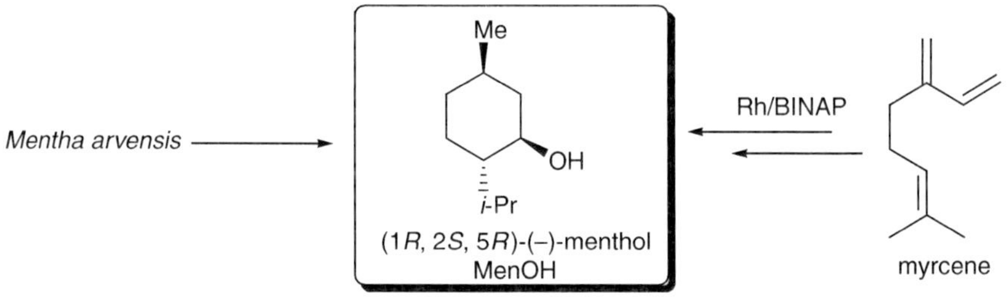

Scheme 2.8 Structure and industrial production of (–)-menthol.

2.3.1.1 *Menthylphosphinates and Related Compounds*

In the late 1960s it was found that diastereomeric, unsymmetrically substituted menthylphosphinates, $R^1R^2(OMen)P(O)$, could be easily prepared and separated into individual isomers by crystallisation or chromatography.[140–143] As an example, Scheme 2.9 depicts the preparation of methylphenylmenthylphosphinate.[144]

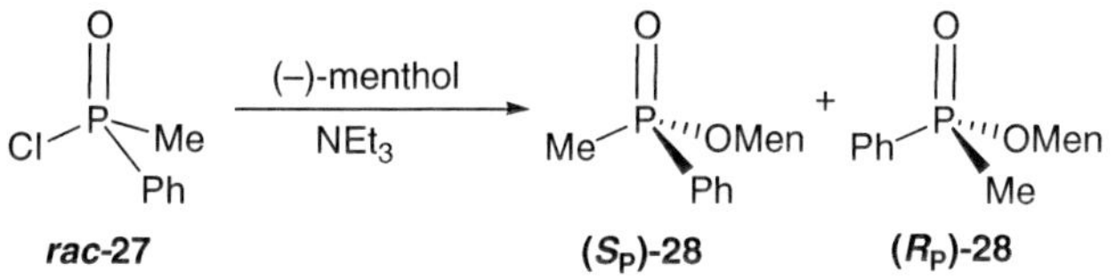

Scheme 2.9 Example of preparation of a *P*-stereogenic menthylphosphinate.

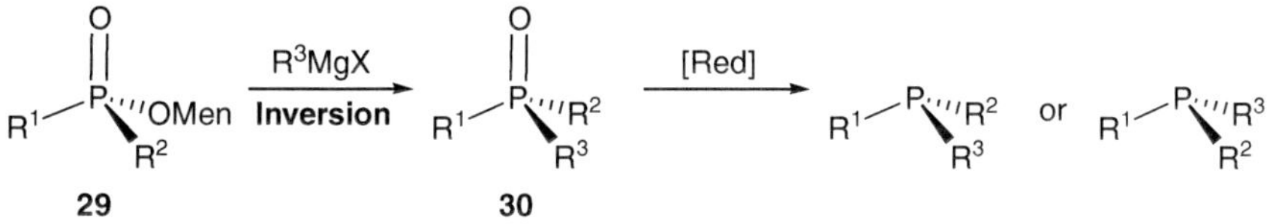

Scheme 2.10 Preparation of *P*-stereogenic phosphines from **29**.

Although some stereoselectivity in the formation of **28** could be expected, a 1:1 epimeric mixture was obtained. Once separated, menthylphosphinates were reacted with Grignard reagents furnishing the corresponding phosphine oxides with high degree of optical purity, with inversion of configuration at the P atom, as rigorously established by Korpiun and Mislow.[140] Since stereoselective reduction methods of phosphine oxides were known by then (Chapter 1, Section 1.3), the overall scheme allowed the preparation of a wide variety of optically enriched *P*-stereogenic phosphines (Scheme 2.10).

This method was described for the first time in the accounts of Mislow and co-workers[140,142] and of Nudelman and Cram.[141,145] These reports are classical papers in *P*-stereogenic chemistry. Table 2.8, which is not comprehensive, lists some of the phosphines prepared in the early reports about this method.

Although the reactions had to be carried out under rather harsh conditions (excess of Grignard reagent, 70 °C in benzene), they attracted the attention of several groups who resolved many tertiary phosphine oxides by this method, although not without difficulties.[1] As can be seen in entries 15–20, the substitution of the menthoxy group is very sensitive to steric hindrance of the groups at the phosphorus atom and on the Grignard reagent.[142] This problem was partially solved by the use of the more nucleophilic organolithium reagents (entries 15 and 16)[143,149] but the phosphine oxides were often (but not always)[150] produced in much lower optical purity.

The most important single application of this chemistry is arguably the first preparation of DiPAMP by Knowles and co-workers[144,151] in 1975 (Scheme 2.11).

The oxidative Cu-promoted dimerisation[152] of PAMPO, obtained by the menthol method, provided enantiopure DiPAMP oxide, which was reduced with inversion of configuration to (*R,R*)-DiPAMP. PAMPO has also been used by Johnson and Imamoto[153] to prepare polydentate phosphines inspired by DiPAMP.

Table 2.8 Some phosphine oxides **30** prepared by separation of menthylphosphinates.

Entry	R^1	R^2	R^3	References
1	Me	Et	*n*-Pr	143
2	Me	Et	Ph	143
3	Me	Ph	Et	140, 142
4	Me	Ph	*n*-Pr	140, 142
5	Me	Ph	*n*-Bu	146
6	Me	Ph	Bn	140–142, 146
7	Me	Ph	*p*-An	140,142
8	Me	Ph	*p*-Tol	147
9	Me	Ph	$p\text{-}CF_3\text{-}C_6H_4$	147
10	Me	Ph	$p\text{-}(t\text{-}Bu)\text{-}C_6H_4$	148
11	Me	Ph	*p*-Biph	142
12	Me	Ph	2-Naphth	142
13	*n*-Pr	Ph	Me	140, 142
14	Ph	2-Naphth	Me	142
15[a,b]	Ph	2-Naphth	*o*-/*p*-An	142, 143
16[a,b]	Me	Cy	*n*-Pr	143
17[a]	Me	Cy	Ph	142
18[a]	Me	Ph	1-Naphth	142
19[a]	Cy	Ph	*p*-Biph	142
20[a]	Cy	Ph	*p*-An	142

[a]No trace of phosphine could be detected.
[b]The reaction worked with the organolithium (R^3Li) reagent.

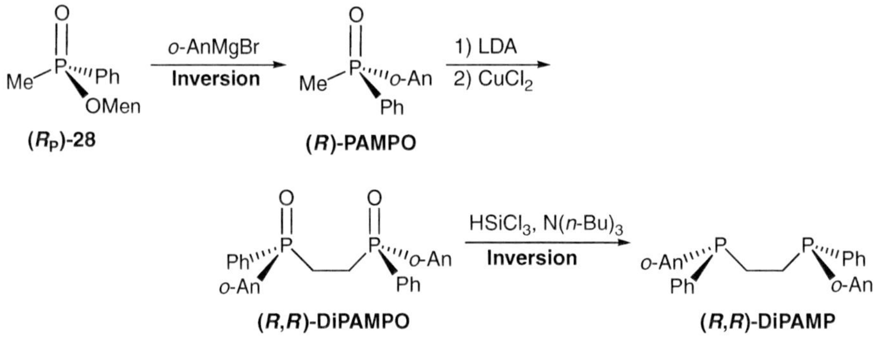

Scheme 2.11 First preparation of DiPAMP.

After these detailed investigations, other methodologies for the preparation of *P*-stereogenic phosphine oxides were little explored until the 1980s, when further advances with the menthol methodology appeared (Scheme 2.12).

Bodalski, Pietrusiewicz and co-workers[154] reacted racemic butyl phosphinite **31** with menthyl bromoacetate to form the expected Michaelis–Arbuzov phosphine oxide **32**. They found, serendipitously, that crystals of pure (**S**$_\mathbf{P}$)-**32** precipitated from the crude oily product and could be easily isolated in 30% yield. This compound is a highly functionalised phosphine oxide which received

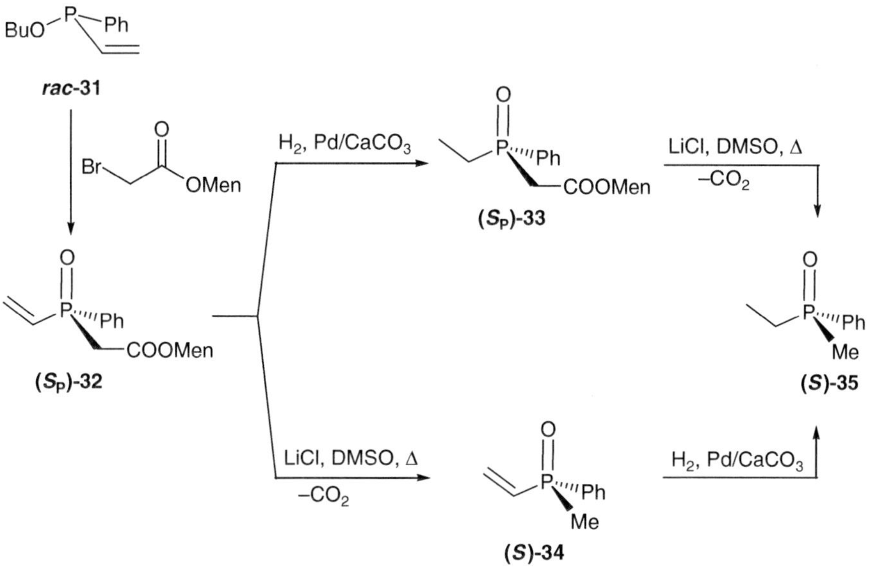

Scheme 2.12 Reactivity of optically pure **32**.

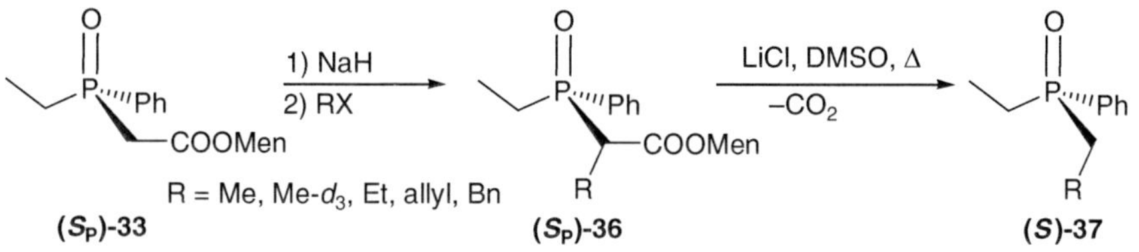

Scheme 2.13 Preparation of (*S*)-alkylethylphenylphosphine oxides from **(S$_P$)-33**.

further attention from the same laboratory.[155,156] The sequence hydrogenation–decarboxylation or *vice versa* afforded the known[142] phosphine oxide **35** in good yields (60–88% in each step) and therefore the absolute configuration of all the products was established by correlation. Interestingly, they also developed a route to alkylethylphenylphosphine oxides, based on the sequence α-alkylation–decarboxylation (Scheme 2.13).[155]

Initially an epimeric mixture of **(S$_P$)-36** was formed with poor diastereoselectivity although that was irrelevant because the stereogenic centre in the α position disappears in the decarboxylation step. The desired oxides **37** were obtained as optically pure compounds in 33–55% of combined yield over the two steps.

Almost at the same time, Imamoto, Johnson and Sato[157] developed a related route starting from a secondary racemic phosphine oxide for the preparation of methylphenylphosphine oxides with encumbered groups (Scheme 2.14).

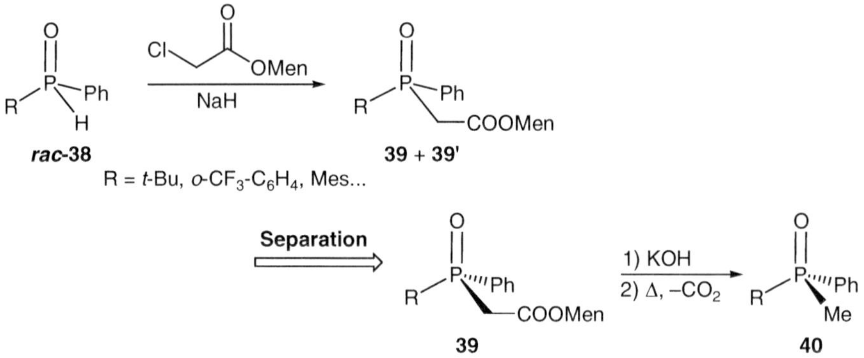

rac-38

R = *t*-Bu, *o*-CF$_3$-C$_6$H$_4$, Mes...

39 + 39'

Separation

39

1) KOH
2) Δ, –CO$_2$

40

Scheme 2.14 Preparation of optically pure methylphenylphosphine oxides.

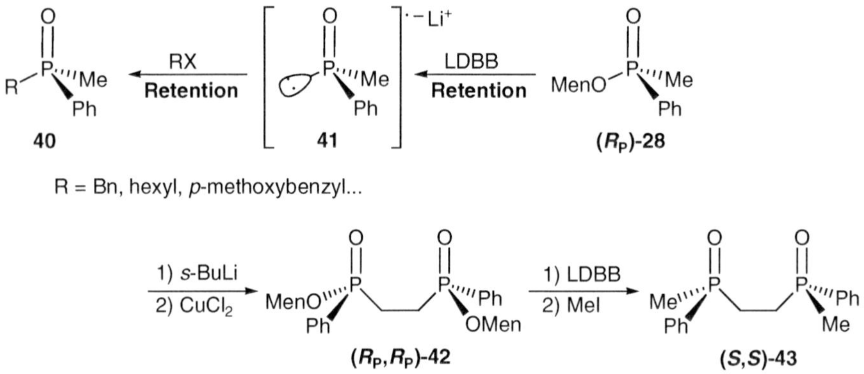

40

RX
Retention

41

LDBB
Retention

(R_P)-28

R = Bn, hexyl, *p*-methoxybenzyl...

1) *s*-BuLi
2) CuCl$_2$

(R_P,R_P)-42

1) LDBB
2) MeI

(*S*,*S*)-43

Scheme 2.15 Substitution of the menthoxy group with retention of configuration.

The alkylation of **38** with menthyl chloroacetate produced a diastereomeric mixture of phosphine oxides **39** and **39'**. These mixtures could be separated into individual diastereomers in 10–41% yields by fractional recrystallisation in hexane. Each epimer was subjected to hydrolysis to the corresponding acid and subsequent decarboxylation to afford the optically pure methylphosphine oxides **40**. Both transformations occurred in very high, often quantitative, yield. This method was an advance because, as seen before, the introduction of encumbered groups by the original Mislow method is cumbersome. Moreover, methylphosphine oxides **40** are key intermediates to prepare diphosphines of the DiPAMP family (see Scheme 2.11). The chemistry of Scheme 2.14 has been used much more recently by Hii and co-workers[158] to prepare series of aminohydroxy phosphine oxide ligands by coupling the acid derived from **39** (R = *t*-Bu) with (–)-norephedrine and (*S*)-valinol.

Another extension is the reductive cleavage of the P–OMen bond with retention of configuration at the phosphorus atom with LDBB developed by Imamoto and co-workers (Scheme 2.15).[159]

After trying several one-electron reductants (Li, lithium naphthalenide and lithium biphenylide) it was found that lithium 4,4'-di-*tert*-butylbiphenylide (LDBB) gave good yields (60–70%) and optical purities (up to 95% ee) for the phosphine oxides **40**, obtained after electrophilic quenching of phosphide **41**. Interestingly, (*R*$_\text{P}$)-**28** was oxidatively dimerised to **42**, which was subjected to reductive cleavage–electrophilic quenching affording (*S,S*)-**43**. The same reactions were performed also with the other epimer of **28**. This method is complementary to the Mislow procedure (Scheme 2.10), in which the displacement of the menthol group proceeds with inversion of configuration.

In 1968 Emmick and Letsinger[160] had reacted unresolved secondary *H*-phenylmenthylphosphinate (**44**) with benzylmagnesium chloride, smoothly affording the expected benzylphenylphosphine oxide. Forty years later Giordano, Buono and co-workers[161] explored the same type of reactions, this time with optically pure (*R*$_\text{P}$)-**44** and an excess of several Grignard and organolithium reagents in order to obtain a variety of secondary phosphine oxides **45** (Scheme 2.16 and Table 2.9).

The results in Table 2.9 show a deep influence of the nucleophilic reagent on the reaction outcome. The Grignard reagents afford high yields of **45** but the optical purity decreases sharply, in parallel with the increase in the bulkiness of R, as entries 1–3 and 4–5 show clearly. With organolithium reagents, the same trend is observed but is much less marked. These results were explained by the rates between two competing processes: the substitution of the menthyloxy by the R group and the racemisation of the anionic species formed upon deprotonation of (*R*$_\text{P}$)-**44** by the first equivalent of the nucleophilic reagent. Apparently, with encumbered Grignard reagents the second process is faster whereas for the more nucleophilic organolithiums the substitution takes place without time for significant racemisation.

In a recent report, Han and co-workers[162] apart from preparing many more SPOs (Scheme 2.16) and tertiary phosphine oxides, studied the mechanism of the nucleophilic substitution on *H*-phosphinates (Scheme 2.17).

They arrived at the conclusion that the substitutions proceeds *via* two competing pathways: direct substitution with inversion of configuration (path **A**) or a two-step process involving first the deprotonation of the *H*-phosphinate followed by substitution and hydrolysis, also with overall inversion of configuration (**B**). Path **B** requires two equivalents of carbanionic reagent, whereas **A** requires only one. They also studied the factors that affect the stereoselectivity of these reactions. They found that anionic intermediates **47** and **48** are configurationally stable but metal alkoxides cause epimerisation of the phosphorus atom on **46** by transesterification.[162]

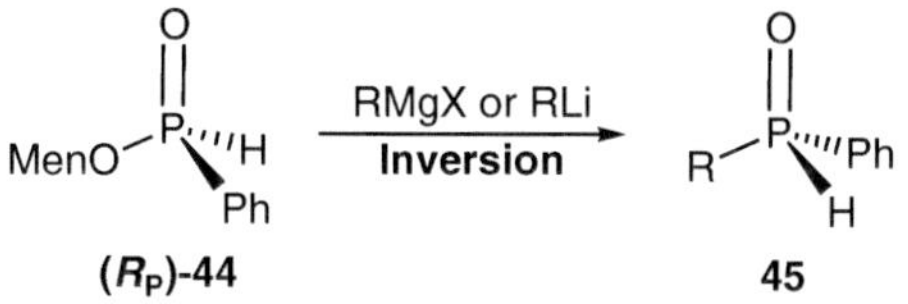

Scheme 2.16 Enantioselective synthesis of SPOs.

Table 2.9   Results in enantioselective synthesis of SPOs (Scheme 2.16).

Entry	RM	Yield of 45 (%)	ee of 45 (%)
1	EtMgCl	60	82
2	*i*-PrMgCl	89	56
3	*t*-BuMgCl	86	0
4	*o*-TolMgCl	84	68
5	*o*-BiphMgCl	88	2
6	1-NaphthMgCl	91	15
7	MeLi	76	97
8	*n*-BuLi	81	96
9	*t*-BuLi	86	86
10	*o*-TolLi	74	98
11	*o*-BiphLi	98	96

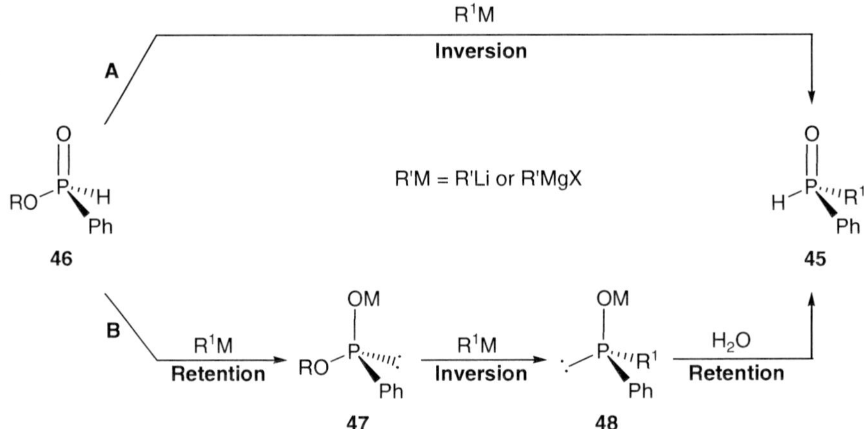

Scheme 2.17   Possible mechanisms for the synthesis of SPOs.

P-stereogenic SPOs are very interesting because although they are formally P(V) compounds, they are in equilibrium with the corresponding P(III) tautomer (phosphinous acid) and can be directly used as ligands for transition metals without racemisation.[163] Interestingly, the P(III) tautomer can be trapped by boronation (Scheme 2.18).[164]

Two procedures were developed: reaction of **44** with two equivalents of the organolithium, followed by quenching of **49** with borane and hydrolysis (method **A**) or the quenching of **49** with TMSCl, borane protection and hydrolysis (method **B**). The results obtained are listed in Table 2.10.

Following procedure **A**, good yields and enantioselectivities were obtained, except for *t*-BuLi and 1-furylLi (entries 3 and 7 respectively). In the former case, it was known that **49** had been obtained with at least 86% ee (Table 2.9, entry 9) and therefore partial racemisation of this compound must have been occurred. To stabilise the configuration of **49**, it was silylated prior to

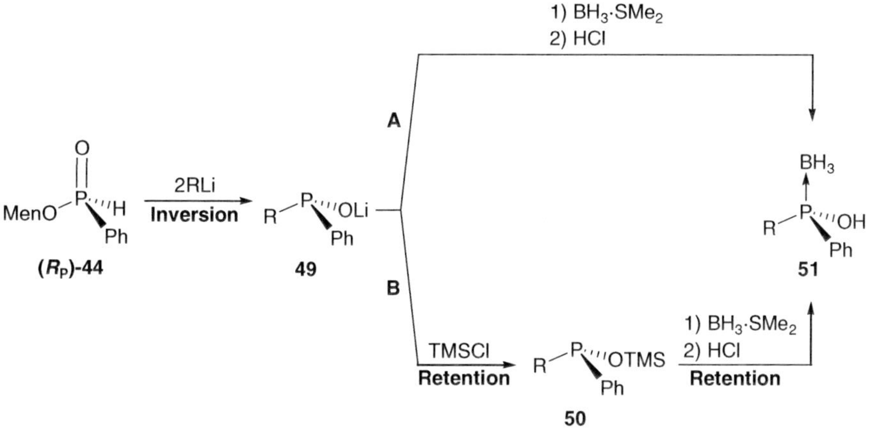

Scheme 2.18 One-pot synthesis of *P*-stereogenic phosphinous acid boranes.

Table 2.10 Synthesis of phosphinous acid boranes (Scheme 2.18).

Entry	R	Yield of 51 (%) A:B	ee of 51 (%) A:B
1	Me	78:78	95:95
2	*n*-Bu	84:70	89:89
3	*t*-Bu	82:70	60:84
4	*o*-Tol	93:75	98:97
5	*o*-Biph	84:85	93:95
6	1-Naphth	88:75	91:99
7	1-furyl	92:72	72:80

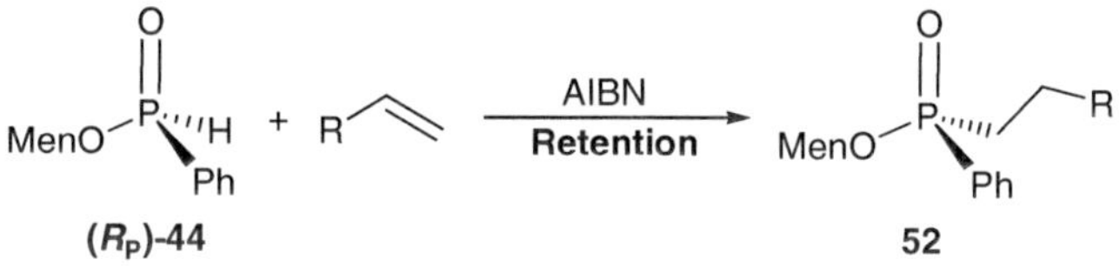

Scheme 2.19 Radical addition of (*R*P)-49 to alkenes.

boronation (method **B**). Following this procedure, the yields were slightly lower but the enantioselectivities improved. With procedure **B**, compound of entry 3 was obtained with the same optical purity as **49** (Table 2.9, entry 9), indicating that the silylation, boronation and acidolysis steps occur without loss of optical purity. The preparation of enantiopure **51** is important because phosphinous acid boranes have been scarcely used in *P*-stereogenic chemistry, in spite of their synthetic potential.[23,24,165]

Another application of (*R*P)-44 was discovered by Han and Zhao,[166] who found that it reacts with alkenes by a radical process (Scheme 2.19).

The reaction proceeds smoothly for monosubstituted electron-rich alkenes but slower for cycloalkenes. In all the cases retention of configuration is observed, therefore this method constitutes an interesting option to prepare a wide variety *P*-stereogenic menthylphosphinates **52**. For electron-poor alkenes, such as acrylonitrile, the reaction was extremely slow. In his case, however,[166] the reaction can be carried out in the presence of alkoxymagnesium halides.

2.3.1.2 *Menthylphosphinite Boranes and Related Compounds*

In the 1990s phosphine boranes began to displace phosphine oxides as synthetic intermediates. Not surprisingly, *P*-stereogenic menthylphosphinite boranes were prepared, separated into pure diastereomers and subjected to analogous reactions as menthylphosphinates discussed in the previous section. These transformations are summarised in Scheme 2.20.

The menthyl group has been substituted, with inversion of configuration, by several organolithium reagents. As will be seen throughout this book, as a rule of a thumb Grignard reagents are the reagents of choice with phosphinates and other P(V) compounds whereas organolithiums are preferred when dealing with phosphine boranes. Most of the phosphine boranes prepared up to date by this method are listed in Table 2.11.

These reactions are only possible at relatively high temperatures (room temperature to 80 °C) in contrast to the very easy substitution of the methoxy group in methylphosphinite boranes (see Chapter 4, Section 4.3.3.1), probably due to the bulkiness of the menthyl group. This may also be the cause of the low optical purity of the products obtained with sterically hindered organolithium reagents compared to the less bulky analogues (compare entries 2 and 3 with entry 1).[150] A remarkable enhancement of the reaction rate is observed in *o*-An-substituted precursors, compared to sterically similar *o*-Me- and *o*-EtC$_6$H$_4$ analogues, which react very sluggishly (entry 4 compared to entries 5 and 6). This acceleration of the substitution step has been attributed to the coordination of the methoxy oxygen to lithium, assisting the nucleophilic attack to the phosphorus.[167] Very bulky and relatively basic menthylphosphinite boranes are unreactive even towards MeLi (entries 9 and 10) highlighting again the sensitivity of the reaction to steric encumbrance. Finally, the menthoxy groups of the diphosphinite–boranes of entries 11–13 were efficiently substituted with MeLi to give the corresponding C_2-symmetric *bis*(phosphine boranes).

The sequence reductive cleavage–electrophilic quenching has often been used to prepare products with retention of configuration after a seminal report of Imamoto and co-workers.[170] This is because these transformations are carried out under much milder conditions giving products of higher optical purity. It was found that the best reducing systems in terms of optical purity of the products was lithium naphthalenide and Li/NH$_3$(l).[169] Care has to be taken because it was also found that the anionic tricoordinated phosphorus species generated racemised extensively *via* pyramidal inversion[168] unless they were

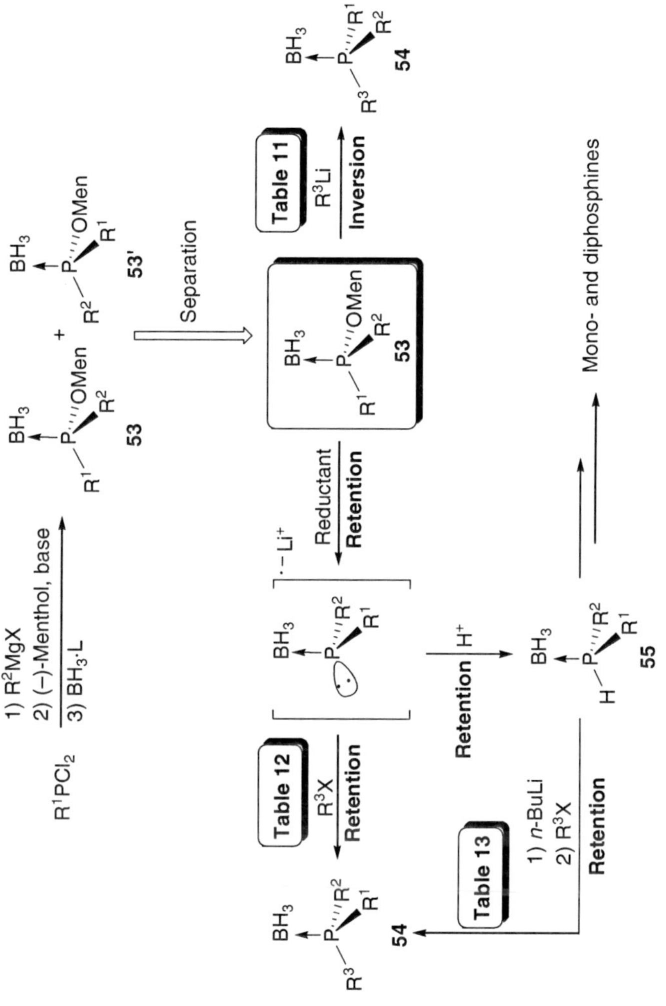

Scheme 2.20 Preparation and reactivity of optically pure menthylphosphinite boranes.

Table 2.11 Phosphine boranes **54** prepared *via* nucleophilic substitution on menthylphosphinite boranes **53** (Scheme 2.20).

Entry	R^1	R^2	R^3	Yield (%)	ee (%)	References
1	Ph	Me	o-An	0–65	0–13	131, 150
2	Ph	Me	m-An	48–57	82–95	150
3	Ph	Me	p-An	54	88–99	150
4	Ph	o-An	Me	95	93–o.p.	131
5	Ph	o-Tol	Me	30	–	167
6	Ph	o-EtC$_6$H$_4$	Me	3	–	167
7	Ph	o-Biph	Me	90	94 (crude) o.p. (recr.)	127
8	o-An	Cy	Me	98	o.p.	168
9	o-Biph	Cy	Me	0	–	127
10	o-Biph	t-Bu	Me	0	–	127
11	(MenO)(Ph)P(BH$_3$)–CH$_2$CH$_2$–P(BH$_3$)(Ph)(OMen)		Me	57	o.p.	169
12	(MenO)(Ph)P(BH$_3$)–CH$_2$CH$_2$CH$_2$–P(BH$_3$)(Ph)(OMen)		Me	57	o.p.	169
13	(MenO)(Ph)P(BH$_3$)–(CH$_2$)$_4$–P(BH$_3$)(Ph)(OMen)		Me	82	o.p.	169

kept at low temperatures. Most of the compounds prepared by this method are listed in Table 2.12.

These reactions often proceed in high yields and enantioselectivities with only a few exceptions (entries 11 and 16). In most of the entries the menthoxy group is substituted either by a hydrogen atom or by methyl group. The products of entries 19–21 are especially interesting because are direct precursors of bidentate phosphines.

Tsuruta and Imamoto[127] disclosed a promising approach to prepare the methylphosphine borane of entry 11 in high optical purity, although it required two more steps (Scheme 2.21).

Deboronation of **56** and methylation afforded the phosphonium salt **57**, which was stereoselectively reduced[175] and reboronated again. Phosphine borane **58** was obtained in a combined yield of 73% and 97% ee.

Many of the secondary phosphine boranes obtained with the menthol methodology have been deprotonated (usually with *n*-BuLi) and quenched with electrophiles leading to a variety of tertiary phosphine boranes (Table 2.13).

The generated phosphide boranes are extremely nucleophilic and reacted smoothly with activated alkyl halides (entries 3–7), epoxides (entries 1, 2 and 8) and molecules with activated multiple bonds such as ethyl acrylate (entry 9), benzyne (10) or even fullerene (entry 14). Some C_2-symmetric diphosphine boranes were also prepared by this method (entries 11, 12 and 15–17). It is noteworthy that phenylmethylphosphine borane epimerises upon deprotonation (entry 1). All these reactions occurred with retention of configuration. Interestingly, the diphosphine boranes obtained in entries 15–17 are the enantiomers of those obtained by the nucleophilic approach (Table 2.11, entries 11–13).

The phosphine borane version of the method presented in Scheme 2.14 has been developed by Imamoto and co-workers[131] as a route to methylphenylphosphine boranes **61** (Scheme 2.22).

With this strategy, *tert*-butylmethylphenylphosphine borane and PAMP·BH_3 were obtained, although the latter presented low optical purity (41% ee) due to racemisation in the decarboxylation step, which had been performed in xylene at 130 °C for 1 h. The menthyl ester intermediate **60**, for R = *o*-An, was also reduced to the alcohol, which was mesylated and iodinated affording the β-iodophosphine borane **62** in 62% combined yield. This compound was reductively dimerised by activated copper to diphosphine borane **63** (64% yield), which is similar to DiPAMP but with a butyl bridge.

Some of the methylphosphine boranes presented in the previous tables have been used to prepare several diphosphine boranes with an ethyl bridge, *i.e.* analogues of DiPAMP *via* Cu(II) promoted dimerisation (Scheme 2.23 and Table 2.14).

All the phosphine boranes in Table 2.14 were obtained as optically pure compounds after removal of the small quantity of *meso* compound, if any. Compounds of entries 1–4 bear an *ortho*-substituted phenyl group and were prepared in order to compare their performance in Rh-catalysed hydrogenation with DiPAMP and to shed light into the role of the methoxy group in the catalysis.[128] The menthyldiphosphinite borane of entry 7 is interesting, because it has been used to prepare other diphosphine boranes (Table 2.11, entry 11, and Table 2.12, entry 19).

2.3.1.3   Other Syntheses Based on Menthol

There are number of other reports on *P*-stereogenic compounds involving separation of diastereomeric menthyl- or menthoxy-containing precursors as a key point of the synthesis.

Mikolajczyk and co-workers[178] reacted racemic ethylphenylchlorophosphine (**rac-65**) with menthol in presence of several bases to afford the corresponding menthylphosphinite **66** as a mixture of diastereomers (Scheme 2.24).

Interestingly, the diastereomeric ratio between **66** and **66′** was dependent upon the amine and the solvent, reaching the maximum value (2:1) for diethylphenylamine in diethyl ether. The diastereomeric mixture of **66** and **66′** was treated with sulfur giving the phosphinothioates **67**. One of the isomers could be obtained in optically pure form by recrystallisation. Its absolute configuration was unequivocally determined by chemical correlation to the known[142] phosphine oxide **69**. With this information several nucleophilic

Table 2.12 Phosphine boranes **54** prepared *via* reductive elimination electrophilic quenching on menthylphosphinite boranes (Scheme 2.20).

Entry	R^1	R^2	R^3	Yield (%)	ee (%)	References
1	Ph	Me	H	95	95	169,170
2	Ph	Me	Bn	100	95	169,170
3	Ph	t-Bu	Me	83	o.p.	168
4	Ph	o-An	H	100	93	169,170
5	Ph	o-An	Me	81–97	77–94	128,169,170
6	Ph	o-EtC$_6$H$_4$	Me	91	88	128,171
7	Ph	o-(i-Pr)C$_6$H$_4$	Me	99	94	128
8			Me	98	84	128
9	o-An	Cy	H	98	o.p.	168
10	o-An	t-Bu	H	94	o.p.	168
11	o-Biph	Cy	Me	–	˜20 75 (crude)	127
12	o-Biph	t-Bu	Me	72	o.p. (recr.)	127
13	1-Naphth	Cy	Me	64	o.p.	172
14	t-Bu	o-An	Me	–	–	173
15	t-Bu	o-(CH$_2$OMe)C$_6$H$_4$	Me	68	o.p.	173
16			H	99	44	174

	Structure	R			
17	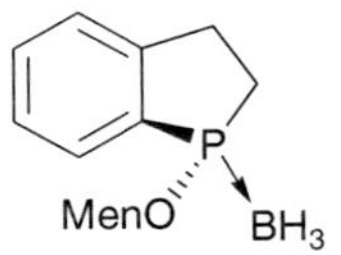	Me	95	99	174
18	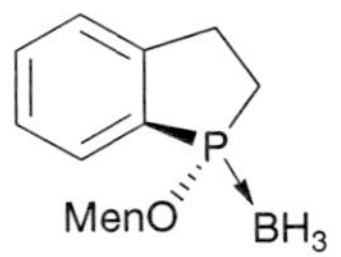	$-CH_2CH_2OMe$	78	99	174
19	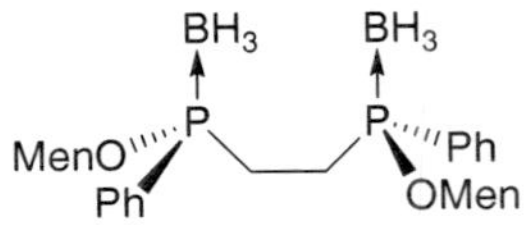	H	91	o.p.	169
20	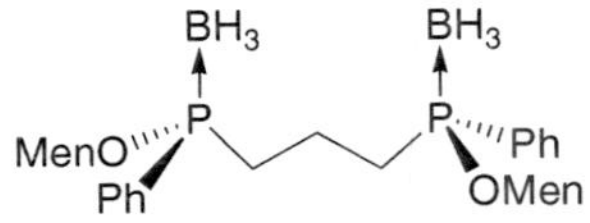	H	88	o.p.	169
21	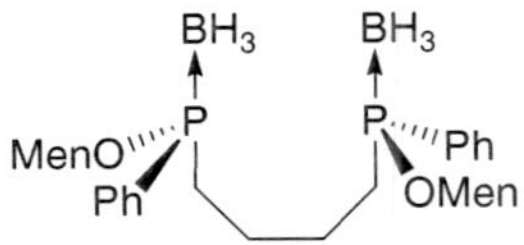	H	83	o.p.	169

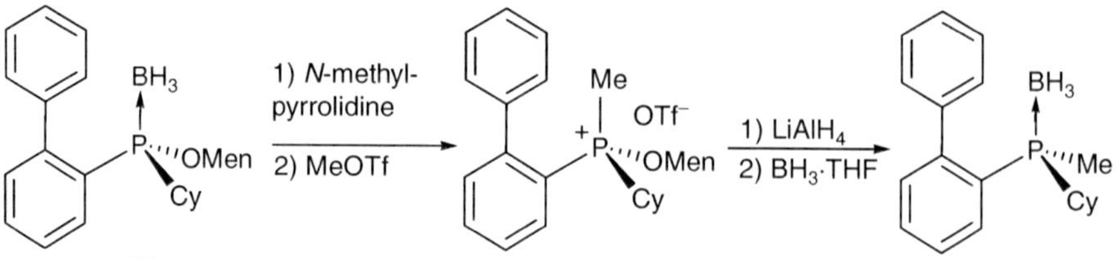

Scheme 2.21 Alternative procedure for the preparation of **58** (Table 2.12, entry 11).

Table 2.13 Phosphine boranes **54** prepared *via* deprotonation of secondary phosphine boranes **55** and electrophilic quenching (Scheme 2.20).

Entry	R^1	R^2	R^3X	Yield (%)	ee (%)	References
1	Ph	Me	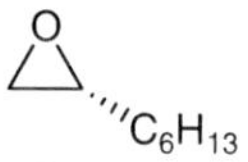 epoxide, C_6H_{13}	95	0^a	168
2	o-An	Cy	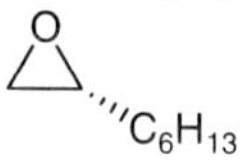 epoxide, C_6H_{13}	61	o.p.^{b}	168
3	o-An	t-Bu	MeI	97	o.p.	168
4	o-An	t-Bu	Allyl bromide	84	o.p.	168
5	o-An	t-Bu	3-butenyl tosylate	85	o.p.	168
6	o-An	t-Bu	$BrCH_2COOMe$	66	o.p.	168
7	o-An	t-Bu	$Br(CH_2)_2COOEt$	79	o.p.	168
8	o-An	t-Bu	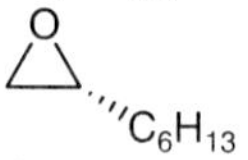 epoxide, C_6H_{13}	69	o.p.^{b}	168
9	o-An	t-Bu	$CH_2=CHCOOEt$	92	o.p.	168
10	o-An	t-Bu	Benzyne^c	44	o.p.	168
11	o-An	t-Bu	$I(CH_2)_3I$	67	o.p.	168
12	o-An	t-Bu	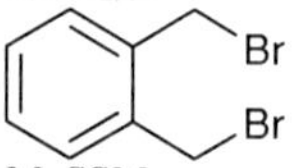 α,α'-dibromo-o-xylene	98	o.p.	168
13	o-An	t-Bu	MeSSMe	89	98	168
14	Ph	OMen	Fullerene	46–71	o.p.	176
15	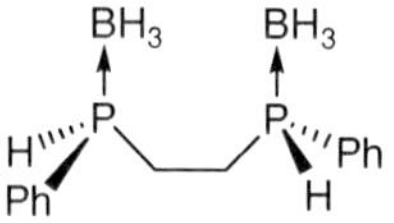 diphosphine (ethylene-bridged)		Me and – CH_2CH_2OMe	79 (Me) / 89	o.p.	169
16	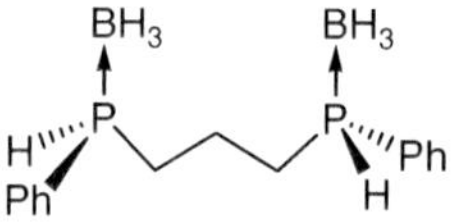 diphosphine (propylene-bridged)		Me and – CH_2CH_2OMe	88 (Me) / 97	o.p.	169
17	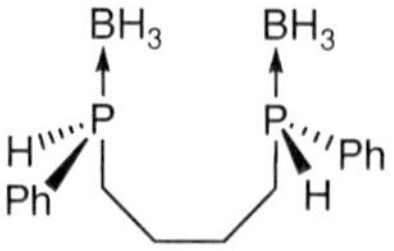 diphosphine (butylene-bridged)		Me and – CH_2CH_2OMe	89 (Me) / 85	o.p.	169

^{a}The phosphorus atom racemised.
^{b}Only one diastereomer detected by ^{1}H NMR.
^{c}Generated by reaction of *o*-bromophenyl triflate with 2 equiv. of *t*-BuLi.

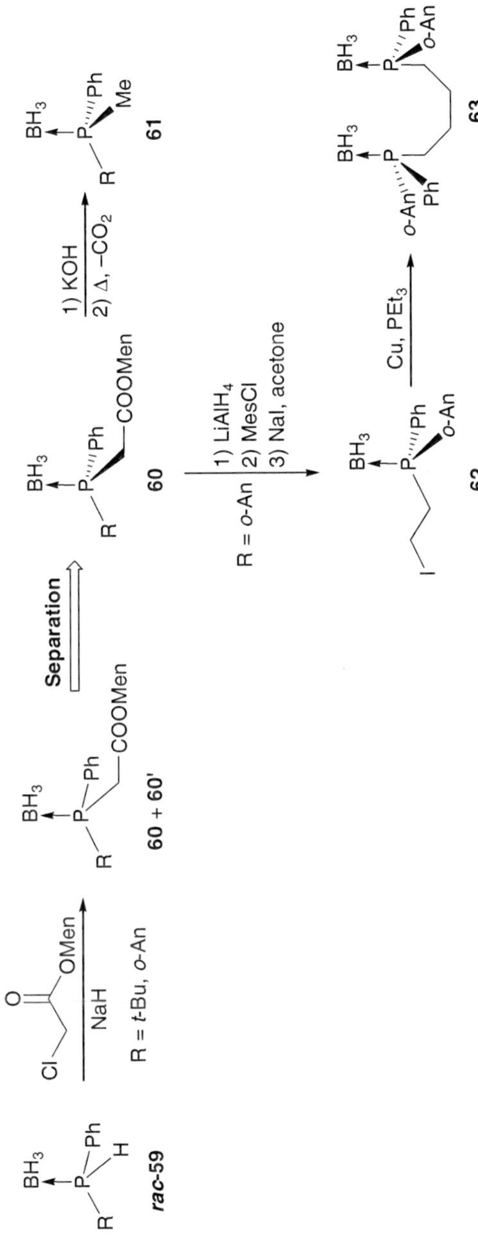

Scheme 2.22 Preparation of optically pure methylphenylphosphine boranes and a diphosphine borane.

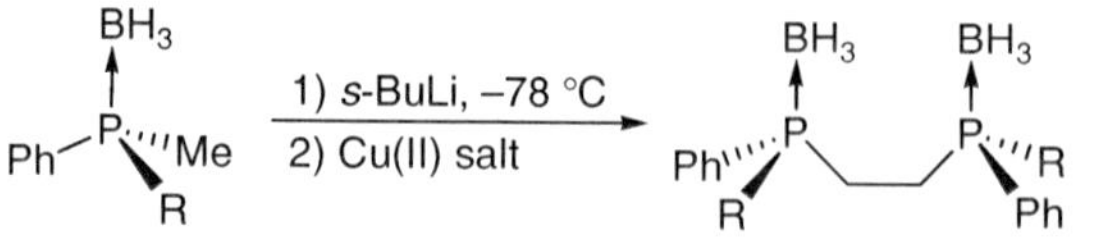

Scheme 2.23   Preparation of DiPAMP-analogues by Cu-promoted oxidative coupling.

substitution reactions on **66**, were studied and were proved that they take place with inversion of configuration at the phosphorus atom.

Richter and Neuffer[179] and more recently Vogt and co-workers[180] studied the mono- and disubstitution of the menthoxy groups on phenylmenthylphosphonite (**70**) by organolithium reagents (Scheme 2.25), without any protecting group involved.

Although no reaction was found for ethyl- or allyllithium, the menthylphosphinites **71** were isolated in good yields (75–82%) and high optical purities (91–96% de) with other reagents. The absolute configuration at the phosphorus atom (R = *t*-Bu) was later determined to be *S* by X-ray analysis.[180] The second nucleophilic substitution with another organolithium reagent afforded the tertiary phosphine **72** in 82 and 64% yield for R = Me and *t*-Bu respectively. The ee values were 73 and 79%, which indicate that partial racemisation occurred in this step.

A considerable number of chiral phosphines bearing menthyl groups directly attached to the phosphorus atom have been prepared. A few of them are *P*-stereogenic, including the first phosphines simultaneously bearing stereogenic carbon and phosphorus atoms, prepared in 1977 by Fisher and Mosher[181] by condensation of sodium methylphenylphosphide and menthyl chloride. A few years later, the method was improved by Valentine and co-workers,[182] who prepared both phosphorus epimers of **73** (Scheme 2.26) by basic decomposition of ammonium salts.

They managed to separate the diastereomers by recrystallisation and determined their absolute configurations at the P atom by X-ray crystallography and chemical correlation, as shown in Scheme 2.26.

Venanzi and co-workers[183] prepared a diphosphine containing a single stereogenic phosphorus atom with a menthyl group (Scheme 2.27).

Treatment of **75** (with a 3:1 epimeric ratio) with vinylmagnesium chloride allowed the isolation of crude **76** as a 2:1 mixture of epimers, which could be separated by column chromatography and recrystallisation (22% yield for the major epimer, which corresponded to the *R*P). Each diastereomer was reacted with lithium diphenylphosphide and borane to afford **77** in good yield. The diphosphine borane was deprotected and complexed to a chiral palladium complex to determine the absolute configuration at the P atom by careful NMR studies.

The menthyl group has also been widely used in the preparation of *P*-stereogenic small ring heterocyclic phosphines, namely phosphiranes and phosphetanes.[184,185]

Table 2.14 Diphosphine boranes **64** with an ethylene bridge by the menthol methodology.

Entry	Diphosphine borane	Yield (%)	References
1	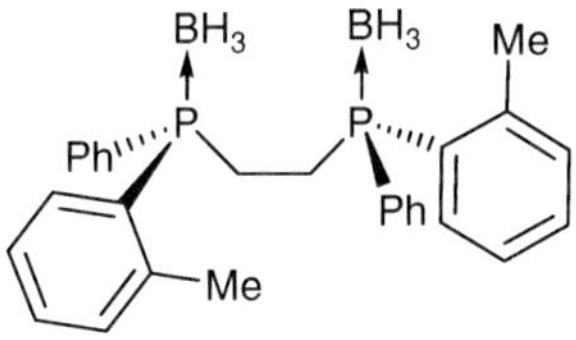	57	128, 177
2	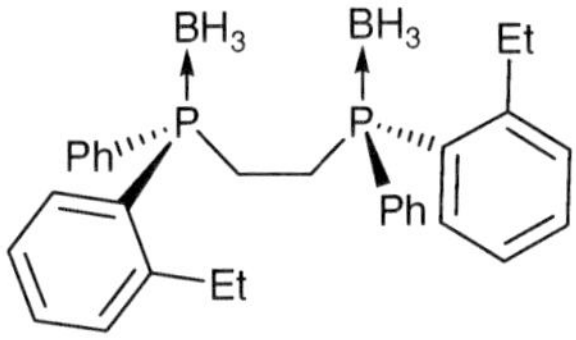	79–88	128, 171
3	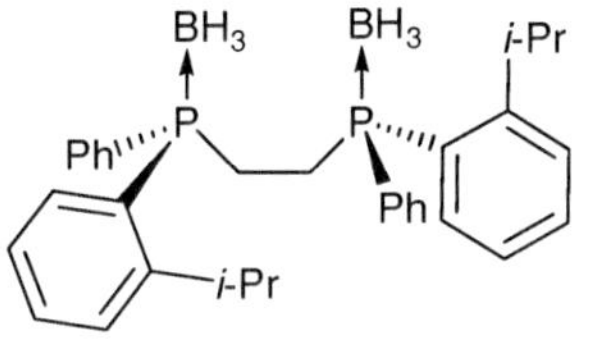	64	128
4	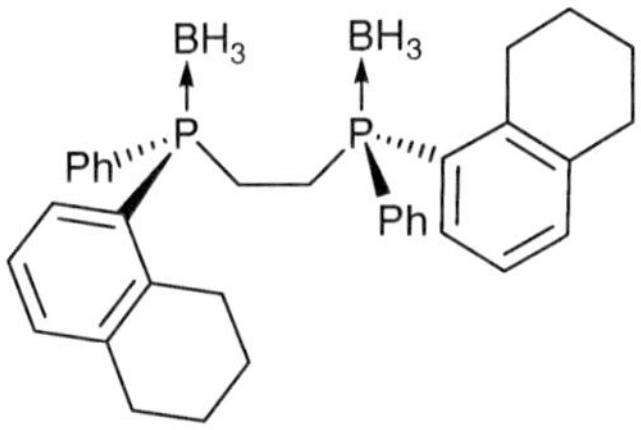	55	128
5	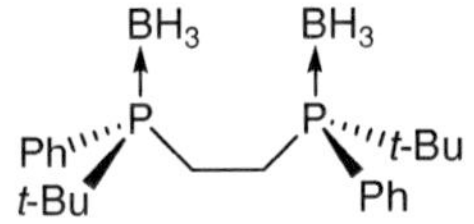	76	131
6	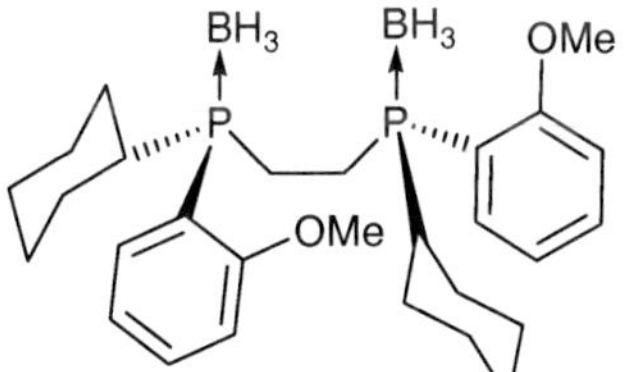 DiCAMP·(BH$_3$)$_2$	75	168
7	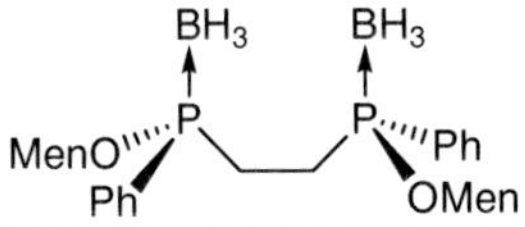	73	169

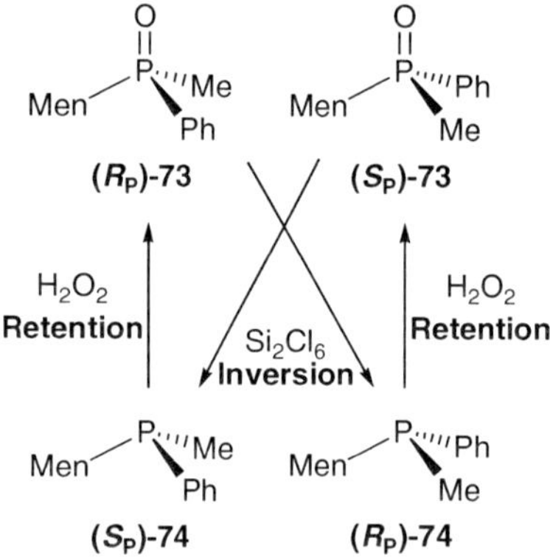

Scheme 2.24 Preparation of **67** and chemical correlation to **(*S*)-69**.

Scheme 2.25 Preparation of *P*-stereogenic menthylphosphinites and phosphines from **70**.

Scheme 2.26 Some of the reported *P*-stereogenic menthylphosphines.

These compounds are interesting as ligands because they present a very high inversion barrier at the phosphorus due to their strained structure.[186]

Marinetti and co-workers[187,188] prepared the first optically active phosphiranes, which contain a menthyl substituent at the phosphorus atom (Scheme 2.28).

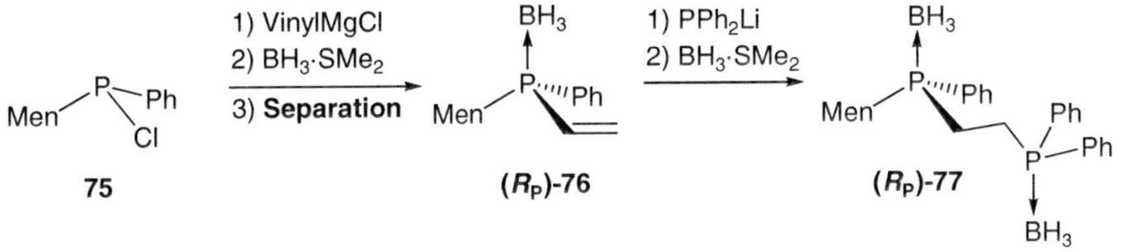

Scheme 2.27 Preparation of a diphosphine borane containing a *P*-stereogenic atom with a menthyl substituent. Only the R_P diastereomer is represented.

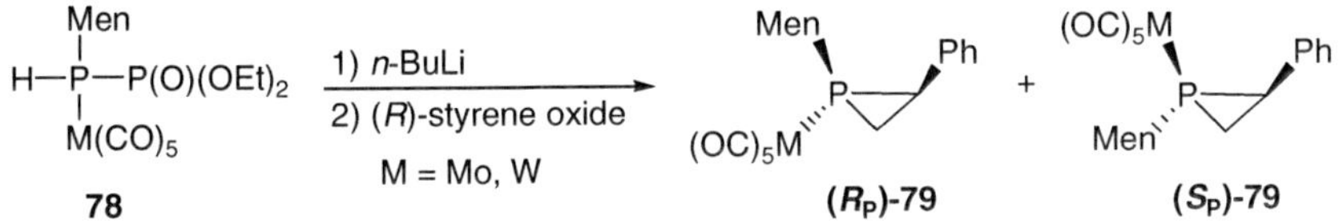

Scheme 2.28 Preparation of *P*-stereogenic phosphirane complexes.

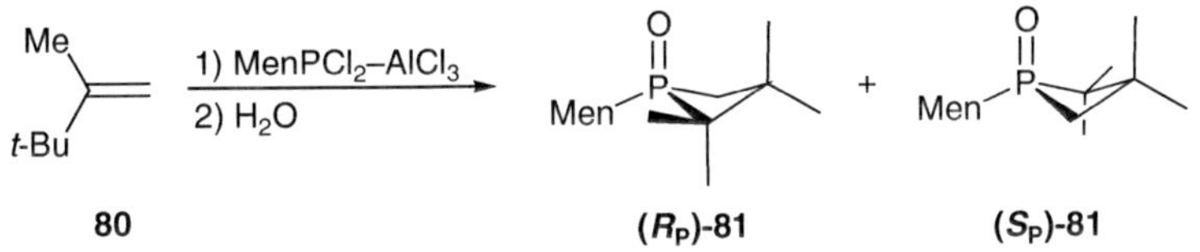

Scheme 2.29 Preparation of a *P*-stereogenic phosphetane oxide.

The alkylation of anions derived from **78** with styrene oxide afforded the diastereomeric mixture of **79** through a 'phospha-Wittig' reaction with very little asymmetric induction from the menthyl group. In these reactions, complete inversion of configuration of the stereogenic oxirane carbon was observed. The individual diastereomers of **79** could be separated by column chromatography or recrystallisation. Decomplexation with dppe furnished the free enantiopure phosphiranes, which were complexed to Rh.[188]

The Marinetti group[189,190] has also prepared many *P*-stereogenic menthylphosphetanes. Although there is one report on phosphetane sulfides,[191] most of the work has been carried out with 1-menthyl-2,2,3,3-tetramethylphosphetane oxide (**81**). This compound is prepared by reaction between menthyldichlorophosphine and alkene **80** followed by hydrolysis of the intermediate phosphetanium salt (Scheme 2.29).[192]

The diastereomeric phosphetane oxides **81** were obtained in equimolar amounts (unlike analogous phosphetane oxides bearing other chiral groups derived from pinene)[193] and could be separated by column chromatography.[194]

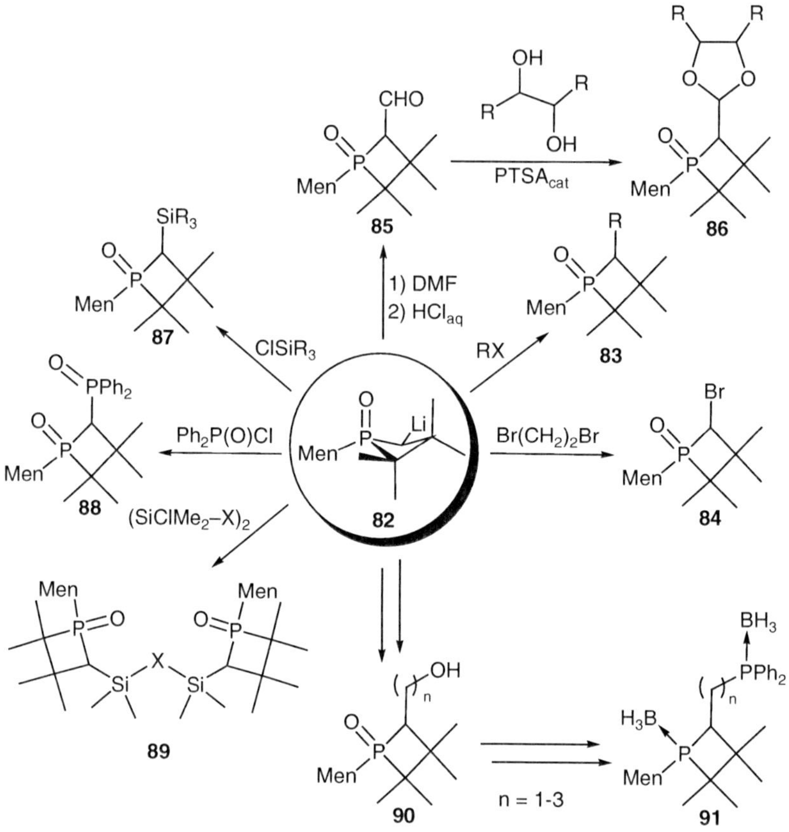

Scheme 2.30 Different α-substituted phosphetanes obtained from anion **82**. The reactions are usually highly stereoselective but for clarity the stereochemistry of the products **83–91** is not represented.

The chemistry of phosphetane oxides **81** has been studied in detail.[189] The most important discovery is that it can be diastereoselectively deprotonated with *n*-BuLi or LDA to give stabilised carbanions **82**, which react with a wide variety of electrophiles to give α-substituted phosphetane oxides (Scheme 2.30).

With this procedure α-alkylated (**83**),[194] brominated (**84**),[194] formylated (**85**),[195] silylated (**87** and **89**),[196] phosphorylated (**88**)[197] and hydroxylated (**90**)[198,199] phosphetane oxides have been prepared. Usually the products are present as a single diastereomer, with retention of configuration at the phosphorus atom and with the new substituent installed in the equatorial position, in *anti* disposition to the menthyl group, as confirmed by X-ray crystallography.[195–197] C_1- and C_2-symmetric phosphetane dioxides or diboranes (**88**, **89** and **91**) have also been obtained. Further elaboration has allowed the preparation of derivatives with multiple stereogenic centres (**86**).[195,198] Most of the phosphetane oxides of Scheme 2.30 have been reduced to the parent phosphines

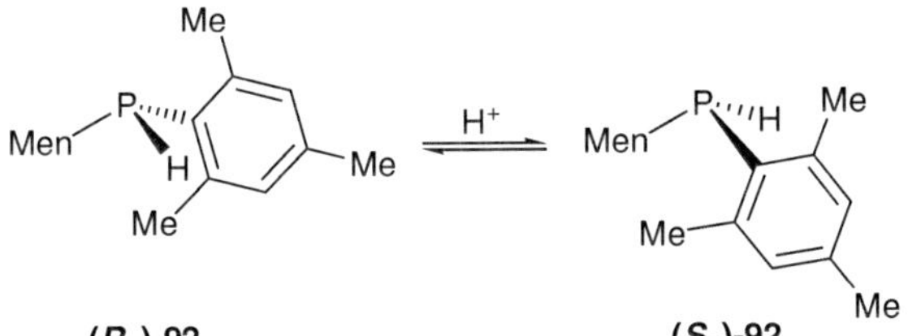

Scheme 2.31 Acid-catalysed epimerisation of **92**.

with trichlorosilane/NEt_3 with net retention of configuration at the P atom and the coordination chemistry of the free phosphetanes towards several metals has been studied.[194–197,199]

The first example of a resolution of a free secondary *P*-stereogenic phosphine was reported by Wild and co-workers.[200,201] They found that the diastereomers of menthylmesitylphosphine (**92**) could be separated by fractional recrystallisation of acetonitrile solutions of the nearly 1:1 epimeric mixture of **92**. The key to their success was the presence of sodium acetylacetonate (0.04%) as proton scavenger. The optical purity of the recovered solids and of the mother liquor was easily monitored by $^{31}P\{^{1}H\}$ because, remarkably, the resonances of the epimers were separated by more than 20 ppm. Dissolution of enantioenriched diastereomer in pure acetonitrile (without proton scavenger) led to immediate epimerisation by an acid-catalysed process (Scheme 2.31).

The epimeric mixture was protected with borane and one of the epimers (the S_P according to X-ray analysis) spontaneously precipitated at low temperature and was obtained in 66% yield and 97% de,[201] which is an example of a second-order asymmetric transformation.

A related enantioselective transformation[202] has been studied by Vedejs and Donde (Scheme 2.32).[203]

The phosphine boranes **93** were obtained by treatment of **75** with fluorenyllithium and borane protection. Optically pure R_P and S_P diastereomers were obtained in 23 and 32% yields after crystallisation and chromatography. The absolute configuration of the former was determined by X-ray crystallography. Both epimers of **93** were deboronated with diethylamine, furnishing the free phosphines **94**. Interestingly, the R_P isomer was found to be an oil whereas the S_P isomer is a crystalline solid. This difference of crystallinity was exploited to perform a crystallisation-induced asymmetric transformation of an equilibrating mixture **94** at high temperature (in refluxing heptane), upon slow evaporation of solvent. After considerable experimentation, the precipitated solid presented a 90% de enrichment of the S_P isomer. This process could also be carried out at room temperature by a I_2-catalysed epimerisation (*via* *P*-iodophosphonium intermediates) of **94** with similar results, although this variation was very sensitive to the purity of **94**. The fluorenyl group in **94** could be reductively cleaved with lithium naphthalenide and the resulting phosphide borane reacted smoothly with sulfuric acid to provide **95** (92% yield) and with

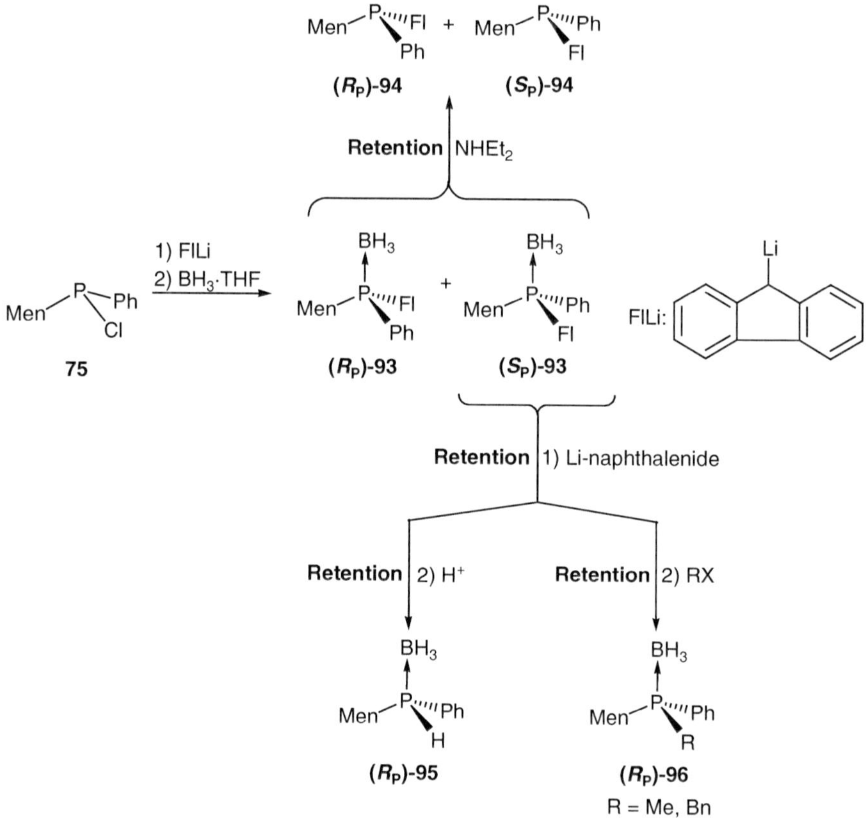

Scheme 2.32 Preparation and resolution of **(S_P)-93**, its reductive cleavage and electrophilic quenching.

alkyl halides to afford the tertiary phosphine boranes **96** in similar yield. Both reactions take place with retention of configuration at the P atom as expected.

Chlorophosphite **97** (as a 1:1 mixture between the two phosphorus epimers) was reacted with 1,2-dicarba-*closo*-dodecaborane(12) by Hey-Hawkins and co-workers[204] giving the expected diphosphonite **98** (Scheme 2.33).

Only the (R_P,R_P)- and (R_P,S_P) isomers of **98** were isolated and separated by fractional crystallisation. They are moderately stable solids stable towards epimerisation and were used to prepare Rh and Mo complexes.[204]

2.3.2 Other Chiral Auxiliaries

Other chiral auxiliaries have been used with similar roles as (–)-menthol. Two them are depicted in Figure 2.4.

An epimer of (–)-menthol, neomenthol, has been sporadically used. For example, King and co-workers[205] managed to obtain an optically pure

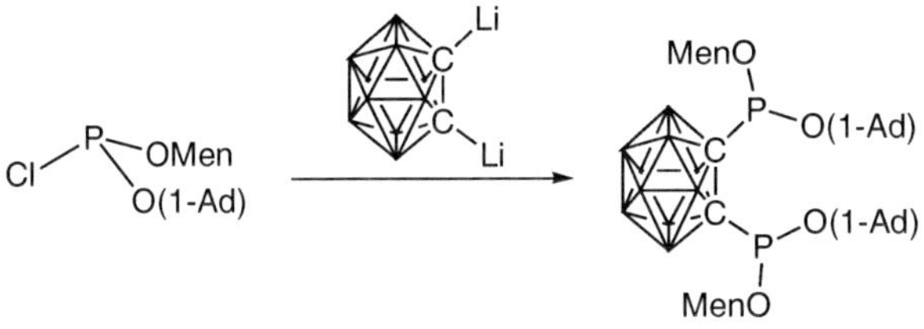

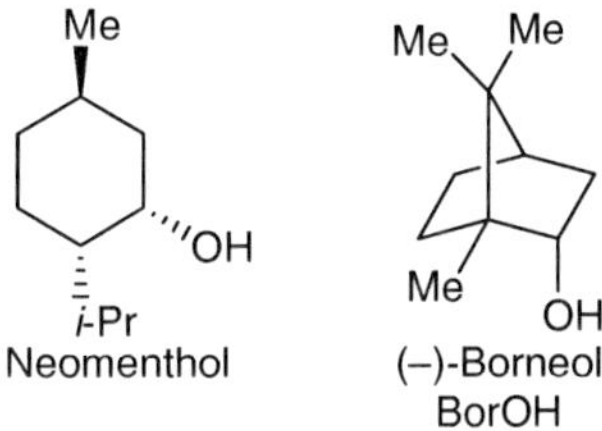

97   98

Scheme 2.33 Synthesis of *P*-stereogenic phosphonites with a dicarbaborane unit. Each vertex in the icosahedron represents a BH unit.

Me
⋮
i-Pr
Neomenthol   (–)-Borneol
                         BorOH

Figure 2.4 Chiral auxiliaries used in similar ways as menthol.

neomenthyldiphosphine bearing one stereogenic phosphorus atom whereas Valentine and co-workers[182] reported the analogous phosphines of Scheme 2.26 but with neomenthyl groups.

The monoterpene borneol has also been employed a few times. Richter and Neuffer[179] prepared the bornylphosphinite analogous to **71** (R = *i*-Pr in Scheme 2.25) but with inferior results (69% yield, 69% de) compared to menthol. Schmidt and co-workers[206] developed the route to DiPAMP depicted in Scheme 2.34.

Although this synthesis seems identical to the one described in Scheme 2.9 and Scheme 2.11 with menthol, (*R*$_P$)-**99** was produced 95% yield and with significant diastereoselectivity (60% de). After separation by chromatography, (*R*,*R*)-DiPAMP was obtained by conventional methods.

More recently, Oohara and Imamoto[207] used bornyl chloroformate to resolve racemic *tert*-butylmethylphosphine borane (**rac-100**), as depicted in Scheme 2.35.

The derivatives **101** were prepared as a crystalline 1:1 epimeric mixture from which the R_P diastereomer could be isolated in 32% yield and 95% de. Alkaline hydrolysis at room temperature provided (*S*)-**100** in 75% yield without any loss of optical purity. Unfortunately, (*S*$_P$)-**101** could not be obtained in high diastereoselectivity. The same route with menthol failed due to the lack of crystallinity of the menthyl esters. An important characteristic of this method is that **rac-100** was obtained by boronation and monomethylation of *tert*-butylphosphine. This is one of the very few occasions where a primary alkyl

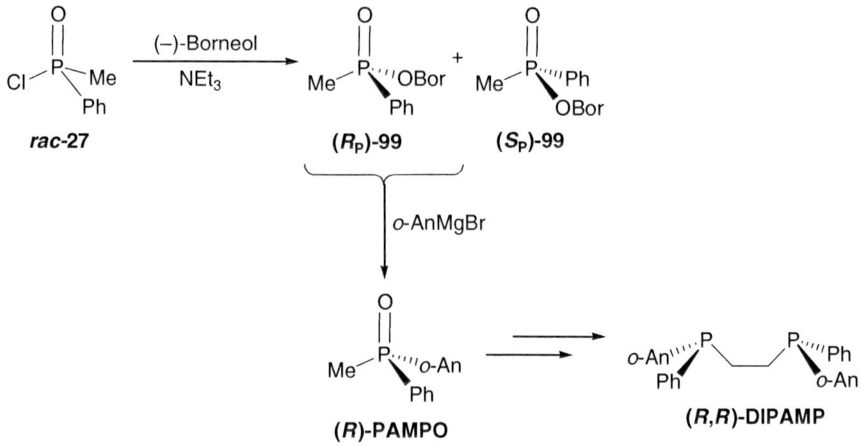

Scheme 2.34 Preparation of DiPAMP *via* bornylphosphinates.

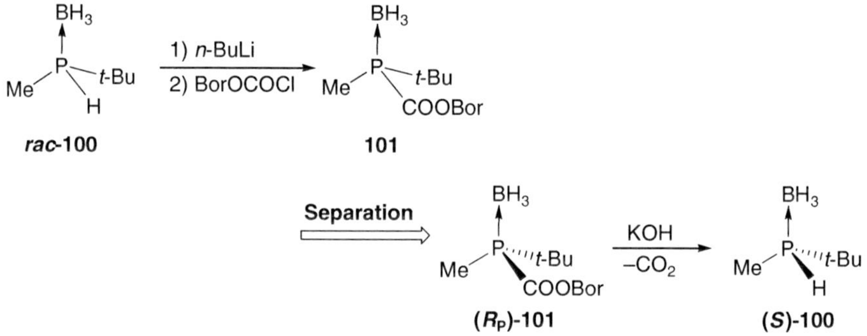

Scheme 2.35 Resolution of *rac*-100 by means of a bornyl ester.

monophosphine has been used in *P*-stereogenic chemistry, despite their potential.

2.4   Chiral Ligands Made by Enantioselective Synthesis

2.4.1   Chiral Alcohols and Amines

Many enantiopure alcohols and some amines have been used to prepare *P*-stereogenic compounds in optically pure form. The most relevant examples are given in this section.

In 1992 Burgess and co-workers[208] prepared DIOP-DiPAMP hybrids to study the matching and mismatching effects between the chiral backbone and the stereogenic phosphorus atoms. These ligands were prepared by substitution on the optically pure ditosylate also used to prepare DIOP (Scheme 2.36).

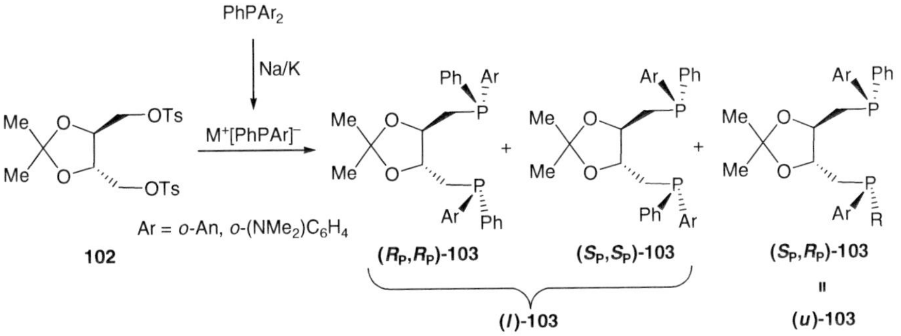

Scheme 2.36 Preparation of DIOP-DiPAMP hybrids.

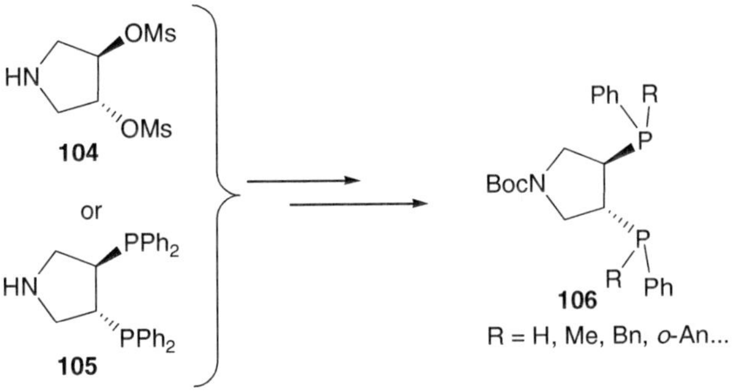

Scheme 2.37 Preparation of pyrrolidine-based *P*-stereogenic phosphines.

The racemic phosphide anion solutions were prepared by selective cleavage of one of the *o*-substituted phenyl rings by sodium/potassium alloy. These solutions were reacted with **102** and furnished a near-statistical mixture of (*l*)- and (*u*)-**103**, indicating no carbon-to-phosphorus chiral induction. The optically pure ligands could only be obtained, after considerable effort, by coordination to molybdenum moieties, separation and decomplexation.

Series of *bis*(phosphines) with a *trans*-1,2-disubstuted pyrrolidine backbone were prepared by Nagel and co-workers by related approaches (Scheme 2.37).[209–211]

Diphosphines **106** were obtained by attack of phosphides to *bis*(mesilate) **104**,[210] by cleavage of phenyl groups in **105** and alkylation of the obtained *bis*(phosphide)[209,210] or by other methods.[211] Usually a mixture of the three possible diastereomers of **106** was formed although they could be often separated after complexation to palladium.[1]

Pellon and co-workers[212] prepared *P*-stereogenic *bis*(1,3-oxaphospholanes), employing phosphine boranes as intermediates (Scheme 2.38).

In this synthesis two cleavages of the phenyl–phosphorus bond by lithium afforded, after hydrolysis, the *bis*(secondary) phosphine borane **109**, which underwent cyclisation with dichloromethane upon deprotonation giving the

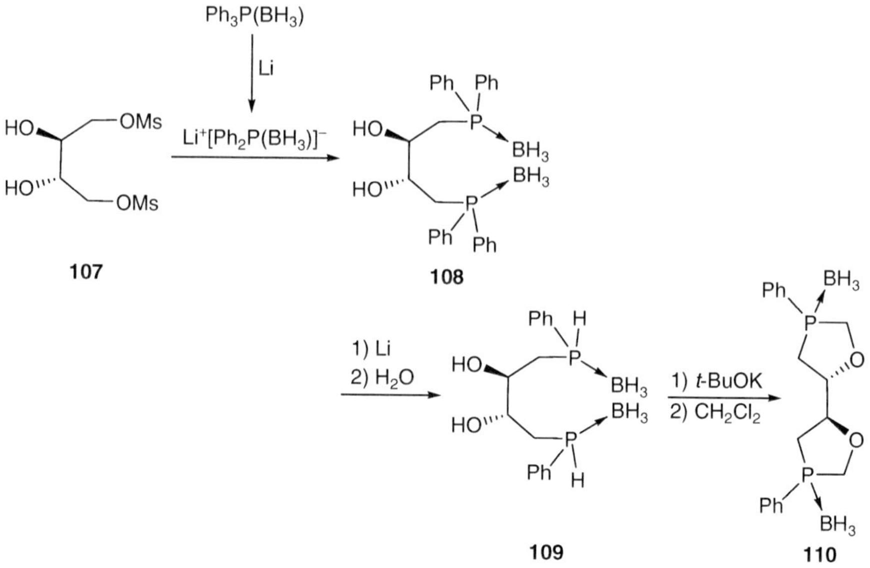

Scheme 2.38 Preparation of *bis*(1,3-oxaphospholanes).

bis-heterocyclic compound **110** as a diastereomeric mixture. The three diastereomers of **110** could be separated by column chromatography.

Many phospholanes have been prepared by reaction of primary phosphines with *bis*mesylates or cyclic sulfates derived from 1,4-diols.[184,213] When the diol is not C_2-symmetric, *P*-stereogenic phospholanes result, which are discussed here.

Series of mono- and *bis*(phospholanes) have been reported by Hoge and co-workers.[214–216] The first example was the synthesis of a *bis*phospholane with an ethyl bridge connecting the phosphorus atoms (Scheme 2.39).[214]

Double deprotonation of methylphosphine borane and addition of **111** produced an equimolar mixture of phospholane boranes **112**. They were separated by 'meticulous' column chromatography, yielding 38% yield of (R_P)-**112**, which was oxidatively dimerised to afford **113** in 50% yield. Although (R_P,R_P)-**113** can obviously be prepared from the other enantiomer of **111**, it was prepared by different protocols (see Scheme 2.40 in this chapter and Chapter 5, Section 5.3.2).

This method has been extended to prepare other mono- and bidentate phospholanes starting from other primary phosphines (Scheme 2.40).[215,216]

Reaction of lithium phosphides with cyclic sulfates **114** provided phospholanes **115–117** in 49–82% yields. Interestingly, the cyclisations took place stereoselectively, affording the *cis* isomers with good selectivity (up to 88% de). For *bis*phospholanes **116** and **117**, no trace of the *trans/trans* isomers was found. As depicted in the scheme, the *cis* isomers could be epimerised to the thermodynamically more stable *trans* isomers by heating the neat compounds

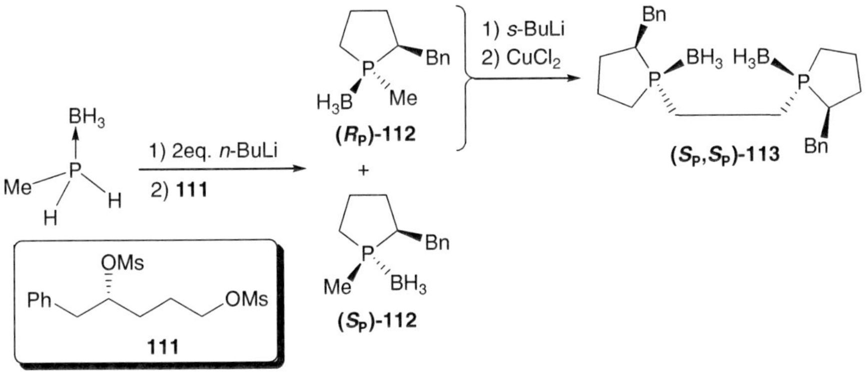

Scheme 2.39 Preparation of *bis*(phospholano)ethane borane **113**.

for several hours. This is one of the few occasions in which pyramidal inversion has been used for synthetic purposes. The degree of epimerisation depended on the substrate. For example, harsher conditions were needed for the more electron-rich phospholane **117**.

Another type of ligand containing phospholane rings, named BeePHOS, was reported by Saito and co-workers.[217] The synthesis of the most important representative is based on cyclisation of mesylate **118** around *bis*(phosphide) **119** (Scheme 2.41).

The desired compound was obtained as a single diastereomer. Other members of the same family were also obtained following a similar methodology.[217]

Vedejs and co-workers[218] prepared bicyclic phospholanes with the 2-phosphabicyclo[3.3.0]octane (PBO) skeleton because they present an exceptional reactivity in enantioselective acylation reactions. The key point in the synthesis is given in Scheme 2.42.[219,220]

The cyclic sulfate **121** was obtained in optically pure form from ketone **120**. Reaction with diphosphides from primary arylphosphines and boronation afforded **122** in 85% yield, good diastereoselectivity (>95%) and even better ee. The absolute configuration of **122** could be ascertained by X-ray crystallography.[220]

Gilbertson and co-workers[221] prepared phosphino oxazoline ligands with a norbornadiene scaffold (Scheme 2.43).

Starting from achiral phospholes, phosphine sulfides **123** were obtained as a racemate. Hydrolysis and coupling to (*S*)-valinol furnished **124** and **124′**, which were easily separated by column chromatography in 93% combined yield. From these precursors, the desired enantiopure ligands **125** could be prepared by conventional methods.

Metallation of enantiomerically pure amines has been used to prepare *P*-stereogenic dihydrobenzazaphosphole boranes (Scheme 2.44).

Wills and co-workers[222] prepared the oxide **127** in 46% yield and good selectivity by cyclisation of dilithiated **126**. Compound **127** was reduced (with inversion of configuration), boronated, desilylated and finally the amino group

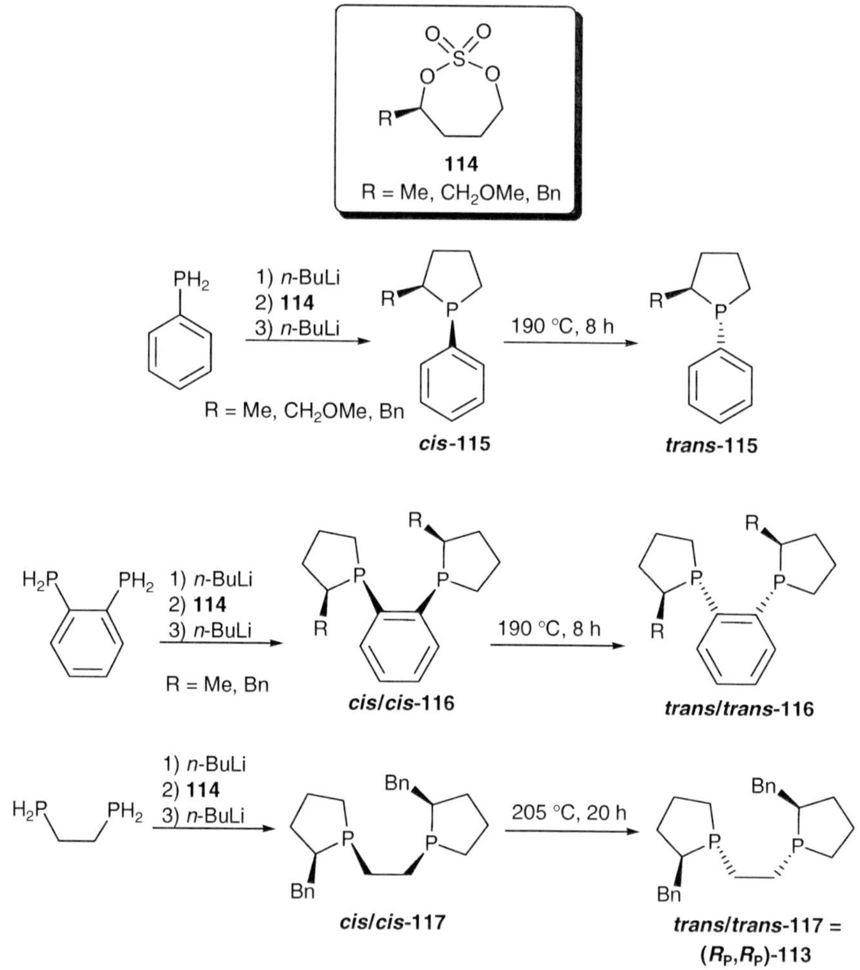

114
R = Me, CH2OMe, Bn
PH2
1) n-BuLi
2) 114
3) n-BuLi
R = Me, CH2OMe, Bn
190 °C, 8 h
cis-115
trans-115
H2P PH2
1) n-BuLi
2) 114
3) n-BuLi
R = Me, Bn
190 °C, 8 h
cis/cis-116
trans/trans-116
1) n-BuLi
2) 114
3) n-BuLi
205 °C, 20 h
cis/cis-117
trans/trans-117 =
(RP,RP)-113

Scheme 2.40 Phospholane boranes prepared from cyclic sulfates **114**.

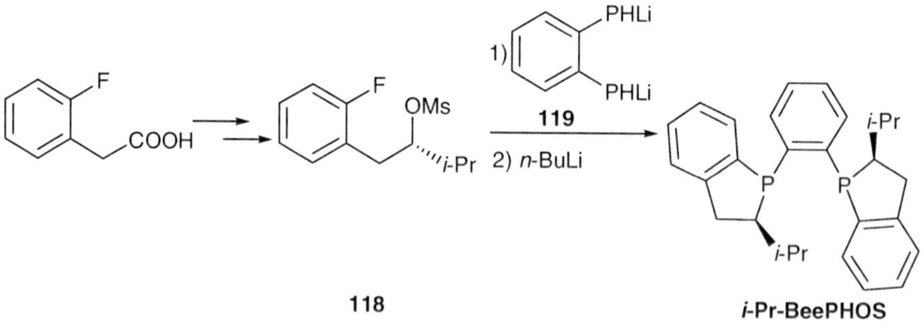

COOH
OMs
i-Pr
PHLi
PHLi
119
1)
2) n-BuLi
118
i-Pr-BeePHOS

Scheme 2.41 Preparation of *i*-Pr-BeePHOS.

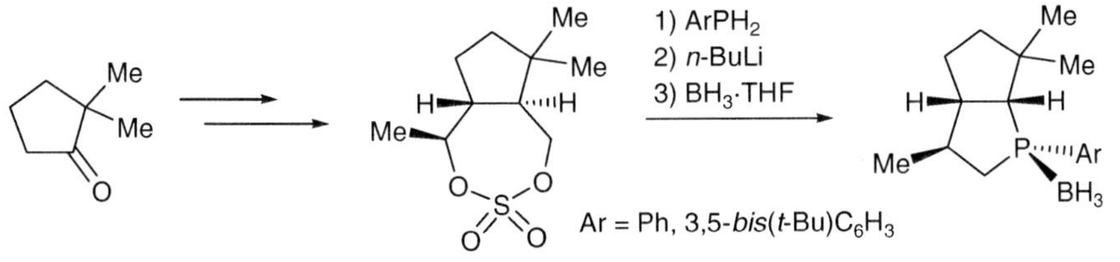

Scheme 2.42 Preparation of a phospholane with the PBO skeleton.

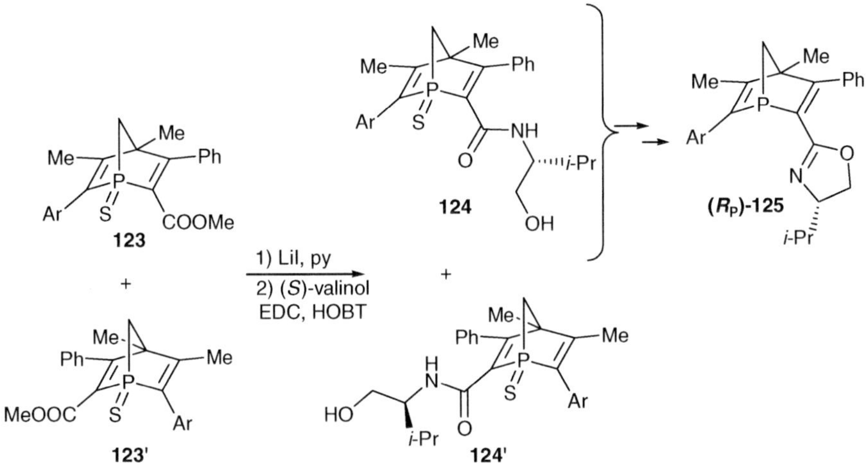

Scheme 2.43 Preparation of phosphanorbornadiene oxazolines. Only one enantiomer of compound **125** is shown.

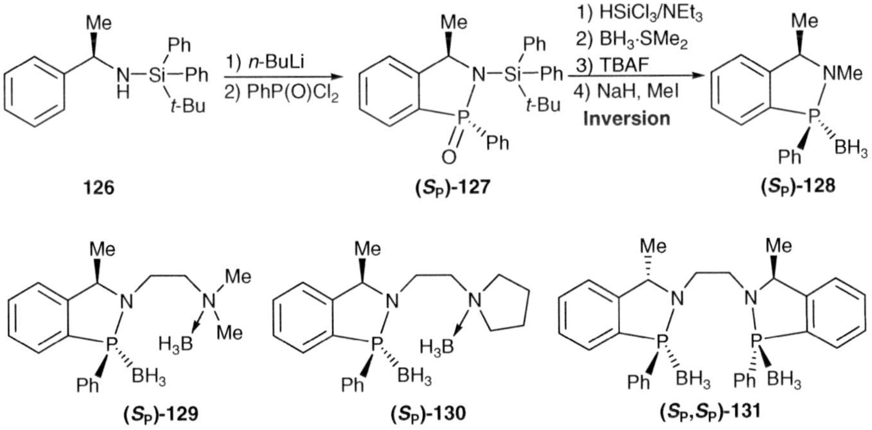

Scheme 2.44 Dihydrobenzazaphosphole boranes prepared from enantiomerically pure amines.

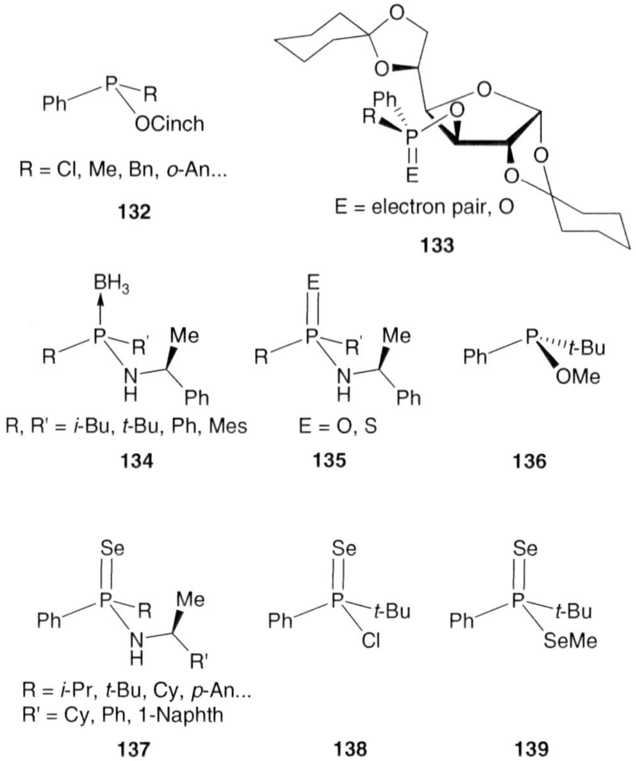

Figure 2.5 *P*-stereogenic compounds with P-heteroatom bonds prepared from chiral amines and alcohols.

methylated to afford **128** in optically pure form. Compound **128** could also be prepared by a shorter route as a mixture of epimers, but they could not be separated. This chemistry has been extended to prepare compounds **129–131** without having to prepare the oxide phosphine oxide intermediate but requiring separation of diastereomers.[223–225]

Several other compounds containing P–hetereoatom bonds have been reported and used as intermediates for other syntheses. Some representative examples are depicted in Figure 2.5.

Early work from Chodkiewicz and co-workers[226,227] described the preparation of cinchonine esters such as **132** as epimeric mixtures, which react with nucleophilies affording phosphonites[227] and tertiary phosphines[226,227] in low to moderate optical purities.

Kolodiazhnyi and Grishkun[228] and simultaneously Fernández, Khiar and co-workers[229] reacted racemic chlorophosphines and their oxides with D-glucoruranose derivatives in the presence of amines to afford the corresponding phosphinites and phosphinates **133** respectively. The nature of the base, solvent and other experimental conditions determined the sense and degree of stereoselectivity[230] but for some cases optically pure compounds were obtained.

Compounds **133** were reacted with organolithium or Grignard reagents under mild conditions to afford phosphines and phosphine oxides with good stereoselectivity.[228,229] Ten years later Hii and co-workers[231] re-examined this chemistry and reacted phosphinates **133** (E = O) with vinylmagnesium bromide affording alkylphenylvinyl phosphine oxides without noticeable polymerisation. The vinylphosphine oxides were used in the same report[231] to prepare families of *P*-stereogenic aminophosphine and aminohydroxyphosphine ligands but they have been used in other syntheses.[232]

Kolodiazhnyi and co-workers[233] reacted racemic chlorophosphines with (*S*)-methylbenzylamine to yield unequal mixtures of epimeric aminophosphines. After boronation, some aminophosphine boranes could be obtained as enantiomerically pure solids by recrystallisation as **134** and used to prepare other compounds such as **135** and **136**.

Series of selenium-containing phosphorus compounds (**137–139**) have been reported by Kimura and Murai.[234–237] Phosphinoselenoic amides **137** were prepared from racemic phosphinoselenoic chlorides (**138**) and chiral amines and the individual isomers could be separated by column chromatography in some cases.[234,237] Reaction of **137** with tributylphosphine afforded the P(III) aminophosphines with retention of configuration at the phosphorus atom. (*S*)-methylbenzylamine was also used to prepare both enantiomers of phosphinoselenoic chlorides **138** in 96% ee.[235] In contrast to the oxygen analogues, they were found to be stable and could be purified by column chromatography. Several compounds, such as **139** for example, were prepared by stereoselective nucleophilic substitutions of the Cl atom.[235]

2.4.2 Chiral Ferrocenes

Chiral ferrocenylphosphines are an important class of ligands in homogeneous catalysis[238–241] and many of them are nowadays commercially available. The introduction of a stereogenic phosphorus atom has been achieved by the standard methods of preparation of *P*-stereogenic compounds (see Chapters 4 and 5) but not all the desired ligands can be made following those methods. This section deals with ferrocene-containing ligands prepared by coupling of an enantiopure ferrocenyl fragment to a racemic or achiral phosphorus unit, followed by separation of diastereomers if required.

Although an *o*-disubstituted ferrocene is chiral, very often (but not always)[242] the ferrocenyl fragments employed contain a group bearing a stereogenic carbon atom. One early example was provided by Togni, Spindler and co-workers[86] (Scheme 2.45).

Reaction of **140** with secondary phosphines at 60 °C gave diphosphines **141** as equimolar epimeric mixtures. Unfortunately, the individual epimers could not be separated by chromatography and the method of the diastereomeric Pd complexes of Section 2.2.3 had to be used (see Table 2.4, entry 9).

Many syntheses of chiral ferrocenes start from the well-known Ugi's amine, and *P*-stereogenic phosphines are not an exception (Scheme 2.46).[243,244]

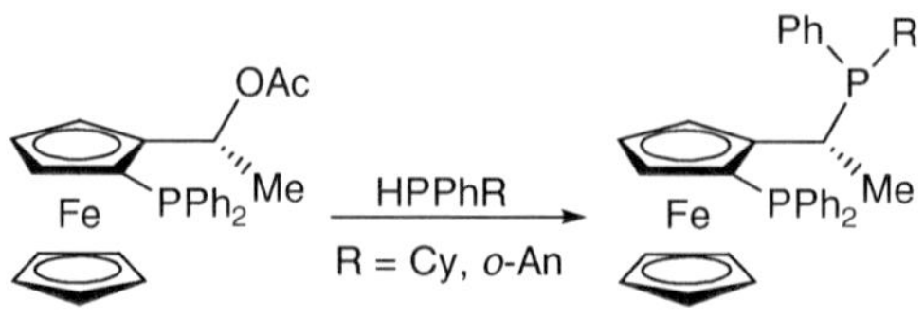

Scheme 2.45 Early syntheses of *P*-stereogenic ferrocenyl diphosphines.

As early as 1986, diastereoselective *ortho*-lithiation of **142** and reaction with racemic *tert*-butylphenylchlorophosphine allowed Cullen and co-workers[243] to prepare **143** as a single isomer. In contrast, when Togni and co-workers[244] prepared **144**, they found an equimolar mixture of epimers, which could be separated by column chromatography. As the objective was the preparation of diphosphines, the second phosphino fragment was introduced by nucleophilic substitution of the dimethylamino group. This step, however, required high temperatures which caused significant racemisation of the stereogenic phosphorus atom in **145**.

The first example of a triphosphine combining a carbon, a phosphorus and a plane as stereogenic elements was reported by Barbaro, Giambastiani and co-workers.[245] The synthesis of this phosphine, named P3Chir, starts with the commercially available phosphine PPFA (Scheme 2.47).

Reaction of PPFA with the racemic secondary phosphine **146** furnished **147** as a 1:1 diastereomeric mixture, which could be separated by column chromatography. Each individual diastereomer of **147** finally afforded P3Chir in good yields after simultaneous reduction/boronation of the phosphine oxide and deprotection of the borane groups. Direct reduction of **147** with silanes epimerised the stereogenic phosphorus atom.

Most of the chiral ferrocenylphosphines are prepared by diastereoselective *ortho*-lithiation of ferrocene and subsequent reaction with a chlorophosphine (Scheme 2.46). Chen and co-workers[246] extended this reaction to dichlorophosphines to have a more flexible approach (Scheme 2.48).

One enantiomer of Ugi's amine was lithiated and reacted successively with several dichlorophosphines and then with organolithiums or Grignard reagents. Remarkably, this method provides monophosphines **149** as single isomers in 80–95% yields. Both epimers at the phosphorus atom are available by using the other enantiomer of Ugi's amine or, more interestingly, exchanging R^1 and R^2. The high diastereoselectivity and the absolute configuration of the products was explained by the formation of the ammonium salt **148'** and attack of the R^2 carbanion from the front as represented in Scheme 2.48. An advantage of this method is that all the transformations are carried out at low or room temperature, preventing the epimerisation of the phosphorus atoms. The same method was applied to the synthesis of the C_2-symmetric diphosphine TriFer[247] as a single isomer after recrystallisation. It was found that this ligand is indefinitely stable in air. A further extension of the method led to the synthesis of C_1-symmetric diphosphines (Scheme 2.49).[248,249]

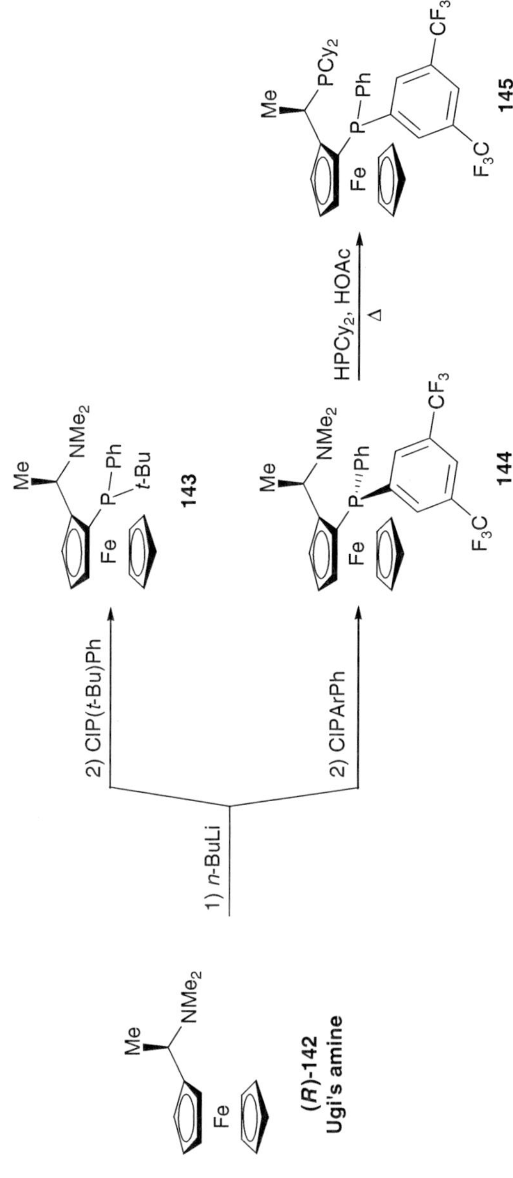

Scheme 2.46 Synthesis of *P*-stereogenic ferrocenyl phosphines starting from Ugi's amine.

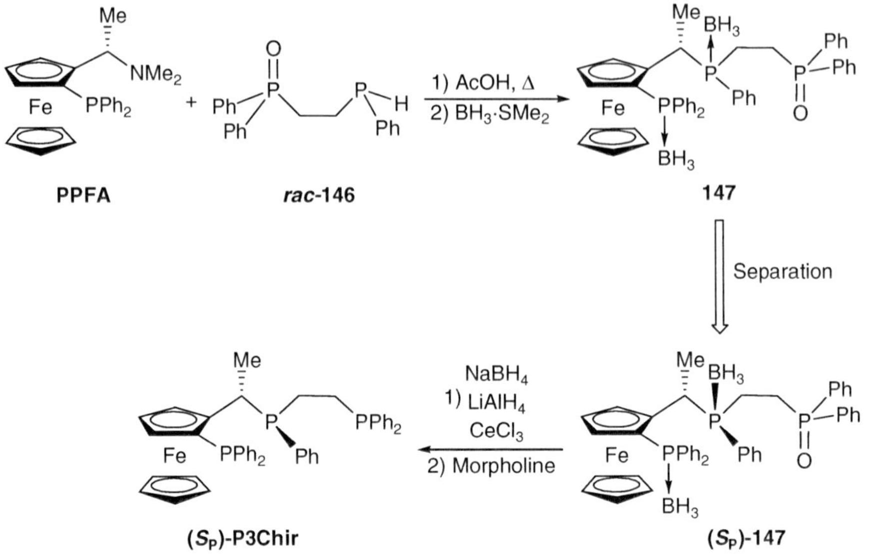

Scheme 2.47 Preparation of one isomer of P3Chir.

Following the sequence seen before, two PingFer diphosphines were prepared from **150** in 90% yield. For R = *o*-An the diastereoselectivity was complete but for R = 1-Naphth a mixture with 80% de was formed although the major diastereomer could be purified by recrystallisation. The selectivity was explained by the formation of a phosphonium salt intermediate. The other route, starting from **151**, was employed to determine the absolute configuration of the products, but is synthetically inferior because it is longer and the diastereoselectivity of aldehydes **152** is lower.[248] In spite of that, Ozawa and co-workers[249] used this aldehyde to prepare the phosphaalkene phosphine **153** in 94% yield.

There is also one example of two successive substitutions on a dichloroferrocenylphosphine, reported by Hey-Hawkins and co-workers (Scheme 2.50).[250]

The first substitution with *o*-carbaboranyl-lithium on enantiomerically pure **154** produces **155** almost exclusively, with only traces of the other epimer. The second chlorine can be substituted in very good yields by reaction with alcohols with clean inversion of configuration at the phosphorus atom. Phosphinites **156** are very stable as the *ortho*-carborane group is bulky and strongly electron-withdrawing, protecting the phosphorus from oxidation.

Another recent advance is the synthesis of *P*-stereogenic ferrocenephospholanes, simultaneously reported by two independent groups. Pugin, Pfaltz and co-workers[251] studied the intramolecular hydrophosphination in a racemic secondary phosphine (Scheme 2.51).

Ortho-substituted bromovinylferrocenes **157** were obtained in few steps from Ugi's amine. Lithiation and reaction with *tert*-butyldichlorophosphine and lithium aluminium hydride afforded secondary phosphines **158** as epimeric

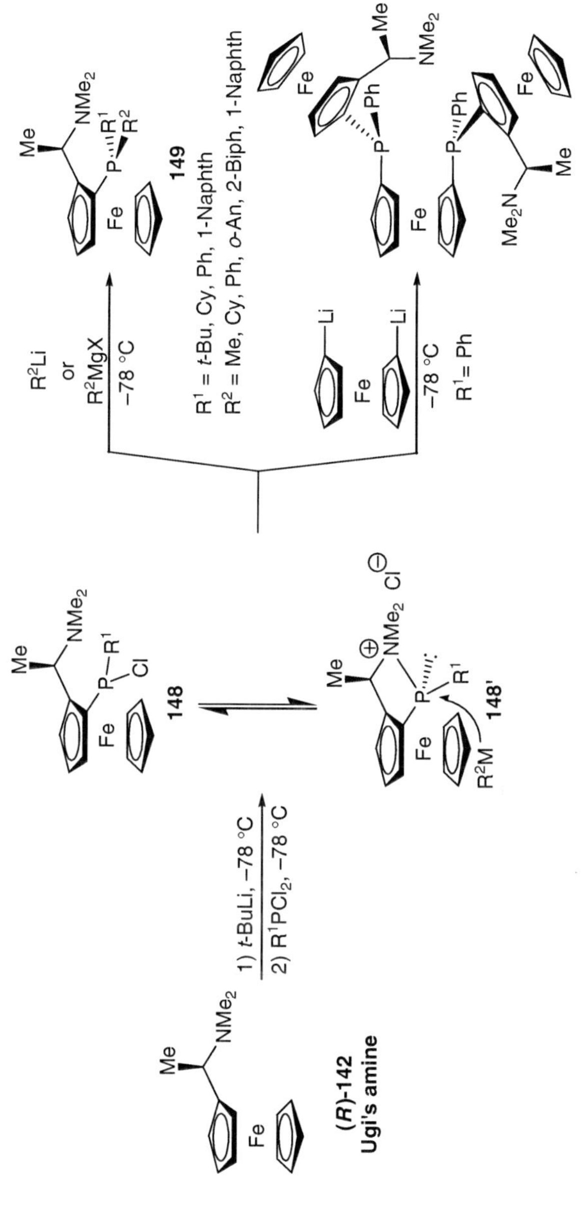

Scheme 2.48　Synthesis of **149** and TriFer using dichlorophosphines.

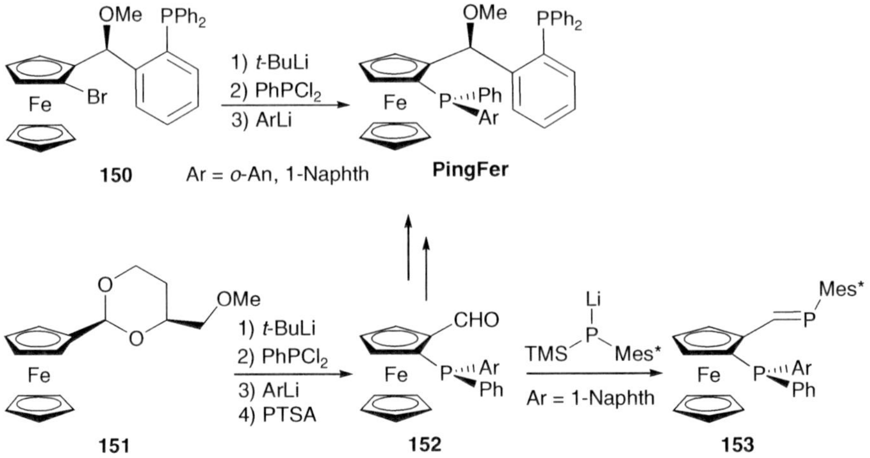

Scheme 2.49 Preparation of C_1-symmetric diphosphines. Mes* is 'supermesityl': 2,4,6-*tris*(*tert*-butyl)phenyl.

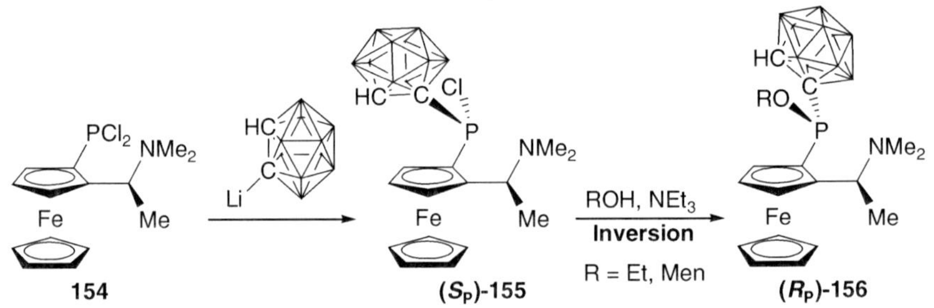

Scheme 2.50 Synthesis of *P*-stereogenic carboranylferrocenylphosphines. Each vertex in the icosahedron represents a BH unit.

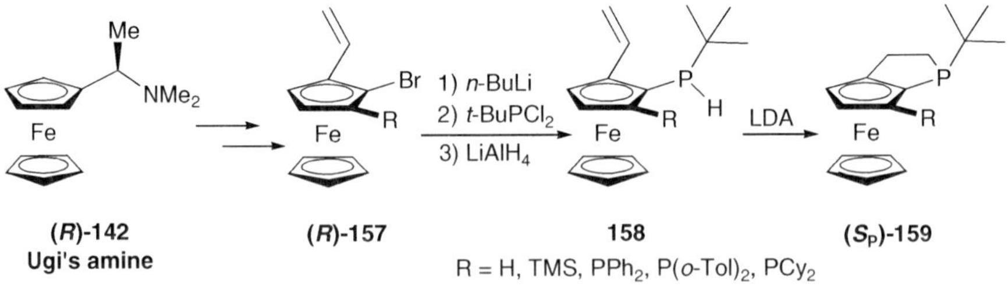

Scheme 2.51 Preparation of ferrocenephospholanes by intramolecular hydrophosphination.

mixtures. Upon deprotonation and intramolecular hydrophosphination, however, resolution of the P atom occurred, affording optically pure products **159** in 50–75% yields. The absolute configuration at the phosphorus atom was determined by X-ray analysis.[251]

Kurita and co-workers[252] prepared a ferrocene-fused benzophosphole by intramolecular cyclisation of a lithiated precursor (Scheme 2.52).

The direct route involved bromine–lithium exchange in **160** followed by intramolecular attack of the carbanion at the phosphorus atom with extrusion of a phenyl group. With this procedure, **162** was obtained in 66% yield as an optically pure compound. This yield could be improved by boronation of the phosphorus atom followed by lithiation with *t*-BuLi, which provoked the cyclisation affording **161**, which was finally deboronated with diethylamine in an overall yield of 87%. The absolute configuration of the phosphorus atom of **162** was determined by X-ray crystallography of a Pd derivative.[252]

An interesting family of *P,N* ligands, known as SiocPhox,[253] was prepared by Hou, Dai and co-workers (Scheme 2.53).[254–256]

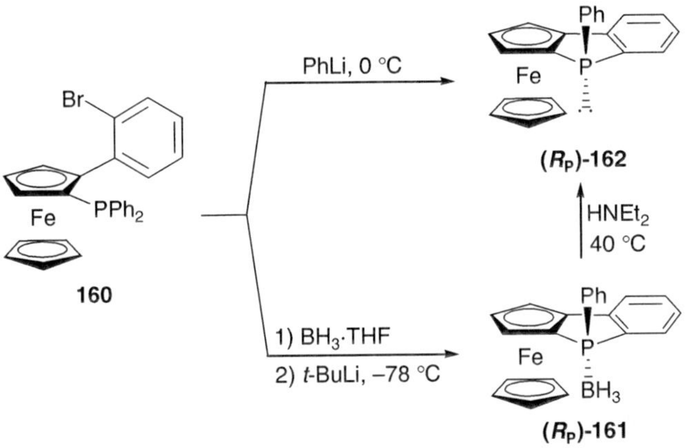

Scheme 2.52 Preparation of a ferrocenephospholane by intramolecular cyclisation.

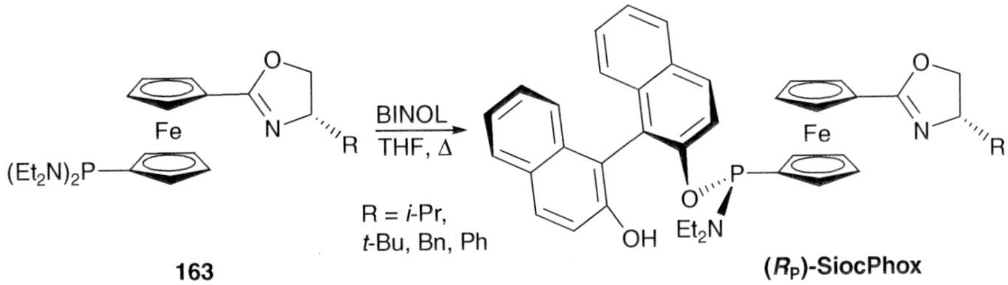

Scheme 2.53 Preparation of SiocPhox ligands.

The original target of these syntheses were phosphonite oxazolines[238] but the second hydroxyl group of the binaphthol did not substitute the diethylamino group of **163** even in refluxing THF. Therefore, phosphonamidate oxazoline ligands were isolated as epimeric mixtures, which could be separated by column chromatography affording 15–45% yields of each pure diastereomer. An unusual feature of this type of ligand is that they possess many stereogenic elements that can be modulated at will. They were found to be excellent ligands for Pd-catalysed allylic substitution reactions.

2.4.3 Chiral Biaryls

Phosphines bearing a stereogenic axis and diphenylphosphino groups, such as MOP and BINAP, are extremely important ligands in enantioselective catalysis.[257] Not surprisingly, there have been attempts to prepare the *P*-stereogenic counterparts. Although there have been successes, it has been found that biaryls are in general poor chiral inductors to the phosphorus atom; therefore, separation of diastereomers is usually needed at some point of the synthesis.

All the work in this area has been carried out with 1,1′-binaphthol, with the exception of the report of Cereghetti and co-workers,[258] which is based on atropoisomeric 1,1′-biphenyls (Scheme 2.54).

Diphosphines with one (**167**) or two (**165**) stereogenic phosphorus atoms were prepared from enantiomerically pure **164** and could be separated into single diastereomers by column chromatography and/or recrystallisation. It was found that the phosphorus atoms epimerised upon heating, reducing the number of diastereomers and sometimes giving a single isomer.

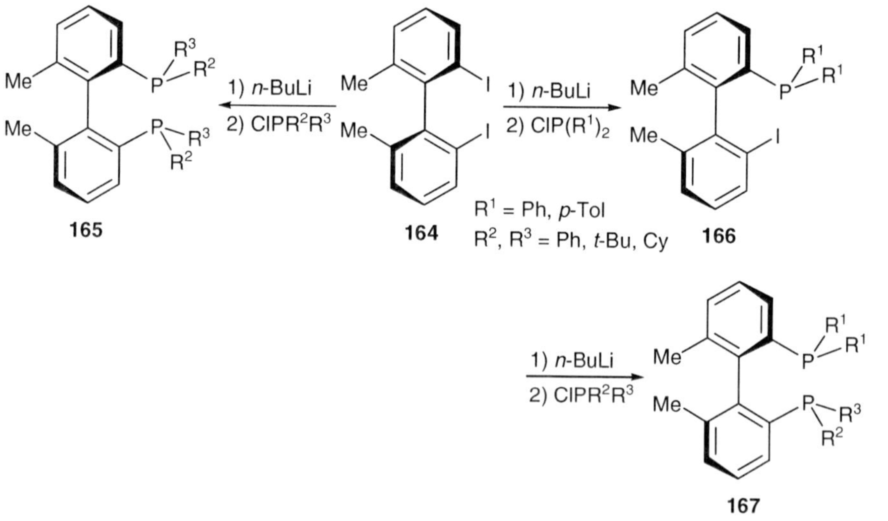

Scheme 2.54 Atropoisomeric *P*-stereogenic biphenyls.

Widhalm and co-workers[259–263] have prepared families of macrocyclic diphosphine ligands with an enantiopure 1,1'-binaphthyl group embedded in the ring (Scheme 2.55).

The multi-step syntheses started from enantiopure **168** or **169** and finished with the closure of the macrocycle by a coupling step with the dilithium salt of 1,2-phenylene*bis*(phenylphosphine), under high-dilution conditions. Starting from **(*S*)-168**, compounds **170** were obtained in good yields with a strong preference for the 'pseudomeso' isomer **(*S*$_a$*S*$_P$*R*$_P$)-170** or even exclusivity for $n = 2$ and 3. The yields of the other isomers were below 15% in all cases.[261] Some of the optically pure isomers could be separated by column chromatography and complexed to Ni and Pd. A similar study was carried out with the nitrogen analogues **171**[262] and more recently other macrocyclic ligands similar to **170** but bearing aryl substitutents at the 3 and 3' positions of the binaphthyl unit have also been prepared.[263]

The same group[264] and others[265] have developed a flexible approach to *P*-stereogenic mono- and bidentate binaphthophosphepines (binepines)[266] by one or two diastereoselective α-deprotonation steps (Scheme 2.56).

Phosphepine sulfides **173** were obtained in 72% yield after ring closure of dilithiated **172** with PhPCl$_2$. The monolithiation and alkylation of **173** proceeded with moderate diastereoselectivity, with preference for **(*R*$_P$)-174**. The mixture could be separated by column chromatography and each isomer alkylated again to afford the α,α'-disubstituted compounds **175**, which were desulfurised with Raney nickel to afford **176** in good yields. This step caused the hydrogenation of the compound with an alkeneic group (R^1 = cinnamyl, R^2 = H) and therefore an alternative synthetic scheme with borane as protective group was developed.[267] More recently, Zhang and co-workers[265] quenched the carbanions derived from **173** with chlorophosphines affording diphosphines **178** after desulfurisation.

A unique example of *P*-stereogenic *bis*(binaphthophosphepine), named BINAPINE, has been reported by Zhang and co-workers[268] (Scheme 2.57).

The protected ligand **180** was obtained as a single diastereoisomer in 25% yield by oxidative coupling of lithiated **179**. Desulfurisation with hexachlorodisilane afforded the desired ligand in 90% yield. Remarkably, the ligand is stable in air for days in spite of being quite electron rich.

Buchwald and Hamada[269] reported in 2002 the first example of a binaphthyl monophosphine (MOP) bearing a *P*-stereogenic atom (Scheme 2.58).

(*R*)-181 was lithiated and reacted with racemic *t*-butylchlorophenylphosphine to afford **(*R*$_P$)-182** and **(*S*$_P$)-182** in a 4:1 ratio. Combining recrystallisation and column chromatography, the pure diastereomers could be separated in 23 and 16% yields for R_P and S_P respectively. Alternatively, the Pd-catalysed phosphination of **(*R*)-181** rendered an equimolar mixture of epimers **183**, which could be separated by column chromatography. Unfortunately, epimerisation at the phosphorus atom occurred upon reduction with a variety of reagents.

Other metal-catalysed couplings were studied by Gilheany and co-workers[270,271] to prepare *P*-stereogenic MOPs. Their results are summarised in Scheme 2.59.

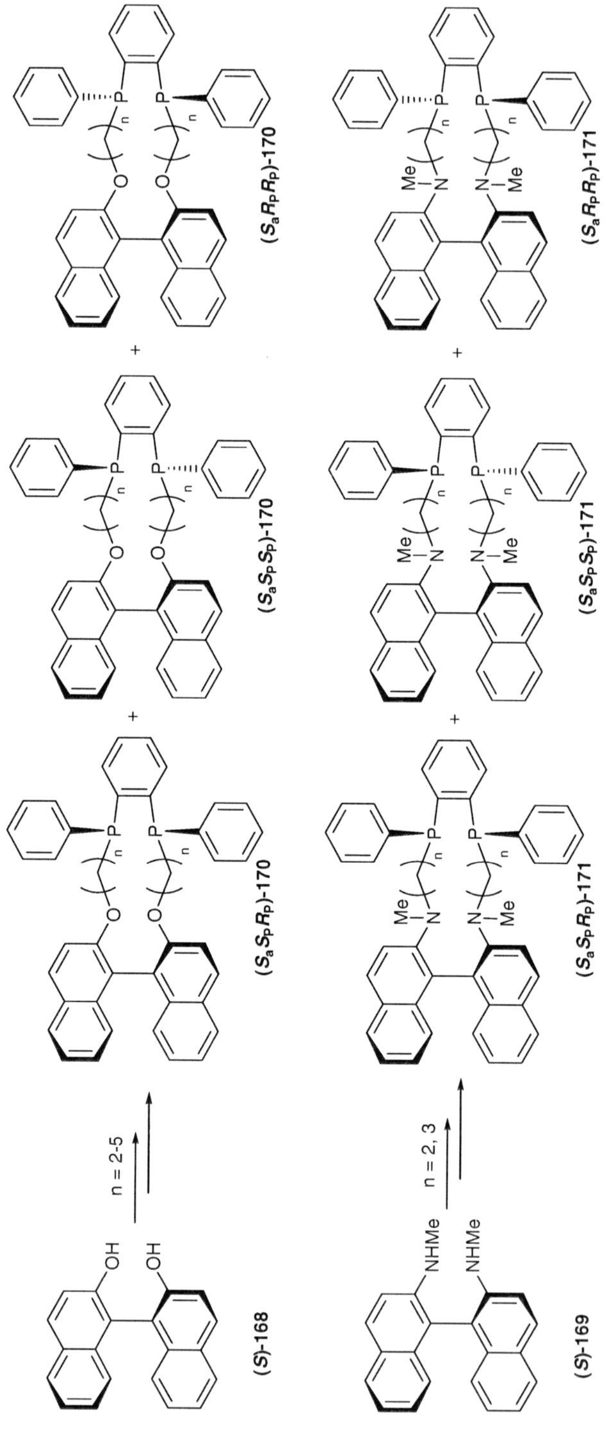

Scheme 2.55 Macrocyclic diphosphines with a 1,1′-binaphthyl group embedded in the ring.

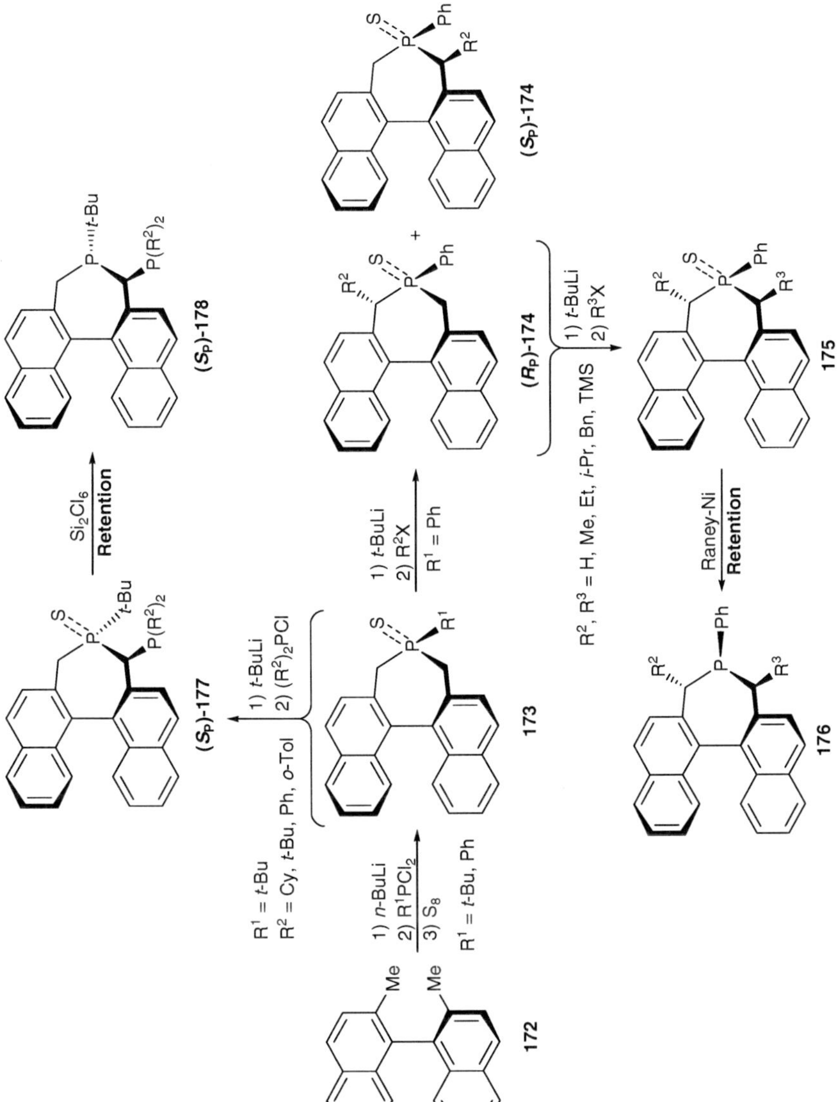

Scheme 2.56 Preparation of *P*-stereogenic binaphthophosphepines.

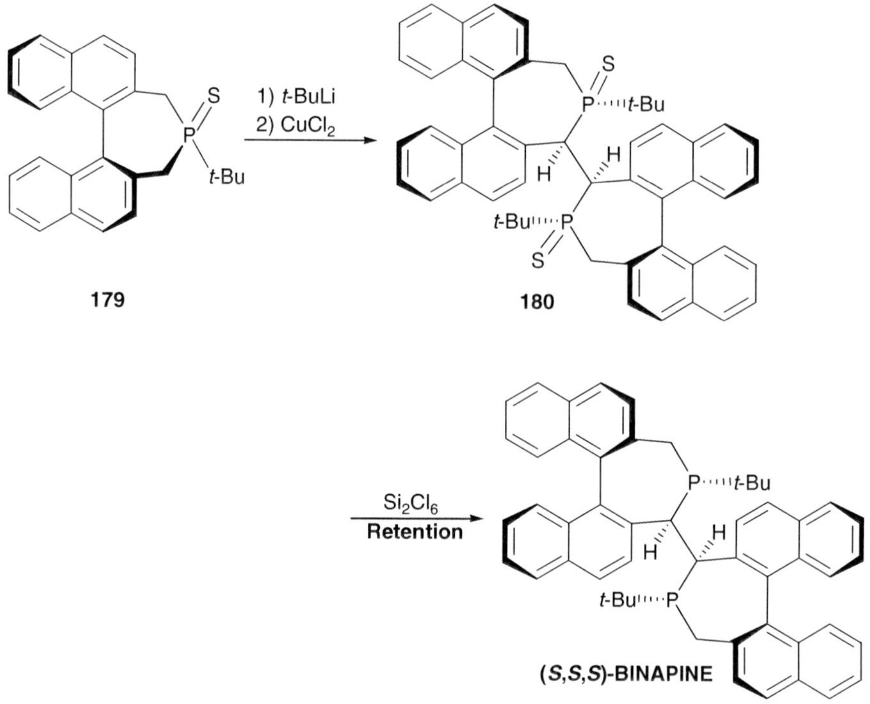

179

1) *t*-BuLi
2) CuCl$_2$

180

Si_2Cl_6
Retention

(*S*,*S*,*S*)-**BINAPINE**

Scheme 2.57   Synthesis of BINAPINE.

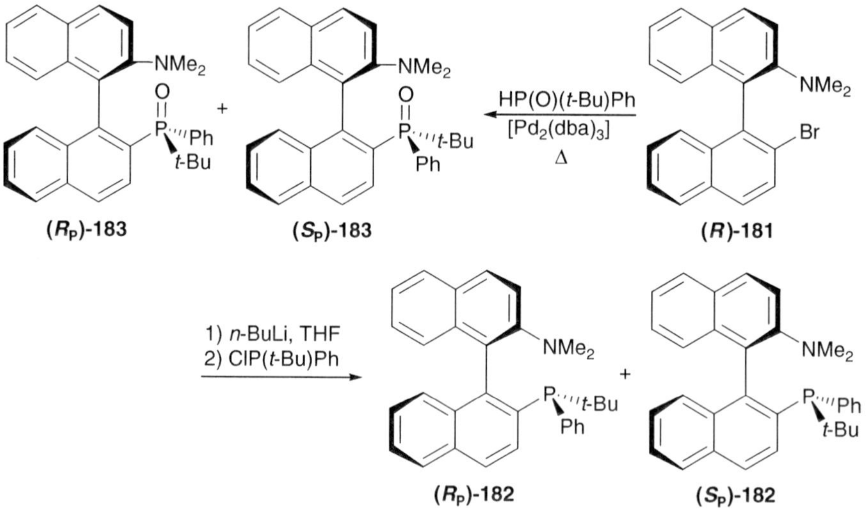

(*R*$_P$)-183

(*S*$_P$)-183

HP(O)(*t*-Bu)Ph
[Pd$_2$(dba)$_3$]
Δ

(*R*)-181

1) *n*-BuLi, THF
2) ClP(*t*-Bu)Ph

(*R*$_P$)-182

(*S*$_P$)-182

Scheme 2.58   Synthesis of a *P*-stereogenic binaphthyl monophosphine.

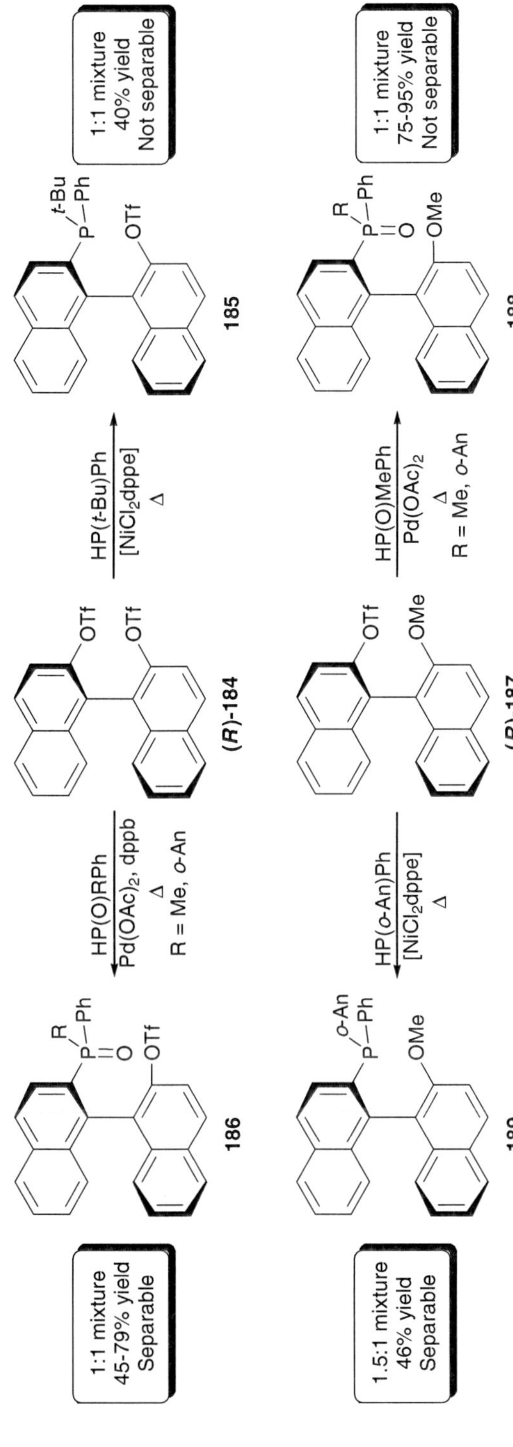

Scheme 2.59 Different metal-catalysed routes to *P*-stereogenic MOPs.

From these examples it is clear that these metal-catalysed couplings are not stereoselective, requiring separation of diastereomers, which seems to be particularly difficult for this type of compound. The phosphine oxide **186** (R = Me) could be converted into **188** by hydrolysis of the triflate group and methylation. Unfortunately, reduction of this phosphine oxide occurred with significant epimerisation at the P atom and only one of the epimers could be isolated in optically pure form.[270] For R = *o*-An, the reduction turned out to be troublesome as well.[271] LiAlH$_4$ reduced the phosphine oxide but also cleaved the phenyl group. Hexachlorodisilane only deprotected the methyl ether without reducing the phosphine. The reaction was successful with phenyl- or trichlorosilane but again was accompanied by considerable epimerisation at the phosphorus atom. The desired phosphines **189** could be finally prepared by Ni-catalysed coupling and separated, with difficulty, either in free form or protected with borane. Similar reactions were studied by Pellet-Rostaing, Lemaire and co-workers,[272] who coupled secondary phosphine boranes with **(*R*)-184** to eventually obtain phosphine oxides bearing an unsubstituted 2-(1,1′-binaphthyl) group. Once again, the problems in the separation of diastereomers and epimerisation in the reduction step appeared.

The conclusion of this section is that the introduction of *P*-stereogenic centres into chiral biaryl moieties is still a synthetic challenge, which remains to be addressed.

2.4.4 Other Chiral Backbones

The chirality of other structures has been used to prepare and separate *P*-stereogenic compounds. Some of these compounds are discussed in this section.

An important example is the preparation of phosphinooxazoline (PHOX) ligands (Scheme 2.60).

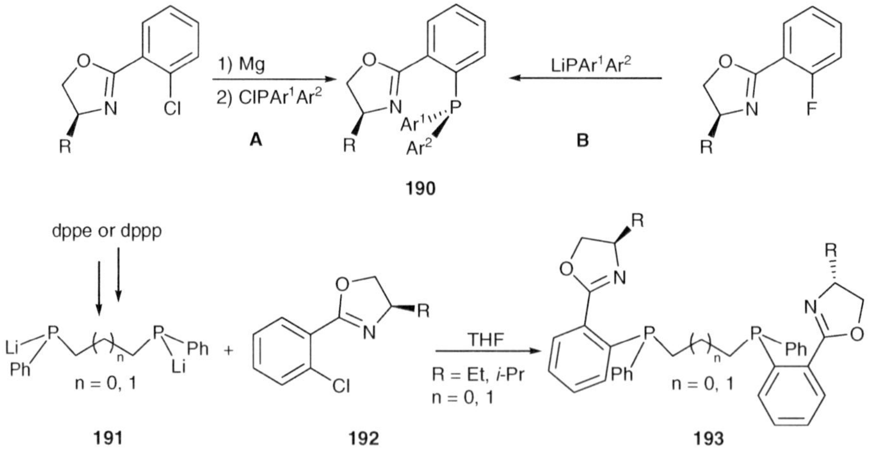

Scheme 2.60 Synthesis of *P*-stereogenic PHOX ligands.

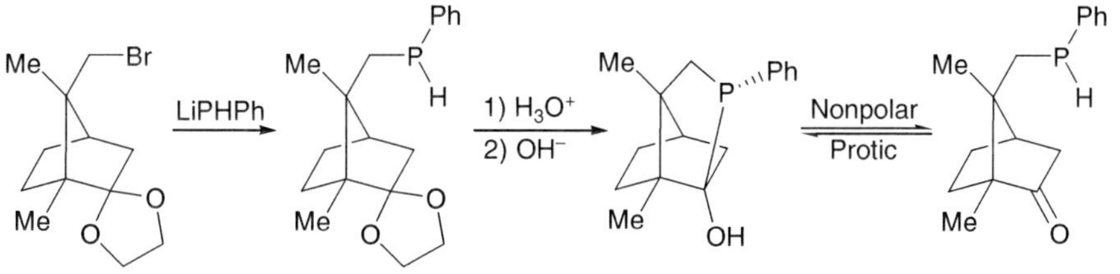

Scheme 2.61 Thermodynamically controlled synthesis of a *P*-stereogenic α-hydroxy-phospholane.

Helmchen and co-workers[273] developed series of bidentate *P*-stereogenic phosphino oxazoline ligands (PHOX) **190** with a stereogenic phosphorus atom. These ligands were prepared *via* either nucleophilic (**A**) or electrophilic (**B**) substitutions at the appropriate phosphorus precursors. The enantiomerically pure oxazoline ring makes the syntheses slightly diastereoselective (diastereomeric ratio of ca. 3:1 in **190**) and optically pure ligands could be obtained after chromatography or recrystallisation. However, only major diastereomers are accessible and the method is obviously not applicable for achiral (R = H) substrates. This limitation can be overcome by Pd-catalysed phosphination (see Chapter 6, Section 6.2.3.1).

Gómez and co-workers[274] reported the preparation of tetradentate NPPN ligands **193**. They were obtained as diastereomeric mixtures by reaction of diphosphides **191** with *o*-chlorooxazolines **192** in good yields. After careful recrystallisations of nickel complexes and decomplexation, two optically pure ligands could be obtained. Their coordination behaviour towards palladium was studied.[274]

A particularly interesting example of a chiral scaffold was given by Komarov, Börner and co-workers.[275] They reported the intramolecular resolution of a *P*-stereogenic centre by a thermodynamically controlled process (Scheme 2.61).

Secondary phosphine **195** was prepared from bromide **194** (derived from (*R*)-camphor) as an equimolar mixture of epimers. Upon liberation of the ketone group under acidic conditions, nucleophilic addition of the secondary phosphine borane took place, giving (***R*_P**)**-196** as the sole compound. Interestingly, in non-polar solvents the phospholane ring opened affording **197** as a mixture of epimers.

2.5 Cyclometallated Pd Complexes as Templates

Ortho-cyclometallated palladium complexes discussed in Section 2.2.2 have found another very important application as chiral templates that promote Diels–Alder and other reactions. This chemistry has been developed by Leung and co-workers[276] and has allowed the synthesis of an impressing number of mono- and diphosphines bearing *P*-stereogenic atoms with peculiar scaffolds bearing at least three more stereogenic centres. These compounds would be very difficult to prepare by other means. The most successful approach has been the activation of DMPP (or its sulfide),[277] depicted in Scheme 2.62.

DMPP itself is not a reactive diene in Diels–Alder reactions,[278] but it is activated by coordination to transition metal ions.[276] Complex **198** contains a labile perchlorato ligand that is easily displaced by the dienophile, which possesses a coordinating atom (O, S, As or P) in the group E. The cycloaddition reaction occurs intramolecularly[278] in a highly organised environment, which leads to the coordinated *exo* cycloadduct **199** exclusively.[279] A standard decoordination step affords the desired enantiopure ligands **200**. Only the *exo-syn* isomers are formed, which bear the lone pair at the phosphorus atom and the dienophile functionalities at the same side of the molecule.[280]

In contrast, complex **201** contains an inert Pd–Cl bond, which is not cleaved by the dienophile. Therefore the cycloaddition reaction takes place inter-molecularly, yielding the thermodynamically more stable *endo-syn* products **203** after decoordination. As can be expected, the stereoselectivity is low (less than 50% de), although the two diastereomers are often separable and the stereoselectivity can be improved by optimising the structure of the cyclo-metallated amine.[276,281,282]

These Diels–Alder cycloadditions have provided many functionalised phosphines as single products, which are listed in Table 2.15.

A disadvantage is that reactions are slow, taking days or even weeks to reach completion although intermolecular cycloadditions are usually faster than the intramolecular ones. Another problem is that the bridgehead phosphorus atoms are not configurationally stable in the free phosphines.

Entries 1–22 list *exo* phosphines **200** whereas entries 23–30 are *endo* phos-phines **203** (Scheme 2.62). The number of *endo* phosphines prepared to date is more limited, probably due to the low stereoselectivity in their formation. Entries 1–10 include *exo* monophosphines with additional oxygen (entries 1–6), sulfur (entries 3, 5–8) and nitrogen (entries 9 and 10) donor atoms. Entries 11–22 contain a second phosphorus or an arsenic atom, which in some cases is also stereogenic (entries 14–16 and 19–21). The compound in entry 19 is the dimerisation of DMPP, which required the use of a platinum complex.[307]

Unsaturated phosphines other than DMPP can also be functionalised upon coordination to Pd cyclometallated complexes (Scheme 2.63).

The coordinated phosphinoalkyne of complex **204** undergoes a hydro-amination reaction with aniline affording iminophosphine complex **(S_P)-205** and its R_P diastereomer (50% de towards the S_P epimer, determined by NMR studies), which could be separated by conventional methods.[314] In solution, complex **205** evolves to the thermodynamically more stable tautomer, the enamino complex **206**. Interestingly, liberation of the ligand from either **205** or **206** furnishes the same phosphine in the imino form, indicating that the tau-tomerism is triggered by complexation.[314]

One of the vinyl groups of the coordinated phosphine of complex **207** can be hydrophosphinated by diphenylphosphine, generating an equimolar mixture of **(R_P)-208** and its S_P epimer in good yields.[315] The diastereomers could be

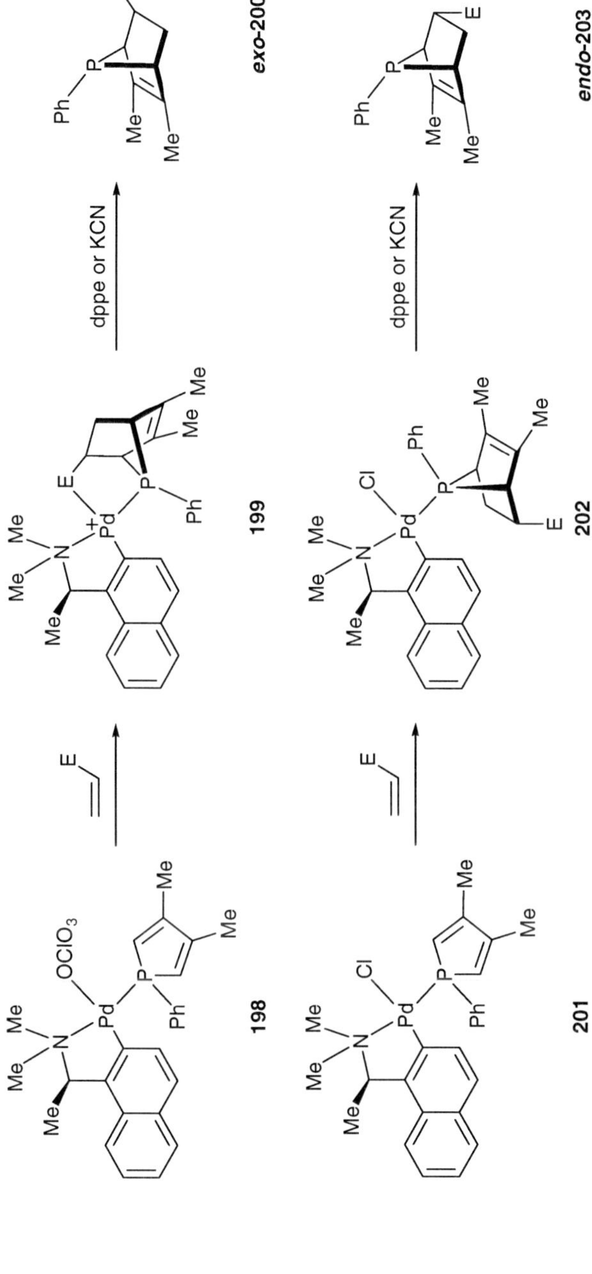

Scheme 2.62 Activation of DMPP towards Diels–Alder reactions by cyclometallated Pd complexes.

Table 2.15 *P*-stereogenic phosphines prepared by Pd-promoted Diels–Alder reactions.

Entry	Phosphine[a]	References
1	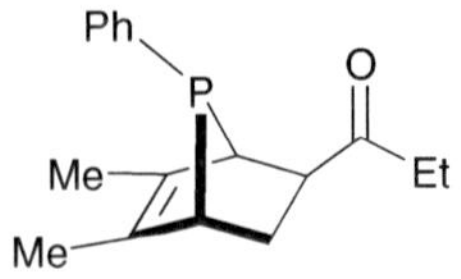	283
2	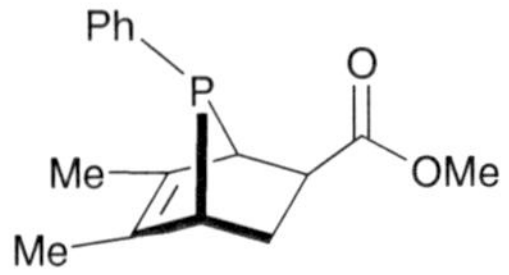	284
3	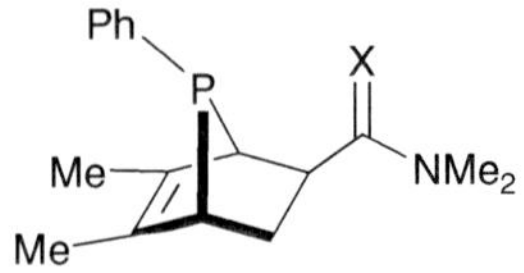 X = O, S	285–288
4	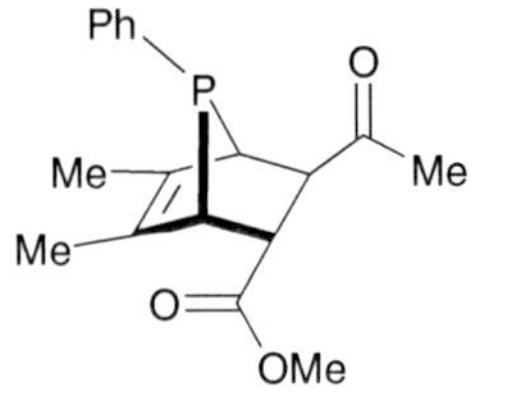	289
5	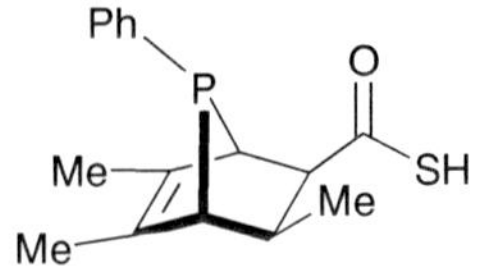	284
6[b]	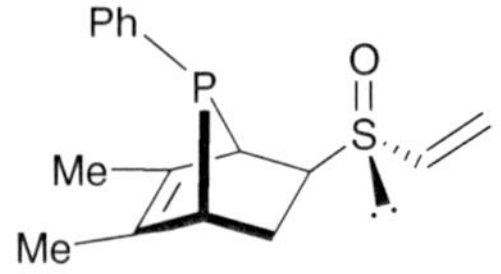	290–292
7	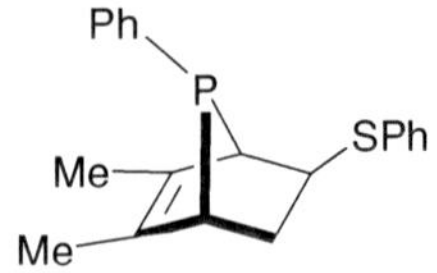	293
8	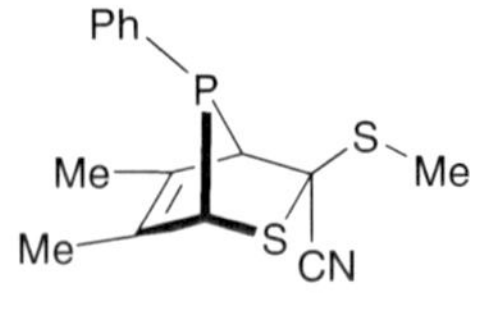	294

Table 2.15 (*Continued*).

Entry	*Phosphine*[a]	*References*
9	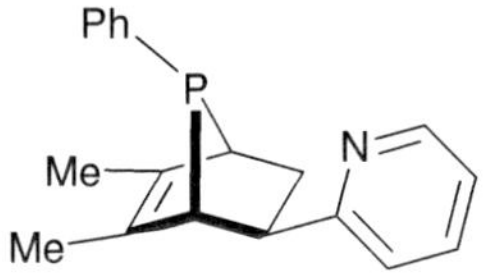	280
10	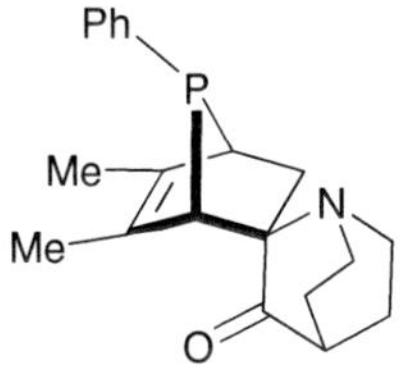	295
11	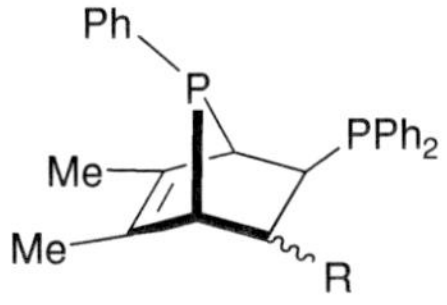 R = H, Me, COOMe, CH_2COOMe	279, 288, 296–300
12	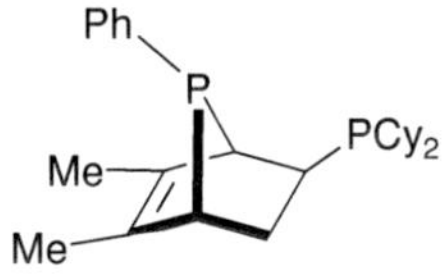	288
13	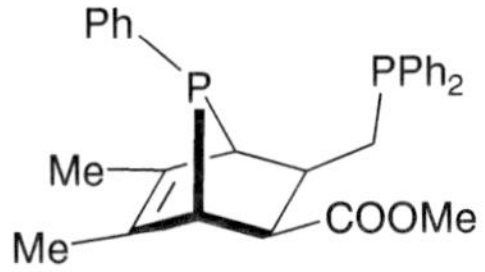	300
14	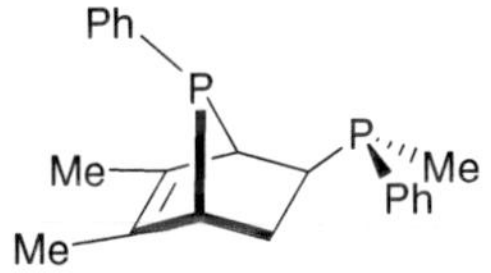	301,302
15[b]	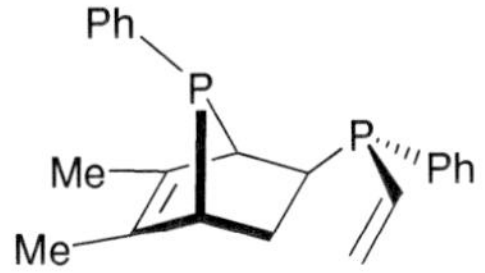	303
16	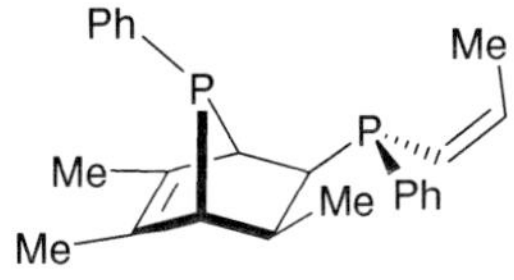	304

Table 2.15   (*Continued*).

Entry	Phosphine[a]	References
17[c]		305
18[a, c]		306
19[c]		307
20		308
21		309
22		310
23		282, 283
24		311

Table 2.15 (*Continued*).

Entry	Phosphine[a]	References
25	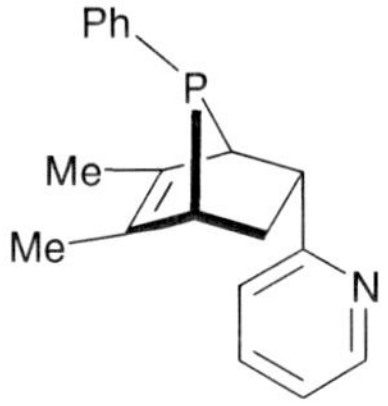 Ph, P, Me, Me, N	280
26	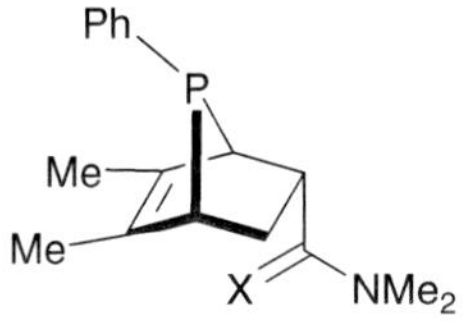 Ph, P, Me, Me, X, NMe$_2$ (X = O, S)	281, 286, 287
27	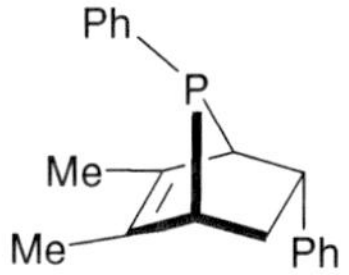 Ph, P, Me, Me, Ph	286, 312
28	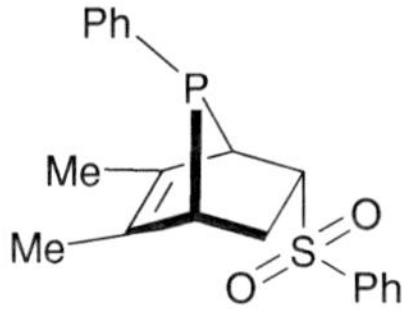 Ph, P, Me, Me, O=S=O, Ph	312
29	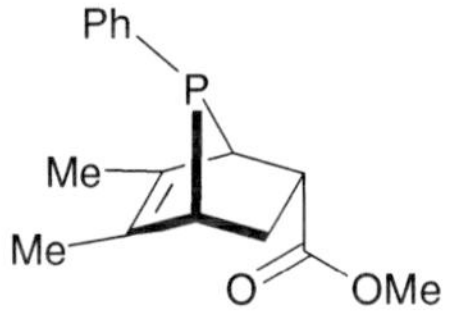 Ph, P, Me, Me, O, OMe	284
30	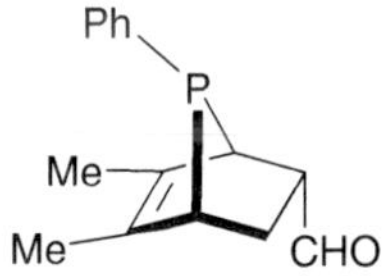 Ph, P, Me, Me, CHO	313

[a]Only one enantiomer is represented.
[b]The free ligand was not obtained.
[c]The platinum analogue of **202** was employed.

separated by fractional crystallisation and the free ligands obtained by standard decomplexation procedures. It has to be noted that although no selectivity was observed in the formation of the stereogenic phosphorus atom, only two of the many other possible products were obtained. Firstly, only one of the vinyl

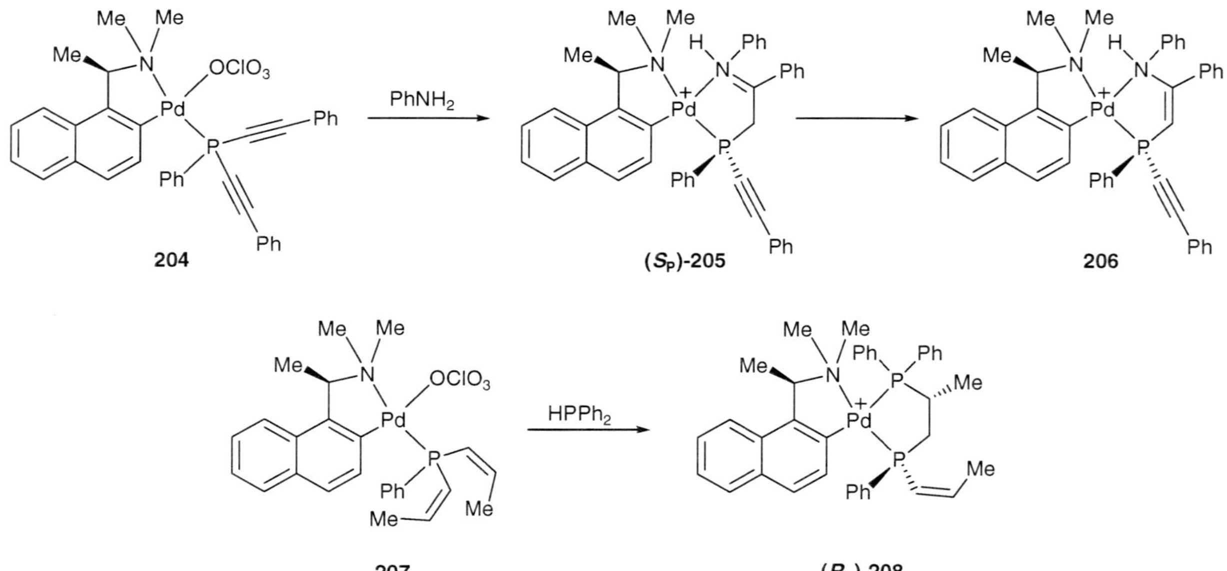

Scheme 2.63 Pd-promoted hydroamination and hydrophosphination of coordinated phosphines. Only one of the possible isomers of compounds **205**, **206** and **207** are shown. None of the reactions occur without the presence of the Pd promoter.

group was hydrophosphinated even with excess of diphenylphosphine, secondly the reaction was regioselective because only the β carbon was hydrophosphinated and thirdly the stereogenic carbon had always *R* absolute configuration.

References

1. K. M. Pietrusiewicz and M. Zablocka, *Chem. Rev.*, 1994, **94**, 1375.
2. J. Meisenheimer and L. Lichtenstadt, *Chem. Ber.*, 1911, **44**, 356.
3. J. Meisenheimer, J. Casper, M. Horing, W. Lauter, L. Lichtenstadt and W. Samuel, *Liebigs Ann. Chem.*, 1926, 213.
4. K. F. Kumli, W. E. McEwen and C. A. VanderWerf, *J. Am. Chem. Soc.*, 1959, **81**, 248.
5. L. Horner, H. Winkler, A. Rapp, A. Mentrup, H. Hoffmann and P. Beck, *Tetrahedron Lett.*, 1961, **2**, 161.
6. L. Horner, *Pure Appl. Chem.*, 1964, **9**, 225.
7. H. J. Bestmann, J. Lienert and E. Heid, *Chem. Ber.*, 1982, **115**, 3875.
8. K. L. Marsi and H. Tuinstra, *J. Org. Chem.*, 1975, **40**, 1843.
9. L. Horner and J. P. Bercz, *Tetrahedron Lett.*, 1966, **7**, 5783.
10. N. Gurusamy and K. D. Berlin, *J. Am. Chem. Soc.*, 1982, **104**, 3114.
11. J. Drabowicz, P. Lyzwa, J. Omelanczuk, K. M. Pietrusiewicz and M. Mikolajczyk, *Tetrahedron: Asymmetry*, 1999, **10**, 2757.
12. I. N. Francesco, A. Wagner and F. Colobert, *Chem. Commun.*, 2010, **46**, 2139.
13. J. Holt, A. M. Maj, E. P. Schudde, K. M. Pietrusiewicz, L. Sieron, W. Wieczorek, T. Jerphagnon, I. W. C. E. Arends, U. Hanefeld and A. J. Minnaard, *Synthesis*, 2009, 2061.
14. R. Freeman, R. K. Haynes, W. A. Loughlin, C. Mitchell and J. V. Stokes, *Pure Appl. Chem.*, 1993, **65**, 647.
15. R. K. Haynes, R. N. Freeman, C. R. Mitchell and S. C. Vonwiller, *J. Org. Chem.*, 1994, **59**, 2919.
16. R. K. Haynes, T. Au-Yeung, W. Chan, W. Lam, Z. Li, L. Yeung, A. S. C. Chan, P. Li, M. Koen, C. R. Mitchell and S. C. Vonwiller, *Eur. J. Org. Chem.*, 2000, 3205.
17. T. Novak, J. Schindler, V. Ujj, M. Czugler, E. Fogassy and G. Keglevich, *Tetrahedron: Asymmetry*, 2006, **17**, 2599.
18. T. Novak, V. Ujj, J. Schindler, M. Czugler, M. Kubinyi, Z. A. Mayer, E. Fogassy and G. Keglevich, *Tetrahedron: Asymmetry*, 2007, **18**, 2965.
19. V. Ujj, P. Bagi, J. Schindler, J. Madarász, E. Fogassy and G. Keglevich, *Chirality*, 2010, **22**, 699.
20. M. Sugiya and H. Nohira, *Bull. Chem. Soc. Jpn.*, 2000, **73**, 705.
21. P. Balcewski, A. Szadowiak, T. Bialas, W. M. Wieczorek and A. Balinska, *Tetrahedron: Asymmetry*, 2006, **17**, 1209.
22. X. Sun, K. Manabe, W. Lam, N. Shiraishi, J. Kobayashi, M. Shiro, H. Utsumi and S. Kobayashi, *Chem. Eur. J.*, 2005, **11**, 361.

23. M. Stankevic and K. M. Pietrusiewicz, *J. Org. Chem.*, 2007, **72**, 816.
24. M. Stankevic and K. M. Pietrusiewicz, *Tetrahedron: Asymmetry*, 2007, **18**, 552.
25. T. Miura and T. Imamoto, *Tetrahedron Lett.*, 1999, **40**, 4833.
26. Y. Hamada, F. Matsuruta, M. Oku, K. Hatano and T. Shiori, *Tetrahedron Lett.*, 1997, **38**, 8961.
27. J. Omelanczuk, A. Karaçar, M. Freytag, P. G. Jones, R. Bartsch, M. Mikolajczyk and R. Schmutzler, *Inorg. Chim. Acta*, 2003, **350**, 583.
28. D. Liu and X. Zhang, *Eur. J. Org. Chem.*, 2005, **1**, 646.
29. T. Imamoto, K. V. L. Crépy and K. Katagiri, *Tetrahedron: Asymmetry*, 2004, **15**, 2213.
30. T. Yoshikuni and J. C. Bailar, *Inorg. Chem.*, 1982, **21**, 2129.
31. F. Leca, F. Fernández, G. Muller, C. Lescop, R. Reau, N. Gabbitas and M. Gómez, *Eur. J. Inorg. Chem.*, 2009, 5583.
32. M. Zanger, C. A. VanderWerf and W. E. McEwen, *J. Am. Chem. Soc.*, 1959, **81**, 3806.
33. A. Blade-Font, C. A. VanderWerf and W. E. McEwen, *J. Am. Chem. Soc.*, 1960, **82**, 2396.
34. W. E. McEwen, K. F. Kumli, A. Blade-Font, M. Zanger and C. A. VanderWerf, *J. Am. Chem. Soc.*, 1964, **86**, 2378.
35. L. Ackermann, in *Phosphorus Ligands in Asymmetric Catalysis: Synthesis and Applications*, ed. A. Börner, Wiley-VCH, Weinheim, 2008, p. 831.
36. W. Lam, R. K. Haynes, L. Yeung and E. W. Chan, *Tetrahedron Lett.*, 1996, **37**, 4733.
37. R. K. Haynes, W. Lam, I. D. Williams and L. Yeung, *Chem. Eur. J.*, 1997, **3**, 2052.
38. T. Au-Yeung, K. Chan, W. Chan, R. K. Haynes, I. D. Williams and L. L. Yeung, *Tetrahedron Lett.*, 2001, **42**, 453.
39. T. A. Yeung, K. Chan, R. K. Haynes, I. D. Williams and L. L. Yeung, *Tetrahedron Lett.*, 2001, **42**, 457.
40. W. Tang, B. Qu, A. G. Capacci, S. Rodriguez, X. Wei, N. Haddad, B. Narayanan, S. Ma, N. Grinberg, N. K. Yee, D. Krishnamurthy and C. H. Senanayake, *Org. Lett.*, 2010, **12**, 176.
41. W. Tang, A. G. Capacci, A. White, S. Ma, S. Rodriguez, B. Qu, J. Savoie, N. D. Patel, X. Wei, N. Haddad, N. Grinberg, N. K. Yee, D. Krishnamurthy and C. H. Senanayake, *Org. Lett.*, 2010, **12**, 1104.
42. S. Lilièvre, F. Mercier, L. Ricard and F. Mathey, *Tetrahedron: Asymmetry*, 2000, **11**, 4601.
43. F. Mercier, F. Brebion, R. Dupont and F. Mathey, *Tetrahedron: Asymmetry*, 2003, **14**, 3137.
44. A. Germoni, B. Deschamps, L. Ricard, F. Mercier and F. Mathey, *J. Organomet. Chem.*, 2005, **690**, 1133.
45. M. Siutkowski, F. Mercier, L. Ricard and F. Mathey, *Organometallics*, 2006, **25**, 2585.
46. N. G. Andersen, P. D. Ramsden, D. Che, M. Parvez and B. A. Keay, *Org. Lett.*, 1999, **1**, 2009.

47. N. G. Andersen, P. D. Ramsden, D. Che, M. Parvez and B. A. Keay, *J. Org. Chem.*, 2001, **66**, 7478.
48. Z. Pakulski, O. M. Demchuk, J. Frelek, R. Luboradzki and K. M. Pietrusiewicz, *Eur. J. Org. Chem.*, 2004, 3913.
49. P. Beak, D. R. Anderson, M. D. Curtis, J. M. Laumer, D. J. Pippel and G. A. Weisenburger, *Acc. Chem. Res.*, 2000, **33**, 715.
50. G. Wittig, H. J. Cristau and H. Braun, *Angew. Chem. Int. Ed.*, 1967, **6**, 700.
51. L. Horner and M. Jordan, *Phosphorus, Sulfur, and Silicon*, 1980, **8**, 225.
52. W. Perlikowska, M. Gouygou, D. J. C. G. Balavoine and M. Mikolajczyk, *Tetrahedron Lett.*, 2001, **42**, 7841.
53. W. Perlikowska, M. Gouygou, M. Mikolajczyk and J. Daran, *Tetrahedron: Asymmetry*, 2004, **15**, 3519.
54. E. Bergin, C. T. O'Connor, S. B. Robinson, E. M. McGarrigle, C. P. O'Mahony and D. G. Gilheany, *J. Am. Chem. Soc.*, 2007, **129**, 9566.
55. J. Granander, F. Secci, D. F. O'Brien and B. Kelly, *Tetrahedron: Asymmetry*, 2009, **20**, 2432.
56. T. H. Chan, *Chem. Commun.*, 1968, 895.
57. A. Grabulosa, J. Granell and G. Muller, *Coord. Chem. Rev.*, 2007, **251**, 25.
58. S. B. Wild, *Coord. Chem. Rev.*, 1997, **166**, 291.
59. D. C. R. Hockless, P. A. Gugger, P. Leung, R. C. Mayadunne, M. Pabel and S. B. Wild, *Tetrahedron*, 1997, **53**, 4083.
60. V. V. Dunina, L. G. Kuz'mina, M. Y. Rubina, Y. K. Grishin, Y. Veits and E. I. Kazakova, *Tetrahedron: Asymmetry*, 1999, **10**, 1483.
61. R. J. Doyle, G. Salem and A. C. Willis, *J. Chem. Soc. Dalton Trans.*, 1997, 2713.
62. J. Albert, R. Bosque, J. Magali Cadena, J. R. Granell, G. Muller and J. I. Ordinas, *Tetrahedron: Asymmetry*, 2000, **11**, 3335.
63. J. Albert, R. Bosque, J. M. Cadena, S. Delgado and J. R. Granell, *J. Organomet. Chem.*, 2001, **634**, 83.
64. J. Albert, R. Bosque, J. Magali Cadena, S. Delgado, J. R. Granell, G. Muller, J. I. Ordinas, M. Font Bardia and X. Solans, *Chem. Eur. J.*, 2002, **8**, 2279.
65. V. V. Dunina, L. G. Kuz'mina, M. Y. Kazakova, Y. K. Grishin, Y. A. Veits and E. I. Kazakova, *Tetrahedron: Asymmetry*, 1997, **8**, 2537.
66. V. V. Dunina and E. B. Golovan, *Tetrahedron: Asymmetry*, 1995, **6**, 2747.
67. S. Otsuka, A. Nakamura, T. Kano and K. Tani, *J. Am. Chem. Soc.*, 1971, **93**, 4301.
68. K. Tani, L. D. Brown, J. Ahmed, J. A. Ibers, M. Yokota, A. Nakamura and S. Otsuka, *J. Am. Chem. Soc.*, 1977, **99**, 7876.
69. V. V. Dunina, E. B. Golovan, N. S. Gulyukina and A. V. Buyevich, *Tetrahedron: Asymmetry*, 1995, **6**, 2731.
70. J. Albert, J. Magali Cadena, J. R. Granell, G. Muller, D. Panyella and C. Sañudo, *Eur. J. Inorg. Chem.*, 2000, 1283.
71. N. Gabbitas, G. Salem, M. Sterns and A. C. Willis, *J. Chem. Soc. Dalton Trans.*, 1993, 3271.

72. J. Albert, J. M. Cadena, J. R. Granell, G. Muller, J. I. Ordinas, D. Panyella, C. Puerta, C. Sañudo and P. Valerga, *Organometallics*, 1999, **18**, 3511.

73. E. Duran, E. Gordo, J. Granell, M. Font-Bardia, X. Solans, D. Velasco and F. Lopez-Calahorra, *Tetrahedron: Asymmetry*, 2001, **12**, 1987.

74. F. Doro, F. Lutz, J. N. H. Reek, A. L. Spek and P. W. N. M. van Leeuwen, *Eur. J. Inorg. Chem.*, 2008, 1309.

75. J. Albert, J. M. Cadena, S. Delgado and J. R. Granell, *J. Organomet. Chem.*, 2000, **603**, 235.

76. A. Bader, G. Salem, A. C. Willis and S. B. Wild, *Tetrahedron: Asymmetry*, 1992, **3**, 1227.

77. M. Pabel, A. C. Willis and S. B. Wild, *Angew. Chem. Int. Ed. Engl.*, 1994, **33**, 1835.

78. M. Pabel, A. C. Willis and S. B. Wild, *Inorg. Chem.*, 1996, **35**, 1244.

79. M. Pabel, A. C. Willis and S. B. Wild, *Tetrahedron: Asymmetry*, 1995, **6**, 2369.

80. J. Leitch, G. Salem and C. R. Hockless, *J. Chem. Soc. Dalton Trans.*, 1995, 649.

81. N. K. Roberts and S. B. Wild, *J. Am. Chem. Soc.*, 1979, **101**, 6254.

82. D. G. Allen, S. B. Wild and D. L. Wood, *Organometallics*, 1986, **5**, 1009.

83. R. J. Doyle, G. Salem and A. C. Willis, *J. Chem. Soc. Dalton Trans.*, 1995, 1867.

84. G. Salem and S. B. Wild, *Inorg. Chem.*, 1983, **22**, 4049.

85. C. E. Barclay, G. Deeble, R. J. Doyle, S. A. Elix, G. Salem, T. L. Jones, S. B. Wild and A. C. Willis, *J. Chem. Soc. Dalton Trans.*, 1995, 57.

86. A. Togni, C. Breutel, M. C. Soares, N. Zanetti, T. Gerfin, V. Gramlich, F. Spindler and G. Rihs, *Inorg. Chim. Acta*, 1994, **222**, 213.

87. F. Robin, F. Mercier, L. Ricard, F. Mathey and M. Spagnol, *Chem. Eur. J.*, 1997, **3**, 1365.

88. F. Mathey, F. Mercier, F. Robin and L. Ricard, *J. Organomet. Chem.*, 1998, **577**, 117.

89. G. Frison, F. Brebion, R. Dupont, F. Mercier, L. Ricard and F. Mathey, *C. R. Chimie.*, 2002, **5**, 245.

90. F. Bienewald, L. Ricard, F. Mercier and F. Mathey, *Tetrahedron: Asymmetry*, 1999, **10**, 4701.

91. G. He, K. F. Mok and P. H. Leung, *Organometallics*, 1999, **18**, 4027.

92. D. G. Allen, G. M. McLaughlin, G. B. Robertson, W. L. Steffen, G. Salem and S. B. Wild, *Inorg. Chem.*, 1982, **21**, 1007.

93. J. W. L. Martin, J. A. L. Palmer and S. B. Wild, *Inorg. Chem.*, 1984, **23**, 2664.

94. P. Leung, A. C. Willis and S. B. Wild, *Inorg. Chem.*, 1992, **31**, 1406.

95. J. A. Ramsden, J. M. Brown, M. B. Hursthouse and A. I. Karalulov, *Tetrahedron: Asymmetry*, 1994, **5**, 2033.

96. N. K. Roberts and S. B. Wild, *Inorg. Chem.*, 1981, **20**, 1892.

97. N. K. Roberts and S. B. Wild, *Inorg. Chem.*, 1981, **20**, 1900.

98. J. A. L. Palmer and S. B. Wild, *Inorg. Chem.*, 1983, **22**, 4054.

99. G. Salem and S. B. Wild, *Inorg. Chem.*, 1984, **23**, 2655.

100. G. Salem, A. Schier and S. B. Wild, *Inorg. Chem.*, 1988, **27**, 3029.

101. A. L. Airey, G. F. Swiegers, A. C. Willis and S. B. Wild, *J. Chem. Soc., Chem. Commun.*, 1995, 693.

102. A. L. Airey, G. F. Swiegers, A. C. Willis and S. B. Wild, *Inorg. Chem.*, 1997, **36**, 1588.

103. V. C. Cook, A. C. Willis, J. Zank and S. B. Wild, *Inorg. Chem.*, 2002, **41**, 1897.

104. P. K. Bowyer, V. C. Cook, N. Gharib-Naseri, P. A. Gugger, A. D. Rae, G. F. Swiegers, A. C. Willis, J. Zank and S. B. Wild, *Proc. Natl. Acad. Sci.*, 2002, **99**, 4877.

105. G. F. Swiegers and S. B. Wild, *Aust. J. Chem.*, 2005, **58**, 831.

106. A. L. Airey, G. F. Swiegers, A. C. Willis and S. B. Wild, *J. Chem. Soc. Chem. Commun.*, 1995, 695.

107. H. J. Kitto, A. D. Rae, A. C. Willis, J. Zank and S. B. Wild, *Z. Naturforsch.*, 2004, **59b**, 1458.

108. H. J. Kitto, A. D. Rae, R. D. Webster, A. C. Willis and S. B. Wild, *Inorg. Chem.*, 2007, **46**, 8059.

109. E. Dunach and H. B. Kagan, *Tetrahedron Lett.*, 1985, **26**, 2649.

110. V. V. Dunina, O. N. Gournova, M. V. Livantsov and Y. K. Grishin, *Tetrahedron: Asymmetry*, 2000, **11**, 2907.

111. Z. Pakulski, O. M. Demchuk, R. Kwiatosz, P. W. Osinski, W. Swierczynska and K. M. Pietrusiewicz, *Tetrahedron: Asymmetry*, 2003, **14**, 1459.

112. A. Brandi, S. Cicchi, F. Gasparrini, F. Maggio, C. Villani, M. Koprowski and K. M. Pietrusiewicz, *Tetrahedron: Asymmetry*, 1995, **6**, 2017.

113. P. Pescher, M. Caude, J. M. Rosset, A. Tambuté and L. Oliveros, *Nouv. J. Chim.*, 1985, **9**, 621.

114. P. Pescher, M. Caude, R. Rosset and A. Tambute, *J. Chromatography.*, 1986, **371**, 159.

115. M. Widhalm, L. Brecker and K. Mereiter, *Tetrahedron: Asymmetry*, 2006, **17**, 1355.

116. X. Jiang, A. J. Minnaard, B. Hessen, B. L. Feringa, A. L. L. Duchateau, J. G. O. Andrien, J. A. F. Boogers and J. G. de Vries, *Org. Lett.*, 2003, **5**, 1503.

117. X. Jiang, M. van den Berg, A. J. Minnaard, B. L. Feringa and J. G. de Vries, *Tetrahedron: Asymmetry*, 2004, **15**, 2223.

118. J. Bigeault, L. Giordano and G. Buono, *Angew. Chem. Int. Ed.*, 2005, **44**, 4753.

119. T. R. Ward, L. M. Venanzi, A. Albinati, F. Lianza, T. Gerfin, V. Gramlich and G. M. Ramos Tombo, *Helv. Chim. Acta*, 1991, **74**, 983.

120. G. Hoge, H. Wu, W. S. Kissel, D. A. Pflum, D. J. Greene and J. Bao, *J. Am. Chem. Soc.*, 2004, **126**, 5966.

121. H. Tsuruta and T. Imamoto, *Tetrahedron: Asymmetry*, 1999, **10**, 877.

122. D. Carmichael, H. Doucet and J. M. Brown, *Chem. Commun.*, 1999, 261.

123. J. A. MacKay and E. Vedejs, *J. Org. Chem.*, 2006, **71**, 498.

124. E. Vedejs, O. Daugulis, L. A. Harper, J. A. MacKay and D. R. Powell, *J. Org. Chem.*, 2003, **68**, 5020.

125. J. Naubron, L. Giordano, F. Fotiadu, T. Burgi, N. Vanthuyne, C. Roussel and G. Buono, *J. Org. Chem.*, 2006, **71**, 5586.

126. T. Miura, H. Yamada, S. Kikuchi and T. Imamoto, *J. Org. Chem.*, 2000, **65**, 1877.

127. H. Tsuruta and T. Imamoto, *Synlett*, 2001, 999.

128. Y. Wada, T. Imamoto, H. Tsuruta, K. Yamaguchi and I. D. Gridnev, *Adv. Synth. Catal.*, 2004, **346**, 777.

129. B. Wolfe and T. Livinghouse, *J. Am. Chem. Soc.*, 1998, **120**, 5116.

130. B. S. Williams, P. Dani, M. Lutz, A. L. Spek and G. Koten, *Helv. Chim. Acta*, 2001, **84**, 3519.

131. T. Imamoto, T. Hoshiki, T. Onozawa, T. Kusumoto and K. Sato, *J. Am. Chem. Soc.*, 1990, **112**, 5244.

132. G. Müller and J. Brand, *Organometallics*, 2003, **22**, 1463.

133. H. Heath, B. Wolfe, T. Livinghouse and S. K. Bae, *Synthesis*, 2001, 2341.

134. C. E. Headley and S. P. Marsden, *J. Org. Chem.*, 2007, **72**, 7185.

135. S. Deerenberg, H. S. Schrekker, G. P. F. van Strijdonck, P. C. J. Kamer, P. W. N. M. van Leeuwen, J. Fraange and K. Goubitz, *J. Org. Chem.*, 2000, **65**, 4810.

136. H. Lebel, M. Berthod and F. Gariépy-Bélanger, *Acta Cryst.*, 2001, **E57**, o282.

137. D. Morales-Morales, R. E. Cramer and C. M. Jensen, *J. Organomet. Chem.*, 2002, **654**, 44.

138. S. Medici, G. M. S. B. Williams, P. A. Chase, S. Gladiali, M. Lutz, A. L. Spek, G. P. M. van Klink and G. van Koten, *Helv. Chim. Acta*, 2005, **88**, 694.

139. M. J. Dearden, M. J. McGrath and D. F. O'Brien, *J. Org. Chem.*, 2004, **69**, 5789.

140. O. Korpiun and K. Mislow, *J. Am. Chem. Soc.*, 1967, **89**, 4784.

141. A. Nudelman and D. J. Cram, *J. Am. Chem. Soc.*, 1968, **90**, 3869.

142. O. Korpiun, R. A. Lewis, J. Chickos and K. Mislow, *J. Am. Chem. Soc.*, 1968, **90**, 4842.

143. R. A. Lewis and K. Mislow, *J. Am. Chem. Soc.*, 1969, **91**, 7009.

144. B. D. Vineyard, W. S. Knowles, M. J. Sabacky, G. L. Bachman and D. J. Weinkauff, *J. Am. Chem. Soc.*, 1977, **99**, 5946.

145. A. Nudelman and D. J. Cram, *J. Org. Chem.*, 1971, **36**, 335.

146. E. P. Kyba, *J. Am. Chem. Soc.*, 1975, **97**, 2554.

147. R. D. Baechler and K. Mislow, *J. Am. Chem. Soc.*, 1970, **92**, 3090.

148. E. P. Kyba, *J. Am. Chem. Soc.*, 1976, **98**, 4805.

149. R. A. Lewis, K. Naumann, K. E. DeBruin and K. Mislow, *Chem. Commun.*, 1969, 1010.

150. T. Oshiki and T. Imamoto, *Bull. Chem. Soc. Jpn.*, 1990, **63**, 3719.

151. W. S. Knowles, M. J. Sabacky, B. D. Vineyard and D. J. Weinkauff, *J. Am. Chem. Soc.*, 1975, **97**, 2567.

152. C. A. Maryanoff, B. E. Maryanoff, R. Tang and K. Mislow, *J. Am. Chem. Soc.*, 1973, **95**, 5839.

153. C. R. Johnson and T. Imamoto, *J. Org. Chem.*, 1987, **52**, 2170.
154. R. Bodalksi, E. Rutkowska-Olma and K. M. Pietrusiewicz, *Tetrahedron*, 1980, **36**, 2353.
155. K. M. Pietrusiewicz, M. Zablocka and J. Monkiewicz, *J. Org. Chem.*, 1984, **49**, 1522.
156. K. M. Pietrusiewicz, M. Kuznikowski and M. Koprowski, *Tetrahedron: Asymmetry*, 1993, **4**, 2143.
157. T. Imamoto, K. Sato and C. R. Johnson, *Tetrahedron Lett.*, 1985, **26**, 783.
158. X. Cheng, P. N. Horton, M. Hursthouse and K. K. Hii, *Tetrahedron: Asymmetry*, 2004, **15**, 2241.
159. Y. Koide, A. Sakamoto and T. Imamoto, *Tetrahedron Lett.*, 1991, **32**, 3375.
160. T. L. Emmick and R. L. Letsinger, *J. Am. Chem. Soc.*, 1968, **90**, 3459.
161. A. Leyris, J. Bigeault, D. Nuel, L. Giordano and G. Buono, *Tetrahedron Lett.*, 2007, **48**, 5247.
162. Q. Xu, C. Zhao and L. Han, *J. Am. Chem. Soc.*, 2008, **130**, 12648.
163. N. V. Dubrovina and A. Börner, *Angew. Chem. Int. Ed.*, 2004, **43**, 5883.
164. D. Moraleda, D. Gatineau, D. Martin, L. Giordano and G. Buono, *Chem. Commun.*, 2008, 3031.
165. M. Stankevic, G. Andrijewski and K. M. Pietrusiewicz, *Synlett*, 2004, 311.
166. L. Han and C. Zhao, *J. Org. Chem.*, 2005, **70**, 10121.
167. T. Imamoto and H. Tsuruta, *Chem. Lett.*, 1996, **1**, 707.
168. T. Imamoto, M. Matsuo, T. Nonomura, K. Kishikawa and M. Yanagawa, *Heteroatom. Chem.*, 1993, **4**, 475.
169. T. Imamoto, T. Oshiki, T. Onozawa, M. Matsuo, T. Hikosaka and M. Yanagawa, *Heteroatom. Chem.*, 1992, **3**, 563.
170. T. Oshiki, T. Hikosaka and T. Imamoto, *Tetrahedron Lett.*, 1991, **32**, 3371.
171. T. Imamoto, H. Tsuruta, Y. Wada, H. Masuda and K. Yamaguchi, *Tetrahedron Lett.*, 1995, **36**, 8271.
172. T. Watanabe, I. D. Gridnev and T. Imamoto, *Chirality*, 2000, **12**, 346.
173. Y. Takahashi, Y. Yamamoto, K. Katagiri, H. Danjo, K. Yamaguchi and T. Imamoto, *J. Org. Chem.*, 2005, **70**, 9009.
174. T. Imamoto, T. Yoshizawa, K. Hirose, Y. Wada, H. Masuda, K. Yamaguchi and H. Seki, *Heteroatom. Chem.*, 1995, **6**, 99.
175. T. Imamoto, S. Kikuchi, T. Miura and Y. Wada, *Org. Lett.*, 2001, **3**, 87.
176. S. Yamago, M. Yanagawa and E. Nakamura, *J. Chem. Soc. Chem. Commun.*, 1994, 2093.
177. H. Tsuruta, T. Imamoto, K. Yamaguchi and I. D. Gridnev, *Tetrahedron Lett.*, 2005, **46**, 2879.
178. M. Mikolajczyk, J. Omelanczuk and W. Perlikowska, *Tetrahedron*, 1979, **35**, 1531.
179. J. Neuffer and W. J. Richter, *J. Organomet. Chem.*, 1986, **301**, 289.
180. R. Bayersdörfer, B. Ganter, U. Englert, W. Keim and D. Vogt, *J. Organomet. Chem.*, 1998, **552**, 187.
181. C. Fisher and H. S. Mosher, *Tetrahedron Lett.*, 1977, **18**, 2487.

182. D. Valentine, J. F. Blount and K. Toth, *J. Org. Chem.*, 1980, **45**, 3691.

183. Q. Jiang, H. Rüegger and L. M. Venanzi, *J. Organomet. Chem.*, 1995, **488**, 233.

184. J. Holz, M. Gensow, O. Zayas and A. Börner, *Curr. Org. Chem.*, 2007, **11**, 61.

185. A. Marinetti, in *Phosphorus Ligands in Asymmetric Catalysis: Synthesis and Applications*, ed. A. Börner, Wiley-VCH, Weinheim, 2008, p. 114.

186. A. Rauk, J. D. Andose, W. G. Frick, R. Tang and K. Mislow, *J. Am. Chem. Soc.*, 1971, **93**, 6507.

187. A. Marinetti, L. Ricard and F. Mathey, *Synthesis*, 1992, 157.

188. A. Marinetti, F. Mathey and L. Ricard, *Organometallics*, 1993, **12**, 1207.

189. A. Marinetti, V. Kruger and F. Buzin, *Coord. Chem. Rev.*, 1998, **178–180**, 755.

190. A. Marinetti and D. Carmichael, *Chem. Rev.*, 2002, **102**, 201.

191. A. Marinetti, F. Buzin and L. Ricard, *Tetrahedron*, 1997, **53**, 4363.

192. J. J. McBride, E. Jungermann, J. V. Killheffer and R. J. Clutter, *J. Org. Chem.*, 1962, **27**, 1833.

193. A. Marinetti, F. Buzin and L. Ricard, *J. Org. Chem.*, 1997, **62**, 297.

194. A. Marinetti and L. Ricard, *Tetrahedron*, 1993, **49**, 10291.

195. A. Marinetti and L. Ricard, *Organometallics*, 1994, **13**, 3956.

196. A. Marinetti, V. Kruger, C. Le Menn and L. Ricard, *J. Organomet. Chem.*, 1996, **522**, 223.

197. A. Marinetti, C. Le Menn and L. Ricard, *Organometallics*, 1995, **14**, 4983.

198. A. Marinetti, V. Kruger and L. Ricard, *J. Organomet. Chem.*, 1997, **529**, 465.

199. A. Marinetti, V. Kruger and B. Couetoux, *Synthesis*, 1998, 1539.

200. A. Bader, M. Pabel and S. B. Wild, *J. Chem. Soc. Chem. Commun.*, 1994, 1405.

201. A. Bader, M. Pabel, A. C. Willis and S. B. Wild, *Inorg. Chem.*, 1996, **35**, 3874.

202. E. Vedejs and Y. Donde, *J. Am. Chem. Soc.*, 1997, **119**, 9293.

203. E. Vedejs and Y. Donde, *J. Org. Chem.*, 2000, **65**, 2337.

204. S. Bauer, S. Tschirschwitz, P. Lonnecke, R. Frank, B. Kirchner, M. L. Clarke and E. Hey-Hawkins, *Eur. J. Inorg. Chem.*, 2009, 2776.

205. R. B. King, J. Bakos, C. D. Hoff and L. Marko, *J. Org. Chem.*, 1979, **44**, 3095.

206. U. Schmidt, B. Riedl, H. Griesser and C. Fitz, *Synthesis*, 1991, 655.

207. N. Oohara and T. Imamoto, *Bull. Chem. Soc. Jpn.*, 2002, **75**, 1359.

208. K. Burgess, M. J. Ohlmeyer and K. H. Whitmire, *Organometallics*, 1992, **11**, 3588.

209. U. Nagel and B. Rieger, *Chem. Ber.*, 1988, **121**, 1123.

210. U. Nagel and T. Krink, *Chem. Ber.*, 1993, **126**, 1091.

211. U. Nagel and T. Krink, *Chem. Ber.*, 1995, **128**, 309.

212. P. Pellon, C. Le Goaster, G. Marchand, B. Martin and L. Toupet, *Heteroatom. Chem.*, 1997, **8**, 123.

213. T. P. Clark and C. R. Landis, *Tetrahedron: Asymmetry*, 2004, **15**, 2123.

214. G. Hoge, *J. Am. Chem. Soc.*, 2003, **125**, 10219.

215. G. Hoge, *J. Am. Chem. Soc.*, 2004, **126**, 9920.

216. G. Hoge and B. Samas, *Tetrahedron: Asymmetry*, 2004, **15**, 2155.

217. H. Shimizu, T. Saito and H. Kumobayashi, *Adv. Synth. Catal.*, 2003, **345**, 185.

218. E. Vedejs, O. Daugulis, J. A. MacKay and E. Rozners, *Synlett*, 2001, 1499.

219. E. Vedejs and O. Daugulis, *J. Am. Chem. Soc.*, 1999, **121**, 5813.

220. E. Vedejs and O. Daugulis, *J. Am. Chem. Soc.*, 2003, **125**, 4166.

221. S. R. Gilberston, D. G. Genov and A. L. Rheingold, *Org. Lett.*, 2000, **2**, 2885.

222. B. Burns, E. Merifield, M. F. Mahon, K. C. Molloy and M. Wills, *J. Chem. Soc., Perkin Trans. 1*, 1993, 2243.

223. G. Brenchley, E. Merifield, M. Wills and M. Fedouloff, *Tetrahedron Lett.*, 1994, **35**, 2791.

224. G. Brenchley, M. Fedouloff, M. F. Mahon, K. C. Molloy and M. Wills, *Tetrahedron*, 1995, **51**, 10581.

225. G. Brenchley, M. Fedouloff, E. Merifield and M. Wills, *Tetrahedron: Asymmetry*, 1996, **7**, 2809.

226. W. Chodkiewicz, D. Jore and W. Wodzki, *Tetrahedron Lett.*, 1979, **12**, 1069.

227. W. Chodkiewicz, *J. Organomet. Chem.*, 1984, **273**, C55.

228. O. I. Kolodiazhnyi and E. V. Grishkun, *Tetrahedron: Asymmetry*, 1996, **7**, 967.

229. A. Benabra, A. Alcudia, N. Khiar, I. Fernández and F. Alcudia, *Tetrahedron: Asymmetry*, 1996, **7**, 3353.

230. I. Fernández, N. Khiar, A. Roca, A. Benabra, A. Alcudia, J. L. Espartero and F. Alcudia, *Tetrahedron Lett.*, 1999, **40**, 2029.

231. M. Oliana, F. King, P. N. Horton, M. Hursthouse and K. K. Hii, *J. Org. Chem.*, 2006, **71**, 2472.

232. A. M. Maj, K. M. Pietrusiewicz, I. Suisse, F. Agbossou and A. Mortreux, *Tetrahedron: Asymmetry*, 1999, **10**, 831.

233. O. I. Kolodiazhnyi, E. V. Gryskun, N. V. Andrushko, M. Freytag, P. G. Jones and R. Schmutzler, *Tetrahedron: Asymmetry*, 2003, **14**, 181.

234. T. Kimura and T. Murai, *Chem. Lett.*, 2004, **33**, 878.

235. T. Kimura and T. Murai, *Chem. Commun.*, 2005, **1**, 4077.

236. T. Kimura and T. Murai, *J. Org. Chem.*, 2005, **70**, 952.

237. T. Kimura and S. Murai, *Tetrahedron: Asymmetry*, 2005, **16**, 3703.

238. L. Dai, T. Tu, S. You, W. Deng and X. Hou, *Acc. Chem. Res.*, 2003, **36**, 659.

239. P. Barbaro, C. Bianchini, G. Giambastiani and S. L. Parisel, *Coord. Chem. Rev.*, 2004, **248**, 2131.

240. W. Chen and H. Blaser, in *Phosphorus Ligands in Asymmetric Catalysis: Synthesis and Applications*, ed. A. Börner, Wiley-VCH, Weinheim, 2008, p. 345.

241. W. Chen and H. Blaser, in *Phosphorus Ligands in Asymmetric Catalysis: Synthesis and Applications*, ed. A. Börner, Wiley-VCH, Weinheim, 2008, p. 359.
242. L. L. Troitskaya, Z. A. Starikova, T. V. Demeshchik and V. I. Sokolov, *Russ. Chem. Bull.*, 1999, **48**, 1738.
243. I. R. Butler, W. R. Cullen and S. J. Rettig, *Organometallics*, 1986, **5**, 1320.
244. C. Gambs, G. Consiglio and A. Togni, *Helv. Chim. Acta*, 2001, **84**, 3105.
245. P. Barbaro, C. Bianchini, G. Giambastiani and A. Togni, *Chem. Commun.*, 2002, 2672.
246. W. Chen, W. Mbafor, S. M. Roberts and J. Whittall, *J. Am. Chem. Soc.*, 2006, **128**, 3922.
247. W. Chen, P. J. McCormack, K. Mohammed, W. Mbafor, S. M. Roberts and J. Whittall, *Angew. Chem. Int. Ed.*, 2007, **46**, 4141.
248. W. Chen, S. M. Roberts, J. Whittall and A. Steiner, *Chem. Commun.*, 2006, 2916.
249. R. Takita, Y. Takada, R. S. Jensen, M. Okazaki and F. Ozawa, *Organometallics*, 2008, **27**, 6279.
250. S. Tschirschwitz, P. Lonnecke and E. Hey-Hawkins, *Organometallics*, 2007, **26**, 4715.
251. B. Gschwend, B. Pugin, A. Bertogg and A. Pfaltz, *Chem. Eur. J.*, 2009, **15**, 12993.
252. S. Yasuike, J. Hagiwara, H. Danjo, M. Kawahata, N. Kakusawa, K. Yamaguchi and J. Kurita, *Heterocycles*, 2009, **78**, 3001.
253. W. Liu, D. Chen, X. Zhu, X. Wan and X. Hou, *J. Am. Chem. Soc.*, 2009, **131**, 8734.
254. S. You, X. Zhu, Y. Luo, X. Hou and L. Dai, *J. Am. Chem. Soc.*, 2001, **123**, 7471.
255. X. Hou and N. Sun, *Org. Lett.*, 2004, **6**, 4399.
256. W. Zheng, N. Sun and X. Hou, *Org. Lett.*, 2005, **7**, 5151.
257. H. Shimizu, I. Nagasaki, N. Sayo and T. Saito, in *Phosphorus Ligands in Asymmetric Catalysis: Synthesis and Applications*, ed. A. Börner, Wiley-VCH, Weinheim, 2008, p. 207.
258. M. Cereghetti, W. Arnold, E. A. Broger and A. Rageot, *Tetrahedron Lett.*, 1996, **37**, 5347.
259. M. Widhalm and G. Klintschar, *Tetrahedron: Asymmetry*, 1994, **5**, 189.
260. M. Widhalm, H. Kalchhauser and H. Kahlig, *Helv. Chim. Acta*, 1994, **77**, 409.
261. M. Widhalm and G. Klintschar, *Chem. Ber.*, 1994, **127**, 1411.
262. M. Widhalm, P. Wimmer and G. Klintschar, *J. Organomet. Chem.*, 1996, **523**, 167.
263. Y. Yan and M. Widhalm, *Tetrahedron: Asymmetry*, 1998, **9**, 3607.
264. P. Kasák, K. Mereiter and M. Widhalm, *Tetrahedron: Asymmetry*, 2005, **16**, 3416.
265. Q. Dai, W. Li and X. Zhang, *Tetrahedron*, 2008, **64**, 6943.

266. S. Gladiali and E. Alberico, in *Phosphorus Ligands in Asymmetric Catalysis: Synthesis and Applications*, ed. A. Börner, Wiley-VCH, Weinheim, 2008, p. 177.

267. P. Kasák, V. B. Arion and M. Widhalm, *Tetrahedron: Asymmetry*, 2006, **17**, 3084.

268. W. Tang, W. Wang, Y. Chi and X. Zhang, *Angew. Chem. Int. Ed.*, 2003, **42**, 3509.

269. T. Hamada and S. L. Buchwald, *Org. Lett.*, 2002, **4**, 999.

270. N. J. Kerrigan, E. C. Dunne, D. Cunningham, P. McArdle, K. Gilligan and D. G. Gilheany, *Tetrahedron Lett.*, 2003, **44**, 8461.

271. L. J. Higham, E. F. Clarke, H. Müller-Bunz and D. G. Gilheany, *J. Organomet. Chem.*, 2005, **690**, 211.

272. M. Jahjah, R. Jahjah, S. Pellet-Rosting and M. Lemaire, *Tetrahedron: Asymmetry*, 2007, **18**, 1224.

273. M. Peer, J. C. de Jong, M. Kiefer, T. Langer, H. Rieck, H. Shell, P. Sennhenn, J. Sprinz, H. Steinhagen, B. Wiese and G. Helmchen, *Tetrahedron*, 1996, **52**, 7547.

274. M. Gómez, S. Jansat, G. Muller, D. Panyella, P. W. N. M. van Leeuwen, P. C. J. Kamer, K. Goubitz and J. Fraange, *Organometallics*, 1999, **18**, 4970.

275. I. V. Komarov, A. Spannenberg, J. Holz and A. Börner, *Chem. Commun.*, 2003, 2240.

276. P. H. Leung, *Acc. Chem. Res.*, 2004, **37**, 169.

277. Y. Qin, S. Selvaratnam, J. J. Vittal and P. H. Leung, *Organometallics*, 2002, **21**, 5301.

278. S. Loh, K. F. Mok, P. H. Leung, A. J. P. White and D. J. Williams, *Tetrahedron: Asymmetry*, 1996, **7**, 45.

279. B. Aw and P. Leung, *Tetrahedron: Asymmetry*, 1994, **5**, 1167.

280. G. He, S. Loh, J. J. Vittal, K. F. Mok and P. H. Leung, *Organometallics*, 1998, **17**, 3931.

281. Y. Li, K. Ng, S. Selvaratnam, G. Tan, J. J. Vittal and P. H. Leung, *Organometallics*, 2003, **22**, 834.

282. Y. Li, S. Selvaratnam, J. J. Vittal and P. H. Leung, *Inorg. Chem.*, 2003, **42**, 3229.

283. P. H. Leung, S. Loh, J. J. Vittal, A. J. P. White and D. J. Williams, *Chem. Commun.*, 1997, 1987.

284. Y. Qin, H. Lang, J. J. Vittal, G. Tan, S. Selvaratnam, A. J. P. White, D. J. Williams and P. H. Leung, *Organometallics*, 2003, **22**, 3944.

285. P. H. Leung, S. Loh, K. F. Mok, A. J. P. White and D. J. Williams, *Chem. Commun.*, 1996, 591.

286. P. H. Leung, G. He, H. Lang, A. Liu, S. Loh, S. Selvaratnam, K. F. Mok, A. J. P. White and D. J. Williams, *Tetrahedron*, 2000, **56**, 7.

287. P. H. Leung, Y. Qin, G. He, K. F. Mok and J. J. Vittal, *J. Chem. Soc., Dalton Trans.*, 2001, 309.

288. K. Y. Ghebreyessus, N. Gül and J. H. Nelson, *Organometallics*, 2003, **22**, 2977.

289. P. H. Leung, H. Lang, A. H. White and D. J. Williams, *Tetrahedron: Asymmetry*, 1998, **9**, 2961.
290. S. Y. Siah, P. H. Leung and K. F. Mok, *J. Chem. Soc., Chem. Commun.*, 1995, 1747.
291. S. Siah, P. Leung, K. F. Mok and M. G. B. Drew, *Tetrahedron: Asymmetry*, 1996, **7**, 357.
292. P. H. Leung, S. Y. Siah, A. J. P. White and D. J. Williams, *J. Chem. Soc., Dalton Trans.*, 1998, 893.
293. P. H. Leung, S. Loh, K. F. Mok, A. J. P. White and D. J. Williams, *J. Chem. Soc., Dalton Trans.*, 1996, 4443.
294. Y. Qin, A. J. P. White, D. J. Williams and P. H. Leung, *Organometallics*, 2002, **21**, 171.
295. Y. Song, J. J. Vittal, S. Chan and P. H. Leung, *Organometallics*, 1999, **18**, 650.
296. B. Aw, S. Selvaratnam, P. H. Leung, N. H. Rees and W. McFarlane, *Tetrahedron: Asymmetry*, 1996, **7**, 1753.
297. S. Selvaratnam, P. Leung, A. J. P. White and D. J. Williams, *J. Organomet. Chem.*, 1997, **542**, 61.
298. B. Aw, T. S. A. Hor, S. Selvaratnam, K. F. Mok, A. J. P. White, D. J. Williams, N. H. Rees, W. McFarlane and P. H. Leung, *Inorg. Chem.*, 1997, **36**, 2138.
299. Y. Song, K. F. Mok, P. H. Leung and S. Chan, *Inorg. Chem.*, 1998, **37**, 6399.
300. M. Yuan, S. A. Pullarkat, C. H. Yeong, Y. Li, D. Krishnan and P. Leung, *Dalton Trans.*, 2009, 3668.
301. A. Liu, K. F. Mok and P. H. Leung, *Chem. Commun.*, 1997, 2397.
302. P. H. Leung, A. Liu and K. F. Mok, *Tetrahedron: Asymmetry*, 1999, **10**, 1309.
303. S. Selvaratnam, K. F. Mok, P. H. Leung, A. J. P. White and D. J. Williams, *Inorg. Chem.*, 1996, **35**, 4798.
304. P. H. Leung, S. Selvaratnam, C. R. Cheng, K. F. Mok, N. H. Rees and W. McFarlane, *Chem. Commun.*, 1997, 751.
305. F. Liu, S. A. Pullarkat, K. Tan, Y. Li and P. Leung, *Organometallics*, 2009, **28**, 6254.
306. W. Yeo, S. Chen, G. Tan and P. K. Leung, *J. Organomet. Chem.*, 2007, **692**, 2539.
307. G. He, Y. Qin, K. F. Mok and P. H. Leung, *Chem. Commun.*, 2000, 167.
308. B. Aw, P. H. Leung, A. J. P. White and D. J. Williams, *Organometallics*, 1996, **15**, 3640.
309. M. Ma, S. A. Pullarkat, Y. Li and P. Leung, *J. Organomet. Chem.*, 2008, **693**, 3289.
310. T. Teo, S. Selvaratnam, J. J. Vittal and P. H. Leung, *Inorg. Chim. Acta*, 2003, **352**, 213.
311. S. Loh, J. J. Vittal and P. H. Leung, *Tetrahedron: Asymmetry*, 1998, **9**, 423.

312. P. H. Leung, H. Lang, X. Zhang, S. Selvaratnam and J. J. Vittal, *Tetrahedron: Asymmetry*, 2000, **11**, 2661.
313. S. Loh, G. Tan, L. L. Koh, S. Selvaratnam and P. H. Leung, *J. Organomet. Chem.*, 2005, **690**, 4933.
314. X. Liu, K. F. Mok and P. H. Leung, *Organometallics*, 2001, **20**, 3918.
315. Y. Zhang, S. A. Pullarkat, Y. Li and P. Leung, *Inorg. Chem.*, 2009, **48**, 5535.

CHAPTER 3
P-Stereogenic Heterocycles

3.1 Introduction

This chapter deals with the species formed upon cyclisation of chiral diamines, amino alcohols or diols around phosphorus moieties to furnish enantiopure or enantioenriched heterocycles containing a trivalent or pentavalent *P*-stereogenic phosphorus atom. Given the tremendous quantity of enantiomerically pure heterobifunctional auxiliaries available there are endless synthetic possibilities. This chapter focuses mainly on five- and six-membered rings, which are by far the most studied (Figure 3.1).

The compounds of this class are easy to prepare and are often obtained in high yields and optical purities. Indeed, they probably constitute the easiest entry to optically pure *P*-stereogenic compounds. Another advantage is that they frequently show marked stability towards moisture and air, in contrast to classic phosphines.

They deserve a chapter on their own for at least three reasons. Firstly, they have been heavily used to study the complicated stereochemistry of substitution reactions at the phosphorus atom.[1] This can be explained by the restriction of rearrangements that occurs when a P atom is placed in a cycle, although it is often found that in cyclic compounds the stereochemical course of the substitution is different to that of their acyclic counterparts.

Secondly, some of the compounds in Figure 3.1 have been used as ligands in transition metal catalysis[2,3] or as organocatalysts.[4] This is especially true for five-membered diaza- and oxazaphospholidines and their corresponding boranes and oxides.

Thirdly, oxazaphospholidines derived from ephedrine are the starting point for one of the principal methods to prepare enantiopure *P*-stereogenic phosphines, discussed in Chapter 4.[5,6]

It would be impossible to give a comprehensive account of all the reported hetereobifunctional *P*-stereogenic heterocycles in a chapter of practical size.

RSC Catalysis Series No. 7
P-Stereogenic Ligands in Enantioselective Catalysis
By Arnald Grabulosa
© Arnald Grabulosa 2011
Published by the Royal Society of Chemistry, www.rsc.org

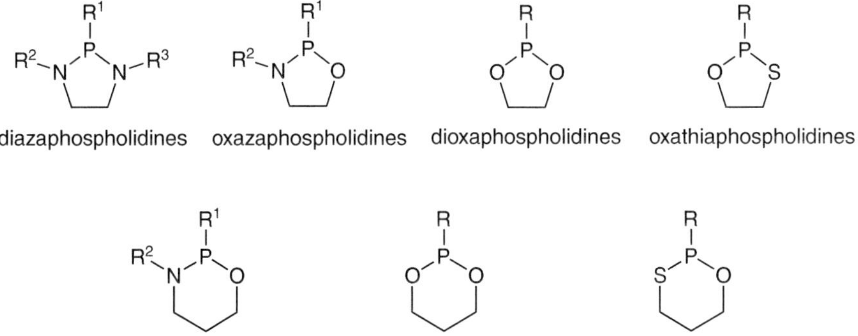

Figure 3.1 Some of the phosphorus heterocycles discussed in this chapter.

Instead, this chapter focuses on diazaphospholidines and oxazaphospholidines, with only a brief discussion of other heterocycles.

The reader must be aware that *P*-stereogenic heterocycles also have many important applications in areas other than catalysis, such as the stereo-controlled synthesis of oligonucleotides[7–12] or as anticancer drugs[13] but these applications lie beyond the scope of this book.

3.2 Diazaphospholidines and Derivatives

3.2.1 P(III) Tertiary Diazaphospholidines

1,3,2-diazaphospholidines are diaminophosphines that bear the N–P–N motif in a five-membered ring. They are easy to prepare and remarkably stable towards oxygen and water.[2,3,14] Their synthesis is based on the displacement of two leaving groups (chlorine or dimethylamino groups) by a chiral secondary 1,2-diamine (Scheme 3.1).

In an overwhelming majority of cases the diamine **1** is (*S*)-2-(anilino-methyl)pyrrolidine (Ar = Ph), easily obtained from L-glutamic acid[15] and now-adays commercially available.

Depending on the absolute configuration at the phosphorus atom, two epimers can be formed, illustrated in Figure 3.2.

The absolute configuration of the phosphorus atom is commonly indicated by the *syn/anti* notation referring to the relative position of the exocyclic substituent at the P atom with respect to pyrrolidine ring.

Figure 3.2 depicts a straightforward way to determine the absolute config-uration at the phosphorus atom of these compounds by ^{13}C NMR spectro-scopy. The value of the C–P coupling constant between C(8) and the P atom is determined by the dihedral angle between the lone pair at the P atom and C(8). For the *syn* isomer this value is close to zero but is over 20 Hz for the *anti* one.[16–18]

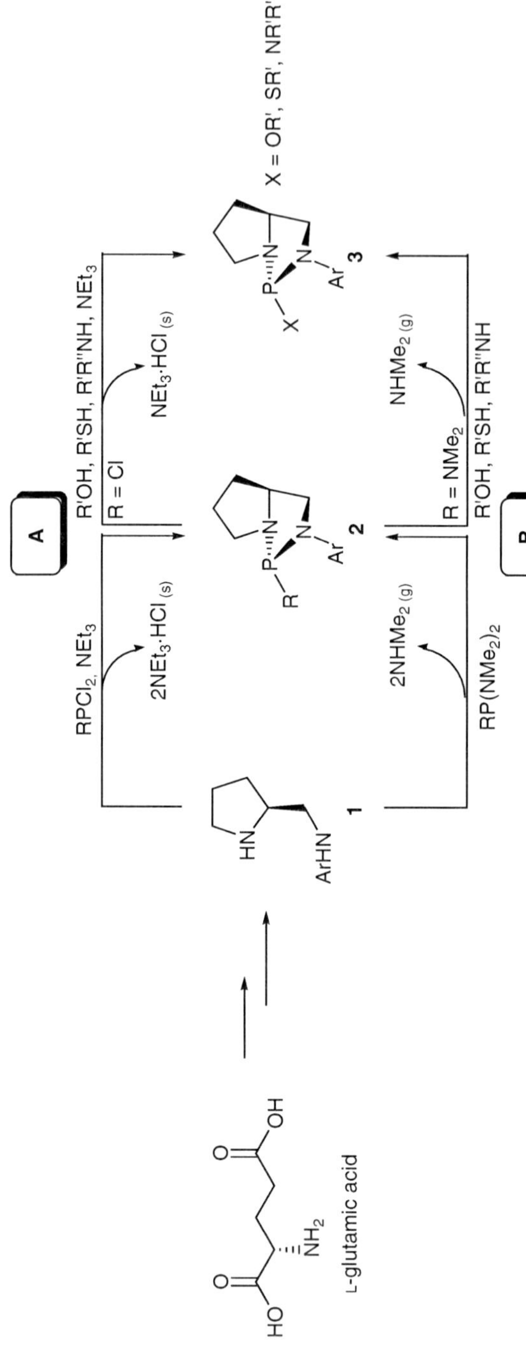

Scheme 3.1 General strategies to prepare diazaphospholidines.

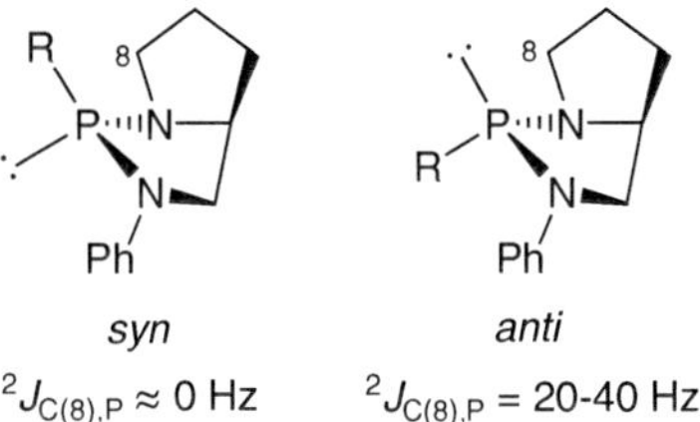

Figure 3.2 Dependence of the C–P coupling constant on the absolute configuration at the P atom.

Depending on the phosphorus precursor, two routes are possible (Scheme 3.1). Route **A** starts from phosphorus trichloride or a dichlorophosphine. The reaction is usually performed in benzene or THF at low temperature and requires the presence of a base, usually triethylamine, to trap the concomitant formation of two equivalents of HCl in the form of triethylammonium chloride, which is removed by filtration. When phosphorus trichloride is employed, compound **2** still has an exchangeable chlorine atom that allows extra functionalisation by similar substitution reactions. Most of the compounds prepared *via* route **A** are listed in Table 3.1.

Entry 1 depicts the key intermediate **4** that has been used in most of the other entries as a phosphitylating reagent for amines (entries 2–8) and alcohols (entries 9–19 and 25–32). This reagent was first reported in 2003 by Bondarev and co-workers[19] as a diastereomerically pure (*anti* isomer) solid. A more detailed study by Leitner, Franciò and co-workers[18] showed that this pholidine is obtained as an epimeric mixture (56% de) in which the *syn* diastereomer predominates. Interestingly, they found by ^{31}P NMR that both epimers are interconverting at a rate that is enhanced by the presence of triethylamine. The reaction of this mixture with different secondary amines furnished *P*-stereogenic triamines in good yields and stereoselectivities towards the *anti* diastereomers (entries 3–8).[18] A few other diamines **1** with R ≠ Ph have also been employed (entries 20, 21) along with different chiral amides (entries 22–24).[29]

From Table 3.1 it can be seen that both monodentate (entries 1–24) and *P,N*-bidentate (entries 25–32) ligands can be prepared. Entry 33 depicts a polymer-bound diazaphospholidine, prepared by reaction of **4** with poly(*p*-hydroxystyrene). The stereoselectivity is often very high towards the thermodynamic *anti* epimer but sometimes significant amounts of the other epimer at the phosphorus atom are obtained (entries 9, 12, 13, 17, 21 and 32). The observed selectivity suggests that a dynamic kinetic resolution takes place during the reaction of **4** with alcohols or amines.[18]

The formation of copious amounts of salts is undesirable, especially when preparing ligands in large quantities. This can be avoided by starting from *bis*(dimethylamino)phosphines or *tris*(dimethylamino)phosphine (HMPT) (Scheme 3.1, route **B**), since no base is required and the only by-product is the volatile dimethylamine. The reaction is typically performed in toluene at reflux (110 °C) for several hours or days, and can be conveniently monitored by titration of the

Table 3.1 *P*-stereogenic diazaphospholidines prepared by route **A** (Scheme 3.1).

Entry	Compound	Yield (%)	de (%)	References
1	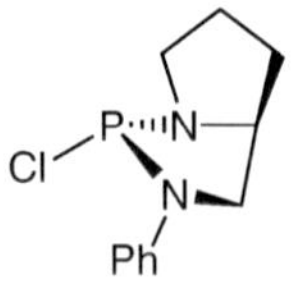 **4** (Cl–P, Ph)	71[19]	o.p.[19] 56[18]	18, 19
2	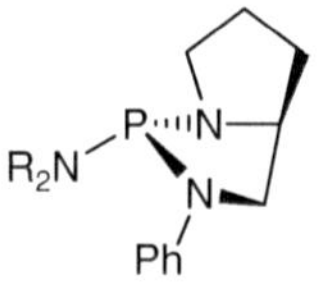 R_2N–P, Ph; $NR_2 = NEt_2, N(CH_2)_5$	88–90	94–o.p.	20
3	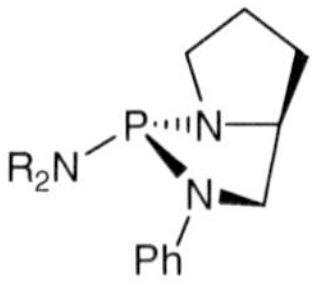 R_2N–P, Ph; R = *i*-Pr, *n*-Bu, Bn, Ph	55–92	96–o.p.	18
4	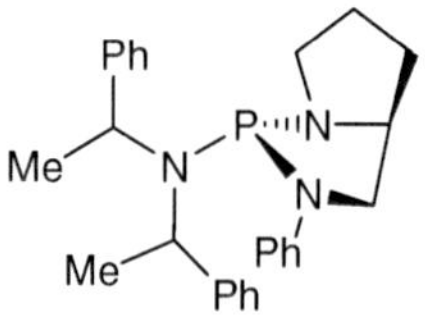 (R,R)- and (S,S) amine	95 (*R,R*)	90 (*R,R*)	18
		83 (*S,S*)	88 (*S,S*)	18
5	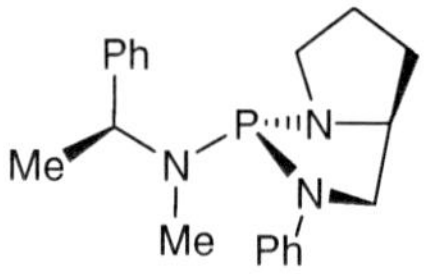	70	98	18
6	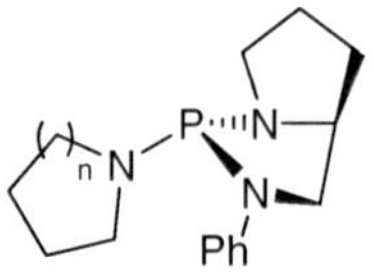 n = 1, 2	59 ($n=1$) 58 ($n=2$)	98 ($n=1$) 96 ($n=2$)	18
7	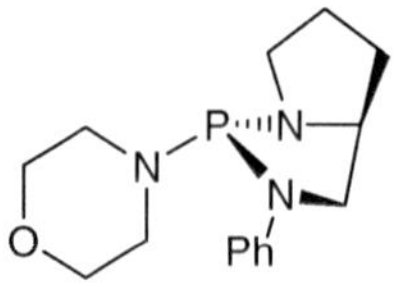	61	98	18

Table 3.1 (*Continued*).

Entry	Compound	Yield (%)	de (%)	References
8	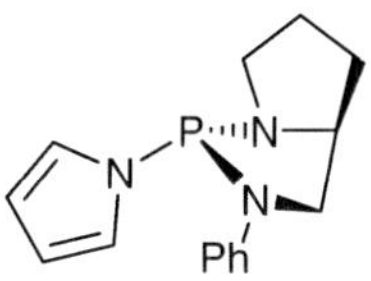	90	88	18
9	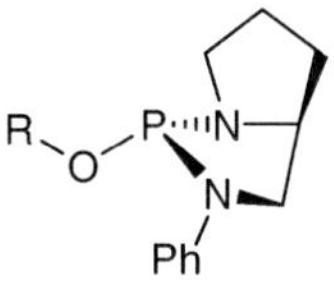 R = Me, *i*-Pr, CH(CF$_3$)$_2$ *t*-Bu, Ad, (−)-Men, Ph	82–95	48–o.p.	20
10	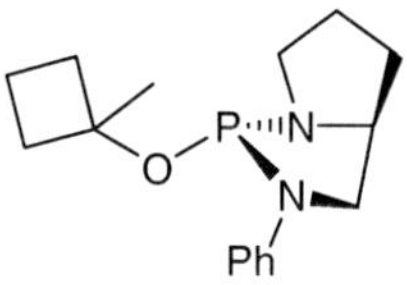	93	82–o.p.	20
11[a]	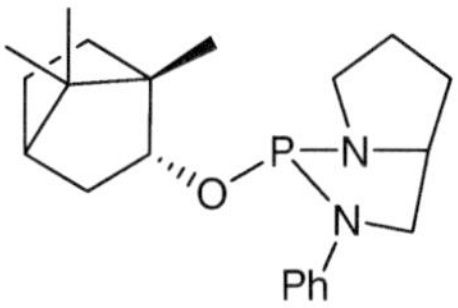	86–89	o.p.	21
12[a]	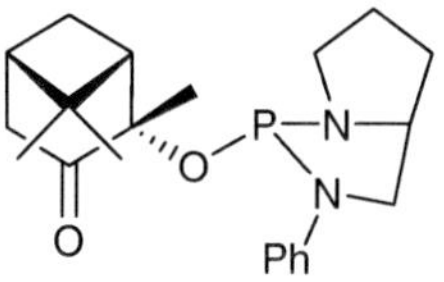	72–75	34–92	21
13	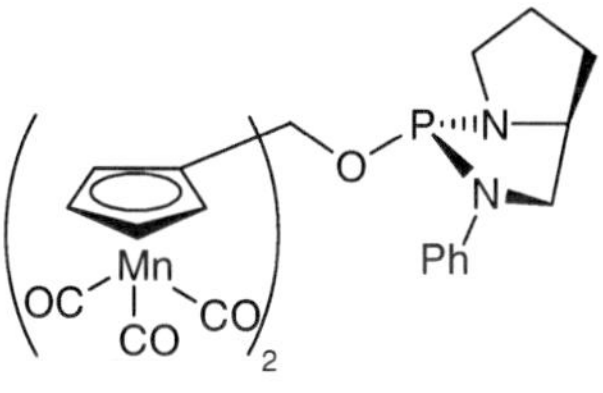	70	54	22
14	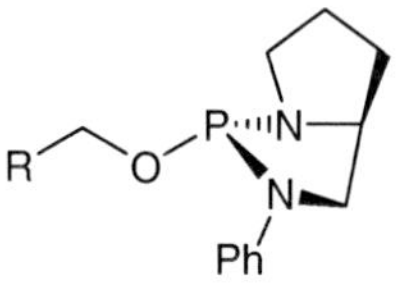 R = Fc, *n*-C$_9$F$_{19}$	66 (Fc) 81 (*n*-C$_9$H$_{19}$)	o.p.	23
15[b]	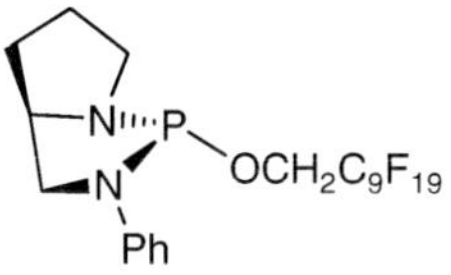	81	o.p.	24

Table 3.1 (*Continued*).

Entry	Compound	Yield (%)	de (%)	References
16[c]	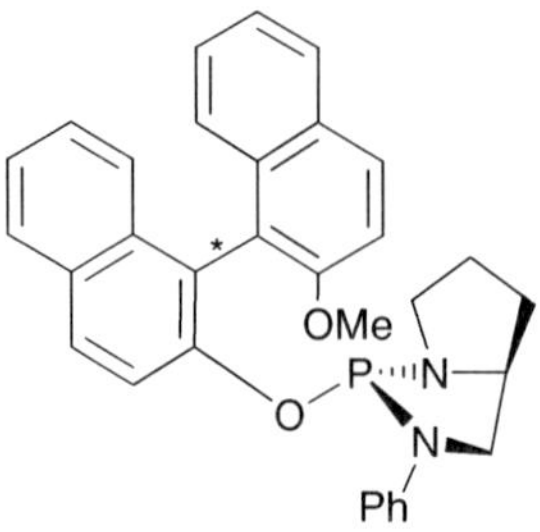	81–84	88–94 (crude) o.p. (recr.)	25
17[c]	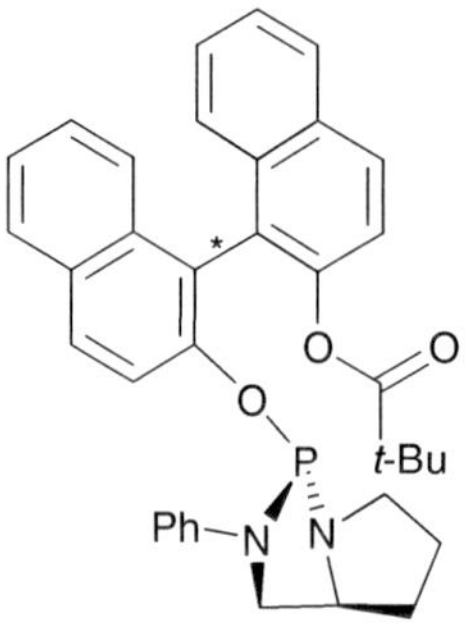	70–75	34–46	23
18	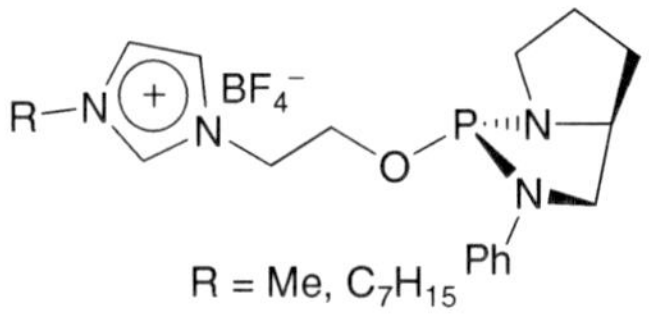 R = Me, C$_7$H$_{15}$	54 (Me) 67 (C$_7$H$_{15}$)	o.p.	26
19	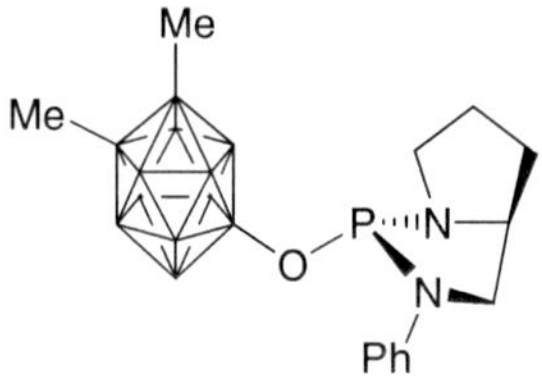	70	o.p.	27
20	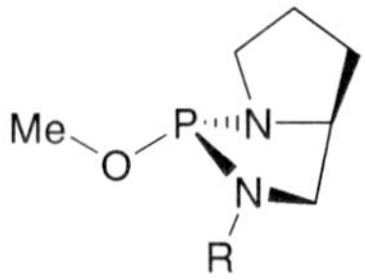 R = *i*-Pr, Cy, (*S*) or (*R*) CH(Ph)Me	65–81	78–90	28
21	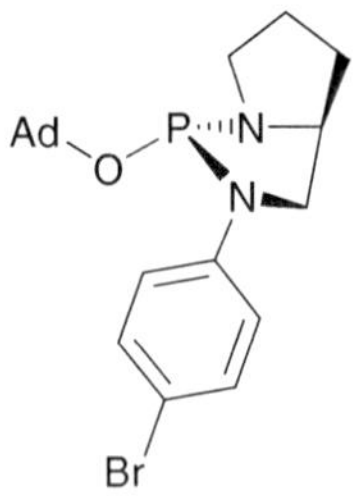	76	68	23

Table 3.1 (*Continued*).

Entry	Compound	Yield (%)	de (%)	References
22	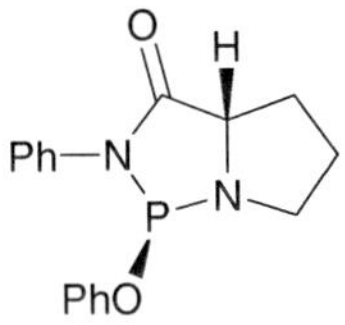	65	o.p.	29
23	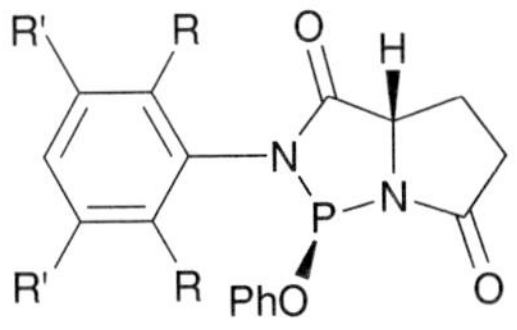 R = H, Me, Et; R' = H R = H, R' = *t*-Bu	63–94	90–o.p.	29
24	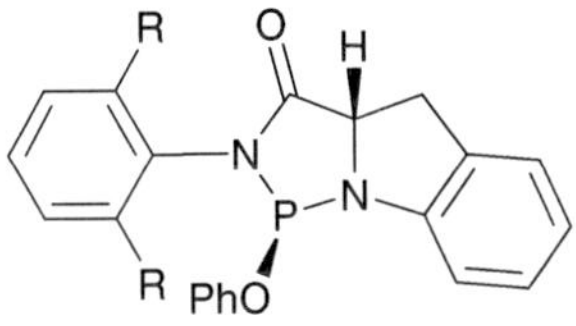 R = H, Me	94 (H) 100 (Me)	96 (H) o.p. (Me)	29
25	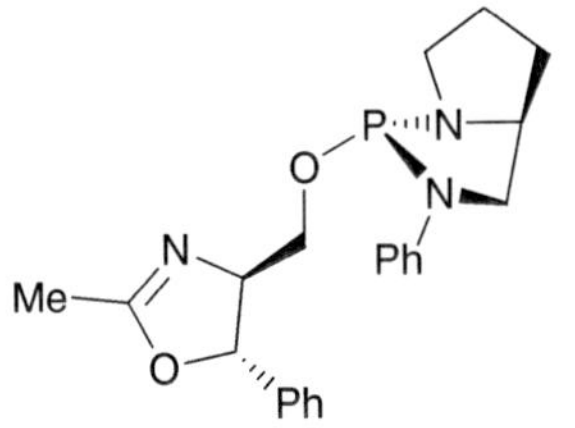	85	o.p.	19
26	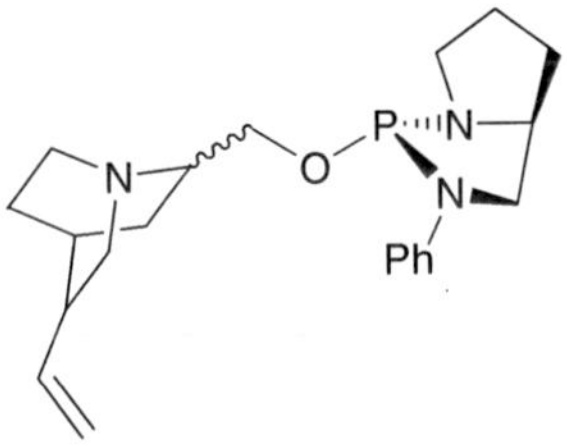	90–91	o.p.	17
27	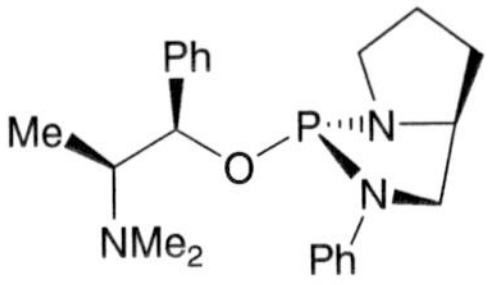	92	o.p.	17
28	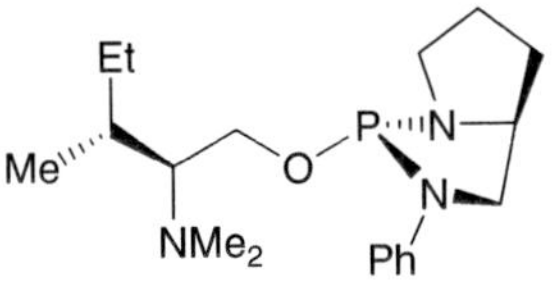	89	96	17

Table 3.1 (*Continued*).

Entry	Compound	Yield (%)	de (%)	References
29	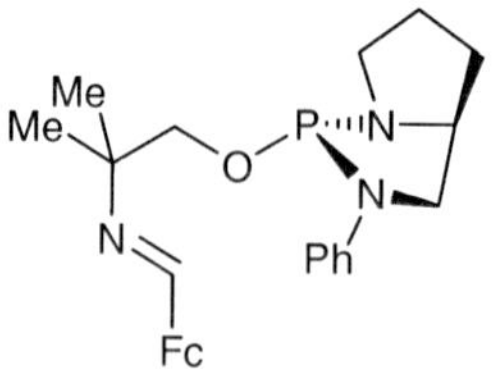	92	o.p.	30
30	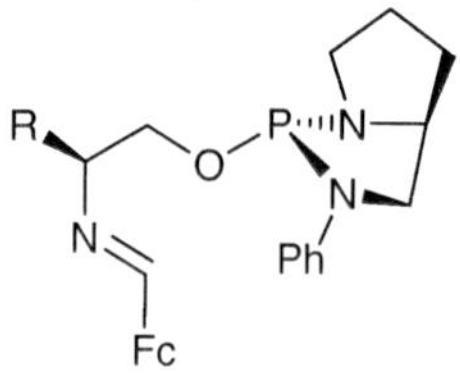 R = (*S*)-*s*-Bu, *t*-Bu	90–91	o.p.	30
31	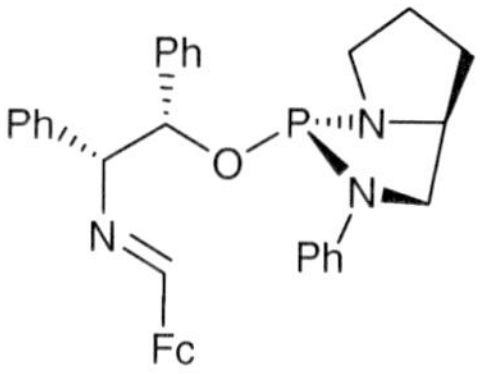	85	o.p.	30
32	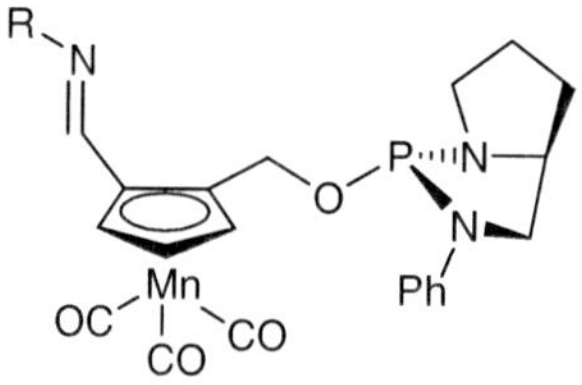 R = *i*-Pr, trimethylbicyclo[3.1.1]heptane	95–98	67–72	31
33	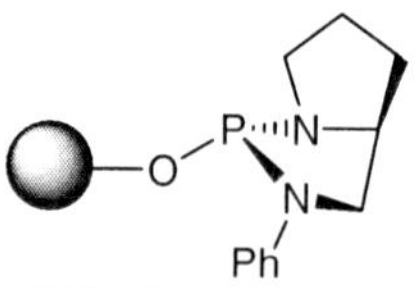	—	o.p.	32

[a] Both enantiomers of 2-(anilinomethyl)pyrrolidine were used.
[b] *R* enantiomer of 2-(anilinomethyl)pyrrolidine was used.
[c] Both enantiomers of the binol moiety were used.

dimethylamine produced. Although route **B** is limited to thermally stable compounds, it represents the usual way to prepare diazaphospholidines.

Given the temperature, **2** is usually obtained as the pure *anti* diastereoisomer following route **B**. One important example of a ligand obtained by this route is SEMI-ESPHOS,[33] depicted in Scheme 3.2.

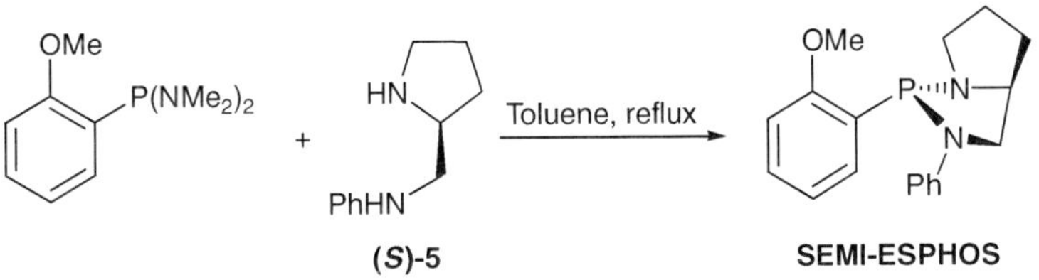

Scheme 3.2   Enantioselective synthesis of SEMI-ESPHOS.

Breeden and Wills[34] reported that SEMI-ESPHOS ligand is obtained as a single isomer in a 69% yield.

X-ray diffraction analysis of the borane adduct of **2** (R = Ar = Ph, Scheme 3.1)[35] revealed the P-bonded extracyclic phenyl group to be in the *trans* position to the pyrrolidine ring, forcing the phosphorus atom to adopt *R* as the absolute configuration in order to minimise steric effects. Most of the diazaphospholidines prepared by route **B** are listed in Table 3.2.

Methodology **B** allows the preparation of monodentate ligands (entries 1–13), *P,N*-bidentate ligands (entries 14–24) and, interestingly, precursors of phosphine–*N*-heterocyclic carbene bidentate ligands (entries 25 and 26). The long reaction times and high temperatures employed in this route allow thermodynamic equilibration of the two epimers at the phosphorus atom and hence only the most stable one is usually obtained, although entries 19–21 form an exception.

Both methodologies have been successfully extended for the preparation of diphosphorus ligands. An important example is ESPHOS (Scheme 3.3), prepared with total stereoselectivity in a 76% yield by Wills and Breeden.[34]

There are only a few examples of similar bi- and polyphosphorus ligands in the literature, listed in Table 3.3.

Notably, the compounds can be prepared in good yields and are reported to be optically pure, including the dendrimeric POSS of entry 7.

3.2.2   P(V) Tertiary Diazaphospholidines

Diazaphospholidines have been successfully applied as ligands in several transition metal-catalysed reactions. In addition, some have been subjected to further transformations (Scheme 3.4) to yield diazaphospholidine boranes **6** and some P(V) compounds: oxides **7**, sulfides **8** and iminophosphines **9**. Most of the reactions have been found to proceed with complete retention of configuration at the phosphorus atom.

Table 3.4 lists of the reported diazaphospholidine derivatives prepared from the P(III) precursors.

To avoid one step, the 'phosphine oxide' version of route **A** (Scheme 3.1) has also been explored for the direct preparation of diazaphospholidine oxides (**7**) from P(V) precursors (Scheme 3.5 and Table 3.5).

Table 3.5 shows these condensations proceed with very low stereoselectivity and hence the prior formation of the diazaphospholidine and subsequent

Table 3.2 *P*-stereogenic diazaphospholidines prepared by route **B** (Scheme 3.1).

Entry	Product	Yield (%)	de (%)	References
1	Me$_2$N, P—N, N—Ph (structure)	75–78	o.p.	36–38
2	Ph, P—N, N—Ph (structure)	–	o.p.	36
3	Me, P—N, N—Ph (structure)	27	o.p.	39
4	OMe, P—N, N—Ph (structure) **SEMI-ESPHOS**	69	o.p.	33, 34
5	OBn, P—N, N—Ph (structure)	–	o.p.	34
6	OSiMe$_3$, R, P—N, N—Ph, R' (structure) R = H, Me, t-Bu; R' = H, Me, t-Bu, Ph, Cl	58–86	o.p.	40
7	OMe, P—N, N—Ph, OTBS (structure)	60	o.p.	41

Table 3.2 (*Continued*).

Entry	Product	Yield (%)	de (%)	References
8	R = Ph, Br	57–63	o.p.	41
9		46	o.p.	41
10	R = Me, 1-naphthyl	60–92 (Me) 42 (1-Naphth)	o.p.	38, 42
11	R = H, Ph, 3,5-Me$_2$C$_6$H$_3$ 2-Biph, 1-NaphthC$_6$H$_4$	92 (H) 87 (Ph) 90 (3,5-Me$_2$C$_6$H$_3$) 90 (2-Biph) 94 (1-NaphthC$_6$H$_4$)	o.p.	42, 43
12	R = Me, *i*-Pr	53 (Me) 31 (*i*-Pr)	o.p.	29

Table 3.2 (*Continued*).

Entry	Product	Yield (%)	de (%)	References
13	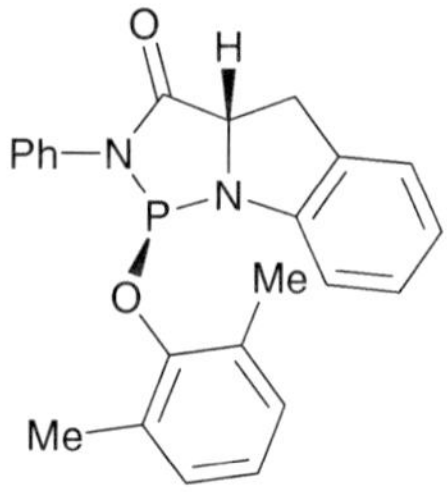	43	o.p.	29
14	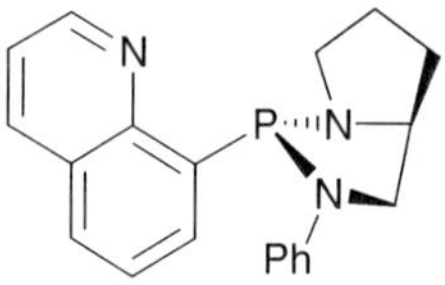 **QUIPHOS-PN5**	61	o.p.	44
15	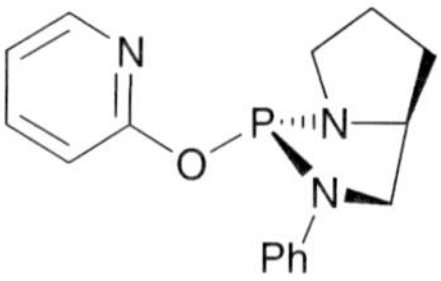	–	o.p.	45
16	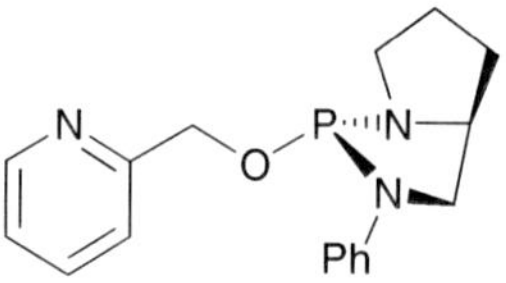	–	o.p.	45
17	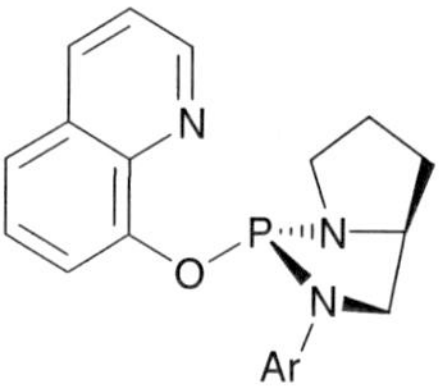 Ar = Ph, 1-naphthyl **QUIPHOS** (Ar = Ph)	72–98 (Ar = Ph)	o.p.	15, 45–47
18	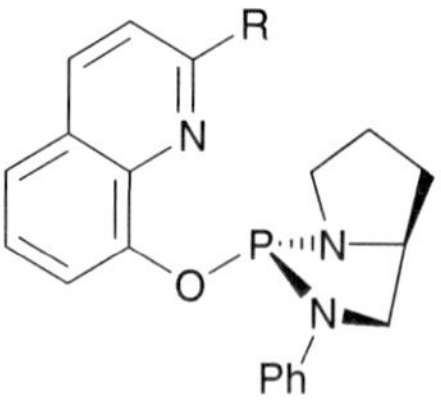 R = Me, *t*-Bu, Ph, CN	49 (Me) 35 (*t*-Bu) 45 (Ph) 72 (CN)	o.p.	48

Table 3.2 (*Continued*).

Entry	Product	Yield (%)	de (%)	References
19	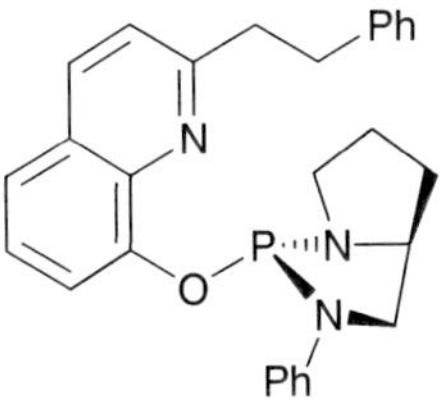	51	94	48
20	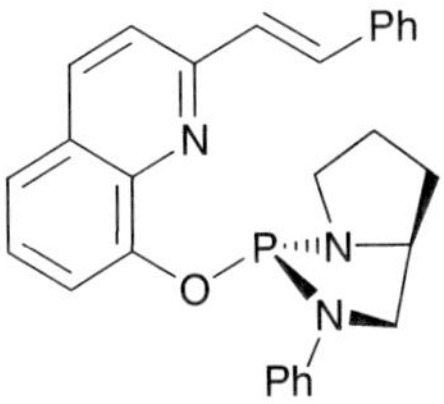	22	84	48
21	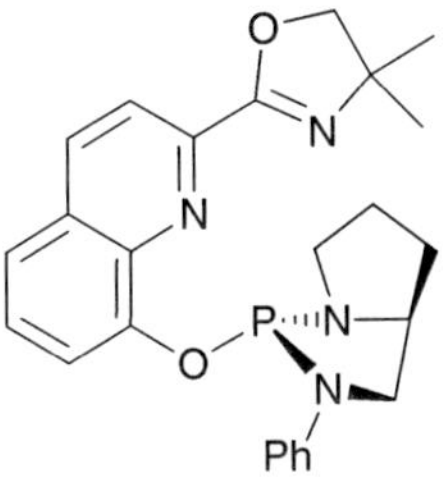	46	54	48
22	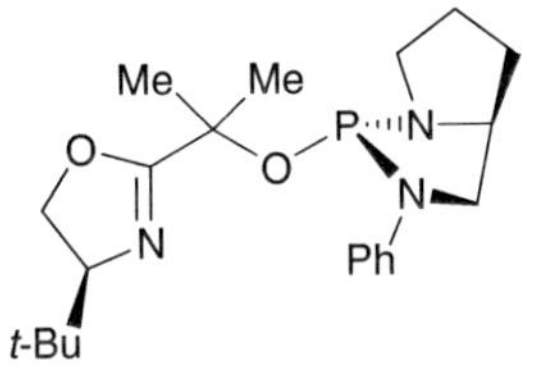	39	o.p.	49
23	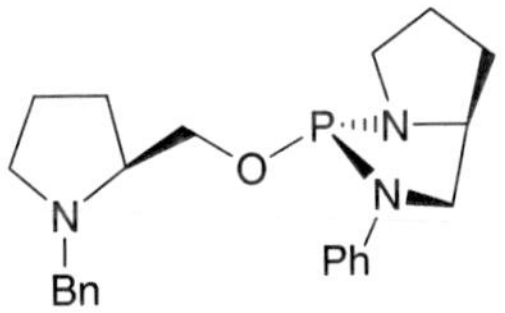	86	o.p.	50
24	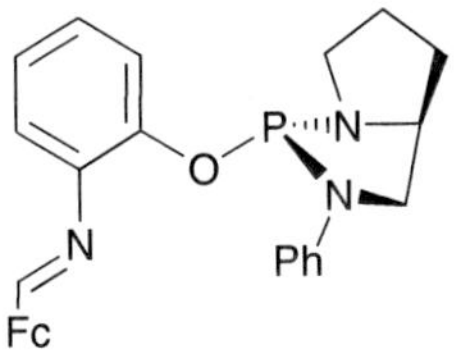	72	o.p.	50
25	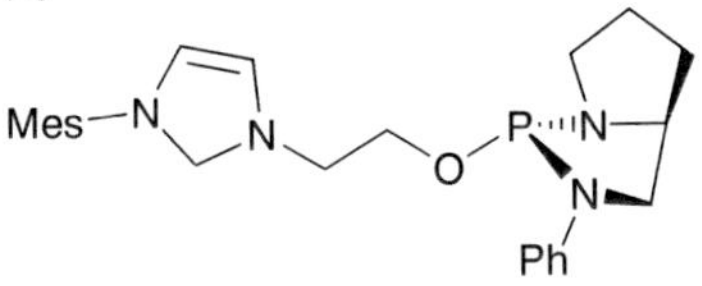	73	o.p.	42

Table 3.2   (*Continued*).

Entry	Product	Yield (%)	de (%)	References
26	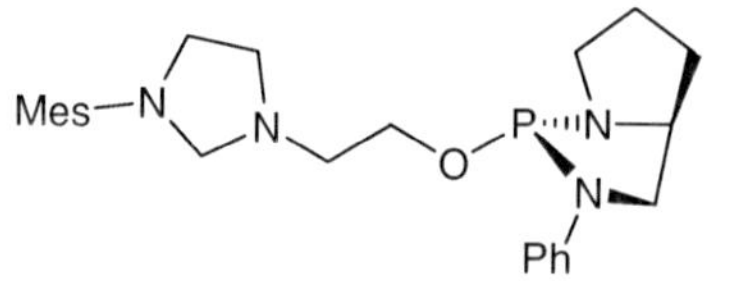	90	o.p.	42

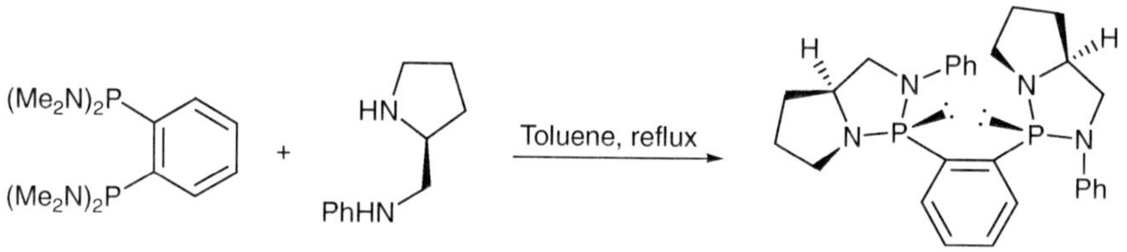

Mes—N
N
O
P
N
N
Ph

(Me$_2$N)$_2$P
(Me$_2$N)$_2$P
+
HN
PhHN
(*S*)-5
Toluene, reflux
H
N—Ph
N—P
H
N
P—N
Ph
ESPHOS

Scheme 3.3   Preparation of ESPHOS.

oxidation seems to be a better route (compare entries 1 and 10 of Table 3.4 with entries 2 and 9 of Table 3.5). In contrast, the relatively stereoselective synthesis of the chloro derivative **11** at 0 °C (entry 1 in Table 3.5) constitutes an exception. This fact was exploited in an early paper by Fiaud and co-workers[60] and more recently by Basavaiah and co-workers,[64,65] who have reacted (*S*$_P$)-**11** with several alcohols and amines to obtain a range of diazaphospholidine oxides, retaining the optical purity at the phosphorus atom (Table 3.6).

Buono and co-workers[43] showed that (aryloxy)diazaphospholidine oxides suffer [1,3] P–O to P–C rearrangement, *i.e.* migration of the phosphine moiety from the oxygen to the *ortho* carbon upon treatment with strong bases such as LDA. For instance, treatment of compound **12** with LDA followed by an aqueous work-up allowed the isolation of the *o*-hydroxyphenyl diazaphospholidine oxide **13** in 94% yield (Scheme 3.6).[43]

The diazaphospholidine oxide group activates the *ortho* position towards metallation (Scheme 3.7) creating the stabilised carbanion **14**, which intramolecularly attacks the phosphorus atom producing the product **13**. It has been suggested that the driving force of the reaction is the extra stabilisation of the lithium cation by two oxygen atoms in **13** compared to only one in **12**.[59]

A remarkable feature of this transformation is its total stereoselectivity, with clean retention of configuration at the phosphorus atom, as confirmed by ^{1}H NMR and X-ray diffraction studies.[43,58] This fact can be explained by considering a mechanism *via* pentacoordinated trigonal bipyramidal (TBP) species and a sequential apical addition–pseudorotation–apical elimination (Scheme 3.8).[67]

Apical attack of the anionic *ortho* carbon in **14** yields intermediate **15**, which rearranges to **15**′ *via* the well-known Berry pseudorotation or, alternatively, the

Table 3.3 *P*-stereogenic *bis*- and polydiazaphospholidines.

Entry	Product	Method	Yield (%)	References
1	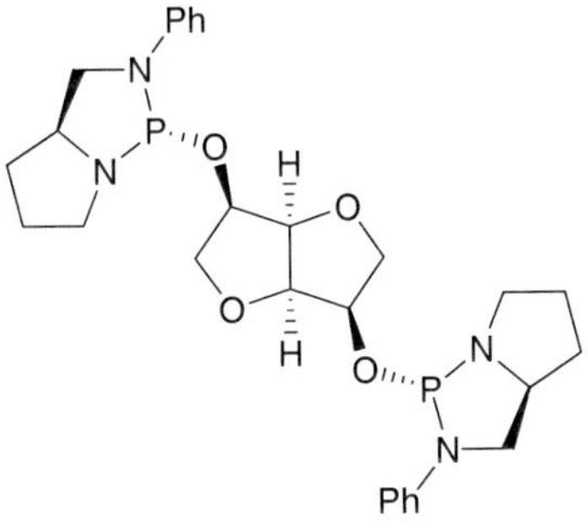	A	73	51
2	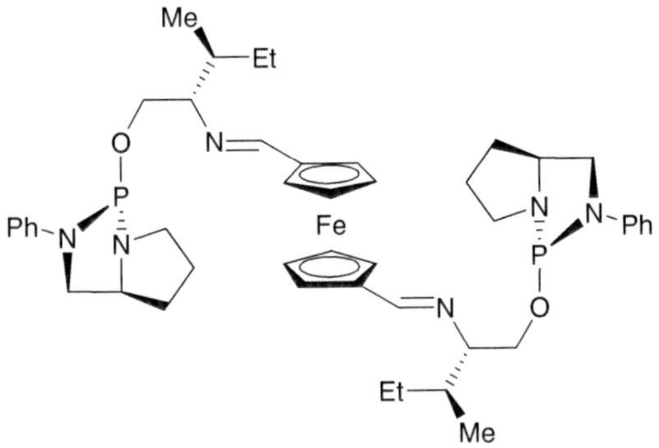	A	95	52
3	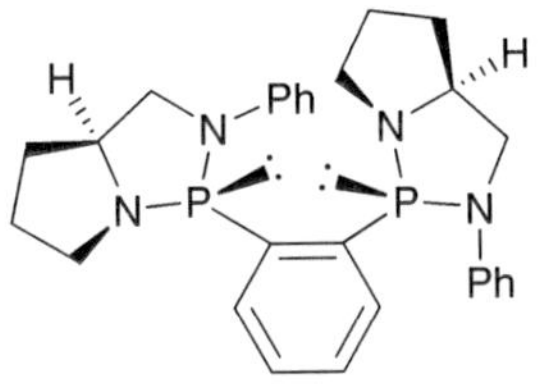 ESPHOS	B	76	34
4	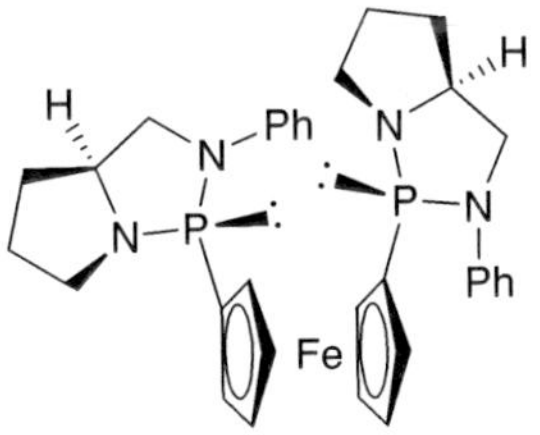 FerriESPHOS	B	77	53

Table 3.3 (*Continued*).

Entry	Product	Method	Yield (%)	References
5	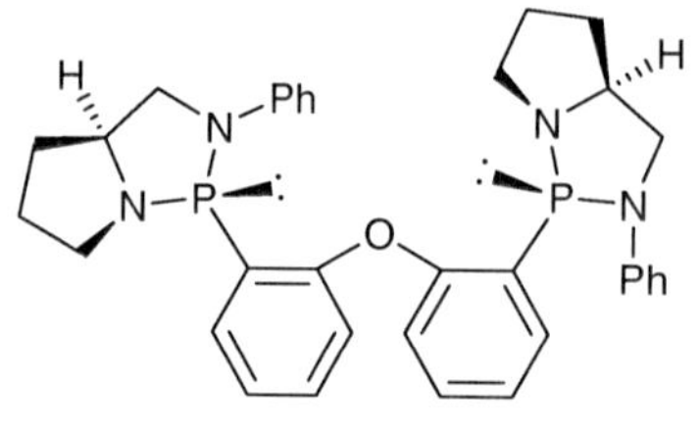 **DiphenESPHOS**	B	quant.	53
6	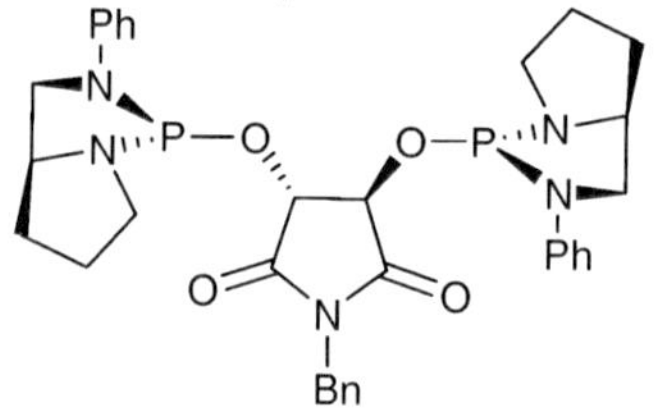	A	70	54
7[a]	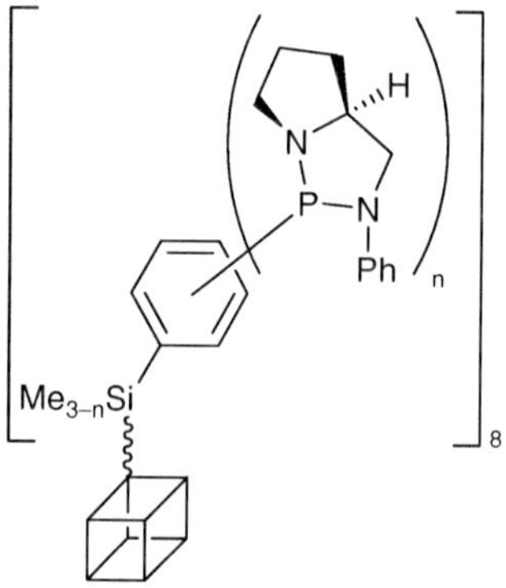	B	52–72	55, 56

[a]The cube represents a Polyhedral Oligomeric SilSesquioxane (POSS); each side is a Si–O–Si linkage.

turnstile rotation. This places the more apicophilic oxygen of the oxaphosphetane ring in an apical position (here the exocyclic oxygen is the pivot) to depart as a leaving group furnishing the product with retention of configuration. In both **15** and **15′** the rings of the oxaphosphetane and diazaphospholidine are in axial-equatorial positions, known to be energetically more stable. In this scenario, inversion of configuration would imply the epimerisation of **15** or **15′** *via* energetically disfavoured intermediates where the oxaphosphetane or the diazaphospholidine would be in diequatorial positions and the oxygen in an apical position.[43,68]

The high yields and the flawless stereoselectivity of this rearrangement prompted the preparation of several *o*-hydroxyaryl diazaphospholidine oxides.

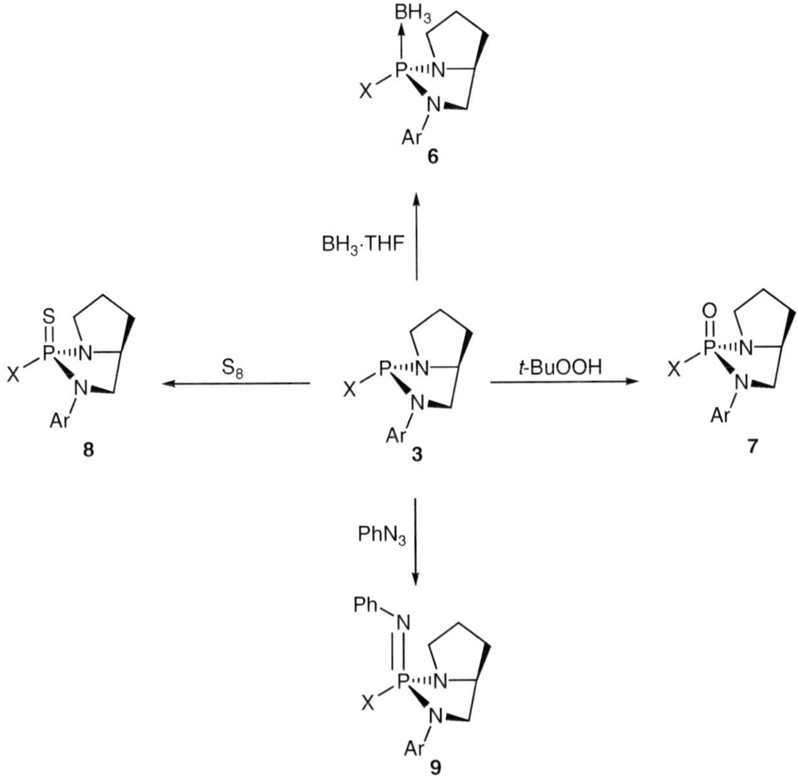

Scheme 3.4   Preparation of diazaphospholidine derivatives.

This class of compounds is interesting because of its bifunctional character, displaying both an acidic OH and a basic phosphoryl group. Table 3.7 lists the *o*-hydroxy diazaphospholidine oxides prepared *via* P–O to P–C rearrangement.

Unsubstituted and *o*/*p*-substituted substrates (entries 1–3) give the expected products in good yields. The reader will have noticed that examples in entries 5–9 were chosen to study the regioselectivity of the rearrangement. It was found to be excellent with the exception of 2-naphthalene substituted substrate (entry 9). In the other cases, electronic (entries 5 and 6) or steric factors (entry 7) govern at which position the carbanion is more likely to be formed.

Further work of Buono and co-workers[62] demonstrated that P–N to P–C carbanionic rearrangement also occurs when excess of LDA is employed (Scheme 3.9).

Compound **16** is formed in 89% yield starting from **12** with clean retention of configuration at the phosphorus atom as determined by X-ray diffraction. This compound is the product of two consecutive (first P–O to P–C followed by P–N to P–C) diastereoselective [1,3] rearrangements. The fact that **16** can also be prepared from **13** suggests that the P–N to P–C migration only occurs once the P–O to P–C step has been completed.

Table 3.4 *P*-stereogenic diazaphospholidine derivatives.

Entry	Product	Yield (%)	de (%)	References
1	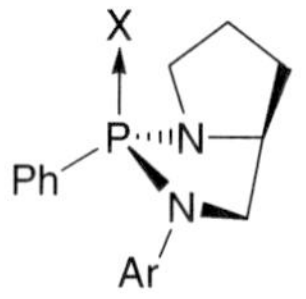 X = lone pair, BH_3 Ar = Ph, *o*-An, *o*-Tol	–	o.p.	35
2	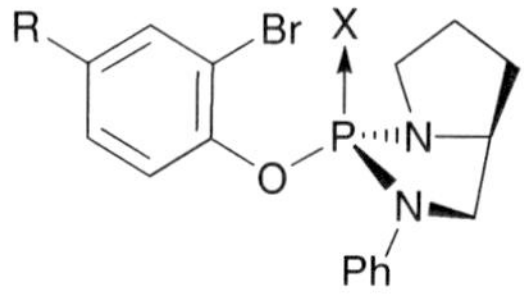 R = H, F, Cl, Me, *t*-Bu, Ph, OMe X = lone pair, BH_3	60–96 (free) 70–92 (BH_3)	o.p.	57
3	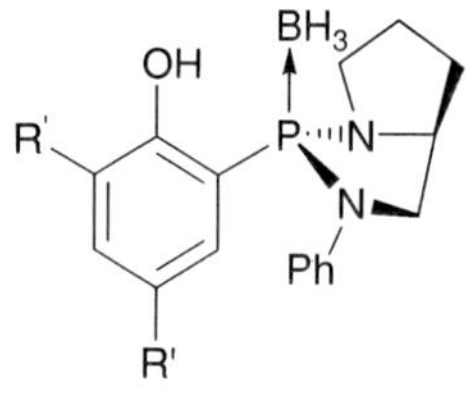 R = H, Me R' = H, Me, Ph, Cl	71–86	o.p.	40
4	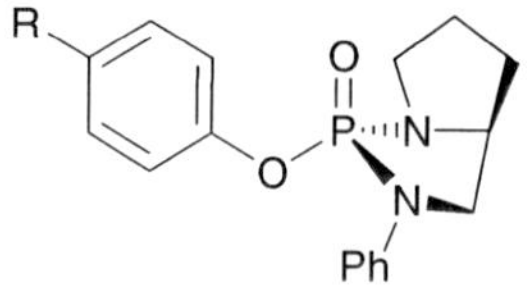 R = OMe, F, Cl	81–84	90 (OMe, F) o.p. (Cl)	58
5	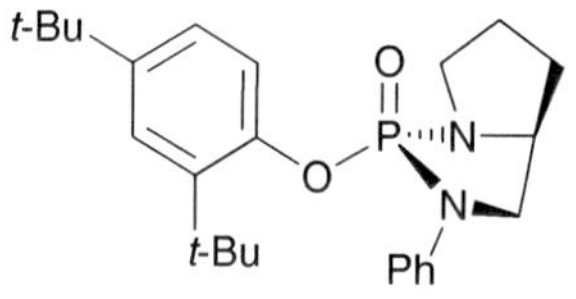	63	o.p.	58
6	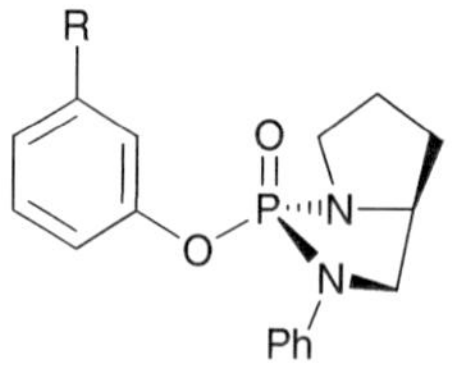 R = OMe, F, Cl, *t*-Bu	63–75	o.p.	58

Table 3.4 (*Continued*).

Entry	Product	Yield (%)	de (%)	References
7	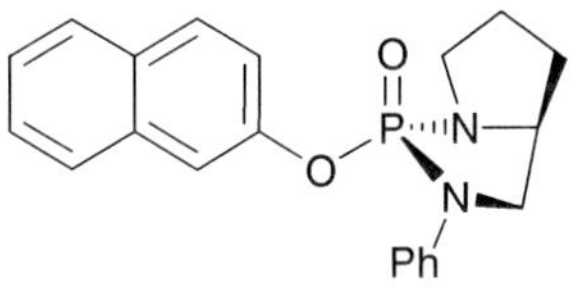	77	88	58
8	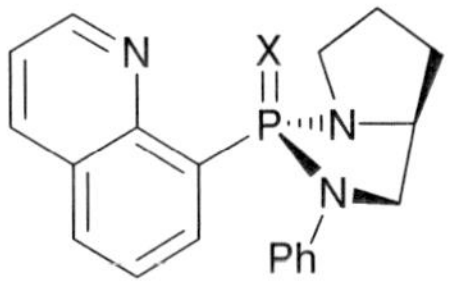 X = O, S, NPh	100 (O) 95 (S) 85 (NPh)	o.p.	44
9	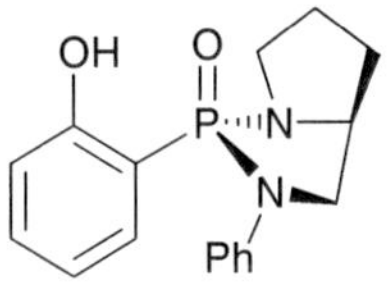	50	o.p.	33
10	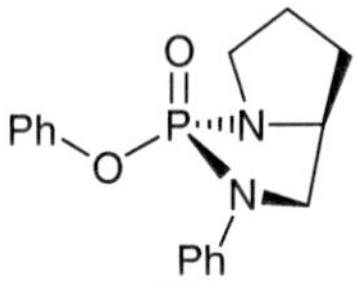 **12**	87	o.p.	59
11	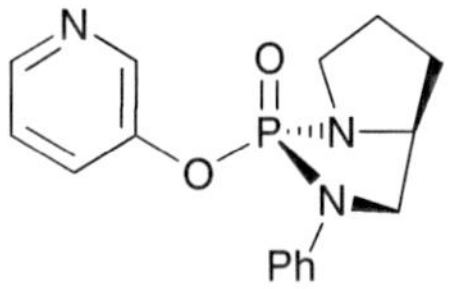	63	o.p.	59
12	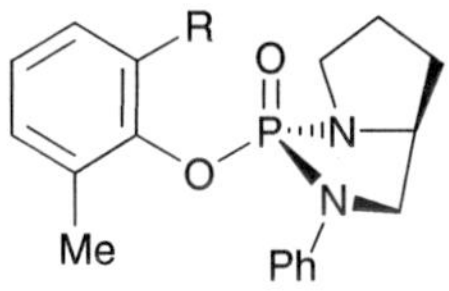 R = H, Me	55	o.p. (H) 80 (Me)	59
13	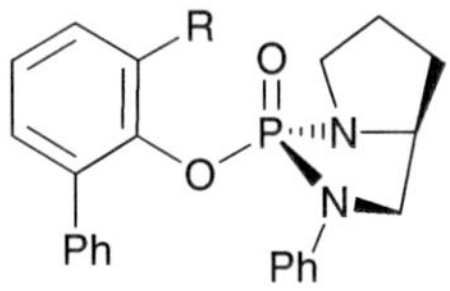 R = H, SiEt$_3$	70 (H) 63 (SiEt$_3$)	90 (H) 80 (SiEt$_3$)	59

Table 3.4 (*Continued*).

Entry	Product	Yield (%)	de (%)	References
14	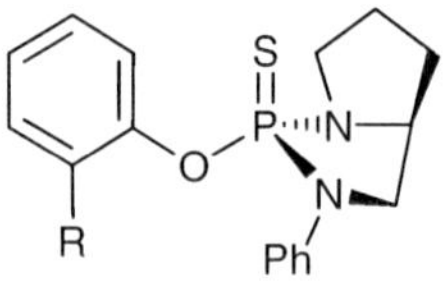	78 (H) 69 (SiEt₃)	o.p.	59
15	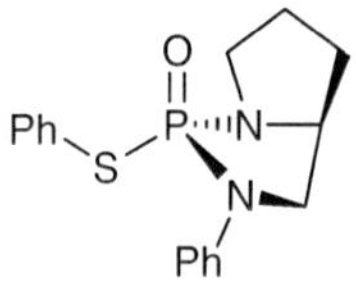	66	54	59
16	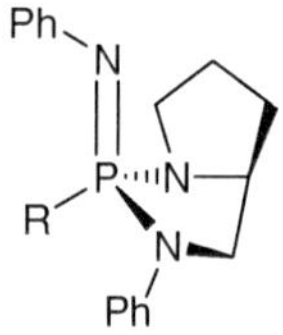	48 (NMe₂) 61 (Ph)	o.p.	36

Scheme 3.5 Synthesis of diazaphospholidine oxides from P(V) precursors.

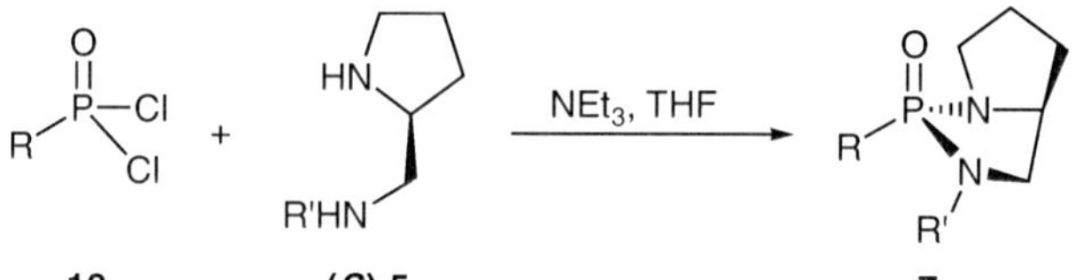

The reaction has been extended to diazaphospholidine boranes (Scheme 3.10).[57]

In this case, the presence of bromine in the *ortho* position is needed to allow the formation of the carbanion, which subsequently attacks the phosphorus centre. In addition, the reaction was found to be completely stereoselective, once more with retention of configuration at the phosphorus atom. It has to be mentioned that the reaction does not work for unprotected diazaphospholidines.

3.2.3 P(V) Secondary Diazaphospholidines

A particular type of secondary *P*-stereogenic diaminophosphine oxide named DIAPHOX (standing for diaminophosphine oxides) was developed by Hamada and co-workers[69,70] starting from aspartic acid. Scheme 3.11 depicts the preparation of the simplest one.

Table 3.5 Diazaphospholidine oxides prepared by direct condensation of **10** with (*S*)-5.

Entry	Product	Yield (%)	de (%)	References
1[a]	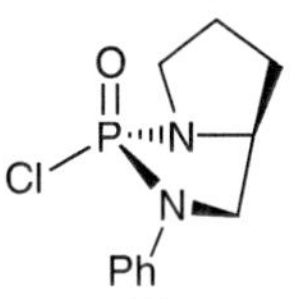 **11**	76[b] 64[c]	86[b] 19[c]	60[b] 61[c]
2[a]	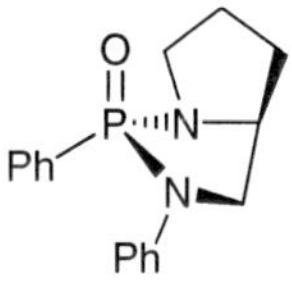	80	0–18	61, 62
3	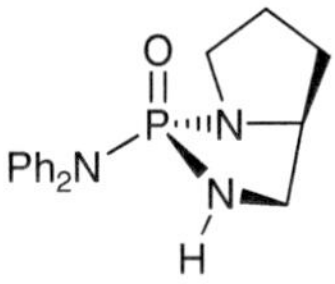	31	40	60
4	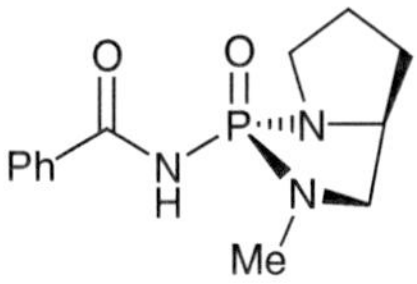	52	38	60
5	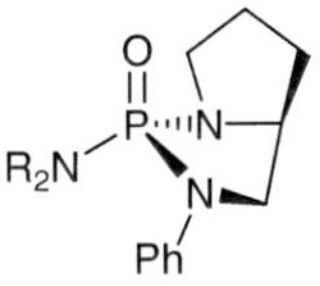  R = Me, X = O, NMe	26–58	0–20	60
6[a]	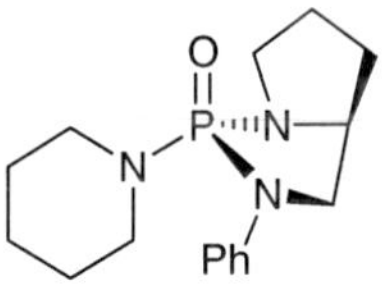	57	16	61
7[d]	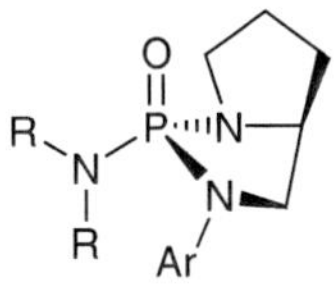 Ar = 1-naphthyl R = Et, Pr, *i*-Bu, –(CH$_2$)$_4$–	15–54	o.p.	63

Table 3.5 (*Continued*).

Entry	Product	Yield (%)	de (%)	References
8[a]	R = Me, Bn, Ph, 1-Naphth, 2-Naphth, 1-Anth *p*-An, *p*-CF$_3$-C$_6$H$_4$, Cy	21–93	0–15	63
9[e]	**12**	91	0	43
10[e]		76	20	43

[a]The epimers were separated.
[b]Conditions: THF, NEt$_3$, 0 °C.
[c]Conditions: toluene, NEt$_3$, 110 °C.
[d]Only one diastereomer was studied.
[e]The epimers were not separated.

Natural L-aspartic acid can be transformed in several steps into triamine **19**, which reacts smoothly with phosphorus trichloride to yield diastereomerically pure **20**. This triaminophosphine is stable in neutral or mildly acidic aqueous conditions but, interestingly, it suffers a S$_N$2-type hydrolysis upon purification by silica gel column chromatography with wet ethyl acetate, yielding the desired secondary diaminophosphine oxide in 60% of overall yield from **19**. The absolute configuration of the phosphorus atom (*R*) could be unequivocally determined by crystallographic analysis.[71]

The synthetic route has been extended for the preparation of several analogues, listed in Table 3.8.

In secondary phosphine oxides the P(V) form (phosphine oxide) predominates but it is always in equilibrium with the P(III) tautomer (phosphinous acid). This equilibrium is represented in Scheme 3.12 for Ph-DIAPHOX. It has to be mentioned that this equilibrium does not affect the stereochemistry of the phosphorus atom.

DIAPHOX ligands are in general air- and moisture-stable since **21** is only formed in minute amounts. ^{31}P NMR in CDCl$_3$ shows, however, that

Table 3.6 Diazaphospholidine oxides prepared from precursor **11**.

Entry	Product	Yield (%)	References
1	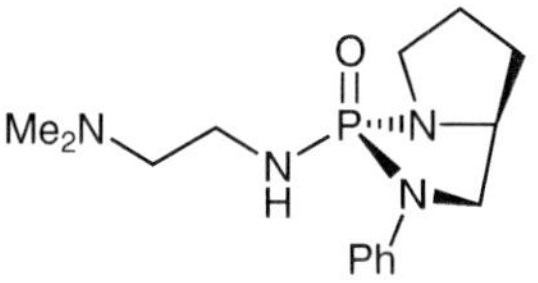	61	60
2	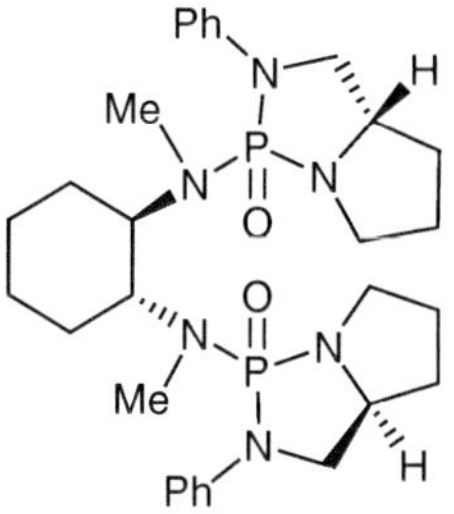	76	64
3	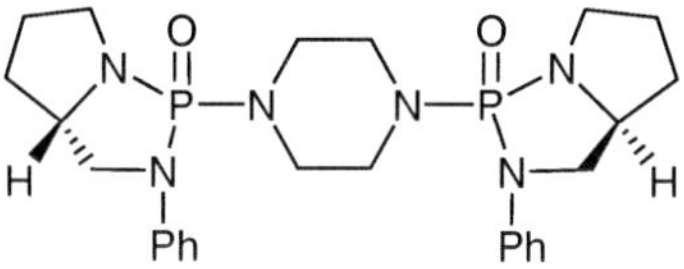	88	64
4	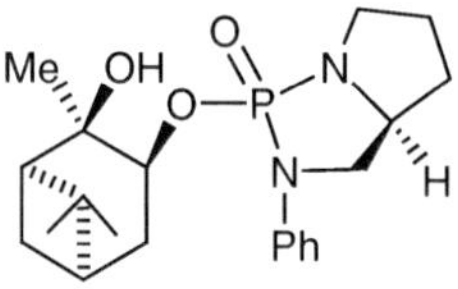	65	65
5	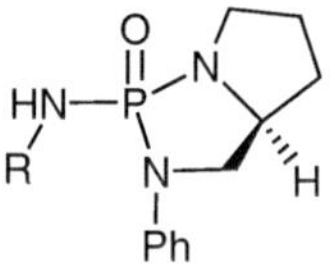 R = *t*-Bu, allyl, Bn	55–60	66
6[a]	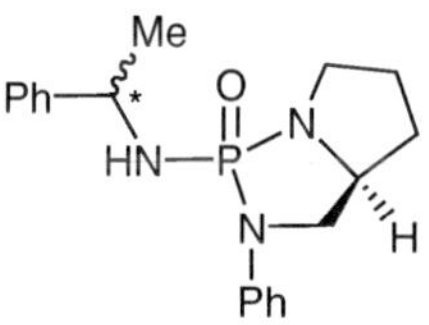	80–81	66

[a] Both products with different absolute configurations at the carbon marked with the asterisk were independently prepared.

N,O-bis(trimethylsilyl)acetamide (BSA, a strong silylating agent) shifts this equilibrium towards the P(III) tautomer **22**.[72] These type of compounds are excellent ligands for Pd and Ir centres and have been used in challenging enantioselective allylic substitution reactions.[69,70]

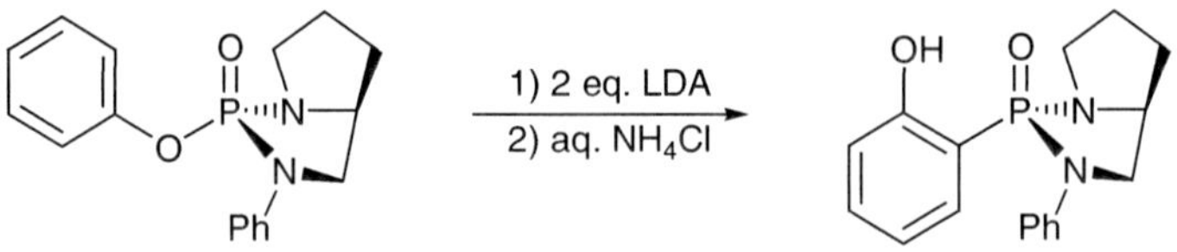

12 13

Scheme 3.6 P–O to P–C [1,3] rearrangement in a diazaphospholidine oxide.

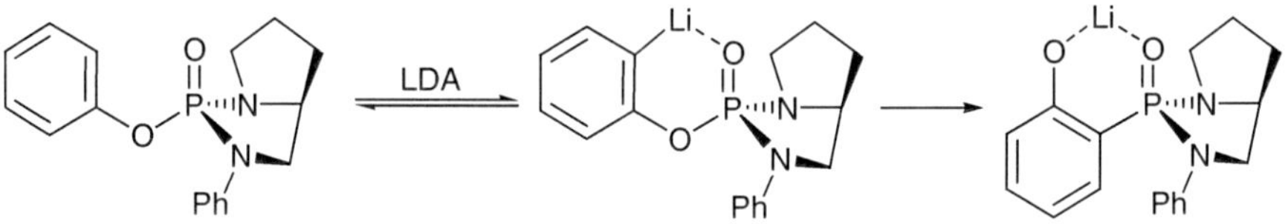

12 14 13

Scheme 3.7 Mechanism of the [1,3] P–O to P–C rearrangement.

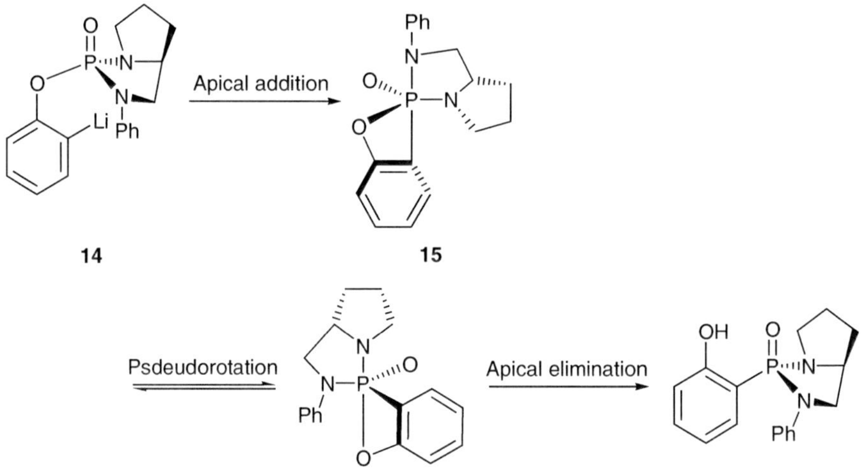

15' 13

Scheme 3.8 Proposed explanation for the retention of configuration in the reaction of Scheme 3.6.

3.3 Oxazaphospholidines and Derivatives

1,3,2-oxazaphospholidines are the N,O counterparts of the 1,3,2-diazaphospholidines discussed in the previous section. Not surprisingly, the preparative

Table 3.7 P–O to P–C rearrangement [1,3] in diazaphospholidine oxides.

Entry	Reactant	Product	Yield (%)	References
1	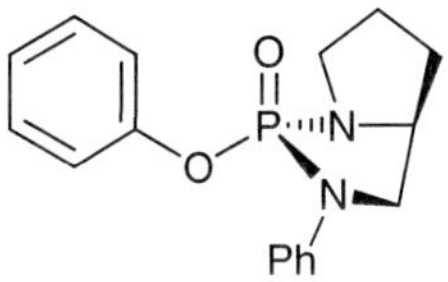	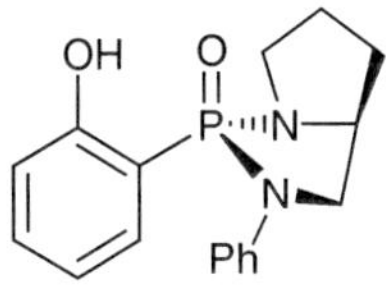	94	43
2	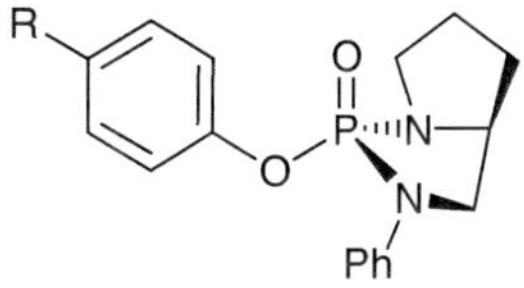 R = OMe, F, Cl	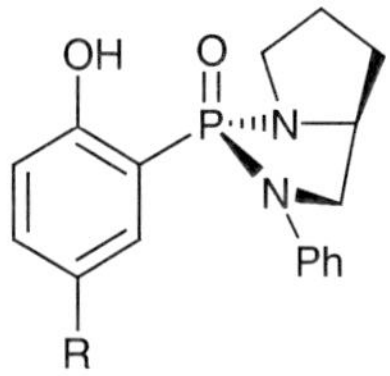 R = OMe, F, Cl	76–86	58
3	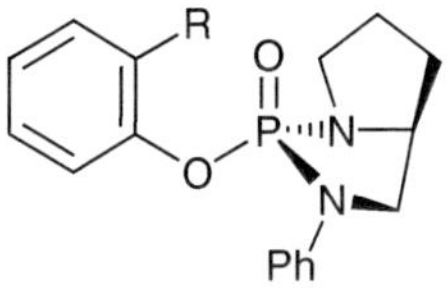 R = Me, Ph	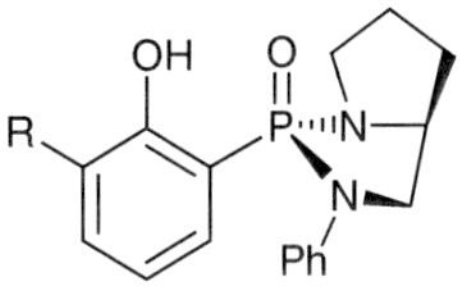 R = Me, Ph	73–82	59
4	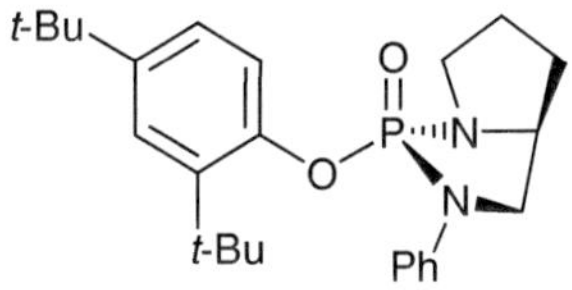	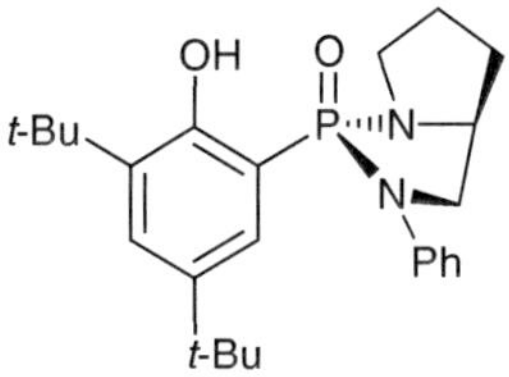	76	58
5	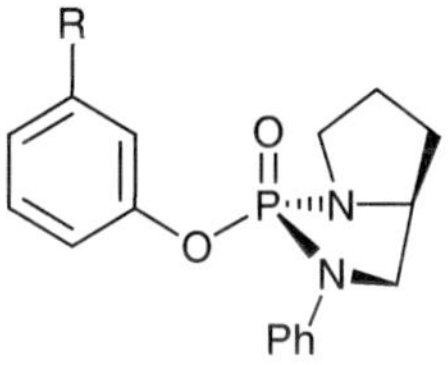 R = OMe, F, Cl	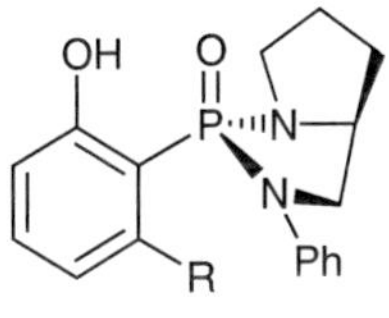 R = OMe, F, Cl	72–92	58
6	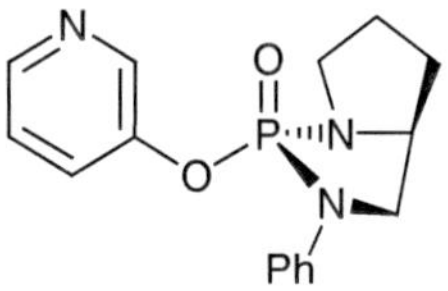	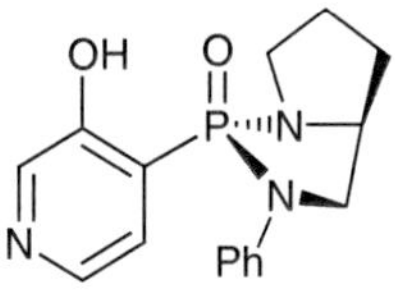	75	59
7	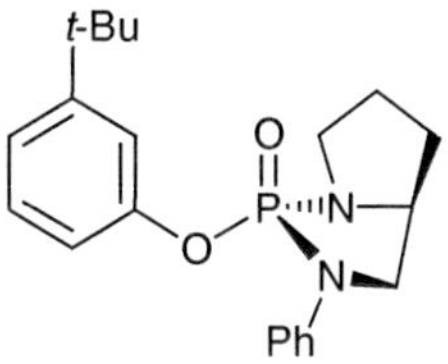	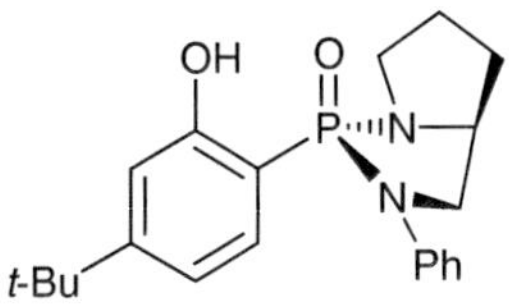	84	58

Table 3.7   (*Continued*).

Entry	Reactant	Product	Yield (%)	References
8	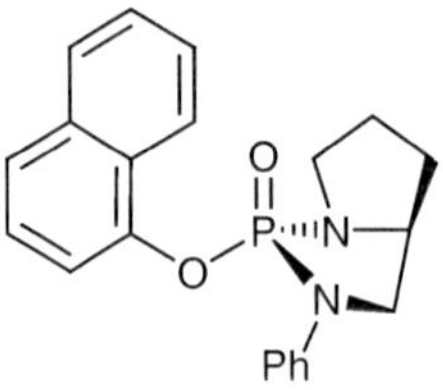	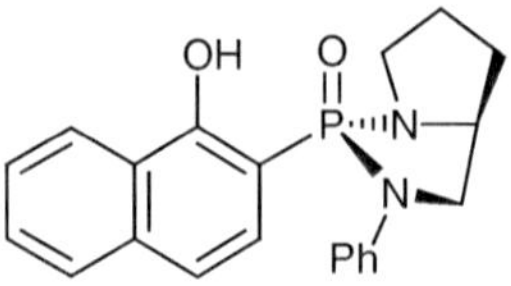	93	43
9[a]	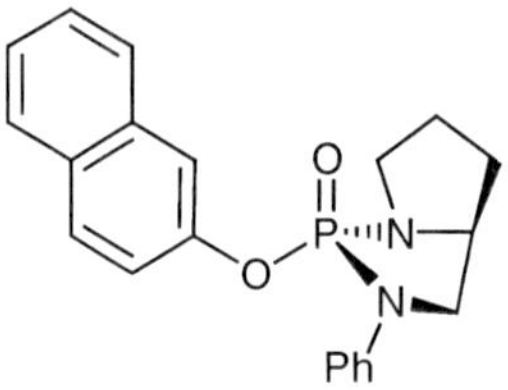	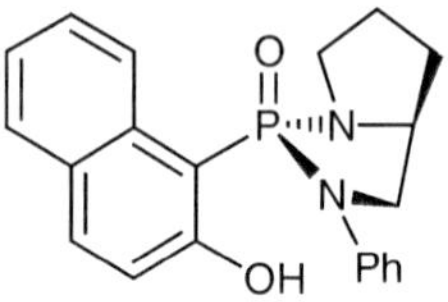 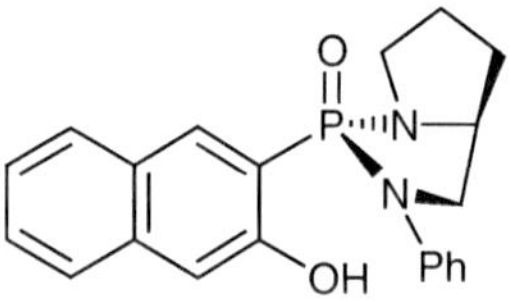	78	58

[a]A 3:1 mixture of regioisomers is obtained.

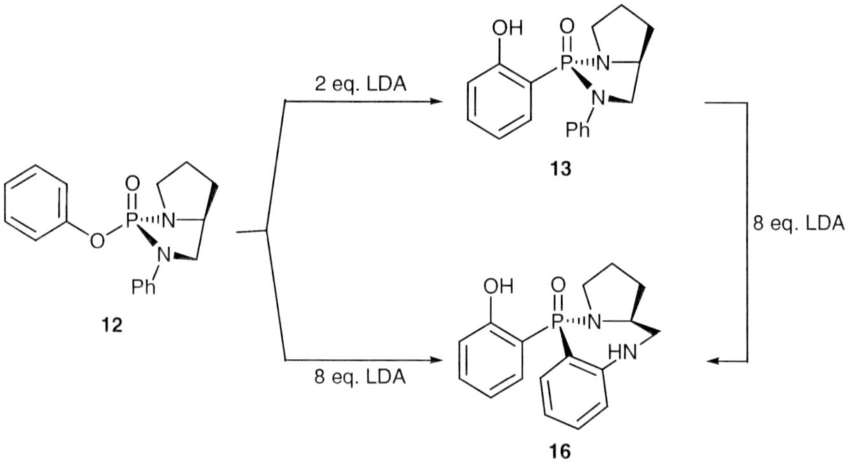

Scheme 3.9   P–O to P–C and P–N to P–C rearrangements in a diazaphospholidine oxide.

methods are basically the same as those shown in Scheme 3.1 but with substitution of the chiral diamine by a chiral β-amino alcohol. Among all the amino alcohols employed, ephedrine and prolinol are the most commonly used.

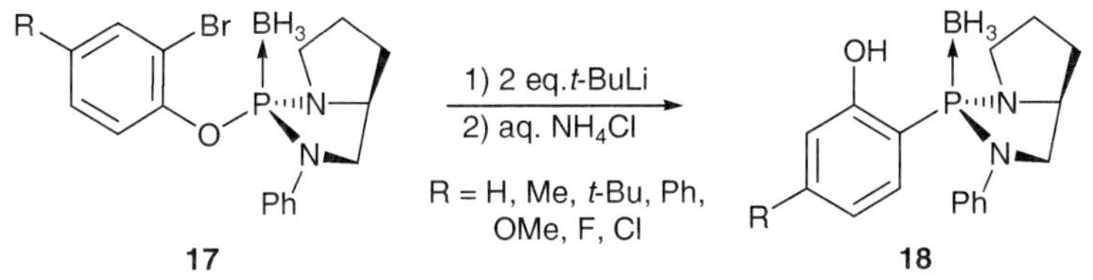

Scheme 3.10   1,3 P–O to P–C rearrangement in diazaphospholidine boranes.

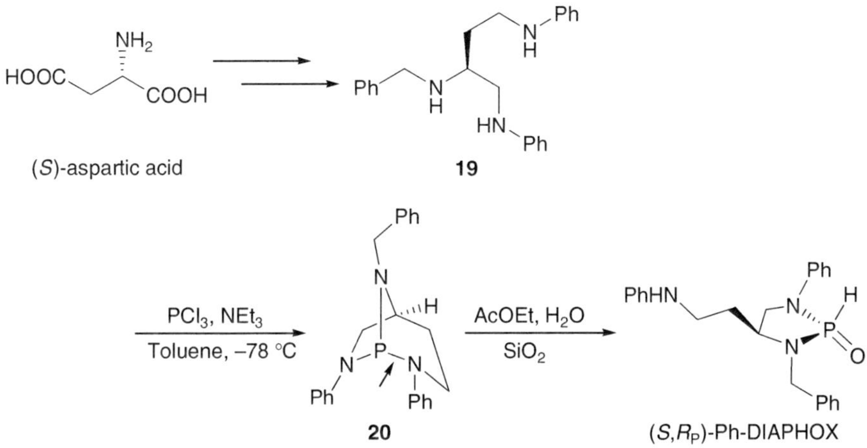

Scheme 3.11   Synthesis of Ph-DIAPHOX. The arrow in **20** indicates the breaking bond.

3.3.1   Derived from Ephedrine

Ephedrine and pseudoephedrine (2-methylamino-1-phenylpropan-1-ol) are natural epimeric amino alcohols extracted from various species of the family of plants *Ephedra*. These plants have been used for millennia in China as stimulants and nasal decongestants. The structures of natural (−)-ephedrine and (+)-pseudoephedrine are depicted in Figure 3.3. The figure also shows norephedrine, which is the nitrogen-unsubstituted counterpart of ephedrine. These compounds are nowadays commercially available as inexpensive crystalline white solids and have been extensively used in phosphorus chemistry, as detailed in the next sections.

3.3.1.1   P(III) Oxazaphospholidines

In 1984 Richter[77] studied the cyclisation reaction of (−)-ephedrine and a few P(III) precursors by displacement of chlorine or dimethylamino groups (Scheme 3.13).

Table 3.8   DIAPHOX ligand precursors.

Entry	Product	Yield (%)[a]	References
1	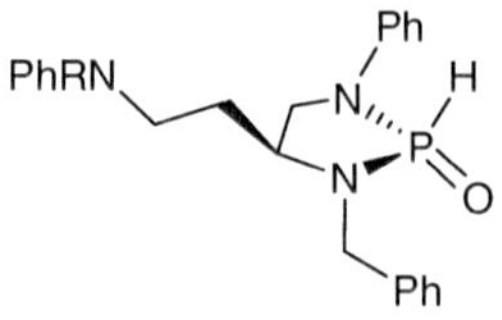 R = H, Me, Ac	60 (H) 86 (Me) 69 (Ac)	71, 72
2	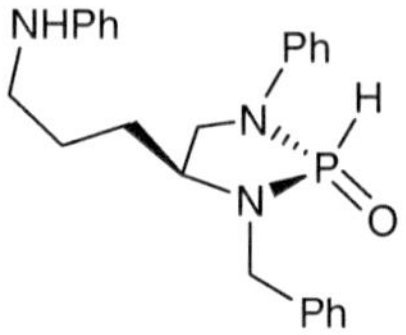	35	72
3[b]	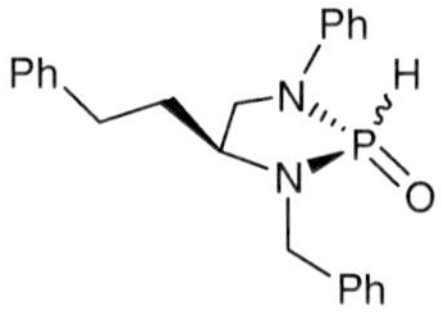	71	72
4	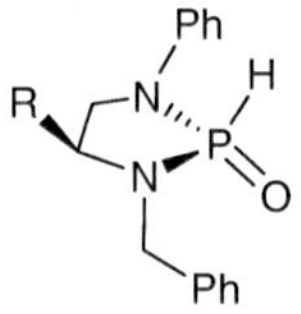 R = *i*-Pr, Bn	20 (Bn)	73
5	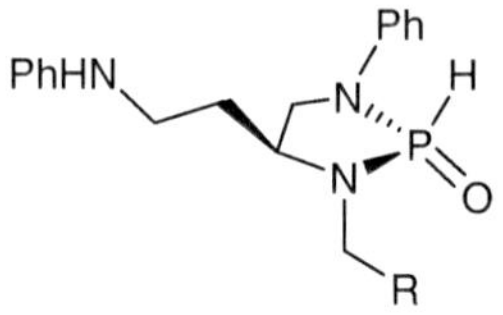 R = 1-Naphth, 2-Naphth, 4-Biph	–	74
6	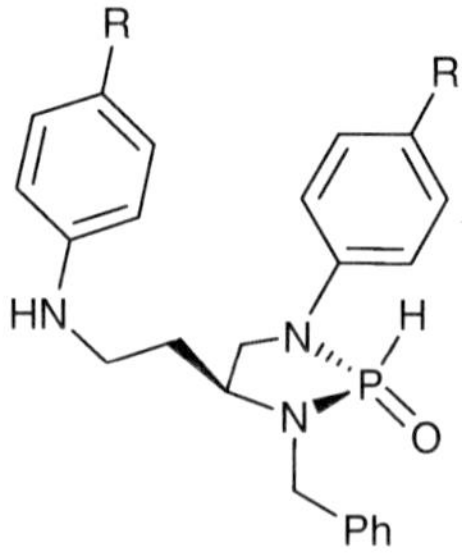 R = *t*-Bu, CF$_3$, OMe	49 (*t*-Bu)	73, 75

Table 3.8 (*Continued*).

Entry	Product	Yield (%)[a]	References
7	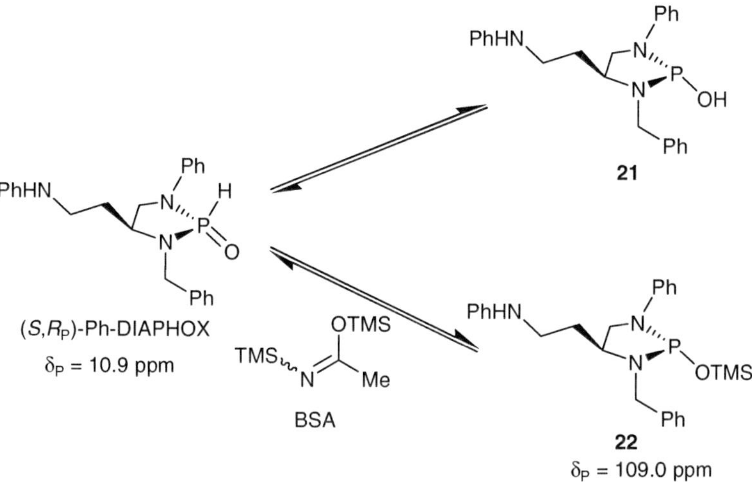	44	76

[a]Combined yield of cyclisation + hydrolysis.
[b]An equimolar mixture of separable epimers was obtained.

Scheme 3.12 Tautomeric equilibrium between P(V) and P(III) forms of Ph-DIAPHOX.

He found that direct cyclisation with dichlorides (method **A**) in the presence of triethylamine was possible with R = *t*-Bu but not for R = Me or Ph. For the latter groups, *bis*(dimethylamino)phosphines proved to be more adequate starting materials (method **B**).

Oxazaphospholidines **23** were obtained in 60–82% yields and with diastereoselectivities in the 85–95% range. No X-ray data was reported, so the absolute configuration at the phosphorus atom of the major diastereomer (*trans*, that corresponds to *R* for alkyl or aryl groups) was determined by careful analysis of the ^{1}H–^{31}P and ^{13}C–^{31}P coupling constants. Other groups managed to prepare **23** with R = Ph using PhPCl$_2$ and *N*-methylmorpholine[78,79] or diisopropylethylamine,[80] suggesting that the base can play a critical role in the success of

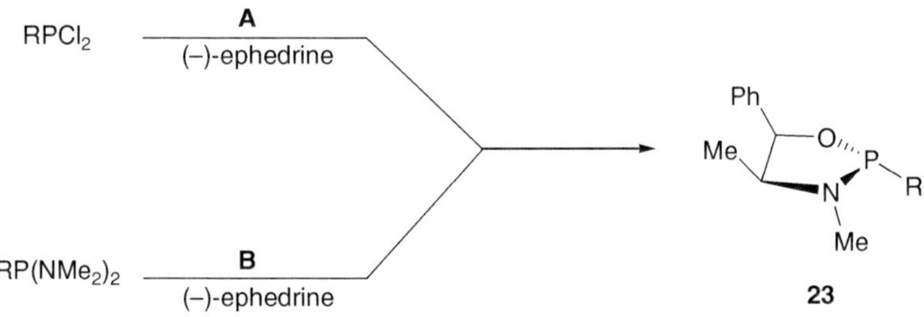

Figure 3.3   Structures of natural ephedrines.

Scheme 3.13   Preparation of ephedrine-derived oxazaphospholidines.

the reaction. Since Richter's report, many others have used similar methods to prepare more *P*-stereogenic oxazaphospholidines and derivatives. Indeed, compounds **23** can be conveniently protected with borane or oxidised to the P(V) species, with complete retention of configuration at the phosphorus atom (Scheme 3.14).[78]

Table 3.9 lists most of the oxazaphospholidines and derivatives prepared from ephedrine reported to date.

Reaction of ephedrine with PCl_3 furnishes **26** (entry 1) with different degrees of stereoselectivity depending on the exact conditions, as determined by ^{31}P NMR. These determinations can be inaccurate, as they depend on many factors such as the resolution of the ^{31}P NMR spectra, the temperature and the presence or absence of impurities that may favour interchange between the epimers.[9] Synthon **26** has been used to prepare other compounds by nucleophilic substitution of the chlorine atom (entries 12, 23, 24, 29–31 and 33). These reactions proceed with predominant retention of configuration,[83] but that depends on the exact structure of the oxazaphospholidine, the temperature and

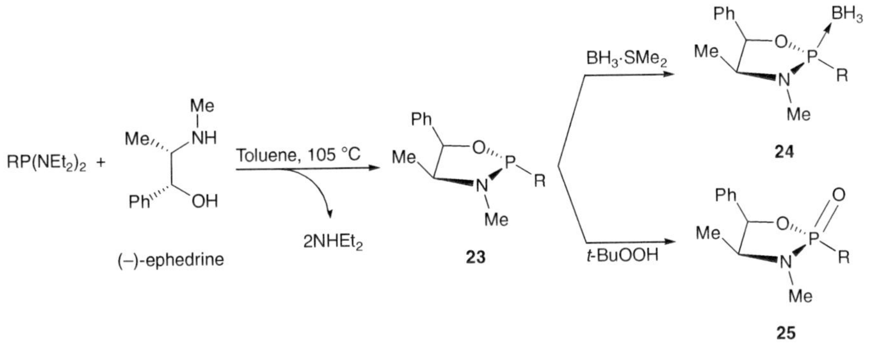

Scheme 3.14 Preparation of oxazaphospholidines by exchange at high temperature.

the base used to quench the HCl produced.[9] In entry 12, the oxazaphospholidine moiety compound was bound to a polystyrene support and was obtained in low optical purity at the phosphorus atom.

In spite of the mentioned stereochemical complexities, in all cases the major diastereomer has the substituent at the P atom and the Ph and Me groups of the ephedrine backbone in *trans* disposition. This has been confirmed by X-ray crystallography[78,79,83,100,101] and can be easily rationalised in terms of thermodynamic preference. The result in entry 27 by Hansen and co-workers,[95] who used pseudoephedrine, sheds more light on this issue. They demonstrated that the phenyl groups at the P atom and in the ephedrine have a *trans* relationship and concluded that the absolute configuration of the P atom is controlled by the carbon atom bearing the phenyl group in the ephedrine or pseudoephedrine.

Product **27** of entry 31 also deserves some comment since it was found that the oxazaphospholidine oxide group turns out to be an excellent *ortho* directing group for the diastereoselective lithiation of ferrocene (Scheme 3.15).

All reactions proceeded to completion, affording optically pure 1,2-disubstituted ferrocenes **28** in 45–95% yields and complete diastereoselectivity when *t*-BuLi was used. With other less basic and more nucleophilic organolithium reagents (*n*-BuLi, LDA) the reaction is less selective because the base also attacks the phosphorus atom yielding opened products such as **30** (Scheme 3.16).

In this case, the desired compound **29** was only isolated in only 37% yield, with 50% of the side product **30**.

Livinghouse and co-workers[102] reported low yields (15–30%) and formation of by-products when performing condensation of (−)-ephedrine with several *bis*(diethylamino)aryl phosphines in hot toluene. He disclosed an interesting variant (Scheme 3.17), based on a stepwise cyclisation that in fact is the hybrid between methods **A** and **B** (see Scheme 3.13).

Ephedrine can be regioselectively *O*-lithiated with *exactly* one equivalent of *n*-BuLi. The resulting alkoxide then reacts with the appropriate (racemic)

Table 3.9   Ephedrine-based oxazaphospholidines and some derivatives.

Entry	Compound	Method	Yield(%)	de (%)	References
1	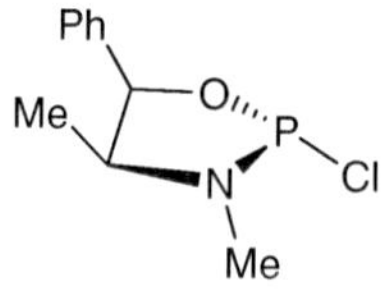 **26**	A	35–80	82–o.p.[a]	7, 9, 81–85
2	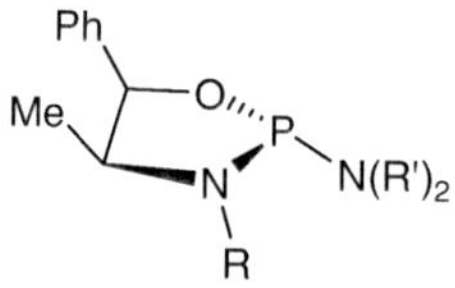 R = Me, *i*-Pr R' = Me, Et	B	43	94–o.p.	84, 86–88
3[b]	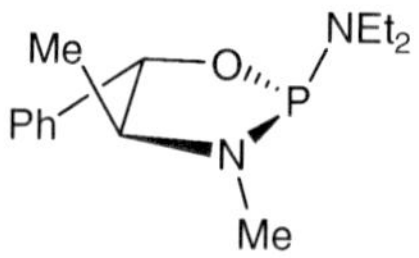	B	–	50	88
4	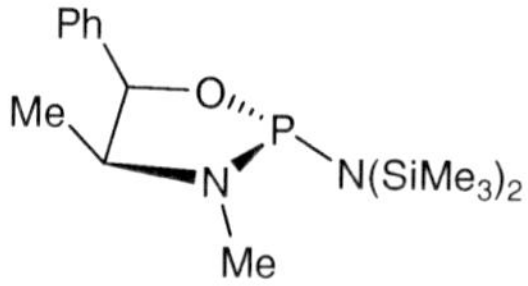	A	80	>96	89
5	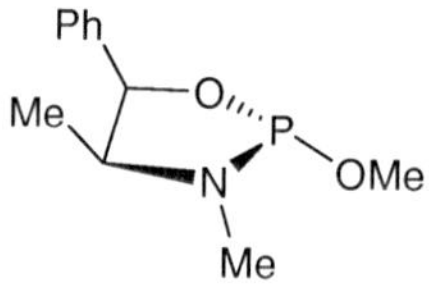	A	61	78	81, 84
6	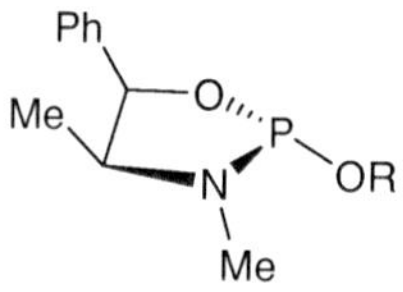 R = *t*-Bu, 2-Naphth...	B	–	92–96	88
7[b]	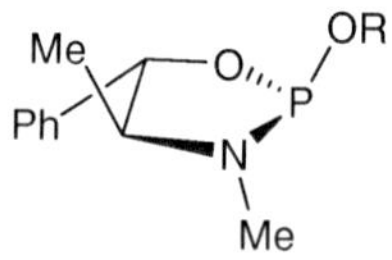 R = *t*-Bu, 2-Naphth...	B	–	82–94	88

Table 3.9 (*Continued*).

Entry	Compound	Method	Yield(%)	de (%)	References
8	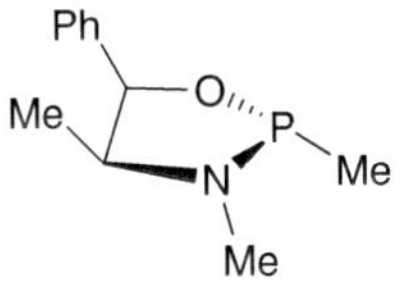	**B**	60	85	77
9	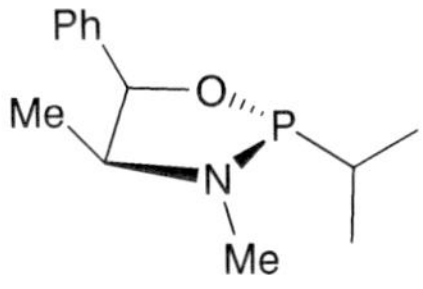	**B**	–	60	90
10	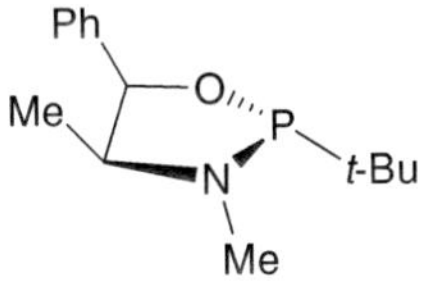	**A**	55	95	77
11	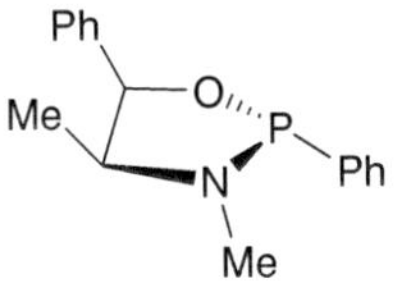	**B**	70–83	93–o.p.	77, 91
12	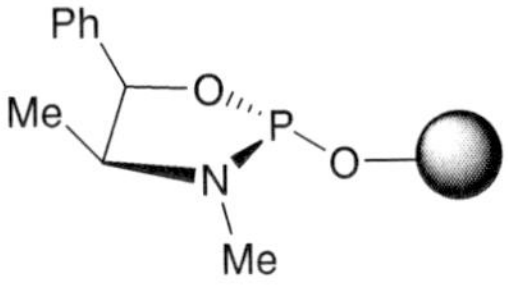	**A**	–	28	32, 92
13	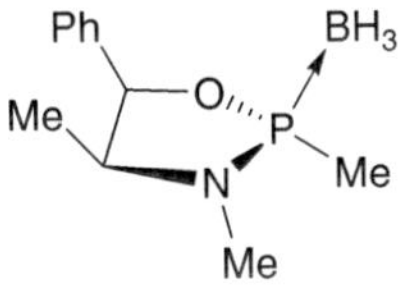	**B**	61	o.p.	92
14	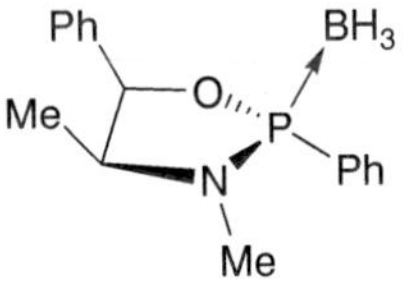	**A**	34	o.p.	78
15	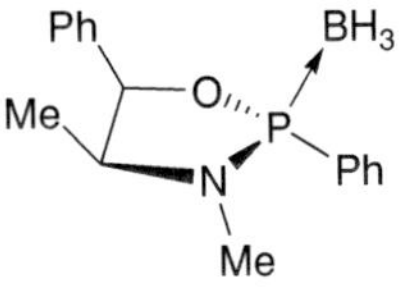	**B**	77–82	o.p.	93–96

Table 3.9 (*Continued*).

Entry	Compound	Method	Yield(%)	de (%)	References
16	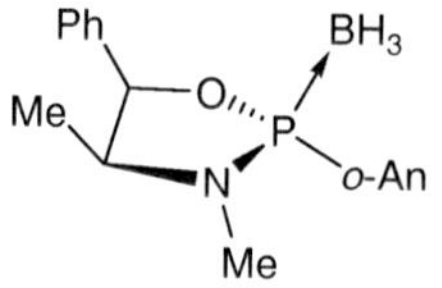	**B**	25	o.p.	92
17	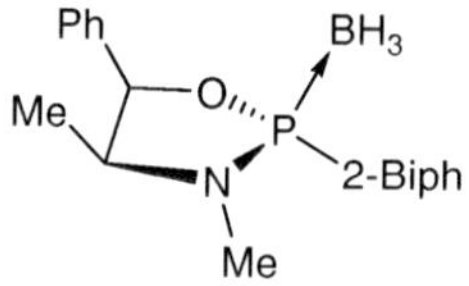	**A**	70	o.p	92
18	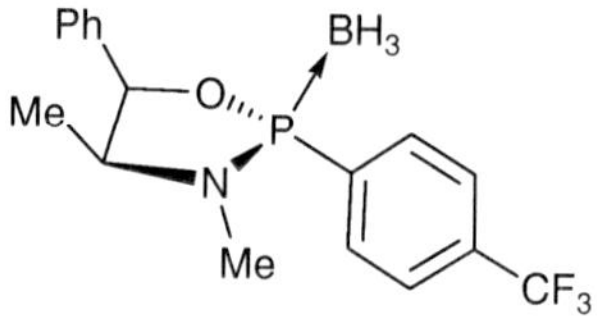	**A**	–	o.p.	97
19	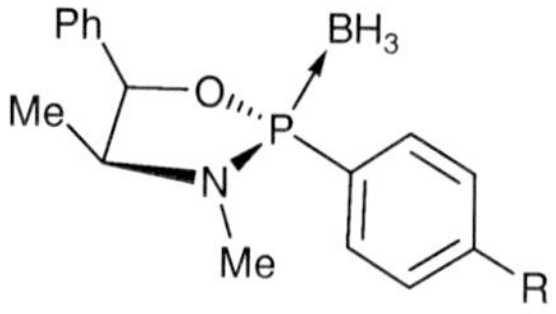	**B**	38 (CF$_3$) 47 (OMe)	o.p.	98
20	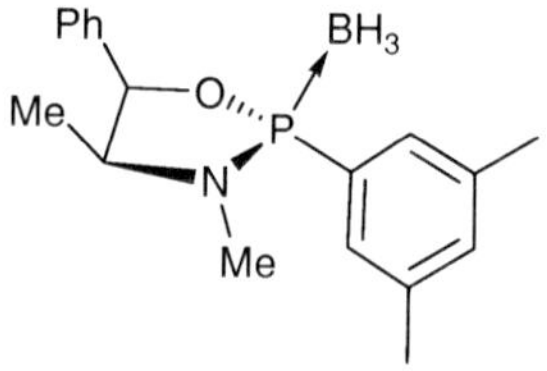	**B**	34	o.p.	90
21	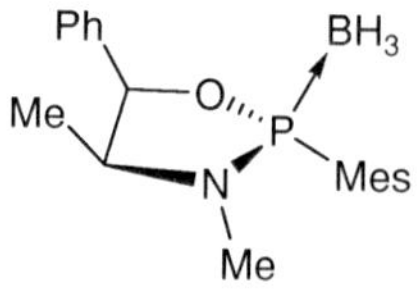	**B**	50	o.p.	99
22	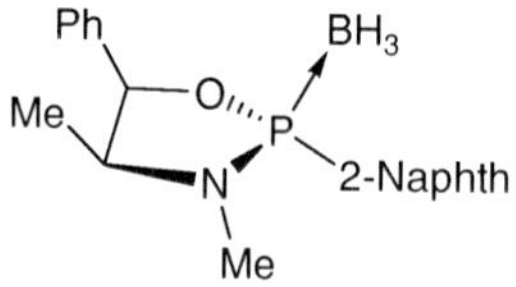	**B**	34	o.p.	92
23[c]	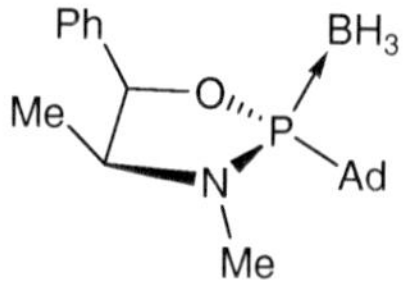	**A**	68	60	100

Table 3.9   (*Continued*).

Entry	Compound	Method	Yield(%)	de (%)	References
24[d]	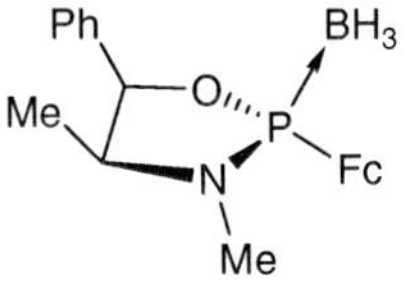	A	64	o.p.	83
25[e]	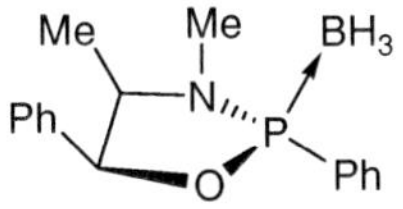	A	38	o.p.	80
26[e]	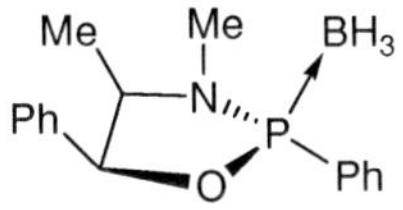	B	77	o.p.	94
27[b]	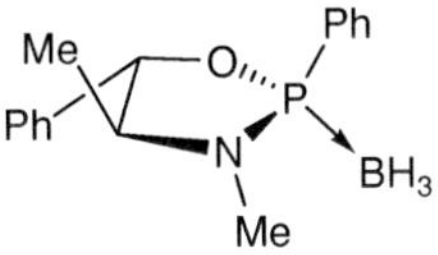	B	95	o.p.	95
28	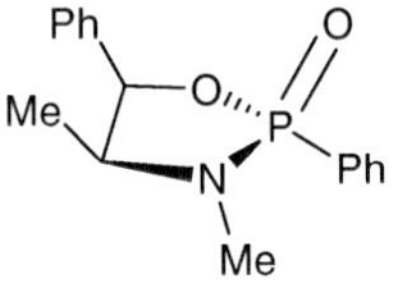	A	33–39	54–o.p.	78, 79
29[f]	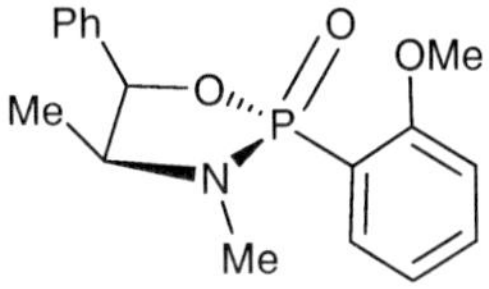	A	61	o.p.[a]	82
30[g]	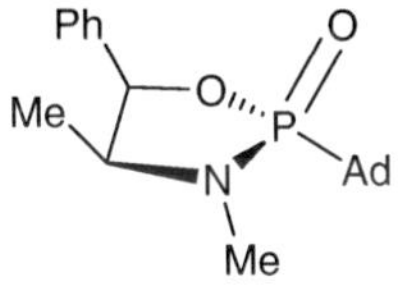	A	61	60	100
31[h]	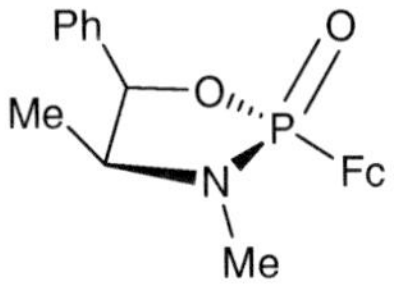	A	61	>99	83
32	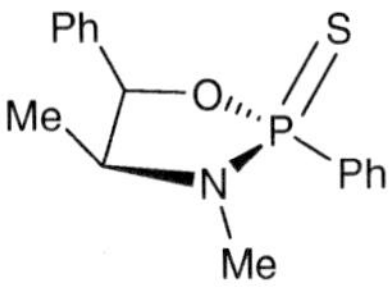	A	35	82	79

Table 3.9 (*Continued*).

Entry	Compound		Method	Yield(%)	de (%)	References
33[i]			**A**	87	o.p.	89

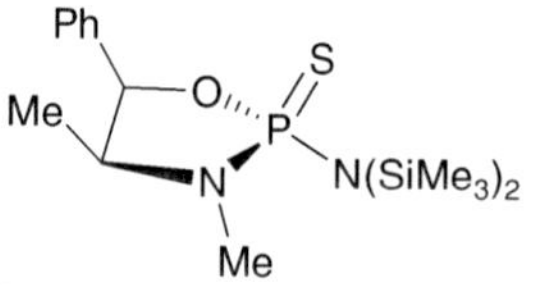

(Structure for entry 33: Ph, Me, O, P(=S), N(SiMe₃)₂, N–Me)

---

[a]In one of the references,[82] both epimers were observed at the beginning of the reaction whereas in another one[85] the epimers are only observed at $-52\,^\circ$C in the NMR.

[b]Prepared from (+)-pseudoephedrine.

[c]Prepared from **26** (entry 1), then AdLi and finally $BH_3 \cdot SMe_2$, both steps with retention of configuration at the P atom.

[d]As in footnote c, but with FcLi.

[e]Prepared from (+)-ephedrine.

[f]Prepared from **26** (entry 1), then o-AnMgBr and finally t-BuOOH, both steps occur with retention of configuration at the P atom.

[g]As in footnote c, but with AdMgBr.

[h]As in footnote g, but with FcLi.

[i]Prepared from **26** (entry 1) and $Li[N(SiMe_3)_2]$.

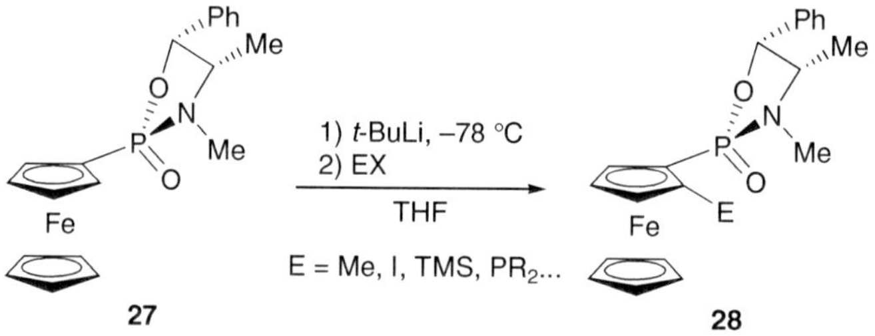

Scheme 3.15   Diastereoselective lithiation and electrophilic quenching of **27**.

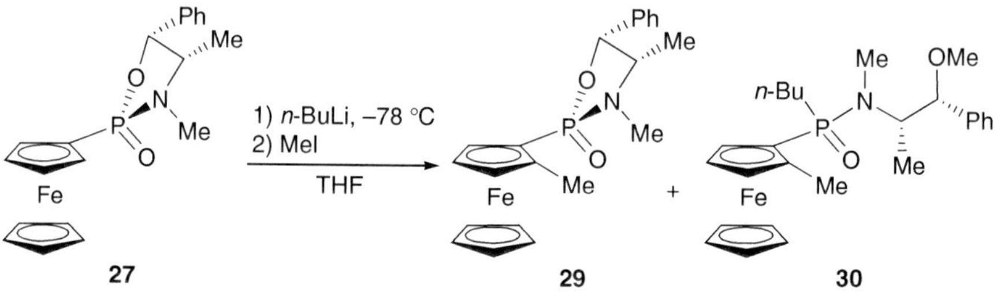

Scheme 3.16   Reaction of **27** with *n*-BuLi and MeI.

chloro(dimethylamino)phosphine **31** to produce a diastereomeric mixture of phosphonamidites **32**, which undergo thermal cyclisation to the desired oxazaphospholidines. Cyclisations were found to be catalysed by proton sources. With this method, the final borane adducts **24** were obtained in 70–82% yield

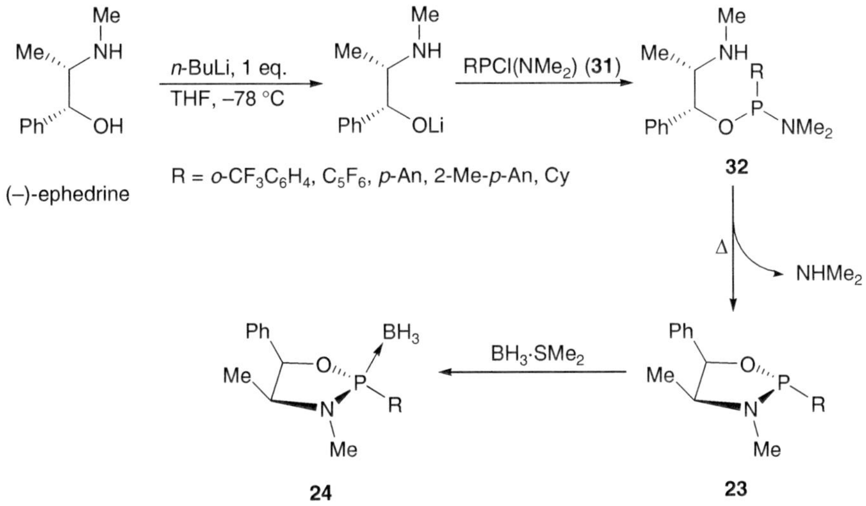

R = *o*-CF₃C₆H₄, C₅F₆, *p*-An, 2-Me-*p*-An, Cy

(–)-ephedrine

32

24

23

Scheme 3.17 Stepwise cyclisation of (–)-ephedrine to prepare oxazaphospholidine boranes (**24**).

from ephedrine and with total diastereoselectivity. A drawback of this strategy is that it requires a two-step synthesis of precursors **31** involving two distillations at reduced pressure.[102]

3.3.1.2 P(V) Oxazaphospholidines

In the last section several oxazaphospholidine oxides, obtained by oxidation of the P(III) precursor with *t*-BuOOH, have already been described. There is also one report by Jugé and co-workers[103] in which they prepare oxazaphospholidine oxides and sulfides by *in situ* deboronation/oxidation of oxazaphospholidine boranes. This section illustrates some more derivatives, prepared directly from P(V) species and ephedrine. Chronologically, these types of compounds were studied earlier than the corresponding P(III) counterparts. Nowadays oxazaphospholidine boranes, not oxides, are the most important precursors used to prepare enantiopure phosphorus ligands. However, apart from historic interest, ephedrine-derived oxazaphospholidine oxides, sulfides and selenides occupy an important place in the study of phosphorus stereochemistry and conformational analysis.[104] Only a few examples are described here.

During the 1970s Inch and co-workers[105–108] published a series of detailed papers where they studied many 1,3,2-oxazaphospholidines derived from (–)-ephedrine, prepared from common P(V) precursors (Scheme 3.18).

(Thio)phosphoryl chloride yields oxides and sulfides **33** which upon treatment with alcohols produces the corresponding alkoxides **34**[106,107] or amino derivatives **35**,[108,109] both with retention of configuration at the P atom.[105,110] Substituted analogues **36** are prepared from (thio)phosphonic dichlorides.

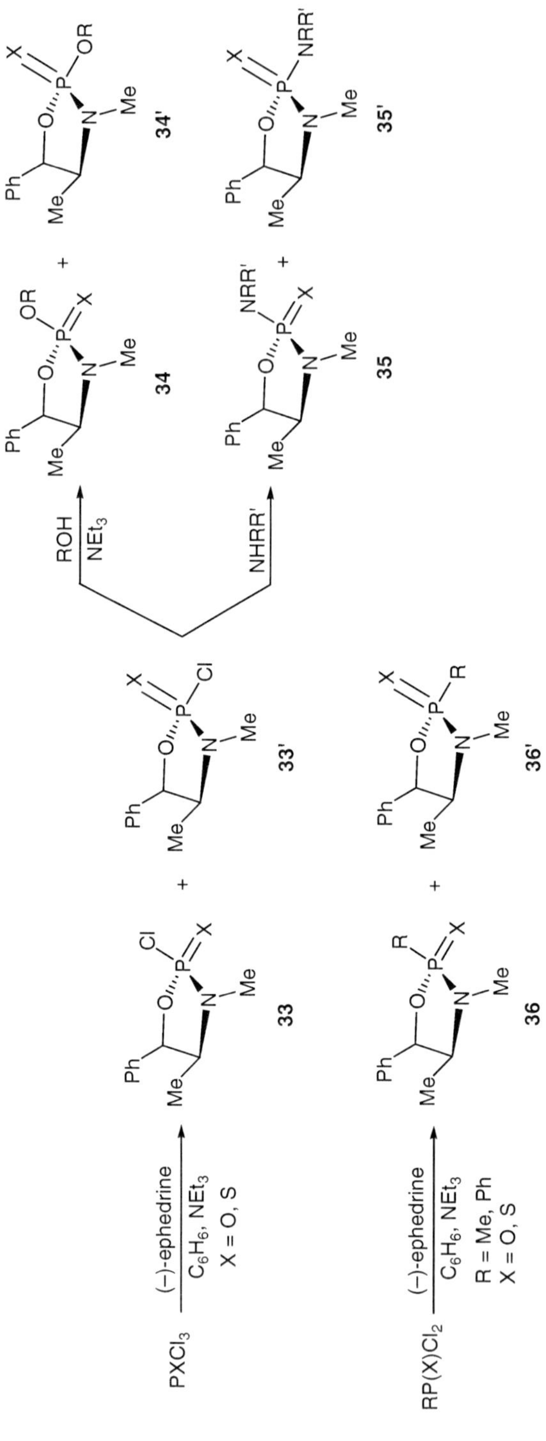

Scheme 3.18 Preparation of oxazaphospholidines from P(V) precursors.

The yields vary considerably (7–91%) and so do the diastereoselectivities (0–84% de), although they tend to be low. Even worse, it was found that they depend upon the exact conditions of the reaction and on the purification method, although the separation of the epimers is often possible. No crystal structures have been reported and the absolute configuration of the products was determined by thorough NMR analysis and comparison with previously reported data on similar compounds.

The mentioned compounds have been used in systematic studies on the regio- and stereoselectivity of substitution reactions at the phosphorus atom.[106–108,111–112] The same authors[113] also reported a similar study on oxazaphospholidine oxides and sulfides derived from (+)-norephedrine (Figure 3.3).

In parallel to P(V) diazaphospholidines (see Section 3.2.2), a totally regio-selective [1,3] P–O to P–C rearrangement was observed upon treatment of P(V) (pseudo)ephedrine-derived oxazaphospholidines with LDA. This was observed first by Welch and co-workers (Scheme 3.19).[114]

Treatment of **37** with LDA affords rearranged product **38** in a 85% yield, with retention of configuration at the phosphorus atom. More recently Buono and co-workers[43] came to the same conclusion with similar substrates.

3.3.2 Derived from Prolinol

The natural amino acid (*S*)-(−)-proline can be easily reduced to the corresponding amino alcohol (+)-prolinol (Scheme 3.20) with total stereoselectivity.[115]

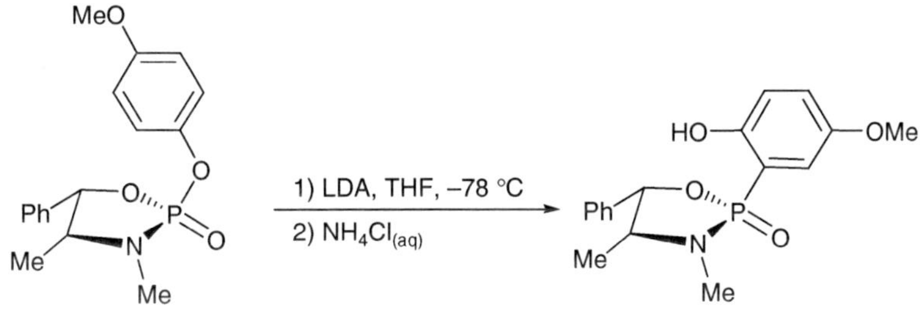

37

38

Scheme 3.19 P–O to P–C 1,3 rearrangement in an oxazaphospholidine oxide.

(*S*)-(–)-proline

(*S*)-(+)-prolinol

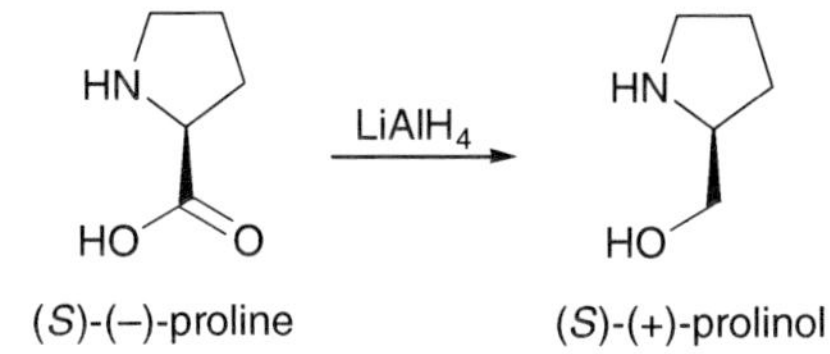

Scheme 3.20 Preparation of (+)-prolinol from (–)-proline.

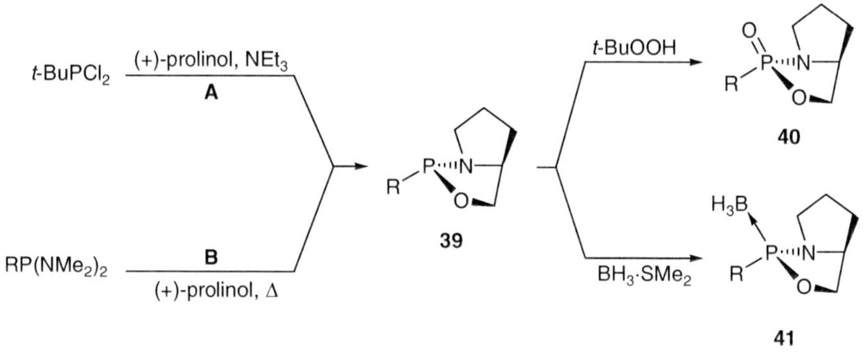

Scheme 3.21 Prolinol-derived oxazaphospholidines **39**, their oxides (**40**) and boranes
(**41**).

The latter compound is the 'amino alcohol' analogue of the diamine (*S*)-2-
(anilinomethyl)pyrrolidine (Scheme 3.1, Section 3.2.1). Not surprisingly, prolinol has also been used to prepare bicyclic *P*-stereogenic oxazaphospholidines
and derivatives thereof. As anticipated, their preparative methods are the same
as those described in the preceding sections (Scheme 3.21).

Richter[77] was the first to study these reactions and was able to prepare the
t-Bu derivative of **39** from *t*-BuPCl$_2$, but not the Me or Ph analogues, which
had to be prepared following method **B**. Since this report others have prepared
a small family of compounds (Table 3.10).

A variety of monodentate ligands (entries 1–9) and P,N-bidentate ligands
(entries 10 and 11) have been successfully prepared. Some have also been
oxidised to the P(V) compounds (entries 12–16) or boronated (entry 17).
Interestingly, entry 18 shows the single example of a C_2 bidentate ligand prepared from prolinol.

In most of the cases (except entries 4 and 8–10), the reactions provide
the desired compound in high or complete stereoselectivity towards the *trans*
diastereomer (represented in the table), which is the thermodynamic product.
Some early papers[16,121] report the formation of much larger quantities of
the *cis* epimer at the beginning of the reaction, suggesting that it is the kinetic
product.

Some oxides and sulfides have also been prepared from phosphoryl or
thiophosphoryl chlorides, but very low or no diastereoselectivity has been
found.[119,122,123] This class of compounds were also found to suffer the [1,3] P–O
to P–C rearrangement upon treatment with strong bases, once again with
retention of configuration at the phosphorus atom.[119]

3.3.3 Derived from Other Amino Alcohols

Ephedrine and prolinol are the most common compounds employed to prepare
P-stereogenic oxazaphospholidines. In spite of that, other 1,2-amino alcohols

Table 3.10 Prolinol-based oxazaphospholidines and some derivatives.

Entry	*Compound*	*Method*	*Yield(%)*	*de (%)*	*References*
1	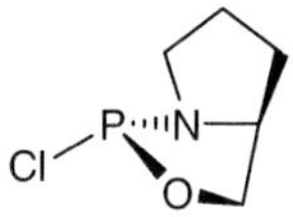	**A**	55–82	o.p.	30, 116
2	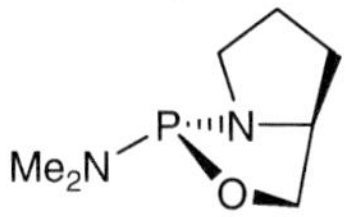	**B**	76	92–o.p.	16, 115
3	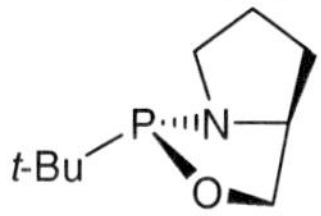	**A**	25	o.p.	77
4	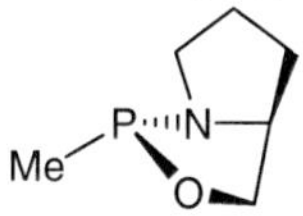	**B**	53	81	77
5	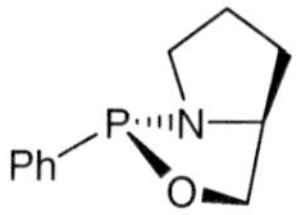	**B**	73–90	74–94	77, 117
6	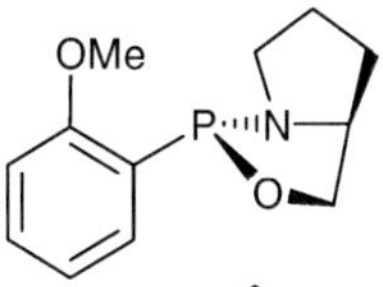	**B**	70	96	16
7	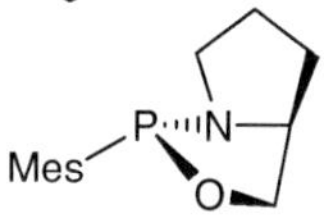	**B**	75	o.p.	115
8	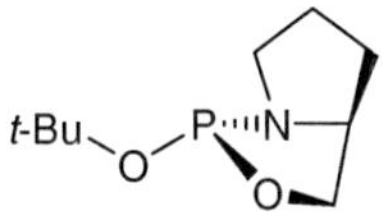	**B**	57	62	115
9	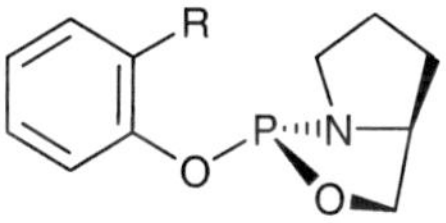 R = H, Me, *t*-Bu, Ph	**B**	74–85	62–82	115
10	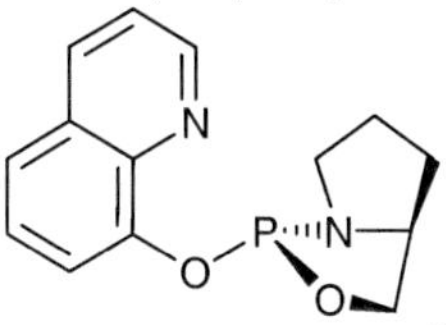	**B**	23	56	48
11	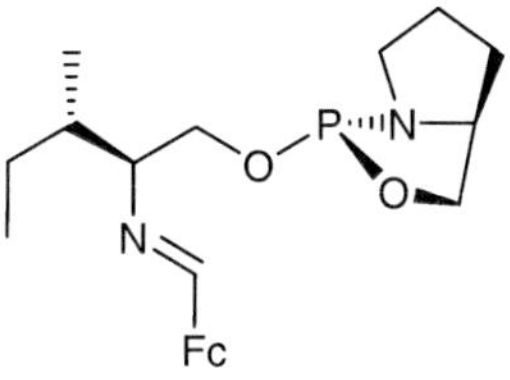	**A**	90	o.p.	30

Table 3.10   (*Continued*).

Entry	Compound	Method	Yield(%)	de (%)	References
12	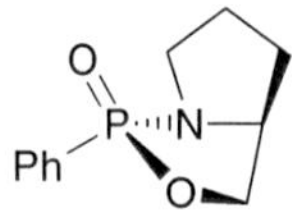 Ph, O, P, N, O	B	–	o.p.	118
13	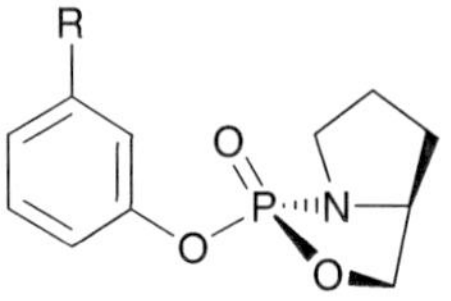 R, O, P, N, O; R = *t*-Bu, F, OMe	B	62–71	o.p.	119
14	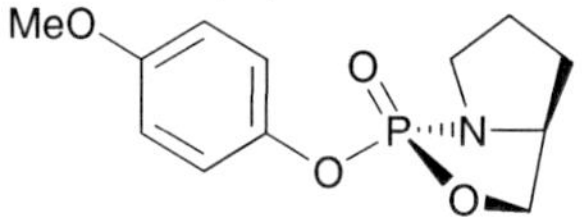 MeO, O, P, N, O	B	82	o.p.	119
15	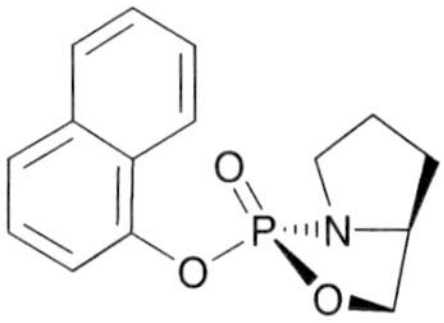 O, P, N, O	B	67	o.p.	119
16	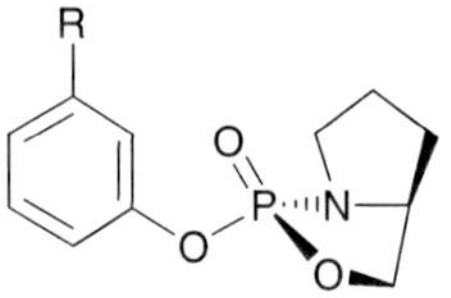 R, O, P, N, O; R = *t*-Bu, F, OMe	B	62–71	o.p.	119
17	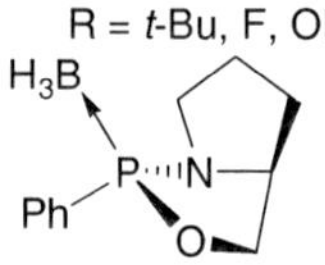 H_3B, Ph, P, N, O	B	97^{120} 71^{117}	o.p.	117, 120
18	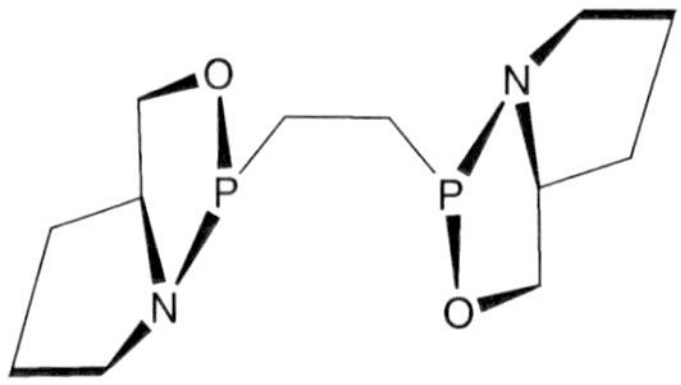 O, P, N, N, P, O	B	50	o.p.	77

have also been reported to furnish such compounds. The preparative methods are basically the same than those reported in the previous sections.

3.3.3.1   *Acyclic Amino Alcohols*

There are some reports on phosphorus heterocycles made from acyclic amino alcohols other than ephedrine.[28,124] Most of these amino alcohols are easily obtained by reduction of the natural amino acids.[117] For example, some oxazaphospholidine oxides and boranes obtained from (*S*)-valinol (**42–45**) and (*S*)-serine (**46, 47**) are depicted in Figure 3.4.

The stereoselectivity in the formation of these heterocycles was null for **42**[125] and **44**,[117] low for **43**[43] and very high for **45**.[126] Interestingly, the analogue of **45** with ephedrine is formed in very low stereoselectivity under the same conditions.[126] Reaction of methyl *N*-benzylserinoate with phosphoryl chloride yields the rather unstable **46**[127] that reacts smoothly with alcohols to yield **47** as nearly equimolar epimeric mixtures.

Pastor and co-workers[128,129] prepared a series of extremely bulky biarylic oxazaphospholidines (Figure 3.5) using chiral and achiral amino alcohols.

Most of the compounds were obtained as single diastereomers and studied by X-ray diffraction and multinuclear NMR. Indeed, some of them present unique spectral characteristics. For example, in the most crowded molecules

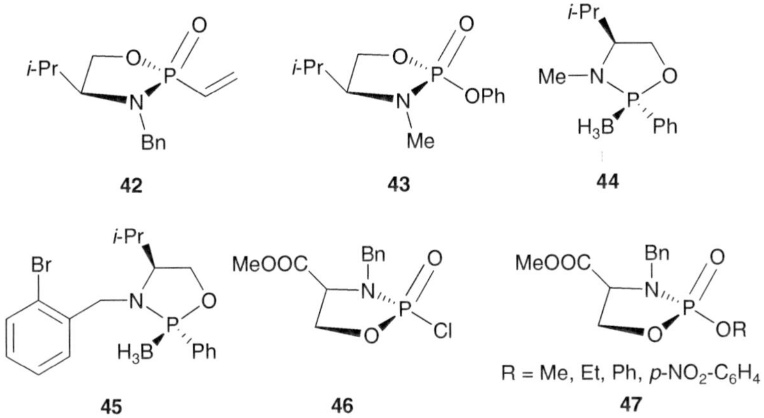

Figure 3.4 Amino acid-derived *P*-stereogenic oxazaphospholidines.

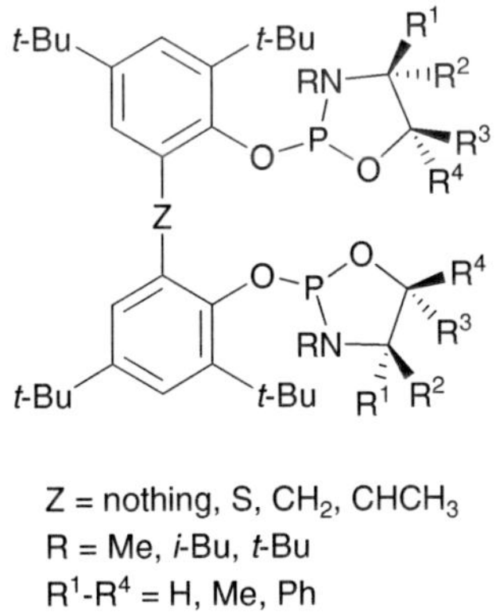

Figure 3.5 Bulky biaryl oxazaphospholidines.

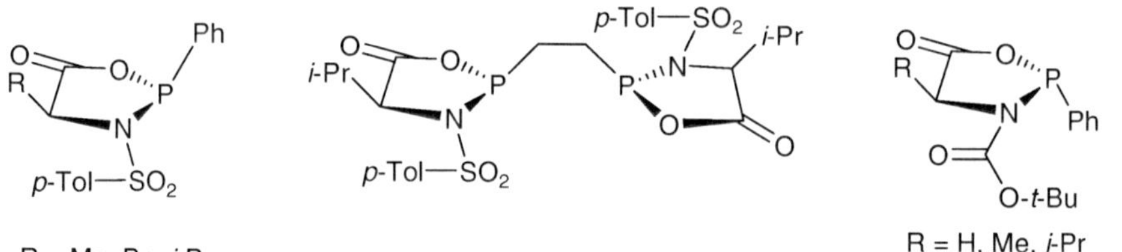

R = Me, Bn, *i*-Pr
48

49

R = H, Me, *i*-Pr
50

Figure 3.6 Amino acid-derived oxazaphospholidines.

both phosphorus atoms appear as non-equivalent in the ^{31}P NMR spectra, but as an AB quartet with a very large ($\sim$30 Hz) coupling constant for two phosphorus separated by seven or eight bonds. The authors suggested[129] that a through-space mechanism was responsible for those large coupling constants.

Finally, it is interesting to mention that Hersh and co-workers[130–132] have prepared oxazaphospholidines directly from α-amino acids in their search for electron-poor phosphorus centres (Figure 3.6).

They first prepared a series of *N*-tolylsulfonylated amino acids and reacted them with PhPCl$_2$ in the presence of NEt$_3$, resulting in the formation of **48** in good yields.[130,131] Whereas for R = Me no diastereoselectivity was observed, in the other cases the isomer bearing the phenyl group at the P atom and the R group in relative *cis* disposition dominated, in sharp contrast with those oxazaphospholidines derived from ephedrine. In fact, for R = *i*-Pr the *cis* isomer is present in up to 80% de in the crude mixture (although the reaction was rather capricious) and was obtained pure by recrystallisation. X-ray diffraction analysis confirmed the *cis* stereochemistry and showed that the isopropyl and phenyl groups only have small steric contact.[131] Consequently, they hypothesised that it is the tolyl fragment of the sulfonyl group that interacts with the isopropyl group and eventually compels the phenyl ring at the phosphorus atom to be *cis* to the isopropyl group. This explanation is in accordance with the observation that for R = Me a 1:1 *cis:trans* ratio is found. Although compounds **48** are stable in the solid state, epimerisation of the phosphorus atom and decomposition were observed in solution.[131] In the same paper, *bis*(oxazaphospholidine) **49** was also described, prepared from Cl$_2$PCH$_2$CH$_2$PCl$_2$ as a mixture of diastereomers, with the *trans/trans* isomer as the major product.

Interestingly, the same authors[130] extended this method to prepare the *N-tert*-butoxycarbonyl derivatives **50** with complete diastereoselectivity and, in contrast to **48**, with *trans* stereochemistry.

3.3.3.2   Cyclic Amino Alcohols

Other cyclic alcohols apart from prolinol have been used to synthesise bicyclic oxazaphospholidines and derivatives. One important example is (*S*)-(−)-α,α-diphenylprolinol (**51**), which is nowadays commercially available (Scheme 3.22).

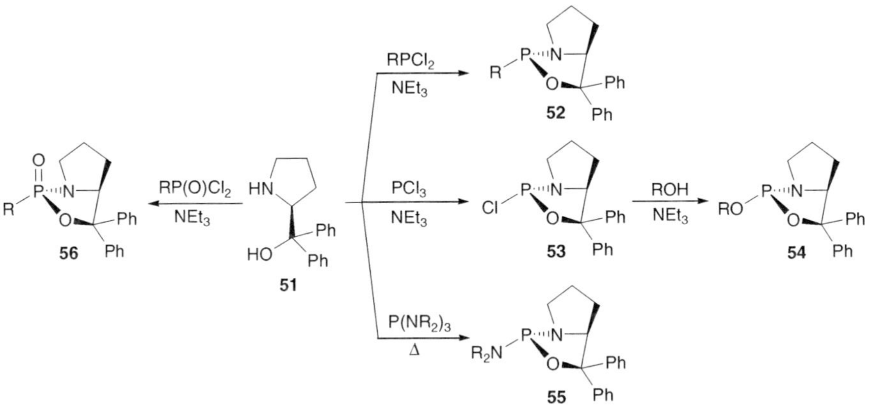

Scheme 3.22   Application of **51** to prepare oxazaphospholidines and oxazaphospholidine oxides.

Schmalz and co-workers[133,134] reported the preparation of key intermediate **53** and a few of its derivatives **54**, but did not discuss the stereochemistry of the phosphorus stereocentre. A later report from Reetz and co-workers[135] showed that reaction of **51** with phosphorus trichloride furnishes exclusively $(\textbf{\textit{R}}_{\textbf{P}})$-**53**. From this precursor, they prepared some bidentate ligands by direct phosphitylation of diols such as ethylene glycol, with retention of configuration at the P atom. The same authors[136] expanded this study to prepare a family of 21 monodentate ligands, including some aminoxazaphospholidines **55**, using the strategies of Scheme 3.22. They also solved the X-ray crystal structure for **54** (R = *o,o*-xylyl), which confirmed the expected *S* configuration at the phosphorus atom. Wills and co-workers[123,137] developed the 'oxide' route, preparing two derivatives **56**. They found good diastereoselectivity for R = Ph, but for R = Me the product was an equimolar epimeric mixture.

Very recently Gebbink and co-workers[138] prepared *bis*(oxazaphospholidines) pincer ligands **57–60** (Figure 3.7).

The ligands were prepared by phosphitylation of diphenols with **53** and its analogue obtained from (*S*)-(+)-indolinemethanol. Compounds **57–60** were obtained in 81–86% yields in high optical purity (>90% de) except for **59**, which was unexpectedly racemic. These compounds were successfully employed to form palladium cyclometallated complexes.[138]

Martens and co-workers[117] prepared two interesting oxazaphospholidines, depicted in Scheme 3.23.

Alcohol **61** is easily prepared from a chemical found in industrial waste and was used to prepare the tricyclic oxazaphospholidine borane **62** in 64% combined yield. The X-ray structure revealed that in this case the phosphorus atom had *S* absolute configuration. In the same paper they described the preparation of **64** in 40% yield and in optically pure form, starting from the azetidinic alcohol **63**. The same authors prepared the 'phosphine oxide' version of **62**[124] *via* oxidation of the P(III) precursor and by direct reaction of **61** with $PhP(O)Cl_2/NEt_3$.

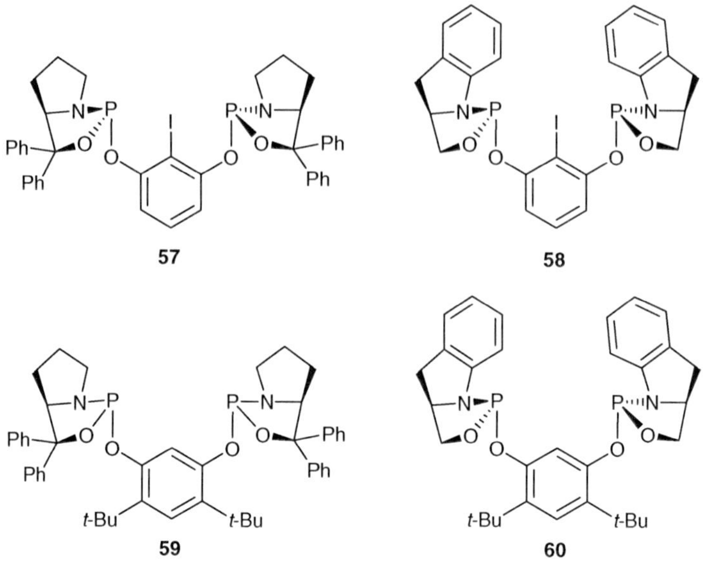

Figure 3.7 *Bis*(oxazaphospholidines) designed to act as pincer ligands.

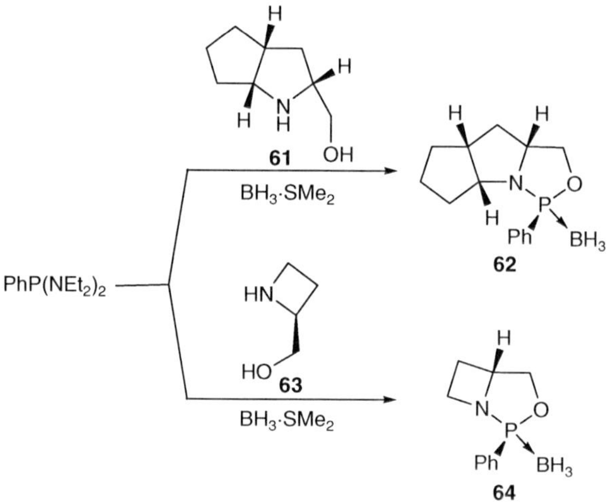

Scheme 3.23 Oxazaphospholidines prepared by Martens and co-workers.[117]

3.4 Other Heterocycles

3.4.1 Five-membered Heterocycles

Many other heterobifunctional chiral auxiliaries have also been employed to afford *P*-stereogenic heterocycles. Some of them are discussed here.

Corey and co-workers[139] used the camphor derivative **65** to form the oxa-thiaphospholidine **66** and its sulfide **67**, the latter obtained in 80% overall yield (Scheme 3.24).

The first step was a thermodynamically controlled process towards the more stable *exo* isomer of **66** (shown in the scheme) although at shorter reaction times the *endo* isomer was also observed. Compound **66** is a liquid that is only stable at low temperatures, whereas **67** is a crystalline solid that was used to prepare *P*-stereogenic phosphines, as detailed in Chapter 4, Section 4.5.2.2. Analogues of **67** with other groups instead of phenyl were also prepared.

There are also reports about 1,3,2-dioxaphospholidines prepared from chiral 1,2-diols; a few of them are represented in Figure 3.8.

Compound **68** was prepared by Stelzer and co-workers[140] from (−)-pina-nediol as a single isomer, but the absolute configuration at the P atom was not determined. Benetsky and co-workers[28] prepared compounds **69–71** from dichlorophosphine precursors and commercially available or easily obtainable diols. Phosphite **71** was obtained in low optical purity (32% de), but for both **69** and **70** the de exceeded 90% and one of them (**70**, R = NEt$_2$) was prepared in

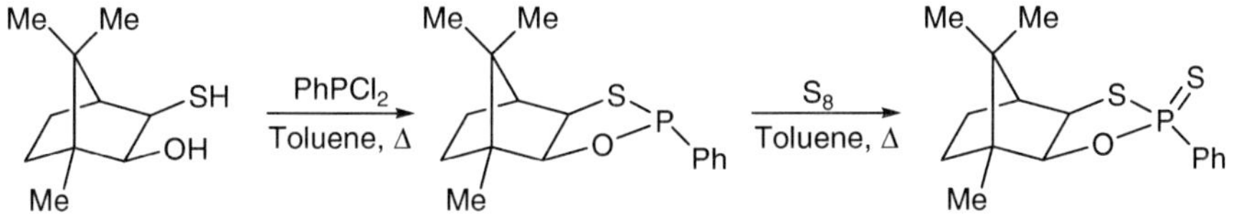

Scheme 3.24 Preparation of an oxathiaphospholidine and its sulfide.

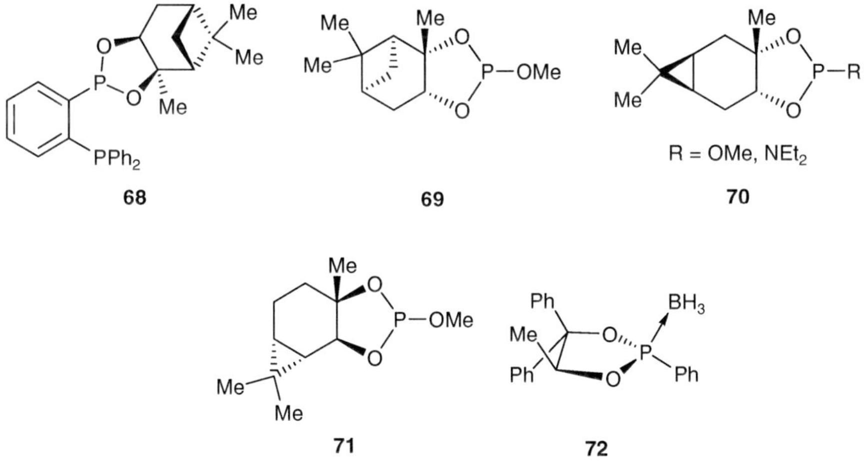

Figure 3.8 Examples of *P*-stereogenic dioxaphospholidines.

optically pure form. Dioxaphospholidine borane **72** was prepared as a single diastereomer by Jugé and co-workers in 55% yield,[141] by reaction of dichlorophenylphosphine with (*S*)-1,1-diphenylpropane-1,2-diol. Compound **72** was used as a model to study the P–O bond cleavage by nucleophilic substitution (see Chapter 4, Section 4.5.2.2).

3.4.2 Six-membered Heterocycles

Inch and co-workers[142–146] pioneered, in the 1970s, the preparation of many phosphorinane oxides and sulfides derived from carbohydrates, confirming their utility in the study of stereochemical aspects in nucleophilic substitutions at phosphorus.[1] Some of the compounds they prepared are shown in Figure 3.9.

Gluco- and galactopyranosides were reacted with phosphonic and phosphoric dihalides to provide dioxaphosphorinane oxides **73** and **74** as mixtures of the two phosphorus epimers, which were separated.[142] A similar method was used to prepare oxathiaphosphorinanes **75** from the corresponding thiopyranoside.[143] Dioxaphosphorinane sulfides **76** were prepared from phosphonothioic dichlorides [RP(S)Cl₂] or by oxidation of the corresponding P(III) precursor with sulfur.[143] Finally, oxazaphosphorinane oxides and sulfides **77** were prepared similarly.[146]

Following these studies, others used chiral diols and amino alcohols to prepare optically active phosphorinanes for different applications. Some selected compounds are depicted in Figure 3.10.

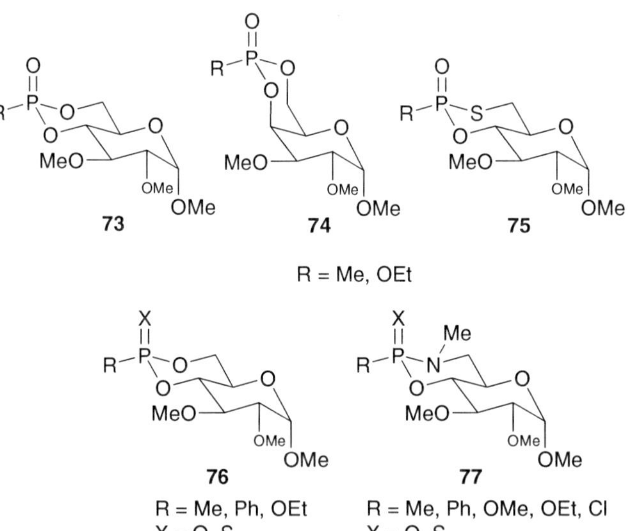

Figure 3.9 *P*-stereogenic phosphorinane oxides and sulfides derived from carbohydrates.

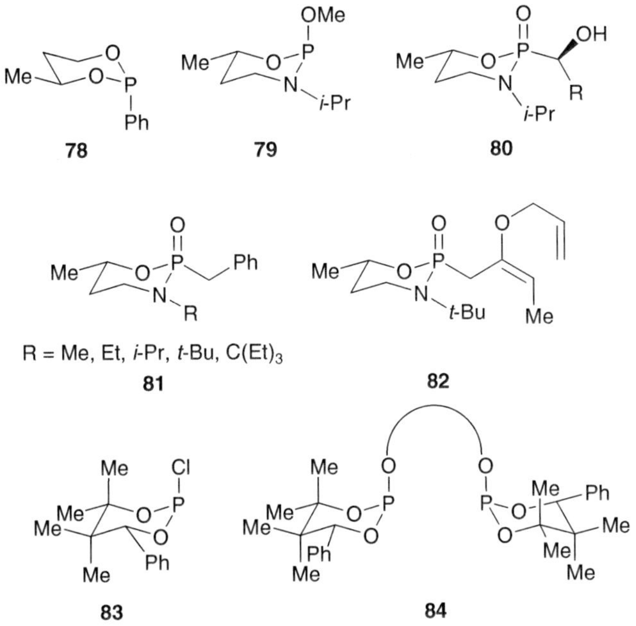

Figure 3.10 Examples of *P*-stereogenic phosphorinanes.

Suga and co-workers[147] prepared dioxaphosphorinane **78** as a single epimer in 62% yield from (*S*)-1,3-butanediol and dichlorophenylphosphine. This compound was used to prepare optically pure phosphine oxides *via* the Arbuzov reaction. Evans and co-workers[148] prepared **79** as an enriched mixture (84% de) of inseparable epimers. They investigated the condensation of **79** with a variety of aldehydes in the presence of BF_3 to give diastereomerically enriched mixtures of α-hydroxyoxazaphosphorinane oxides (**80**). Denmark and co-workers[149,150] prepared oxazaphosphorinane oxides **81** and studied the alkylation of the benzylic position as a function of several parameters. The same group[151] prepared **82** and its epimer that were separated by column chromatography. These compounds were employed to study the sense of asymmetric induction in carbanion-accelerated Claisen rearrangements. *P*-stereogenic phosphochloridite **83** was prepared by Kellogg and co-workers[152] and reacted with several diols to produce diphosphites **84**. These compounds were applied to self-recognition studies.

Many other phosphorinane oxides have been prepared due to their pharmacological applications. Of particular importance are those containing fused bicyclic and tricyclic systems. They can be readily prepared by reaction of the chiral cyclic amino alcohol and the appropriate phosphonic dichloride and they are usually obtained as mixtures of epimers, often separable by column chromatography. A few examples are displayed in Figure 3.11.

Fülöp and co-workers[153,154] prepared tricycles **85** as mixtures of separable epimers and studied their conformation in solution in detail. More recently the same group did a similar study, complemented with theoretical calculations, on

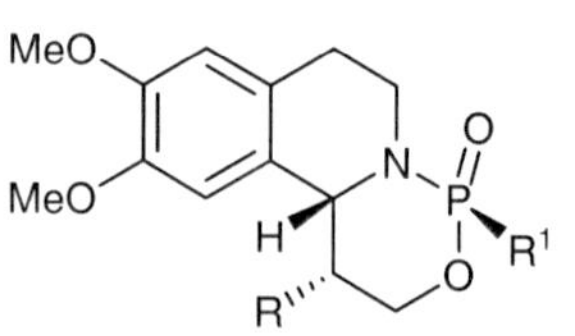

R = H, Me
R^1 = Cl, Ph, $N(CH_2CH_2Cl)_2$
85

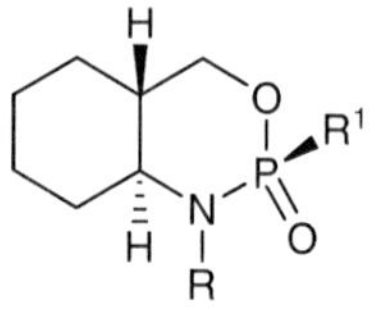

R = H, Me, Bn
R^1 = Ph, OPh, $N(CH_2CH_2Cl)_2$
86

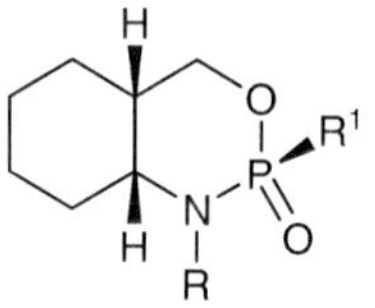

R = H, Me, Bn
R^1 = Ph, OPh, $N(CH_2CH_2Cl)_2$
87

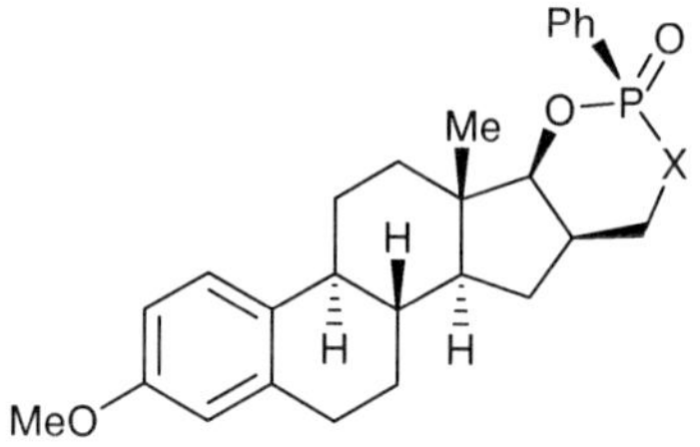

X = O, NR
R = H, *n*-Pr, Bn
88

Figure 3.11 Examples of phosphorinane oxides with pharmacological interest.

bicyclic systems **86** and **87**.[155] Similarly, Frank and co-workers[156] reported the fused pentacyclic oxazaphosphorinane and dioxaphosphorinane oxides **88**, prepared from derivatives of the steroid estrone and phenylphosphonic dichloride in the presence of triethylamine.

More recently Keay and co-workers[157] prepared some 1,3,2-oxazaphosphorinanes from chiral spiro 1,3-amino alcohols with the aim to use them as ligands in Rh-catalysed hydrogenation (Scheme 3.25).

Spiro amino alcohols **89** were reacted with dichlorophenylphosphine and after treatment with borane provided protected oxazaphosphorinanes **90**. The diastereoselectivity of the cyclisation was found to be low (0% for R = H and 43% for R = Me) but the epimers could be separated by crystallisation or column chromatography. In the case of R = H, bidentate ligands (**91**) could be prepared *via* deprotonation of the NH group and double nucleophilic substitution on dibromomethane. X-ray crystallography of **90** (R = Me) and **91** was used to determine the absolute configuration of the phosphorus atom.

Upon deboronation of **90** and **91** with DABCO, however, it was found that the free oxazaphosphorinanes are extremely sensitive to oxidation. Although Rh complexes could be detected by ^{31}P NMR, they are also very sensitive and give the corresponding oxides when exposed to air.

The extreme sensitivity of oxazaphosphorinanes is surprising and shows that there can be discrepancies between basicity and oxygen sensitivity, pointing out that factors affecting oxidation rates in tricoordinated phosphorus compounds are still not clearly understood.

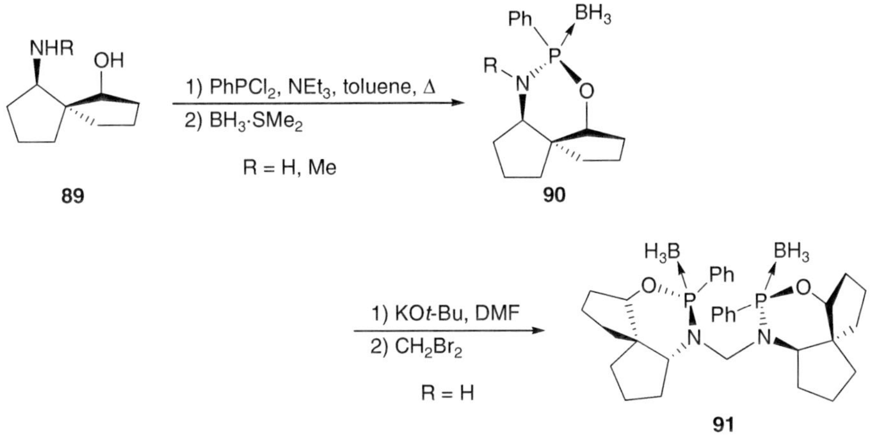

Scheme 3.25 1,3,2-oxazaphosphorinanes from spiro 1,3-amino alcohols.

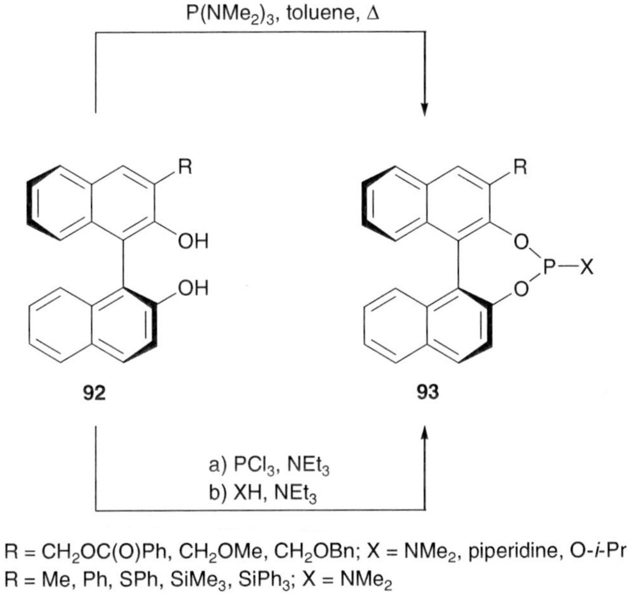

Scheme 3.26 *P*-stereogenic binaphthol-derived phosphites and phosphoramidites.

3.4.3 Seven-membered Heterocycles

Reetz and co-workers[158] published the only examples of binaphthol-based *P*-stereogenic phosphites and phosphoramidites. They were prepared from known *ortho*-monosubstituted binaphthols (**92**) *via* classical phosphitylation methods (Scheme 3.26).

The phosphitylation reactions are high yielding (79–97%) and furnish **93** as the expected mixtures of epimers with different absolute configuration at the phosphorus atom. The final diastereomeric distribution varies from 33% to 100% de and was found to change upon heating the compounds for prolonged periods of time. The diastereomers could be separated by HPLC or crystallisation and their absolute configurations were ascertained by X-ray structural analysis and ^{31}P NMR data.

References

1. C. R. Hall and T. D. Inch, *Tetrahedron*, 1980, **36**, 2059.
2. J. Ansell and M. Wills, *Chem. Soc. Rev.*, 2002, **31**, 259.
3. K. N. Gavrilov, O. G. Bondarev and A. I. Polosukhin, *Russ. Chem. Rev.*, 2004, **73**, 671.
4. G. Buono, O. Chiodi and M. Wills, *Synlett*, 1999, 377.
5. S. Jugé, *Phosphorus, Sulfur, and Silicon*, 2008, **183**, 233.
6. C. Darcel, J. Uziel and S. Jugé, in *Phosphorus Ligands in Asymmetric Catalysis: Synthesis and Applications*, A. Börner (ed.), Wiley-VCH, Weinheim, 2008, p. 1211.
7. R. P. Iyer, D. Yu, N. Ho, W. Tan and S. Agrawal, *Tetrahedron: Asymmetry*, 1995, **6**, 1051.
8. Y. Lu and G. Just, *Angew. Chem. Int. Ed.*, 2000, **39**, 4521.
9. N. Oka, T. Wada and K. Saigo, *J. Am. Chem. Soc.*, 2003, **125**, 8307.
10. A. Wilk, A. Grajkowski, L. R. Phillips and S. L. Beaucage, *J. Am. Chem. Soc.*, 2000, **122**, 2149.
11. B. Whittaker, M. Ruiz and C. J. Hayes, *Tetrahedron Lett.*, 2008, **49**, 6984.
12. N. Oka, T. Kondo, T. Fujiwara, Y. Maizuru and T. Wada, *Org. Lett.*, 2009, **11**, 967.
13. M. D. Sorensen, L. K. A. Blaehr, M. K. Christensen, T. Hoyer, S. Latini, P. V. Hjarnaa and F. Bjorkling, *Bioorg. Med. Chem.*, 2003, **11**, 5461.
14. G. Buono, N. Toselli and D. Martin, in *Phosphorus Ligands in Asymmetric Catalysis*, A. Börner, (ed.), Wiley-VCH, Weinheim, 2008, p. 529.
15. J. M. Brunel, T. Constantieux and G. Buono, *J. Org. Chem.*, 1999, **64**, 8940.
16. H. Arzoumanian, G. Buono, M. Choukrad and J. F. Petrignani, *Organometallics*, 1988, **7**, 59.
17. K. N. Gavrilov, V. N. Tsarev, A. A. Shiryaev, O. G. Bondarev, S. E. Lyubimov, E. B. Benetsky, A. A. Korlyukov, M. Y. Antipin, V. A. Davankov and H. Gais, *Eur. J. Inorg. Chem.*, 2004, 629.
18. K. Barta, M. Holscher, G. Franciò and W. Leitner, *Eur. J. Org. Chem.*, 2009, 4102.
19. K. N. Gavrilov, O. G. Bondarev, V. N. Tsarev, A. A. Shiryaev, S. E. Lyubimov, A. S. Kucherenko and V. A. Davankov, *Russ. Chem. Bull. Int. Ed.*, 2003, **52**, 122.

20. V. N. Tsarev, S. E. Lyubimov, A. A. Shiryaev, S. V. Zheglov, O. G. Bondarev, V. A. Davankov, A. A. Kabro, S. K. Moiseev, V. N. Kalinin and K. N. Gavrilov, *Eur. J. Org. Chem.*, 2004, 2214.

21. K. N. Gavrilov, E. B. Benetsky, T. B. Grishina, S. V. Zheglov, E. A. Rastorguev, P. V. Petrovskii, F. Z. Macaev and V. A. Davankov, *Tetrahedron: Asymmetry*, 2007, **18**, 2557.

22. S. E. Lyubimov, V. A. Davankov, P. V. Petrovskii and N. M. Loim, *Russ. Chem. Bull. Int. Ed.*, 2007, **56**, 2094.

23. K. N. Gavrilov, E. B. Benetskiy, T. B. Grishina, E. A. Rastorguev, M. G. Maksimova, S. V. Zheglov, V. A. Davankov, B. Schaffner, A. Börner, J. M. Rosset, G. Bailat and A. Alexakis, *Eur. J. Org. Chem.*, 2009, 3923.

24. E. B. Benetskii, V. A. Davankov, P. V. Petrovskii, E. A. Rastorguev, T. B. Grishina, K. N. Gavrilov, S. Rosset, G. Bailat and A. Alexakis, *Russ. J. Org. Chem.*, 2008, **44**, 1846.

25. K. N. Gavrilov, S. E. Lyubimov, S. V. Zheglov, E. B. Benetsky, P. V. Petrovskii, E. A. Rastorguev, T. B. Grishina and V. A. Davankov, *Adv. Synth. Catal.*, 2007, **349**, 1085.

26. K. N. Gavrilov, S. E. Lyubimov, O. G. Bondarev, M. G. Maksimova, S. V. Zheglov, P. V. Petrovskii, V. A. Davankov and M. T. Reetz, *Adv. Synth. Catal.*, 2007, **349**, 609.

27. S. E. Lyubimov, I. V. Kuchurov, A. A. Vasil'ev, A. A. Tyutyunov, V. N. Kalinin, V. A. Davankov and S. G. Zlotin, *J. Organomet. Chem.*, 2009, **694**, 3047.

28. E. B. Benetsky, S. V. Zheglov, T. B. Grishina, F. Z. Macaev, L. P. Bet, V. A. Davankov and K. N. Gavrilov, *Tetrahedron Lett.*, 2007, **48**, 8326.

29. M. Kimura and Y. Uozumi, *J. Org. Chem.*, 2007, **72**, 707.

30. V. N. Tsarev, S. E. Lyubimov, O. G. Bondarev, A. A. Korlyukov, M. Y. Antipin, P. V. Petrovskii, V. A. Davankov, A. A. Shiryaev, E. B. Benetsky, P. A. Vologzhanin and K. N. Gavrilov, *Eur. J. Org. Chem.*, 2005, 2097.

31. K. N. Gavrilov, V. N. Tsarev, S. I. Konkin, N. M. Loim, P. V. Petrovskii, E. S. Kelbyscheva, A. A. Korlyukov, M. Y. Antipin and V. A. Davankov, *Tetrahedron: Asymmetry*, 2005, **16**, 3224.

32. B. H. G. Swennenhuis, R. Chen, P. W. N. M. van Leeuwen, J. G. de Vries and P. C. J. Kamer, *Eur. J. Org. Chem.*, 2009, 5796.

33. J. Brunel, T. Constantieux, O. Legrand and G. Buono, *Tetrahedron Lett.*, 1998, **39**, 2961.

34. S. Breeden and M. Wills, *J. Org. Chem.*, 1999, **64**, 9735.

35. J. M. Brunel, O. Chiodi, B. Faure, F. Fotiadu and G. Buono, *J. Organomet. Chem.*, 1997, **529**, 285.

36. J. M. Brunel, O. Legrand, S. Reymond and G. Buono, *J. Am. Chem. Soc.*, 1999, **121**, 5807.

37. S. Reymond, J. M. Brunel and G. Buono, *Tetrahedron: Asymmetry*, 2000, **11**, 1273.

38. T. Pfretzschner, L. Kleemann, B. Janza, K. Harms and T. Schrader, *Chem. Eur. J.*, 2004, **10**, 6048.

39. V. V. Dunina, O. N. Gorunova, V. A. Stepanova, P. A. Zykov, M. V. Livantsov, Y. K. Grishin, A. V. Churakov and L. G. Kuz'mina, *Tetrahedron: Asymmetry*, 2007, **18**, 2011.

40. C. J. Ngono, T. Constantieux and G. Buono, *J. Organomet. Chem.*, 2002, **643–644**, 237.

41. C. W. Edwards, M. R. Shipton, N. W. Alock, H. Clase and M. Wills, *Tetrahedron*, 2003, **59**, 6473.

42. N. Toselli, D. Martin and G. Buono, *Org. Lett.*, 2008, **10**, 1453.

43. O. Legrand, J. M. Brunel, T. Constantieux and G. Buono, *Chem. Eur. J.*, 1998, **4**, 1061.

44. G. Delapierre, M. Achard and G. Buono, *Tetrahedron Lett.*, 2002, **43**, 4025.

45. J. M. Brunel, T. Constantieux, A. Labande, F. Lubatti and G. Buono, *Tetrahedron Lett.*, 1997, **38**, 5971.

46. J. M. Brunel, B. Del Campo and G. Buono, *Tetrahedron Lett.*, 1998, **39**, 9663.

47. T. Constantieux, J. Brunel, A. Labande and G. Buono, *Synlett*, 1998, 49.

48. G. Delapierre, J. M. Brunel, T. Constantieux and G. Buono, *Tetrahedron: Asymmetry*, 2001, **12**, 1345.

49. R. Hilgraf and A. Pfaltz, *Adv. Synth. Catal.*, 2005, **347**, 61.

50. O. G. Bondarev, K. N. Gavrilov, V. N. Tsarev, V. A. Davankov, R. V. Lebedev, S. K. Moiseev and V. N. Kalinin, *Russ. Chem. Bull. Int. Ed.*, 2002, **51**, 521.

51. K. N. Gavrilov, S. V. Zheglov, P. A. Vologzhanin, M. G. Maksimova, A. S. Safronov, S. E. Lyubimov, V. A. Davankov, B. Schaffner and A. Börner, *Tetrahedron Lett.*, 2008, **49**, 3120.

52. V. N. Tsarev, S. I. Konkin, A. A. Shyryaev, V. A. Davankov and K. N. Gavrilov, *Tetrahedron: Asymmetry*, 2005, **16**, 1737.

53. G. J. Clarkson, J. R. Ansell, D. J. Cole-Hamilton, P. J. Pogorzelec, J. Whittell and M. Wills, *Tetrahedron: Asymmetry*, 2004, **15**, 1787.

54. K. N. Gavrilov, S. V. Zheglov, E. B. Benetsky, A. S. Safronov, E. A. Rastorguev, N. N. Groshkin, V. A. Davankov, B. Schaffner and A. Börner, *Tetrahedron: Asymmetry*, 2009, 20.

55. N. R. Vautravers and D. J. Cole-Hamilton, *Chem. Commun.*, 2009, 92.

56. N. R. Vautravers, P. Andre and D. J. Cole-Hamilton, *Dalton Trans.*, 2009, 3413.

57. C. J. Ngono, T. Constantieux and G. Buono, *Eur. J. Org. Chem.*, 2006, 1499.

58. O. Legrand, J. M. Brunel and G. Buono, *Eur. J. Org. Chem.*, 1999, 1099.

59. O. Legrand, J. M. Brunel and G. Buono, *Tetrahedron*, 2000, **56**, 595.

60. J. Peyronel, O. Samuel and J. Fiaud, *J. Org. Chem.*, 1987, **52**, 5320.

61. D. Basavaiah, V. Chandrashekar, U. Das and G. J. Reddy, *Tetrahedron: Asymmetry*, 2005, **16**, 3955.

62. O. Legrand, J. M. Brunel and G. Buono, *Angew. Chem. Int. Ed.*, 1999, **38**, 1479.

63. K. Iseki, Y. Kuroki, M. Takahashi, S. Kishimoto and Y. Kobayashi, *Tetrahedron*, 1997, **53**, 3513.
64. D. Basavaiah, G. J. Reddy and V. Chandrashekar, *Tetrahedron: Asymmetry*, 2001, **12**, 685.
65. D. Basavaiah, G. J. Reddy and V. Chandrashekar, *Tetrahedron: Asymmetry*, 2004, **15**, 47.
66. D. Basavaiah, G. J. Reddy and K. V. Rao, *Tetrahedron: Asymmetry*, 2004, **15**, 1881.
67. K. Mislow, *Acc. Chem. Res.*, 1970, **3**, 321.
68. R. P. Holmes, *J. Am. Chem. Soc.*, 1978, **100**, 433.
69. T. Nemoto and T. Hamada, *Chem. Rec.*, 2007, **7**, 150.
70. T. Nemoto, *Chem. Pharm. Bull.*, 2008, **56**, 1213.
71. T. Nemoto, T. Matsumoto, T. Masuda, T. Hitomi, K. Hatano and Y. Hamada, *J. Am. Chem. Soc.*, 2004, **126**, 3690.
72. T. Nemoto, T. Masuda, T. Matsumoto and Y. Hamada, *J. Org. Chem.*, 2005, **70**, 7172.
73. T. Nemoto, T. Harada, T. Matsumoto and Y. Hamada, *Tetrahedron Lett.*, 2007, **48**, 6304.
74. T. Nemoto, T. Masuda, Y. Akimoto, T. Fukuyama and Y. Hamada, *Org. Lett.*, 2005, **7**, 4447.
75. T. Nemoto, T. Sakamoto, T. Matsumoto and Y. Hamada, *Tetrahedron Lett.*, 2006, **47**, 8737.
76. T. Nemoto, T. Fukuyama, E. Yamamoto, S. Tamura, T. Fukuda, T. Matsumoto, Y. Akimoto and Y. Hamada, *Org. Lett.*, 2007, **9**, 927.
77. W. J. Richter, *Chem. Ber.*, 1984, **117**, 2328.
78. J. M. Brown, J. V. Carey and M. J. H. Russell, *Tetrahedron*, 1990, **46**, 4877.
79. C. H. Schwalbe, G. Chopra, S. Freeman, J. M. Brown and J. V. Carey, *J. Chem. Soc., Perkin Trans.*, 1991, **2**, 2081.
80. E. A. Colby and T. F. Jamison, *J. Org. Chem.*, 2003, **68**, 156.
81. C. R. Hall and T. D. Inch, *Tetrahedron Lett.*, 1976, **40**, 3645.
82. J. V. Carey, M. D. Barker, J. M. Brown and M. J. H. Russell, *J. Chem. Soc., Perkin Trans. 1*, 1993, **1**, 831.
83. D. Vinci, N. Mateus, X. Wu, F. Hancock, A. Steiner and J. Xiao, *Org. Lett.*, 2006, **8**, 215.
84. J. Nielsen and O. Dahl, *J. Chem. Soc., Perkin Trans. 2*, 1984, **2**, 553.
85. V. Sum, A. J. Davies and T. P. Kee, *J. Chem. Soc., Chem. Commun.*, 1992, 1771.
86. A. Alexakis, S. Mutti and J. F. Normant, *J. Am. Chem. Soc.*, 1991, **113**, 6332.
87. A. Alexakis, J. Frutos and P. Mangeney, *Tetrahedron: Asymmetry*, 1993, **4**, 2427.
88. M. T. Reetz and M. Surowiec, *Heterocycles*, 2006, **67**, 567.
89. V. Sum, C. A. Baird, T. P. Kee and M. Thornton-Pett, *J. Chem. Soc., Perkin Trans. 1*, 1994, **1**, 3183.
90. A. Grabulosa, PhD Thesis, Universitat de Barcelona, Barcelona, 2005.

91. S. Jugé and J. P. Genêt, *Tetrahedron Lett.*, 1989, **30**, 2783.

92. R. den Heeten, B. H. G. Swennenhuis, P. W. N. M. van Leeuwen, J. G. de Vries and P. C. J. Kamer, *Angew. Chem. Int. Ed.*, 2008, **47**, 6602.

93. S. Jugé, M. Stephan, J. A. Laffitte and J. P. Genêt, *Tetrahedron Lett.*, 1990, **31**, 6357.

94. E. B. Kaloun, R. Merdès, J. P. Genêt, J. Uziel and S. Jugé, *J. Organomet. Chem.*, 1997, **529**, 455.

95. A. J. Rippert, A. Linden and H. Hansen, *Helv. Chim. Acta*, 2000, **83**, 311.

96. R. Ewalds, E. B. Eggeling, A. C. Hewat, P. C. J. Kamer, P. W. N. M. van Leeuwen and D. Vogt, *Chem. Eur. J.*, 2000, **6**, 1496.

97. M. J. Johansson, N. Kann and K. Larsson, *Acta Cryst. A*, 2004, **E60**, o287.

98. U. Nettekoven, P. C. J. Kamer and P. W. N. M. van Leeuwen, *Organometallics*, 2000, **19**, 4596.

99. F. Maienza, F. Spindler, M. Thommen, B. Pugin, C. Malan and A. Mezzetti, *J. Org. Chem.*, 2002, **67**, 5239.

100. J. M. Brown and J. C. P. Laing, *J. Organomet. Chem.*, 1997, **529**, 435.

101. S. Jugé, M. Stephan, J. P. Genêt, S. Halut-Desportes and S. Jeannin, *Acta Cryst. Sect. C*, 1990, **46**, 1869.

102. S. K. Sheeban, M. Jiang, L. McKinstry, T. Livinghouse and D. Garton, *Tetrahedron*, 1994, **50**, 6155.

103. J. Uziel, C. Darcel, D. Moulin, C. Bauduin and S. Jugé, *Tetrahedron: Asymmetry*, 2001, **12**, 1441.

104. W. N. Setzer, B. G. Black, B. A. Hovanes and J. L. Hubbard, *J. Org. Chem.*, 1989, **54**, 1709.

105. D. B. Cooper, J. M. Harrison and T. D. Inch, *Tetrahedron Lett.*, 1974, **31**, 2697.

106. D. B. Cooper, R. C. Hall, J. M. Harrison and T. D. Inch, *J. Chem. Soc., Perkin Trans.*, 1977, **1**, 1969.

107. R. C. Hall and T. D. Inch, *J. Chem. Soc., Perkin Trans. 1*, 1978, **1**, 1104.

108. R. C. Hall and T. D. Inch, *J. Chem. Soc., Perkin Trans. 1*, 1979, **1**, 1646.

109. G. Boche and W. Schrott, *Tetrahedron Lett.*, 1982, **23**, 5403.

110. P. M. Cullis, A. Iagrossi, A. J. Rous and M. B. Schilling, *J. Chem. Soc., Chem. Commun.*, 1987, 996.

111. R. C. Hall and T. D. Inch, *J. Chem. Soc., Perkin Trans. 1*, 1981, **1**, 2368.

112. C. R. Hall, T. D. Inch, G. Peacock, C. Pottage and N. E. Williams, *J. Chem. Soc., Perkin Trans.*, 1984, **1**, 669.

113. C. R. Hall, T. D. Inch and N. E. Williams, *J. Chem. Soc., Perkin Trans. 1*, 1982, **1**, 639.

114. S. C. Welch, J. A. Levine, I. Bernal and J. Cetrullo, *J. Org. Chem.*, 1990, **55**, 5991.

115. P. Cros, G. Buono, G. Peiffer, A. Denis, A. Mortreux and F. Petit, *New Jour. Chem.*, 1987, **11**, 574.

116. R. P. Iyer, M. Guo, D. Yu and S. Agrawal, *Tetrahedron Lett.*, 1998, **39**, 2491.

117. V. Peper, K. Stingl, H. Thumler, W. Saak, D. Haase, S. Pohl, S. Jugé and J. Martens, *Liebigs Ann.*, 1995, 2123.

118. O. Chiodi, F. Fotiadu, M. Sylvestre and G. Buono, *Tetrahedron Lett.*, 1996, **37**, 39.

119. J. M. Brunel, O. Legrand and G. Buono, *Eur. J. Org. Chem.*, 2000, 3313.

120. J. Brunel, O. Pardigon, B. Faure and G. Buono, *J. Chem. Soc., Chem. Commun.*, 1992, 287.

121. B. Faure, A. Archavlis and G. Buono, *J. Chem. Soc., Chem. Commun.*, 1989, 805.

122. T. Koizumi, R. Yanada, H. Takagi, H. Hirai and E. Yoshii, *Tetrahedron Lett.*, 1981, **22**, 477.

123. M. P. Gamble, J. R. Studley and M. Wills, *Tetrahedron Lett.*, 1996, **37**, 2853.

124. V. Peper and J. Martens, *Tetrahedron Lett.*, 1996, **37**, 8351.

125. K. Katagiri, M. Yamamoto, T. Iwaoka and C. Kaneko, *J. Chem. Soc., Chem. Commun.*, 1991, 1429.

126. W. Leung, S. Cosway, R. H. V. Jones, H. McCann and M. Wills, *J. Chem. Soc., Perkin Trans. 1*, 2001, **1**, 2588.

127. C. M. Thompson, J. A. Frick and D. L. C. Green, *J. Org. Chem.*, 1990, **55**, 111.

128. S. D. Pastor, J. L. Hyun, P. A. Odorisio and R. K. Rodebaugh, *J. Am. Chem. Soc.*, 1988, **110**, 6547.

129. S. D. Pastor, R. K. Rodebaugh, P. A. Odorisio, B. Pugin, G. Rihs and A. Togni, *Helv. Chim. Acta*, 1991, **74**, 1175.

130. W. H. Hersh, P. Xu, C. K. Simpson, T. Wood and A. L. Rheingold, *Inorg. Chem.*, 1998, **37**, 384.

131. W. H. Hersh, P. Xu, C. K. Simpson, J. Grob, B. Bickford, M. S. Hamdani, T. Wood and A. L. Rheingold, *J. Org. Chem.*, 2004, **69**, 2153.

132. W. H. Hersh, L. Klein and L. J. Todaro, *J. Org. Chem.*, 2004, **69**, 7355.

133. R. Kranich, K. Eis, O. Geis, S. Muhle, J. W. Bats and H. Schmalz, *Chem. Eur. J.*, 2000, **6**, 2874.

134. F. Blume, S. Zemolka, T. Fey, R. Kranich and H. Schmalz, *Adv. Synth. Catal.*, 2002, **344**, 868.

135. M. T. Reetz, G. Mehler and O. Bondarev, *Chem. Commun.*, 2006, 2292.

136. O. G. Bondarev and R. Goddard, *Tetrahedron Lett.*, 2006, **47**, 9013.

137. M. P. Gamble, A. R. Smith and M. Wills, *J. Org. Chem.*, 1998, **63**, 6068.

138. J. Li, M. Lutz, A. L. Spek, G. P. M. van Klink, G. van Koten and R. J. M. K. Gebbink, *Organometallics*, 2010, **29**, 1379.

139. E. J. Corey, Z. Chen and G. J. Tanoury, *J. Am. Chem. Soc.*, 1993, **115**, 11000.

140. W. K. Kottsieper, U. Küher and O. Stelzer, *Tetrahedron: Asymmetry*, 2001, **12**, 1159.

141. J. Uziel, M. Stephan, E. B. Kaloun, J. P. Genêt and S. Jugé, *Bull. Soc. Chim. Fr.*, 1997, **134**, 379.

142. D. B. Cooper, T. D. Inch and G. J. Lewis, *J. Chem. Soc., Perkin Trans.*, 1974, 1043.

143. D. B. Cooper, J. M. Harrison, T. D. Inch and G. J. Lewis, *J. Chem. Soc., Perkin Trans.*, 1974, **1**, 1049.

144. J. M. Harrison, T. D. Inch and G. J. Lewis, *J. Chem. Soc., Perkin Trans. 1*, 1974, **1**, 1053.

145. D. B. Cooper, J. M. Harrison, T. D. Inch and G. J. Lewis, *J. Chem. Soc., Perkin Trans. 1*, 1974, **1**, 1058.

146. J. M. Harrison, T. D. Inch and G. J. Lewis, *J. Chem. Soc., Perkin Trans.*, 1975, **1**, 1892.

147. M. Segi, Y. Nakamura, T. Nakajima and S. Suga, *Chem. Lett.*, 1983, 913.

148. N. J. Gordon and S. A. Evans, *J. Org. Chem.*, 1993, **58**, 5293.

149. S. E. Denmark and R. L. Dorow, *J. Org. Chem.*, 1990, **55**, 5926.

150. S. E. Denmark and C. Chen, *J. Org. Chem.*, 1994, **59**, 2922.

151. S. E. Denmark and J. E. Marlin, *J. Org. Chem.*, 1987, **52**, 5742.

152. A. C. Dros, A. Meetsma and R. M. Kellogg, *Tetrahedron*, 1999, **55**, 3071.

153. F. Fulop, E. Forro, T. Martinek, G. Gunther and R. Sillanpaa, *J. Mol. Struc.*, 2000, **554**, 119.

154. T. Martinek, E. Forro, G. Gunther, R. Sillanpaa and F. Fulop, *J. Org. Chem.*, 2000, **65**, 316.

155. H. Kivela, Z. Zalan, P. Tahtinen, R. Sillanpaa, F. Fulop and K. Pihlaja, *Eur. J. Org. Chem.*, 2005, 1189.

156. E. Frank, B. Kazi, K. Ludányi and G. Keglevich, *Tetrahedron Lett.*, 2006, **47**, 1105.

157. W. L. Benoit, M. Parvez and B. A. Keay, *Tetrahedron: Asymmetry*, 2009, **20**, 69.

158. M. T. Reetz, J. Ma and R. Goddard, *Angew. Chem. Int. Ed.*, 2005, **44**, 412.

P-Stereogenic Compounds Derived from Enantiopure Heterocycles

4.1 Introduction

Some of the heterocycles discussed in Chapter 3 have been found to be ideally suited to prepare a wide variety of *P*-stereogenic compounds and tertiary phosphines in particular.[1,2] This is described in this chapter in detail. The most important precursors are oxazaphospholidine boranes derived from ephedrine, which have given rise to one of the main modern strategies to prepare *P*-stereogenic phosphines in high stereoselectivities. This method, sometimes referred to as the 'ephedrine' or 'Jugé–Stephan' method, is therefore discussed step by step in this chapter. Some previous versions of this method and extensions providing interesting ligands for enantioselective catalysis have been included as well. Finally, other similar procedures with chiral auxiliaries other than ephedrine are also briefly considered.

4.2 Overview of Different Methodologies

The methods discussed in this chapter are based on substitutions on optically pure *P*-stereogenic heterocycles prepared from bifunctional chiral auxiliaries, discussed in Chapter 3. The starting heterocycle, apart from containing the first desired substituent, bears two leaving groups displaying very different reactivity allowing stepwise substitutions to take place. When these reactions are sufficiently regio- and stereoselective, the strategy depicted in Scheme 4.1 becomes a powerful tool to prepare *P*-stereogenic phosphines and other organophosphorus compounds.

RSC Catalysis Series No. 7
P-Stereogenic Ligands in Enantioselective Catalysis
By Arnald Grabulosa
© Arnald Grabulosa 2011
Published by the Royal Society of Chemistry, www.rsc.org

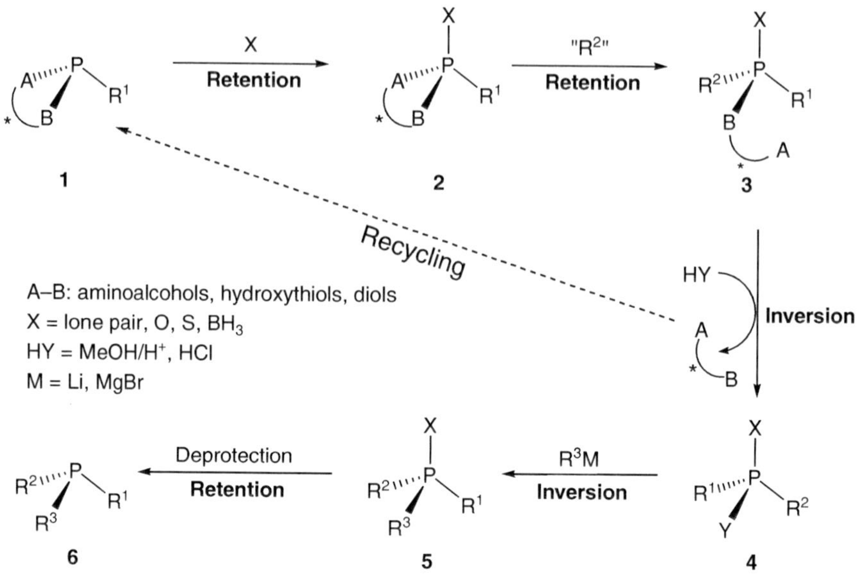

Scheme 4.1 Preparation of *P*-stereogenic phosphines *via* stepwise substitutions on **1**.

Although unprotected heterocycles **1** have been used[3,4] most of the work in this area has been carried out with protected P(III) or P(V) derivatives **2**, easily obtainable from **1** with retention of configuration. The second desired group is then installed by nucleophilic or electrophilic attack, yielding **3**. This reaction usually proceeds with retention of configuration, but not always.[5] The chiral auxiliary is then recovered *via* acidolysis of **3**, which also yields **4**, still possessing a leaving group. Nucleophilic substitution installs the last substituent with inversion of configuration. Finally, **5** can be deprotected by conventional methods (see Chapter 1, Section 1.3) to eventually afford the desired *P*-stereogenic phosphine **6**, habitually with retention of configuration. The advantage of this stepwise approach is that most of the intermediates are quite stable in air and can be purified by conventional methods. Furthermore, they are often solids and therefore their optical purity can be easily improved by recrystallisation.

4.3 The Jugé–Stephan Method

In the original Jugé–Stephan method A–B = ephedrine, X = BH$_3$, HY = MeOH/ H$_2$SO$_4$ and M = Li.[1,2,6] This method is depicted in Scheme 4.2.

The first step is the nucleophilic ring opening of **7** with organolithiums at low temperature to produce aminophosphine boranes **8**. This step occurs regioselectively (P–O bond cleavage) and diastereoselectively (retention of configuration). The selective cleavage of the P–N bond is then performed in methanol with the presence of one equivalent of sulfuric acid. The products are

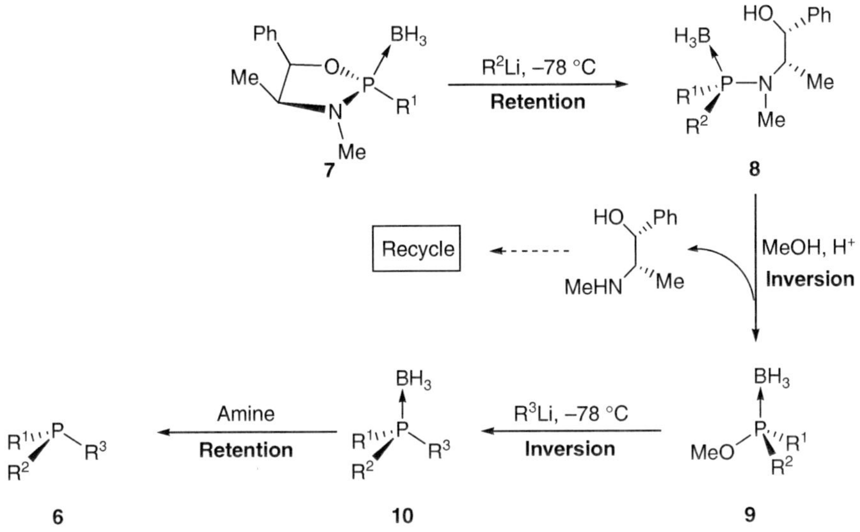

Scheme 4.2 The Jugé–Stephan method.

methylphosphinites **9** and ephedrine, which can be recycled. This step occurs with inversion of configuration. Phosphinite boranes **9** react with organolithiums to afford tertiary phosphine boranes **10**, also with inversion of configuration.

4.3.1 Introduction of R^2 by Nucleophilic Ring Opening

4.3.1.1 Ring Opening with Mono(organolithiums)

The first step in the Jugé–Stephan methodology is the introduction of R^2 by nucleophilic ring opening of the starting oxazaphospholidine borane. This reaction is completely regioselective, with exclusive P–O bond cleavage, and usually very stereoselective, with retention of configuration at the phosphorus atom (Scheme 4.3).

Organolithium reagents react smoothly with **7** (with $R^1 = Ph$ in most of the cases) at low temperature (often $-78\,^{\circ}C$) to afford, after aqueous work-up, ring opened alcohols **8** in high yields and high stereoselectivities. Methyl-, *n*-butyl, 1-naphthyl and *o*-anisyllithium were used in the early papers of Jugé and co-workers[6,7] but since then many other aminophosphines **8** have been synthesised (see Table 4.1 below). In contrast, it has been found that Grignard reagents require much higher temperatures to react, resulting in serious erosion of stereoselectivity. For example, *o*-anisylmagnesium reagents provide **8** at $80\,^{\circ}C$ with only 20% de.[7]

The retention of configuration at the P atom has been confirmed by X-ray analysis of some aminophosphines **8** prepared from alkyl- and aryllithiums.[7–9] This stereochemistry can be rationalised by a mechanism implying a kinetically

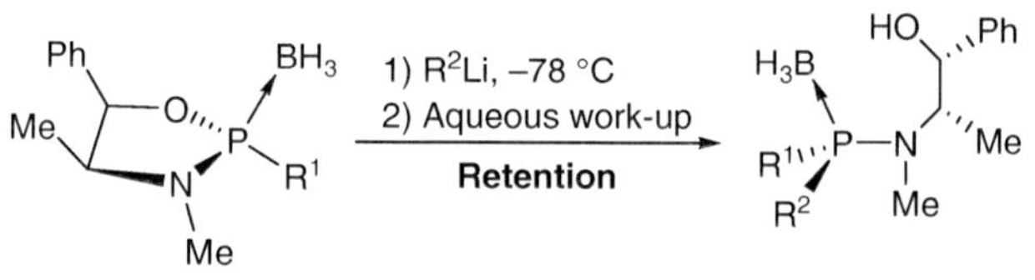

Scheme 4.3 Nucleophilic ring opening of **7** by organolithium reagents.

controlled nucleophilic attack and a reorganisation in the pentacoordinated phosphorus atom (Scheme 4.4).[2,7]

The oxygen atom assists the attack of the organolithium to the less hindered face of the P–O bond, opposite to the NMe group, affording pentacoordinated compound **11**, which can suffer a turnstile rotation (TR) producing **11′**, with the O atom in an apical position. Finally, ring opening in this intermediate yields **12** with retention of configuration, which eventually produces **8** after protonation. Interestingly, a Berry pseudorotation in either **11** or **11′** could afford products with inversion of configuration, but this is energetically unfavourable in five-membered rings.

Many aminophosphines **8** have been prepared in the last 20 years. Most of these are listed in Table 4.1.

Table 4.1 shows the variety of substituents that have been introduced in this step of the Jugé–Stephan method. In the vast majority of cases, the starting oxazaphospholidine borane **7** bears a Ph group, although there are some exceptions (entries 57–61). In general both alkyl substituents (entries 1–13) and aryl substituents (entries 15–61) can be introduced in good yields and good diastereoselectivities. In fact, aminophosphines **8** are commonly obtained optically pure after purification by column chromatography and/or recrystallisation. Usually compounds **8** are moderately air and moisture sensitive but can be stored for a long time under inert atmosphere. A few cases of low stereoselectivities (entries 11, 55–58 and 61), however, must be noted. Entry 11 shows that allyllithium is surprisingly unselective in the ring opening of **7**.[18] This entry constitutes also a rare example where optically inactive (±)-ephedrine has been used to prepare **7**. Phosphines containing ferrocenyl groups form an important group of ligands. As a consequence, the introduction of the ferrocene group into a stereogenic phosphorus atom is an important goal, which has been achieved in entry 55 with relatively low stereoselectivity, although both epimers of **8** could be separated by column chromatography.[29] In a similar study, Stephan and co-workers[9] prepared the analogous ruthenocene compound (entry 56). *p*-Substituted phenyl oxazaphospholidines give unexpected lower stereoselectivities in the reaction with 1-naphthyllithium compared to unsubstituted starting materials (compare entries 57 and 58 with entries 40 and 43), although separation of isomers could be achieved by column chromatography.[30] The extremely bulky 2-adamantyl group severely restricts the diastereoselectivity of the reaction (entry 61).[31]

Table 4.1 Aminophosphines **8** prepared from **7** and organolithium reagents.

Entry	R^1	R^2	Yield (%)	de (%)	References
1	Ph	Me	80–97	o.p.	6, 10–12
2[a]	Ph	Me	79–95	80–o.p.	13–15
3	Ph	*i*-Pr	59	o.p.	16
4	Ph	*n*-Bu	97	o.p.	6, 12
5[a]	Ph	*n*-Bu	80	88	15
6[a,b]	Ph	*s*-Bu	97	o.p.	9
7	Ph	*t*-Bu	9–93	o.p.	8, 11, 17
8[a]	Ph	*t*-Bu	82	o.p.	14
9[c]	Ph	*t*-Bu	94	o.p.	8
10[a]	Ph	Cy	87–88	o.p.	14, 15
11[d]	Ph	Allyl	77	60	18, 19
12	Ph	Bn	50	o.p.	16
13[a]	Ph	Neophyl[e]	72	o.p.	15
14	Ph	Vinyl	96–quant.	o.p.	19, 20
15[f]	Ph	Ph	—	o.p.	11
16	Ph	*o*-An	93	o.p.	6, 8, 10, 11, 21, 22
17[a]	Ph	*o*-An	95	o.p.	14
18[c]	Ph	*o*-An	93–97	o.p.	8
19[a]	Ph	*m*-An	99	o.p.	23
20[a]	Ph	*p*-An	74–80	o.p.	15, 23
21[a]	Ph	2,3-(OMe)$_2$C$_6$H$_3$	87	o.p.	24
22[a]	Ph	2,4-(OMe)$_2$C$_6$H$_3$	92	o.p.	24
23[a]	Ph	2,5-(OMe)$_2$C$_6$H$_3$	81	o.p.	24
24[a]	Ph	2,6-(OMe)$_2$C$_6$H$_3$	35	o.p.	9

Table 4.1 (*continued*).

Entry	R^1	R^2	Yield (%)	de (%)	References
25^a	Ph	3,4-$(OMe)_2C_6H_3$	87	o.p.	23
26^a	Ph	3,5-$(OMe)_2C_6H_3$	91	o.p.	23
27^a	Ph	2,4,6-$(OMe)_3C_6H_2$	24	o.p.	9
28^a	Ph	2,3,4-$(OMe)_3C_6H_2$	70	o.p.	24
29^a	Ph	2,3,4,5-$(OMe)_4C_6H$	76	o.p.	24
30^a	Ph	o-Tol	80–92	o.p.	14, 15
31^a	Ph	2-(t-BuO)C_6H_4	99	o.p.	25
32^a	Ph	2,6-$(Me)_2C_6H_3$	46	o.p.	9
33^a	Ph	Mesityl	57	o.p.	9
34	Ph	o-(methoxyethyl)C_6H_4	57	o.p.	26
35^a	Ph	o-MEMC$_6$H$_4{}^g$	—	o.p.	27
36^a	Ph	3-Ph-2-An	98	o.p.	24
37^a	Ph	3-(i-PrO)-2-An	97	o.p.	23
38^a	Ph	3-TMS-2-An	99	o.p.	24
39^a	Ph	3,5-(t-Bu)$_2$-2-An	96	o.p.	24
40	Ph	2-biphenylyl	85	o.p.	22
41^a	Ph	2-biphenylyl	81	o.p.	14, 15
42	Ph	2-p-terphenylyl	82	o.p.	16
43	Ph	1-naphthyl	92–99 (crude) 61–82 (purified)	o.p.	11, 12, 22, 28
44^a	Ph	1-naphthyl	87	o.p.	14

45[a]	Ph	2-methoxy-1-naphthyl	38	o.p.	9
46[a]	Ph	8-methoxy-1-naphthyl	59	o.p.	9
47	Ph	2-naphthyl	89	o.p.	6, 11, 22
48[a]	Ph	2-naphthyl	89	o.p.	14
49[a]	Ph	1-methoxy-2-naphthyl	92	o.p.	24
50[a]	Ph	3-methoxy-2-naphthyl	98	o.p.	24
51[a]	Ph	9-phenanthryl	83	o.p.	16
52[a]	Ph	9-phenanthryl	86	o.p.	22
53[a]	Ph	9-anthryl	4	o.p.	9
54	Ph	9-pyrenyl	85	o.p.	16
55	Ph	Ferrocenyl	81	82–96 (crude) o.p. (purified)	9, 29
56	Ph	'Ruthenocyl'	27	92	9
57	p-An	1-naphthyl	81	90 (crude) o.p. (purified)	30
58	p-CF$_3$C$_6$H$_4$	1-naphthyl	72	90 (crude) o.p. (purified)	30
59	3,5-Me$_2$C$_6$H$_3$	2-biphenylyl	85	o.p.	16
60	Mesityl	Ph	9	o.p.	17
61	2-Ad	Ph	–	50	31

[a]Prepared from (+)-ephedrine.
[b]No details about any possible stereoselection at the stereogenic carbon atom of the *s*-Bu group were reported.
[c]Prepared from (+)-(1*S*,2*S*)-pseudoephedrine.
[d]Complex 7 prepared from (±)-ephedrine.
[e]Neophyl = 2-methyl-2-phenyl-1-propyl.
[f]The phosphorus atom is not stereogenic.
[g]MEM = 2-[(2-methoxy)ethoxy]methoxy.

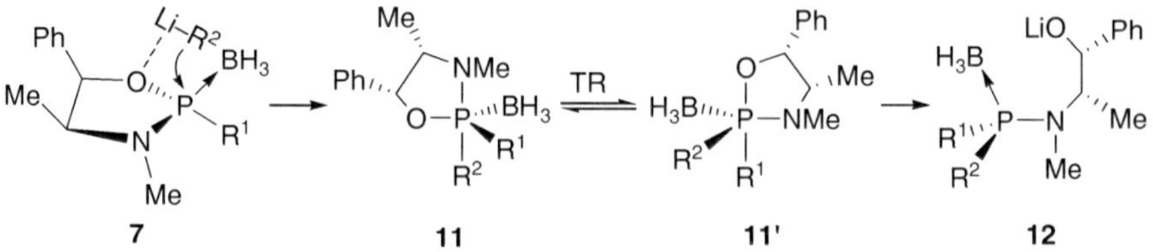

Scheme 4.4 depicts the proposed mechanism with structures **7**, **11**, **11'**, and **12**.

Scheme 4.4 Proposed mechanism to explain the stereochemistry of the formation of **8**.

o-Substituted aryl groups can be easily introduced, as many of the entries in Table 4.1 demonstrate. In contrast, the introduction of much bulkier *o,o*-disubstituted aryl groups is more troublesome, although phosphines bearing highly symmetric, bulky substituents at the stereogenic phosphorus atom are an interesting synthetic target.[17] Towards this end, Mezzetti and co-workers[17] reacted **7** ($R^1 = Ph$) with 2,6-disubstituted aryllithiums (mesityllithium, 2,4,6-trimethoxyphenllithium and 9-anthryllithium) but no reaction was observed, which was attributed to steric factors. They also prepared **7** with $R^1 = Mes$ and treated it with methyl- and phenyllithium, but a reaction was only observed in the second case with very low yield (entry 60). More recently, Stephan and co-workers[9] re-examined the reaction of **7** and 2,6-disubstituted aryllithiums in more detail (Scheme 4.5).

They found that the expected products **13** are indeed formed, but are accompanied by an unexpected compound (**14**). The yield of **13** was reasonably good for 2,6-dimethylphenyl- and mesityllithium, whereas it was extremely low for 9-anthryllithium and the limit was exceeded for supermesityllithium since only 'supermesitylene' was recovered after hydrolysis. Compound **14** (characterised by X-ray diffraction) was obtained exclusively as the isomer depicted in Scheme 4.5. Its formation can be explained by deprotonation of the benzylic proton of the ephedrine moiety by the aryllithium and subsequent elimination. This reaction is usually slow compared to the nucleophilic attack of the aryl group to the P atom, but when this attack is sterically hindered, the deprotonation–elimination reaction becomes competitive or even prevalent. With 1-lithio-8-methoxynaphthalene, a mixture of products was also obtained (Scheme 4.6).

^{1}H NMR of the crude showed the presence of a 90:10 mixture of the expected product **15** and its isomer **15'**, which could be separated by column chromatography to finally afford **15** in a 65% yield.

4.3.1.2 Ring Opening with Bis(organolithiums)

Despite the good results of the ring opening reaction with mono(organolithiums), much less effort has been devoted to the use of *bis*(organolithiums) notwithstanding the potential of this reaction for the preparation of diphosphine ligands. An exception is the condensation of **7** with 1,1'-dilithioferrocene, which has been studied in some detail (Scheme 4.7).

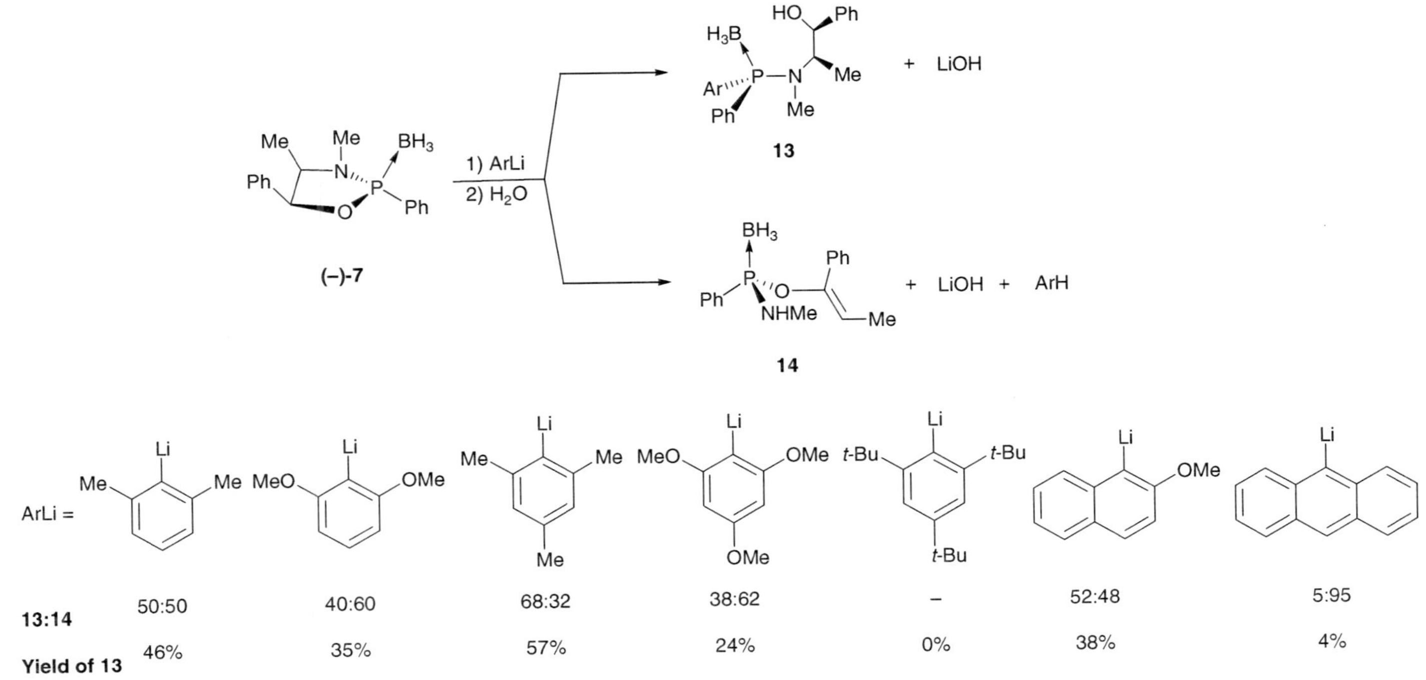

Scheme 4.5 Reaction of (−)-**7** with bulky aryllithiums.

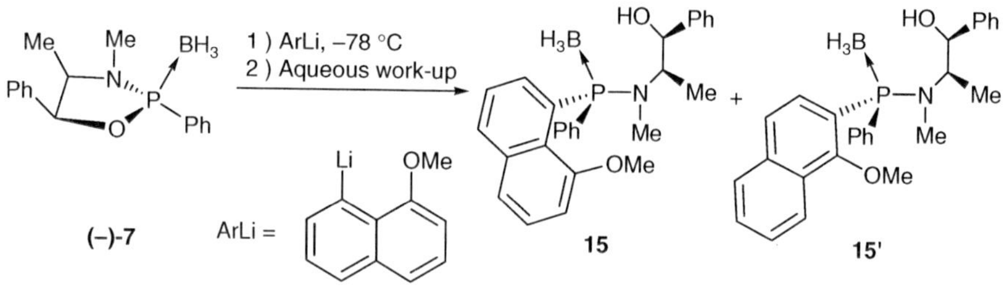

Scheme 4.6   Reaction of (−)-**7** with 1-lithio-8-methoxynapthalene.

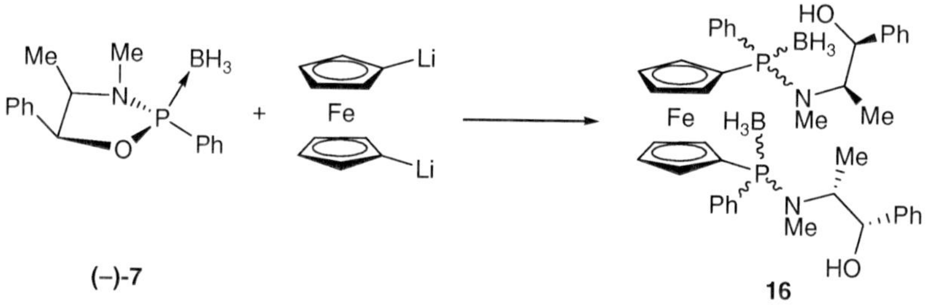

Scheme 4.7   Reaction of 1,1′-dilithioferrocene with (−)-**7**.

Jugé and co-workers[13] investigated the reaction of 2 equivalents of (−)-**7** with 1,1′-dilithioferrocene in THF, at 45 °C. Under these conditions, the expected product **16** was obtained only in 10% yield as a mixture of diastereomers. The same reaction was also attempted by van Leeuwen, Widhalm and co-workers,[22] who reported the synthesis of **16** in only 30% de as a mixture of the (R_P,R_P) and (R_P,S_P) diastereomers. More recently, Stephan and co-workers[9] re-examined the same reaction in THF at −20 °C and managed to obtain crude **16** (impurified with 5% of monosubstituted analogue) in 64% de. The separation of diastereomers by column chromatography was successful and the individual diastereomers (R_P,R_P) and (R_P,S_P) could be obtained in 64% and 10% yields respectively. Interestingly, they carried out the same reaction with 1,1′-dilithioruthenocene obtaining a distribution of 19:72:9 of monosubstituted:disubstituted:trisubstituted products. After chromatographic purification, the major disubstituted (R_P,R_P)-diastereomer was obtained in a 34% yield.

The low diastereoselectivity of these reactions has been attributed to the second nucleophilic attack due to increased steric shielding since it is known that the first attack occurs with a de higher than 96%.[9]

2,2′-dilithio-1,1′-biaryls (biaryl = biphenylyl or binaphthyl) have also been reacted with **7** with the aim to obtain, for example, *P*-stereogenic analogues of BINAP (Scheme 4.8).[9]

It was found that the reactions afford complex mixtures, which contain phospholanes **18** along with the non *P*-stereogenic diaminophosphine **19**. These

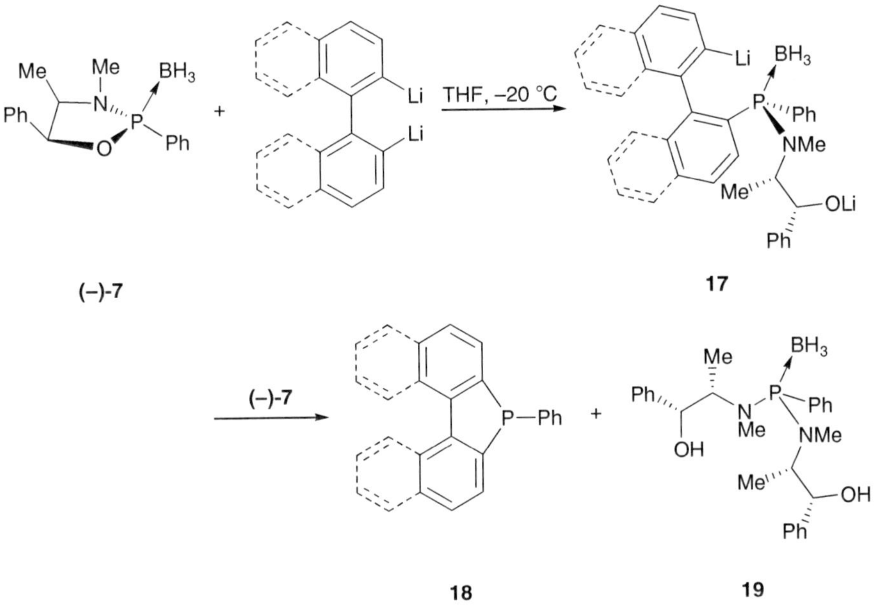

Scheme 4.8   Reaction of 2,2′-dilithioarenes with (−)-7.

compounds can be produced from intermediate **17** by ring closure to form **18** with extrusion of *N,O*-dilithiated ephedrine, which reacts with another equivalent of **7** to form compound **19** after hydrolysis (its dimethyl ether derivative was characterised by X-ray diffraction).[9] The ring-closing reaction is thought to be strongly favoured due to the proximity of the second carbanion to the phosphorus atom, in contrast to the slow attack of **17** on another molecule of **7**.[9] Interestingly, the borane group spontaneously decoordinates from the phosphorus atom during reaction or work-up, although traces of the borane adducts of **18** could be detected by ^{31}P NMR spectroscopy.

4.3.2   P–N Bond Cleavage by Acidolysis

4.3.2.1   *Formation of Phosphinite Boranes*

The cleavage of the P–N bond of **8** in the original Jugé–Stephan method is performed by acidic methanolysis (Scheme 4.9).

Reaction conditions are very straightforward, consisting of stirring the aminophosphines **8** in freshly distilled methanol in the presence of 1 equivalent of sulfuric acid for several hours at 0 °C or room temperature. Methylphosphinite boranes **9** are either clear liquids or crystalline solids, which can be easily purified by column chromatography or recrystallisation and, in contrast to **8**, are indefinitely stable in air. Another advantage is that the reaction yields one equivalent of unchanged ephedrine, which can be recycled. The reaction is

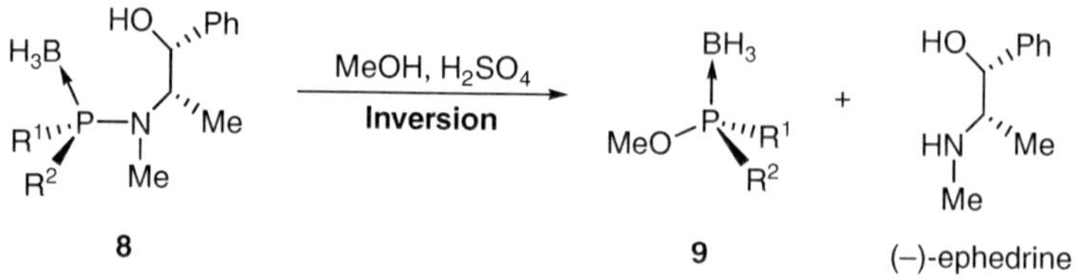

Scheme 4.9 Preparation of phosphinite boranes **9** by acidic methanolysis of **8**.

considered to be an S_N2 substitution and therefore occurs with inversion of configuration at the phosphorus atom. Most of the methylphosphinite boranes reported to date are listed in Table 4.2.

Table 4.2 shows that very often the reactions proceed in high yields and very high enantioselectivities. Nonetheless, the reader must be warned that the ee values usually refer to recrystallised compounds and therefore the optical purity of the crude products could be lower. This detail, however, is rarely discussed in the literature, although an exception can be found in entry 33. The methanolysis of aminophosphine boranes **8** with two aromatic groups (entries 12–46) does not seem to be particularly sensitive to steric hindering since it is not prevented by *o,o*-disubstituted groups (entries 21 and 46).[17,24] On the other hand, alkylphenylaminophosphine boranes (entries 1–11) show a decrease in reactivity upon increasing the bulkiness and/or basicity of the alkyl group. That is clearly illustrated by the decreasing yield in the series $P(BH_3)(R)Ph(N$-ephedrine) for R = Me, *n*-Bu, *i*-Pr and Cy (entries 1, 5, 3 and 6 respectively). For R = Cy (entry 6) even after heating to 45 °C for 18 h only low yields of product are obtained.[15] Indeed, the reaction becomes impossible for the very basic and bulky *tert*-butyl[8,17] or 2-adamantyl[31] groups, even at reflux in methanol for prolonged periods of time.[8]

The reaction is also troublesome when the precursor contains a ferrocenyl group (entries 41 and 42). Here the reaction is sluggish and low yielding, with around 35% of starting material recovered.[15] In contrast, the methanolysis of a diastereomeric mixture of the *bis*(aminophosphine) boranes derived from 1,1′-dilithioferrocene affords **20** in 90% yield but with only 20% of ee (Scheme 4.10).[13]

A virtually unexplored extension is the use of alcohols other than methanol. There is only one report by Beak and co-workers,[26] where they employ a iodinated primary alcohol (Scheme 4.11).

The reaction of aminophosphine borane **21** with *o*-iodophenethyl alcohol yields the phosphinite borane **22**, with the expected inversion of configuration, although only in 24% yield.

Although methylphosphinite boranes are usually used as electrophilic precursors to phosphine boranes (see Section 4.3.3.1), they can also be deprotected by amines, affording the corresponding optically pure methylphosphinites. The deprotection reaction is very enantioselective, with retention of configuration at the P centre. Only a few free phosphinites have been prepared with the Jugé–Stephan method; these are listed in Figure 4.1.

Table 4.2 Methylphosphinite boranes **9** from aminophosphine boranes **8**.

Entry	R^1	R^2	Yield (%)	ee (%)	References
1	Ph	Me	80	o.p.	6, 10
2[a]	Ph	Me	75	o.p.	15
3	Ph	i-Pr	56	–	16
4	Ph	n-Bu	–	o.p.	6
5[a]	Ph	n-Bu	62	o.p.	15
6[a]	Ph	Cy	29	o.p.	15
7	Ph	$-CH_2CH_2PPh_2(BH_3)$	80	o.p.	20
8[b]	Ph	Allyl	79	0	19
9	Ph	Bn	74	–	16
10[a]	Ph	Neophyl[c]	65	o.p.	15
11[b]	Ph	Vinyl	87	0	19
12	Ph	o-An	80–97	o.p.	6, 8, 10, 21
13[a]	Ph	o-An	77–85	99.3–o.p.	15, 22
14[d]	Ph	o-An	97	o.p.	8
15[a]	Ph	m-An	96	o.p.	23
16[a]	Ph	p-An	77–86	o.p.	15, 23
17[a]	Ph	2,3-$(OMe)_2C_6H_3$	95	o.p.	24
18[a]	Ph	3,4-$(OMe)_2C_6H_3$	90	o.p.	23
19[a]	Ph	2,4-$(OMe)_2C_6H_3$	98	o.p.	24
20[a]	Ph	2,5-$(OMe)_2C_6H_3$	96	o.p.	24
21[a]	Ph	2,6-$(OMe)_2C_6H_3$	90	o.p.	24
22[a]	Ph	3,5-$(OMe)_2C_6H_3$	94	o.p.	23
23[a]	Ph	2,3,4-$(OMe)_3C_6H_2$	96	o.p.	24
24[a]	Ph	2,3,4,5-$(OMe)_4C_6H$	79	o.p.	24
25[a]	Ph	o-Tol	87	o.p.	15
26[a]	Ph	o-$(t$-BuO)C_6H_4	93	o.p.	25
27	Ph	o-(methoxyethyl)C_6H_4	54	o.p.	26
28[a]	Ph	3-Ph-2-An	96	o.p.	24
29[a]	Ph	3-(i-PrO)-2-An	89	o.p.	23
30[a]	Ph	3,5-$(t$-Bu)$_2$-2-An	85	o.p.	24
31	Ph	2-biphenylyl	82	o.p.	16
32[a]	Ph	2-biphenylyl	77–87	98.1	15, 22
33	Ph	1-naphthyl	75 (crude) 60 (recr.)	92 (crude) o.p. (recr.)	28
34[a]	Ph	1-naphthyl	84	o.p.	22
35	Ph	2-naphthyl	94	o.p.	6, 22
36[a]	Ph	1-methoxy-2-naphthyl	92	o.p.	24
37[a]	Ph	3-methoxy-2-naphthyl	92	o.p.	24
38	Ph	9-phenanthryl	42	o.p.	16
39[a]	Ph	9-phenanthryl	64	o.p.	22
40	Ph	9-pyrenyl	42	o.p.	16
41	Ph	Ferrocenyl	31	–	29
42[a]	Ph	Ferrocenyl	30	–	15
43	p-An	1-naphthyl	84	o.p.	30
44	p-$CF_3C_6H_4$	1-naphthyl	75	o.p.	30
45	3,5-$Me_2C_6H_3$	2-biphenylyl	76	–	16
46	Mesityl	Ph	55	–	17

[a]Precursor **8** prepared from (+)-ephedrine.
[b]Precursor **8** prepared from (±)-ephedrine so **9** was obtained racemic.
[c]Neophyl = 2-methyl-2-phenyl-1-propyl.
[d]Precursor **8** was prepared from (+)-(1S,2S)-pseudoephedrine.

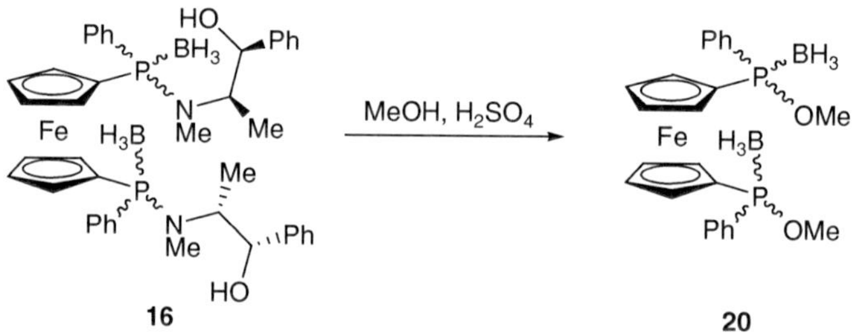

Scheme 4.10   Acidic methanolysis of *bis*(aminophosphine) **16**.

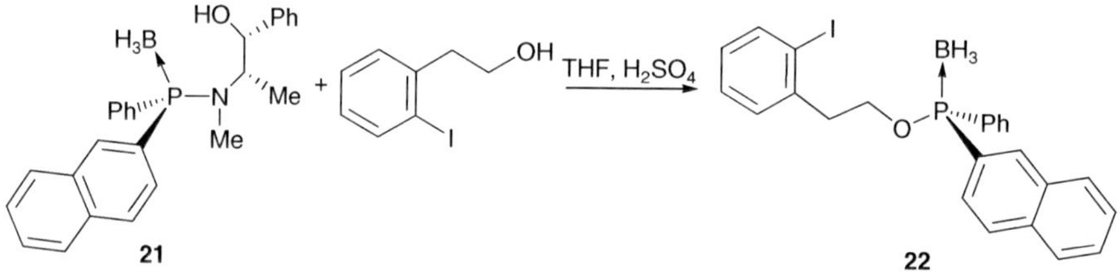

Scheme 4.11   Acidolysis of **21** with *o*-iodophenethyl alcohol.

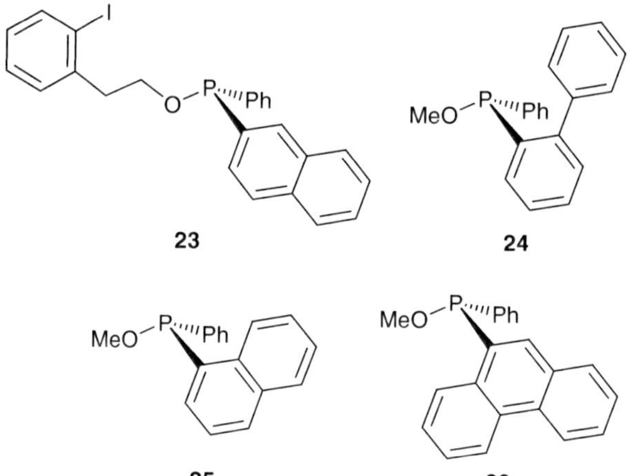

Figure 4.1   Enantiomerically pure free phosphinites prepared from the parent boranes.

Deboronation of **22** in refluxing diethylamine affords **23** in quantitative yield[26] whereas methylphosphinites **24–26** have been reported by Muller and co-workers[16,32] after borane decomplexation with morpholine at room temperature. The absolute configuration of **24** was confirmed by the X-ray structure of a derived palladium complex.[32]

4.3.2.2 Formation of Chlorophosphine Boranes

Methylphosphinite boranes react smoothly with organolithium reagents to afford the corresponding tertiary phosphine boranes, as will be discussed in the next section. However, phosphinite boranes are not electrophilic enough to react with other weaker nucleophiles such as alcohols, amines or thiols. More reactive precursors, capable of producing a wide variety of phosphorus compounds, were needed. In phosphorus chemistry halophosphines, and chlorophosphines **27** in particular, are essential synthons (Scheme 4.12) as nucleophilic (after transformation into metal phosphides **28**) and electrophilic building blocks.

The widespread use of achiral or racemic chlorophosphines **27** in synthetic phosphorus chemistry (PClPh$_2$ in particular is an ubiquitous diphenylphosphinating agent) contrasts with their rare application in *P*-stereogenic chemistry. The reason is the lack of methods to produce enantioenriched *P*-stereogenic chlorophosphines and their marked tendency to racemise. Indeed, the first example of enantiomerically enriched chlorophosphine was reported by Omelanczuk[33] in 1992, who prepared (*S*)-P(*t*-Bu)PhCl in 49% ee by decomposition of an optically enriched thiophosphonium salt. The same compound has been prepared more recently, albeit in low ee, by kinetic resolution.[34]

The configurational instability of chlorophosphines is not due to low pyramidal inversion barriers (estimated to be around $168\,\text{kJ}\,\text{mol}^{-1}$), which are $20\,\text{kJ}\,\text{mol}^{-1}$ higher than tertiary phosphines (see Chapter 1, Section 1.2).[35] This issue has been studied by Humbel, Jugé and co-workers[36] using a mixture of computational (DFT) and experimental methods. They envisaged several possible mechanisms of racemisation and concluded that the most reasonable one involves HCl-catalysed racemisation of the phosphorus atom by formation of an achiral pentacoordinated intermediate **29** (Scheme 4.13).

The formation of **29** can be explained by a double nucleophilic attack: the phosphorus atom to the H$^+$ and the chloride anion to the phosphorus centre. Other isomers of **29** are very disfavoured due to the high apicophilicity of the chloro substituent. The easy substitution of Cl by Br when reacting **27** with HBr

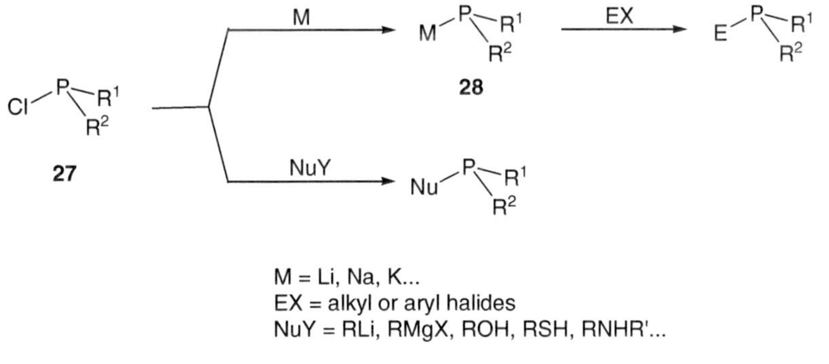

M = Li, Na, K...
EX = alkyl or aryl halides
NuY = RLi, RMgX, ROH, RSH, RNHR'...

Scheme 4.12 Basic reactivity of chlorophosphines.

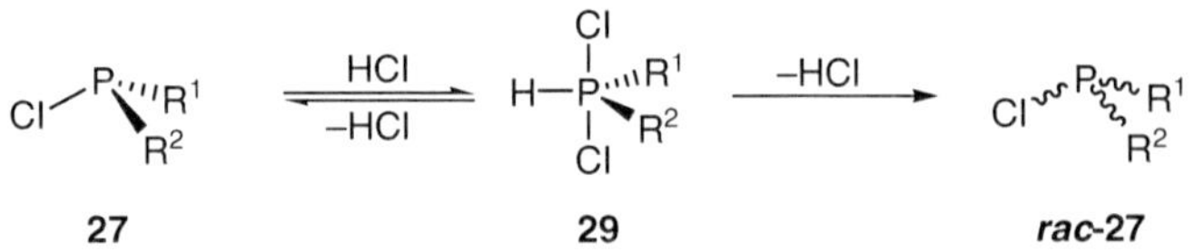

Scheme 4.13 Mechanism for racemisation of optically active **27**.

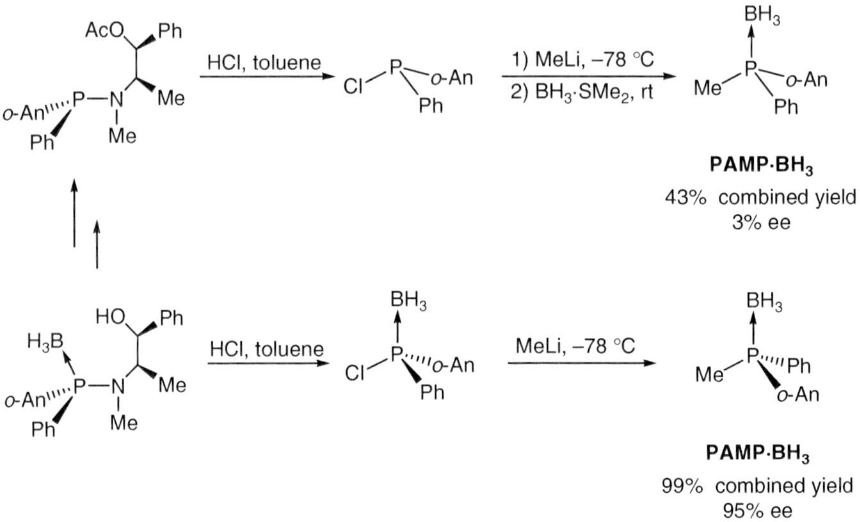

Scheme 4.14 Preparation of PAMP·BH$_3$ *via* a chlorophosphine and its borane derivative.

supports this mechanism.[36] From the present mechanism two important conclusions can be drawn. Firstly, pure chlorophosphines should be configurationally stable, but traces of hydrogen (or metal) halides, often present under experimental conditions, preclude their use as valuable synthons in *P*-stereogenic chemistry. Secondly, as the racemisation mechanism involves the electron pair at the phosphorus atom it should not be operative in P(V) compounds or, more interestingly, in chlorophosphine boranes. The latter compounds have been prepared in optically enriched form, from the aminophospine boranes **7**, and used to prepare several organophosphorus compounds.

The difference in the optical purities with or without the presence of the borane protecting group is properly illustrated in the preparation of PAMP·BH$_3$ (Scheme 4.14).[36]

The first report of an optically pure chlorophosphine borane appeared in 1997 (Scheme 4.15).[13]

Treatment of **8** (derived from (+)-ephedrine) with two equivalents of a toluenic solution of HCl (previously titrated) produced **30** in 90% yield, with inversion of configuration at the P atom. Purification by column chromatography, although

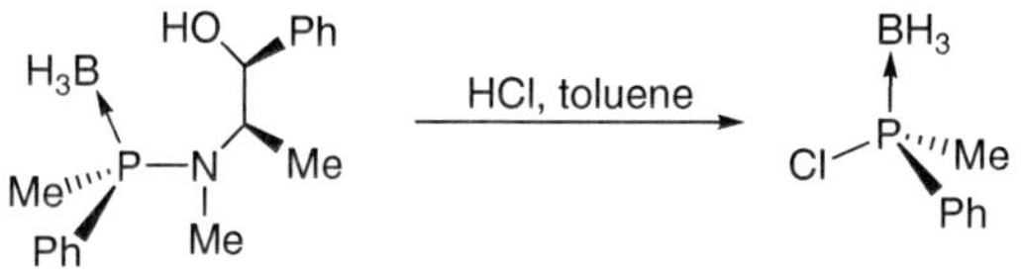

Scheme 4.15   The first report of an optically pure chlorophosphine borane.

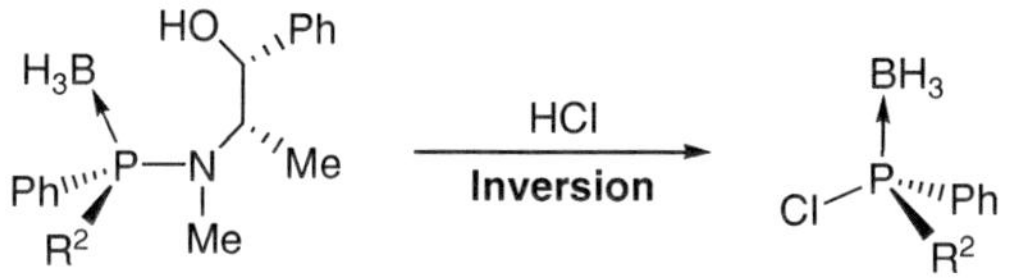

Scheme 4.16   Preparation of chlorophosphine boranes.

possible, resulted in poor yield, even using oven-dried silica, highlighting the sensitivity chlorophosphine boranes. In spite of that, the ephedrine was filtrated and the crude solution of **30** could be used immediately for further transformations.

Since this report, a small number of chlorophosphine boranes have been prepared. They are very moisture-sensitive and still prone to racemisation in spite of the borane protection. Their optical purities can only be assessed after transformation into other less reactive compounds, such as tertiary phosphine boranes, assuming that the substitution reaction is completely stereoselective. Table 4.3 lists the compounds reported so far, along with the best conditions for their preparation. As depicted in Scheme 4.16, in all the reported cases the phosphorus atom bears a phenyl group.

The acidolysis reactions are usually performed at room temperature, adding a toluenic HCl solution to **8** (which is also dissolved in toluene). Toluenic HCl solutions are not commercially available and must be prepared by bubbling dry HCl gas through toluene and calculating the concentration of HCl by titration. To avoid this inconvenience, in some occasions[17,18] commercial solutions of HCl in diethyl ether have been used. It has to be mentioned that chlorophosphine boranes are much more sensitive and difficult to handle than the phosphinite boranes of the last section. In Table 4.3 the best enantioselectivities are reported, but they are only attained with careful control of the HCl equivalents with respect to **8**, the concentrations and the reaction times. In general, higher loadings of HCl and/or longer reaction times assure higher yields of **31** but at the expense of the enantioselectivity.[14] Also purification of **31** involves loss of optical purity, sometimes severe (entries 1, 8 and 9) and therefore chlorophosphine boranes are usually employed *in situ*, after filtration of the ephedrine salt. In parallel with the methanolysis, the acidolysis reaction is

Table 4.3 Reported chlorophosphine boranes.

Entry	R^2	Yield (%)	Conditions	ee (%)[a]	References
1[b]	Me	80–85	Toluene, 2.1 eq. HCl	90 (crude) 63 (purified)	13, 14, 37
2	Me	99	Et$_2$O, 2.1 eq. HCl	99 (crude)	17
3[b]	Cy	74	Toluene, 6.0 eq. HCl	80 (crude)	14
4[c]	Allyl	–	Et$_2$O	–	18
5[b]	o-An	99	Toluene, 2.1–5.8 eq. HCl	98 (crude) 95 (purified)	11, 14, 36, 37
6[b]	o-Tol	87	Toluene, 2.1–6.0 eq. HCl	98 (crude) 95 (purified)	14, 37
7[b]	1-naphthyl	68	Toluene, 6.0 eq. HCl	0	14
8[b]	2-naphthyl	61	Toluene, 3.0 eq. HCl	85 (crude) 68 (purified)	14
9[b]	2-biphenylyl	45	Toluene, 6.0 eq. HCl	99 (crude) 59 (purified)	14
10[b]	Ferrocenyl	76	Toluene, 6.0 eq. HCl	95 (crude)	2

[a]Values determined by HPLC after reaction with MeLi, assuming that this reaction is totally stereoselective.
[b]Prepared from (+)-ephedrine.
[c]Precursor **8** prepared from (±)-ephedrine so **31** was obtained racemic.

sensitive to the steric bulk of the substituents at the phosphorus atom (compare entry 1 with 3 and entries 5 and 6 with 9). In the case of the *tert*-butyl substituted **8**, no reaction is observed, analogously to the methanolysis reaction. The methanolysis of the 1-naphthyl derivative of **8** produces **31** as a racemate (entry 7). This is unexpected since the related compound with 2-naphthyl reacts with good enantioselectivity (entry 8). It has been suggested that this racemisation is due to a particularly low energy barrier for the stereopermutation of the pentacoordinated species formed from the attack of HCl on **8** when $R^2 =$ 1-naphthyl.[14]

4.3.3   Introduction of R^3 by Nucleophilic Substitution

4.3.3.1   *Nucleophilic Substitution on Phosphinite Boranes*

The reaction of methylphosphinite boranes with organolithiums at low temperature affords the corresponding *P*-stereogenic tertiary phosphine boranes by substitution of the methoxy group and inversion of configuration at the phosphorus atom (Scheme 4.17).

These reactions proved to be quite general, providing a wide range of phosphine boranes. Most of them are listed in Table 4.4.

Table 4.4 gives a clear overview of the type of phosphine boranes that can be prepared with the Jugé–Stephan method. The method seems particularly well suited for the synthesis of phosphines containing at least one aryl group in extremely high enantioselectivities. These include a few dialkylarylphosphine boranes (entries 2–4 and 7), triarylphosphine boranes (entries 17, 18 and 35)

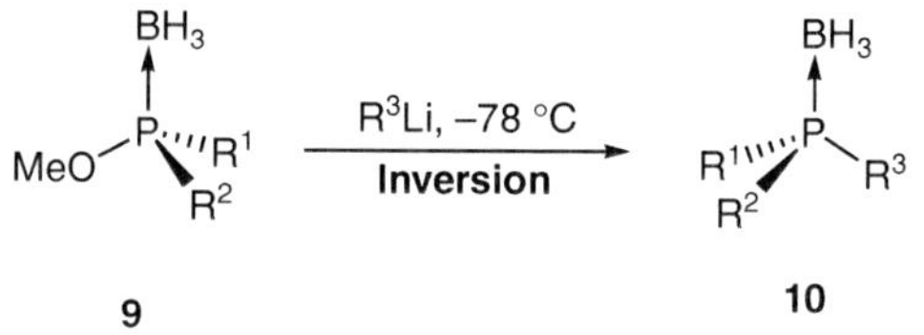

Scheme 4.17 Preparation of phosphine boranes.

and a large number of alkyldiarylphosphine boranes. A prominent example of the latter class of compounds is PAMP·BH$_3$ (entries 1 and 12). Indeed the Jugé–Stephan method constitutes the best way to prepare this important compound in enantiomerically pure form. Almost exclusively (except in entries 50 and 51) one of the aryl groups is a phenyl although this is only due to the accessibility of PhP(NEt$_2$)$_2$, which is nowadays commercially available.

The optical purity of the compounds, determined by HPLC, is usually very high and can be brought to perfection with a simple recrystallisation (entries 3, 4 and 42). In a few cases, the expected absolute configuration of the phosphorus atom has been confirmed by X-ray analysis.[8,10,28,32]

One of the most interesting features of the Jugé–Stephan method is that it allows the synthesis of both enantiomers of phosphine boranes (Scheme 4.18).

This can be done using either (−)- or (+)-ephedrine in the synthesis of **7** or, more interestingly, reverting the order of addition of the organolithium reagents from one of the precursors. An example is found in Table 4.4 (entries 1 and 12) in the synthesis of both enantiomers of PAMP·BH$_3$ from (+)-**7** (derived from (−)-ephedrine). Although promising, this approach has rarely been used, *i.e.* in most cases only one of the enantiomers of the phosphine borane has been obtained.

Although the substitution of the methoxy group by organolithiums has a broad scope, some limitations have been found. Mezzetti and co-workers[17] suggested that *o,o*-substituted aryllithiums do not react with (*S*)-P(BH$_3$)Me(OMe)Ph. In contrast, the bulky *t*-BuLi reacts smoothly (Table 4.4, entries 3 and 4), probably due to its higher nucleophilicity.

Muller and co-workers[16,32] studied the reaction of increasingly hindered alkyllithiums with the bulky diarylmethylphosphinite boranes displayed in Figure 4.2.

Compounds **32–35** react smoothly with MeLi to furnish the expected aryl-methylphenylphosphine boranes.[16,28,32] In contrast, they are unreactive towards *t*-BuLi.[16] With *i*-PrLi, **32** and **33** give the expected substitution product[32] whereas no reaction is observed for **35**.[16] For the 9-phenanthryl substituted derivative **34** an unexpected product was formed (**37** in Scheme 4.19).[32]

Phosphine borane **37** has three stereogenic centres, but a thorough analysis, including the X-ray crystal structure,[32] indicated that it was present as a racemic mixture. Its formation was interpreted *via* the formation of an unde-tected intermediate **36** (Scheme 4.19) with a P–C double bond, which suffers a second attack of *i*-PrLi to form an anionic phosphine borane and the final

Table 4.4 Phosphine boranes **10** prepared from methylphosphinite boranes **9**.

Entry	R^1	R^2	R^3	Yield (%)	ee (%)	References
1	Ph	Me	o-An	–	98	6
2	Ph	Me	n-Bu	–	o.p.	6
3	Ph	Me	t-Bu	66 (crude)	92 (crude) o.p. (recr.)	17
4[a]	Ph	Me	t-Bu	66 (crude)	91 (crude) o.p. (recr.)	17
5	Ph	Me	2-naphthyl	–	o.p.	6
6[a]	Ph	Me	Ferrocenyl	58	83	15
7	Ph	n-Bu	Me	–	o.p.	6
8[a]	Ph	n-Bu	Ferrocenyl	90	77	15
9	Ph	$-CH_2CH_2PPh_2(BH_3)$	o-An	44	o.p.	20
10[a]	Ph	Neophyl[b]	Ferrocenyl	79	98	15
11[a]	Ph	Cy	Ferrocenyl	66	97	15
12	Ph	o-An	Me	–	o.p.	6
13	Ph	o-An	2-picolyl	70	o.p.	10, 21
14	Ph	o-An	2,6-lutidyl	55	o.p.	10
15[a]	Ph	o-An	t-Bu	51	≥97	31
16[a]	Ph	o-An	1-Ad	37	–	31
17	Ph	o-An	3,4-$(OMe)_2C_6H_3$	80	o.p.	8
18[c]	Ph	o-An	3,4-$(OMe)_2C_6H_3$	80	o.p.	8
19[a]	Ph	o-An	Ferrocenyl	67–92	≥92	15, 31
20[a]	Ph	m-An	Me	97	o.p.	23
21[a]	Ph	p-An	Me	93	o.p.	23
22[a]	Ph	p-An	Ferrocenyl	85	95	15
23[a]	Ph	2,3-$(OMe)_2C_6H_3$	Me	91	o.p.	24
24[a]	Ph	2,5-$(OMe)_2C_6H_3$	Me	99	o.p.	24
25[a]	Ph	2,6-$(OMe)_2C_6H_3$	Me	92	o.p.	24
26[a]	Ph	3,4-$(OMe)_2C_6H_3$	Me	92	o.p.	23

27[a]	Ph	3,5-(OMe)$_2$C$_6$H$_3$	Me	93	o.p.	23
28[a]	Ph	3-(i-PrO)-2-An	Me	91	o.p.	23
29[a]	Ph	3-TMS-2-An	Me	80	o.p.	24
30[a]	Ph	2,3,4-(OMe)$_3$C$_6$H$_2$	Me	99	o.p.	24
31[a]	Ph	2,3,4,5-(OMe)$_4$C$_6$H	Me	79	o.p.	24
32[a]	Ph	o-Tol	Me	70[d]	o.p.	38
33[a]	Ph	o-Tol	Ferrocenyl	70	83	15
34[a]	Ph	o-(t-BuO)C$_6$H$_4$	Me	92	o.p.	25
35	Ph	o-(MeOEt)C$_6$H$_4$	2-naphthyl	28	o.p.	26
36[a]	Ph	3-Ph-2-An	Me	90	o.p.	24
37[a]	Ph	3,5-(t-Bu)$_2$-2-An	Me	76	o.p.	24
38	Ph	2-biphenylyl	Me	55	o.p.	16
39	Ph	2-biphenylyl	i-Pr	31	o.p.	32
40[a]	Ph	2-biphenylyl	Ferrocenyl	69	98	15
41	Ph	2-biphenylyl	Ferrocenyl	66[e]	o.p.	39
42	Ph	1-naphthyl	Me	96 (crude) 81 (recr.)	93 (crude) o.p. (recr.)	28
43	Ph	1-naphthyl	i-Pr	61	o.p.	32
44	Ph	1-naphthyl	Ferrocenyl	66[e]	o.p.	39
45	Ph	2-naphthyl	Me	–	o.p.	6
46[a]	Ph	1-methoxy-2-naphthyl	Me	97	o.p.	24
47[a]	Ph	3-methoxy-2-naphthyl	Me	96	o.p.	24
48	Ph	9-phenanthryl	Me	44	o.p.	32
49	Ph	9-pyrenyl	Me	37	o.p.	16
50	p-An	1-naphthyl	(1′-Br)-1-Fc	41	o.p.	30
51	p-CF$_3$C$_6$H$_4$	1-naphthyl	Lithiated product of entry 44	70[e]	o.p.	30

[a]Precursor **9** prepared from (+)-ephedrine.
[b]Neophyl = 2-methyl-2-phenyl-1-propyl.
[c]Precursor **9** prepared from (+)-(1S,2S)-pseudoephedrine.
[d]Overall yield starting from (+)-**7**.
[e]Yields correspond to the free phosphine borane, after deprotection with HNEt$_2$ or CF$_3$SO$_3$H.

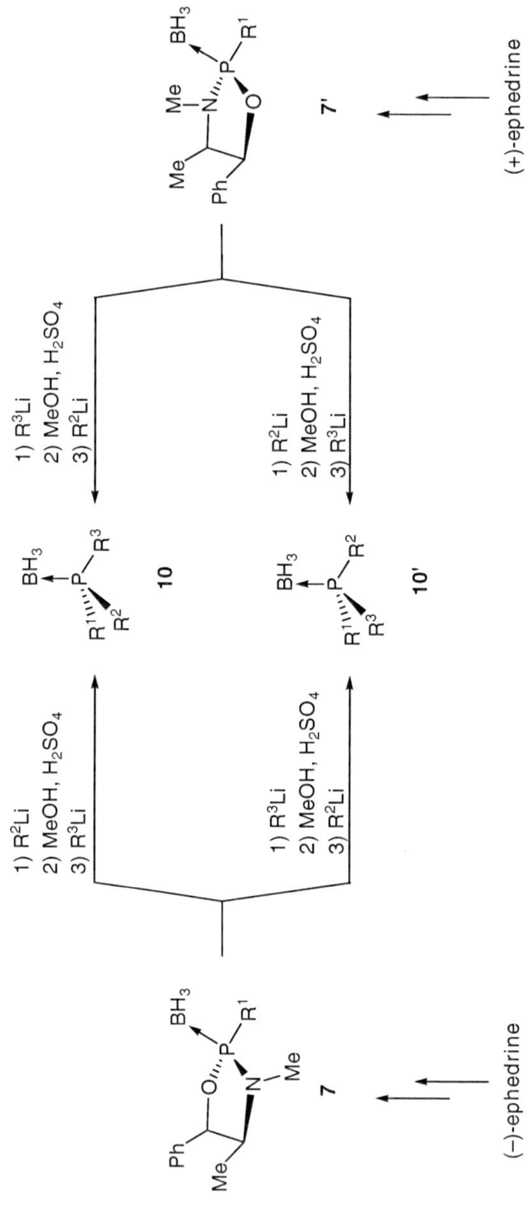

Scheme 4.18 Different ways to obtain both enantiomers of phosphine boranes *via* the Jugé–Stephan method. Usually R^1 = Ph.

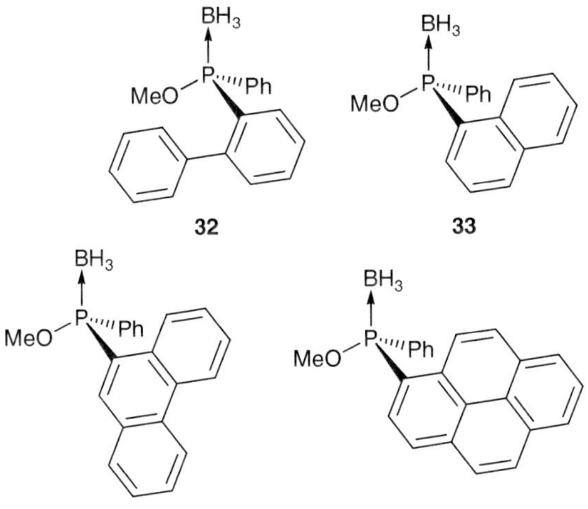

Figure 4.2 Bulky diarylmethylphosphinite boranes.

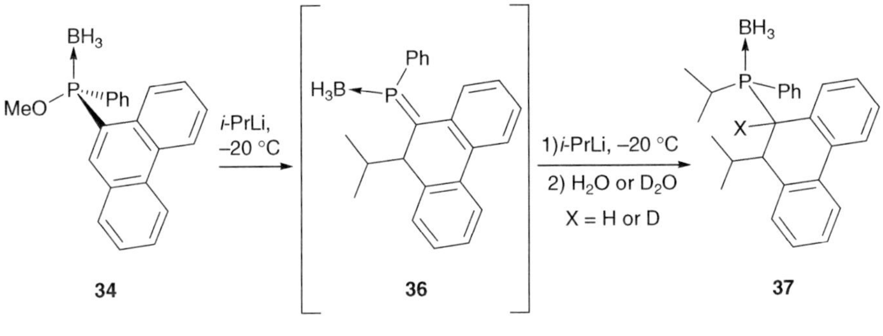

Scheme 4.19 Possible mechanism for the formation of **37**.

product **37** after washing with water. High diastereoselectivity (but no enantioselectivity) in those reactions explain the formation **37** exclusively as a racemate.

Introduction of ferrocene met with more success at this stage of the method, in contrast to the troublesome introduction in the nucleophilic ring opening of **7**. Brown and Laing[31] (Scheme 4.20 and Table 4.4, entry 19) reacted the *o*-anisylphosphinite borane **38** with monoferrocenyllithium and successfully obtained **39** in 67% yield and with an ee over 92%.

Monoferrocenyllithium was prepared by direct metallation of ferrocene with *t*-BuLi, a procedure that resulted in the formation of traces of 1,1′-dilithioferrocene, explaining the obtention of the interesting diphosphine borane **40** in 3% yield. This has been exploited to prepare bidentate ligands, as detailed below. Other groups[15,30,39] prepared other ferrocenyl monophosphines as can be

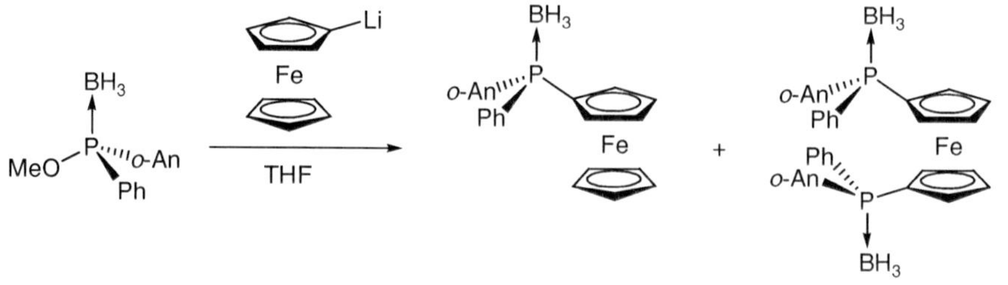

Scheme 4.20 First ferrocenylphosphines prepared by the Jugé–Stephan method.

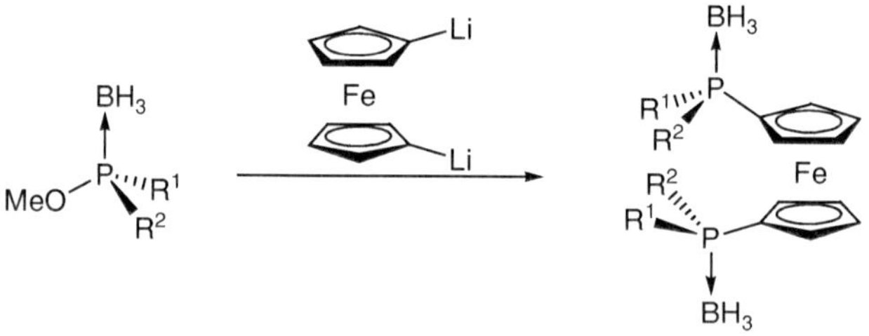

Scheme 4.21 Preparation of C_2-symmetric diphosphine boranes based on the ferrocenyl backbone.

seen in several entries of Table 4.4. It is worth mentioning that Colby and Jamison[15] used the halogen–metal exchange on commercially available bromoferrocene to generate monoferrocenyllithium free from 1,1′-dilithioferrocene.

As suggested by Brown and Laing,[31] the reaction of phosphinite boranes with 1,1′-dilithioferrocene produces C_2-symmetric diphosphine boranes **41** (Scheme 4.21).

Mezzetti's[40] and van Leeuwen's[22,29,30,41] groups followed this method to prepare a small family of diphosphine boranes. For $R^1 = Ph$, diphosphines with $R^2 = o$-An and $R^2 = 1$-naphthyl have been named BPAF and BPNF respectively.[41] Table 4.5 lists the diphosphine boranes of this type reported to date.

Table 4.5 shows that the Jugé–Stephan methodology is well suited for the preparation of C_2-symmetric ferrocenyldiphosphines, including the compound in entry 8,[29,42] which is remarkable because it contains three ferrocenyl groups connected through stereogenic phosphorus atoms. It also seems clear that deferring the introduction of the ferrocene fragment until the nucleophilic substitution on methylphosphinite boranes is the best option to prepare such diphosphines. However, some loss of optical purity is sometimes observed (entries 1, 6 and 8) and seems unavoidable. The same happens when the 1,1′-ferrocenyl unit is introduced earlier in the Jugé–Stephan method, although more severely (Section 4.3.1.2). Therefore, special care and monitoring of the

Table 4.5 Ferrocenyl-based diphosphine boranes **41** prepared from **9**.

Entry	R^1	R^2	Yield (%)	ee (%)	References
1	Ph	*o*-An	93 (crude)	56 (crude)	40
			72 (purif.)	o.p. (purif.)	
2[a,b]	Ph	*o*-An	73	o.p.	22, 31, 41
3	Ph	1-naphthyl	48	o.p.	40
4[a,b]	Ph	1-naphthyl	74	o.p.	22, 41
5[b]	Ph	2-naphthyl	78	o.p.	22
6[a,b]	Ph	2-biphenylyl	85 (crude)	92 (crude)	22
			81 (purif.)	o.p. (purif.)	
7[a]	Ph	9-phenanthryl	72	o.p.	22
8	Ph	Ferrocenyl	54 (purif.)	90 (crude)	29, 42
				o.p. (purif.)	
9[b]	*p*-An	1-naphthyl	51	o.p.	30
10[b]	$p\text{-}CF_3\text{-}C_6H_4$	1-naphthyl	36	o.p.	30

[a]Precursor **9** prepared from (+)-ephedrine.
[b]Data for the free diphosphine.

optical purity of each intermediate is advised when preparing *P*-stereogenic ferrocenylphosphines. This configurational instability probably stems from the particular steric and electronic features of ferrocene. Luckily, it has been found that most of the intermediates are quite air-stable and easy to purify by column chromatography and/or recrystallisation. That greatly facilitates obtaining analytically and optically pure diphosphine boranes and their derived free diphosphines.

The preparation of a very unusual diphosphine borane, reported by van Leeuwen and co-workers,[30] is depicted in Scheme 4.22.

Reaction of phosphinite borane **42** (Table 4.2, entry 43) with 1-bromo-1′-lithioferrocene afforded the monophosphine borane **43** (Table 4.4, entry 50), which still possesses a bromine atom susceptible to metallation on the non-phosphinated Cp ring. After lithiation, the ferrocenyl anion reacted with the second phosphinite borane **44** (Table 4.2, entry 44), affording the desired optically pure dissymmetric diphosphine borane **45** in 29% yield from **42**. During the synthesis impurities such as undesired monosubstituted ferrocene derivatives could be separated by column chromatography.

The first 1,1′-biferrocenyls containing *P*-stereogenic units were also reported by van Leeuwen and co-workers[39] using a diastereoselective *ortho*-metallation of ferrocene ring as the key step (Scheme 4.23).

Optically pure diarylferrocenylphosphine oxides **47** were prepared by reacting ferrocenyllithium with methylphosphinite boranes **46**, followed by deboronation and oxidation. It was found that *ortho*-lithiation of **47** was both ineffective and unselective (compare this with the report of Xiao and co-workers,[43] described in Chapter 3, Section 3.3.1.1). In contrast, *ortho*-magnesiation with diisopropylamidomagnesium bromide was more selective and afforded, after electrophilic quenching with iodine, the desired iodinated phosphine oxides **48** in good yields. The absolute configuration of the phosphorus atom (R_P) is not affected by this reaction, but a stereogenic plane is

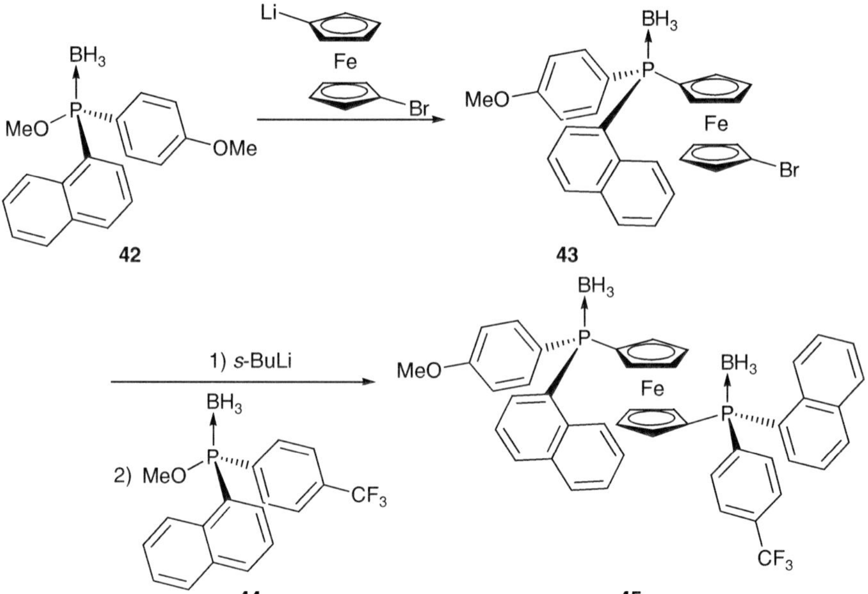

Scheme 4.22   Preparation of a dissymmetric ferrocenyldiphosphine borane.

created (whose absolute configuration is denoted by R_m or S_m) and hence two diastereomers can be formed. The *P*-stereogenic atom controlled the diastereoselectivity of the reaction, favouring the formation of the (R_P,S_m) isomer. The de values were 50% and 94% for Ar $=$ 1-naphthyl and 2-biphenylyl respectively; in both cases the isomers could be separated by column chromatography. The Ullmann coupling of the major diastereomers (although the same reaction also worked for the minor ones)[29] afforded the expected 1,1′-biferrocenyl-2,2′-diphosphine oxides **49** in good yields and without any loss of optical purity. The reduction of the oxides to the free phosphines with trichlorosilane required rather harsh conditions (130 °C, 72 h), which drastically reduced the optical purity of the products. After work-up, only 60–65% of retention of configuration in the phosphines was found, the rest being mono- and doubly epimerised products. Luckily, the reduction step left the stereointegrity of the stereogenic plane untouched and in consequence all the products were diastereomeric. This allowed the separation of the different isomers by column chromatography after their protection with borane. Eventually, the desired phosphines **50** were obtained enantiomerically pure after a deboronation step.

4.3.3.2   *Nucleophilic Substitution on Chlorophosphine Boranes*

Chlorophosphine boranes are more reactive than phosphinite boranes and therefore are easily attacked by carbanionic reagents such as organolithiums, organosodiums or Grignard reagents.[2,13,14,17,18,36] However, they can lead to

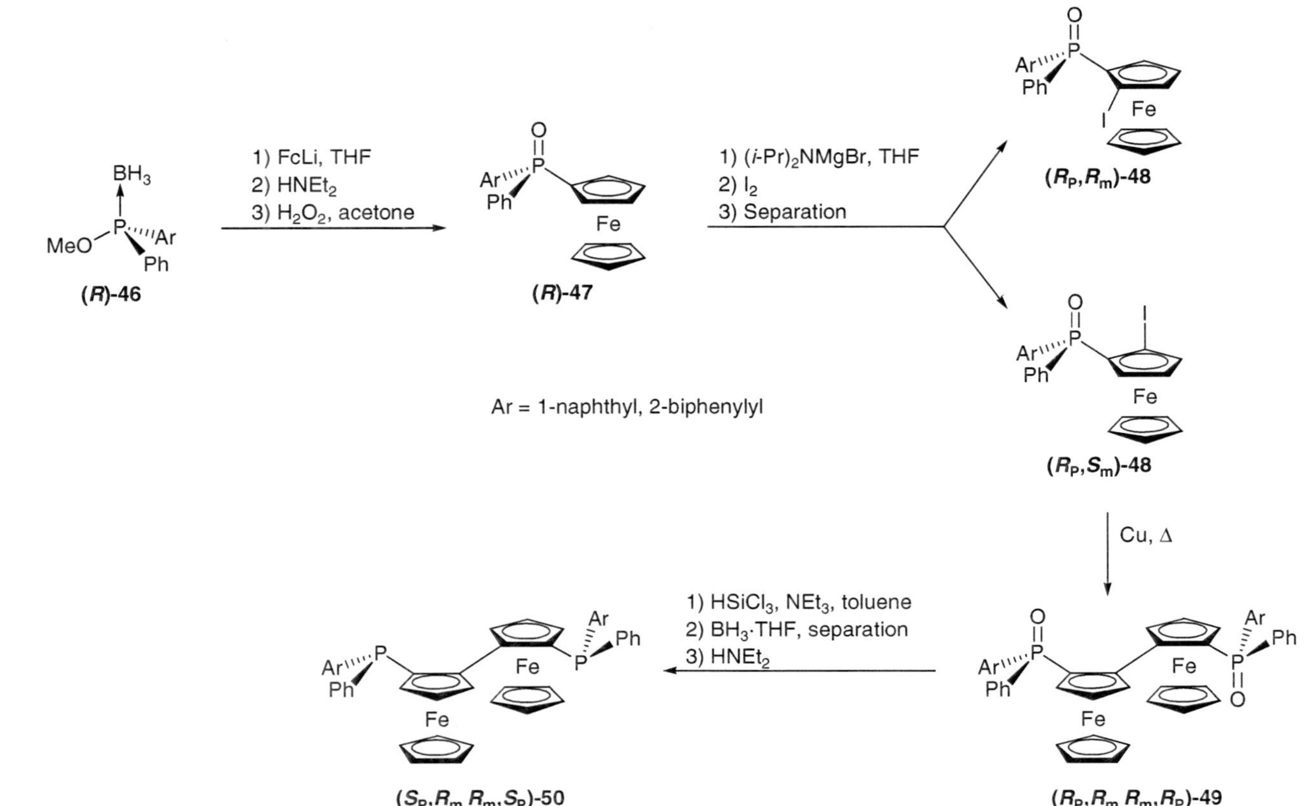

Scheme 4.23 Preparation of 2,2′-*bis*(diarylphosphino)-1,1′-biferrocenyls.

products with lower optical purity or even racemic[2,14] and therefore methylphosphinite boranes are preferred, unless the higher reactivity of chlorophosphine boranes is required.

Jugé and co-workers[13] were the first to use chlorophosphine boranes as valuable synthons in phosphine synthesis, as Scheme 4.24 depicts.

Chlorophosphine borane **30** was prepared and immediately reacted with cyclopentadienyl sodium at low temperature. After work-up and purification a mixture of isomeric phosphine boranes **51** (major component) and **51'** was isolated in 83% combined yield and 85% ee for **51**. A minor amount (10%) of the diphosphine borane **52** was also produced as a mixture of diastereomers. The deprotonation of the mixture **51/51'** forms the expected cyclopentadienylphosphine borane anion, which reacts with Fe(II) to form the desired diphosphine borane **53** as a single enantiomer. Here the chlorophosphine borane **30** is superior to the analogous methylphosphinite borane, since the latter does not react cleanly with NaCp.[13]

Another example was reported by Gouverneur and co-workers,[18] who used allylphenylchlorophosphine borane for the preparation of a few rather unusual *bis*(alkenyl)phosphine boranes (Scheme 4.25).

The desired products **55** were obtained in 29–68% yields after purification. Unfortunately, the precursor **54** was prepared racemic (on purpose) from (±)-ephedrine, so the enantioselectivity in of the reaction is unknown.

Mezzetti and co-workers,[17] in their exploration of phosphines bearing highly symmetric and bulky substituents, used (*S*)-chloromethylphenylphosphine borane (Scheme 4.26) to prepare a rare example of non-C_2 symmetric methylene-bridged *P*-stereogenic diphosphine borane.

Deprotonation of two equivalents of **56** with *s*-BuLi formed a racemic mixture of α-carbanions **57**, which were reacted with one equivalent of **(*S*)-30**. After work-up, 66% of the desired diphosphine borane **58** was isolated. HPLC analysis revealed that it was a mixture of *l:u* diastereomers, in a 83:17 ratio and with 86% and 82% ee respectively. The major isomer was, as expected, **(2*R*,4*R*)-58**, resulting from inversion of configuration at the phosphorus atom of **(*S*)-30**. The enantioselectivities are in the range of those expected for **30**, suggesting that the reaction is completely enantioselective. Interestingly, using stoichiometric amounts of **57** gave lower yields of product but the same isomeric distribution, suggesting a kinetic resolution in **57**, with the *S* enantiomer reacting preferentially.

In contrast to phosphinite boranes, the high electrophilicity of chlorophosphine boranes make them reactive towards a variety of non-carbanionic nucleophiles such alcohols, amines and thiols (Scheme 4.27).[2,14]

As expected, predominant inversion of configuration at the phosphorus atom was observed. It has to be kept in mind that heterophosphine boranes **59** are very difficult to obtain by other means, particularly in optically pure form.[2]

Reaction of 2-bromophenolates and 2-bromonaphtholates with chlorophosphine boranes affords the expected 2-bromoarylphosphinite boranes, which undergo [1,3] P–O to P–C rearrangement upon treatment with excess of *t*-BuLi (Scheme 4.28, compare this with Chapter 3, Sections 3.2.2 and 3.3.1.2).[37]

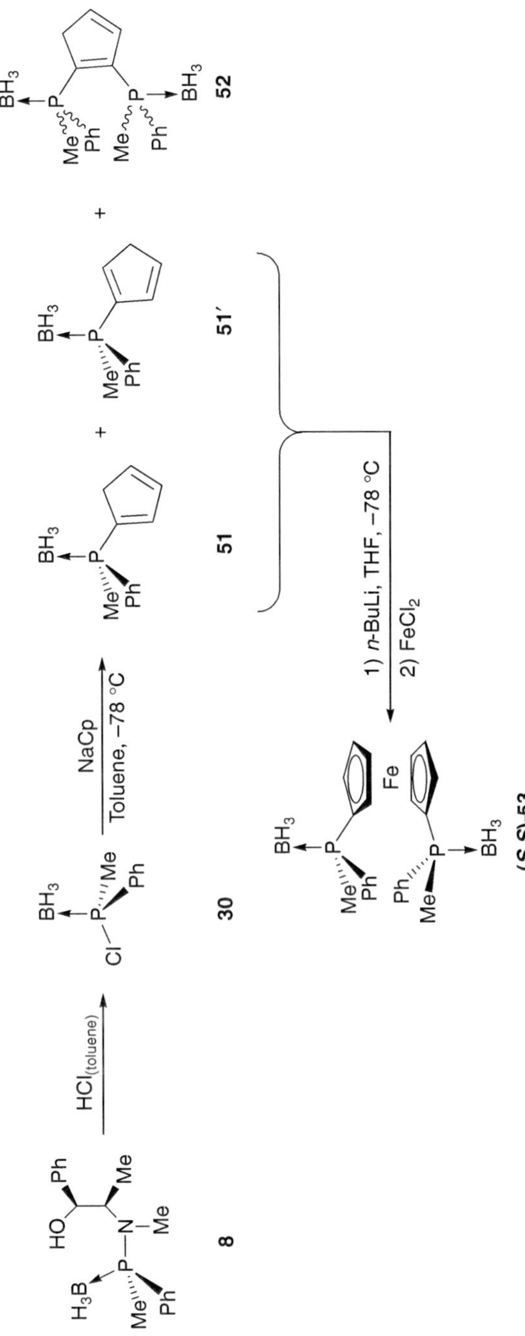

Scheme 4.24 Utility of a chlorophosphine borane in the synthesis of a ferrocenyldiphosphine borane.

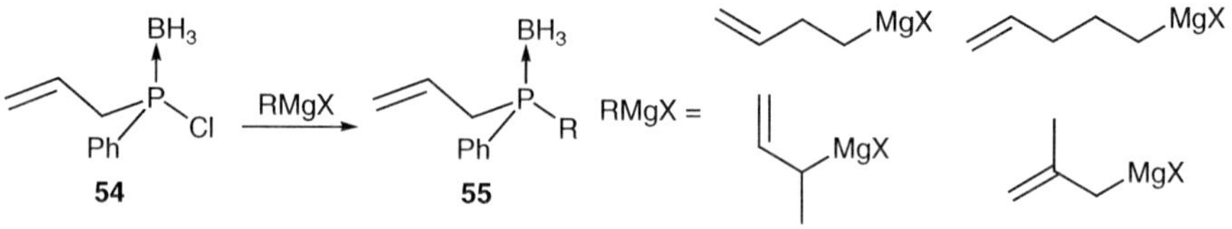

Scheme 4.25 *P*-stereogenic *bis*(alkenyl)phosphine boranes.

Phosphinite boranes **61** were obtained in around 70% yield with 71–100% ee, the latter being limited by the optical purity of **31**. They rearranged to afford the phosphine boranes **62** in 48–96% yields. As can be expected (see Chapter 3, Section 3.2.2) the rearrangement occurs with clean retention of configuration at the phosphorus atom, explained by the formation of an oxaphosphetane ring blocking the racemisation of the pentacoordinated intermediate.[37] A final quenching with water yielded 2-hydroxyphosphines or interestingly, quenching with chlorodiphenylphosphine afforded phosphinophosphinite boranes. A similar rearrangement has been reported to occur with 2-bromoarylaminophosphine boranes, although it requires the protection of the NH group during the lithiation step.[2]

More recently Jugé[2] reported an unexpected reaction of chlorophosphine boranes with *t*-BuLi (Scheme 4.29).

Reaction of **31** with *t*-BuLi at very low temperature did not afford the expected *tert*-butylphosphine borane but prompted the metallation of **31** to form a solution of phosphide boranes **63**. The steric bulk of the *t*-Bu carbanion probably accounts for this reactivity. Phosphide boranes are very nucleophilic, and can be protonated to form secondary phosphine boranes **64** or alkylated to produce tertiary phosphines, **65**. The reported yields for these compounds are in the range of 21–75% with ee values up to 99%. Although no details of this reaction have been yet published,[2] it is likely that this transformation will attract considerable attention because **64** are very versatile synthons for the preparation of many other *P*-stereogenic compounds.

4.4   Extensions of the Jugé–Stephan Method

4.4.1   *P*-Stereogenic AMPP Ligands

Aminophosphine phosphinite (AMPP) ligands constitute an important class of non-C_2 diphosphorus ligands.[44–46] Some of them are prepared from ephedrine and therefore the Jugé–Stephan method constitutes an ideal procedure for the synthesis of *P*-stereogenic AMPP ligands (Scheme 4.30).[45,46]

After ring opening of **7** by an organolithium, the opened aminophosphine alkoxide **12** can be trapped with an equivalent of a chlorophosphine to afford **66** after boronation. Alternatively, attack of **12** on a highly enantioenriched chlorophosphine borane **31** affords the doubly *P*-stereogenic compounds **67**. Both **66** and **67** can be easily deprotected with DABCO in toluene, for example, with complete retention of configuration at the phosphorus atom.

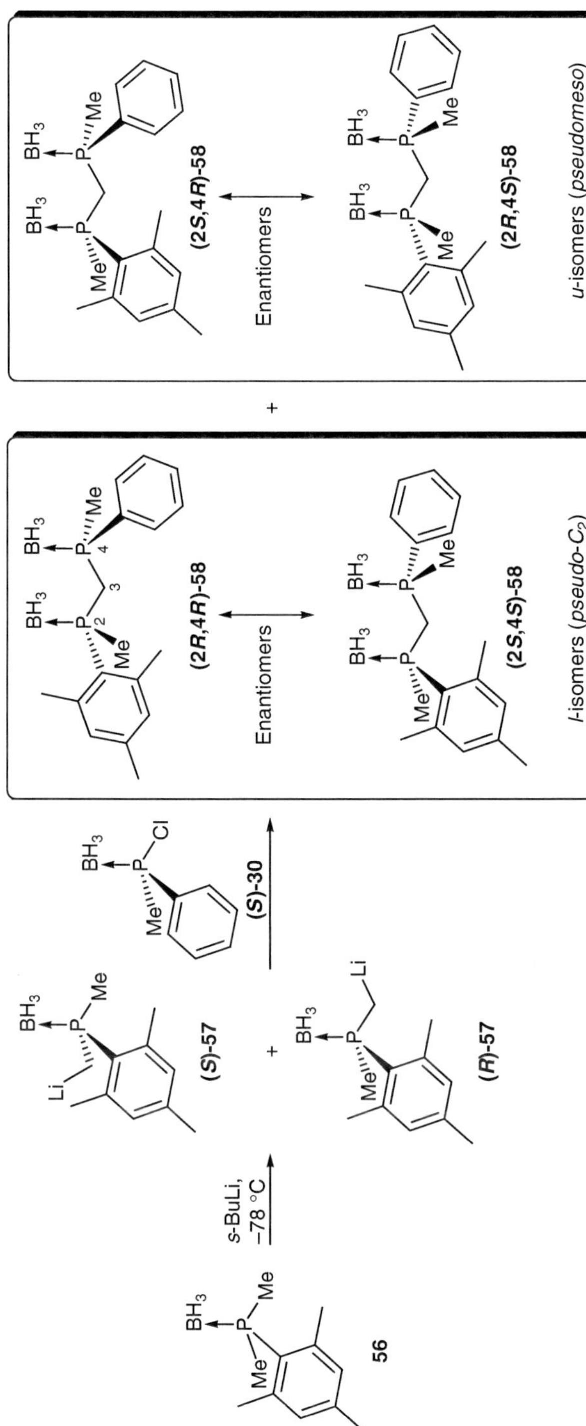

Scheme 4.26 Preparation of a dissymmetric, methylene-bridged diphosphine borane.

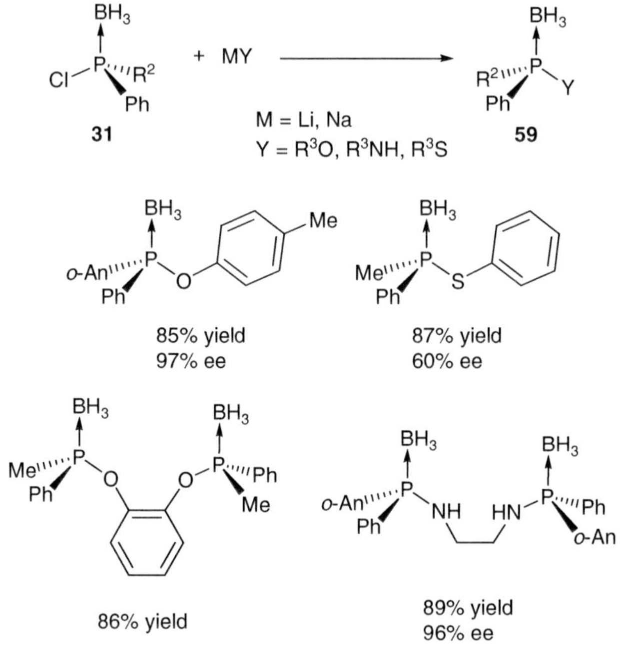

Scheme 4.27   Non-carbanionic attack on chlorophosphine boranes.

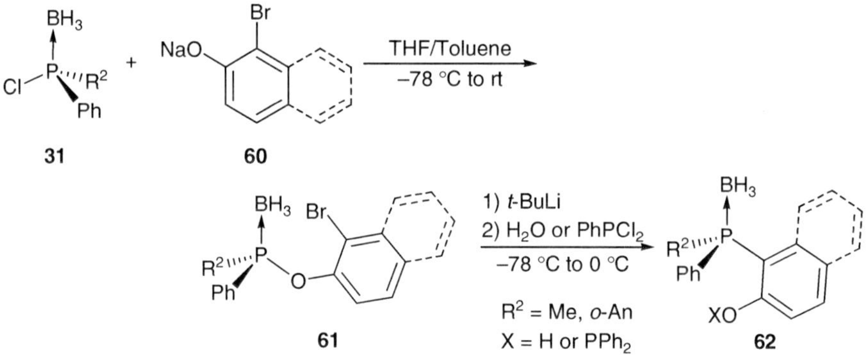

Scheme 4.28   [1,3] P–O to P–C rearrangement in 2-bromoarylphosphinite boranes.

Although prepared differently, all these ligands can be formally considered as derivatives of the non *P*-stereogenic EPHOS ligand ($R^2 = R^3 = Ph$) (Table 4.6).

A variety of substituents have been introduced in good yields with high stereoselectivity ($\geq 92\%$ de),[12] proving the versatility of the synthetic scheme. It has to be kept in mind that the absolute configuration of each of the stereogenic centres of the molecules **67** and **68** can be individually selected simply by changing the ephedrine enantiomer used in the preparation of **7** and/or **31**. This

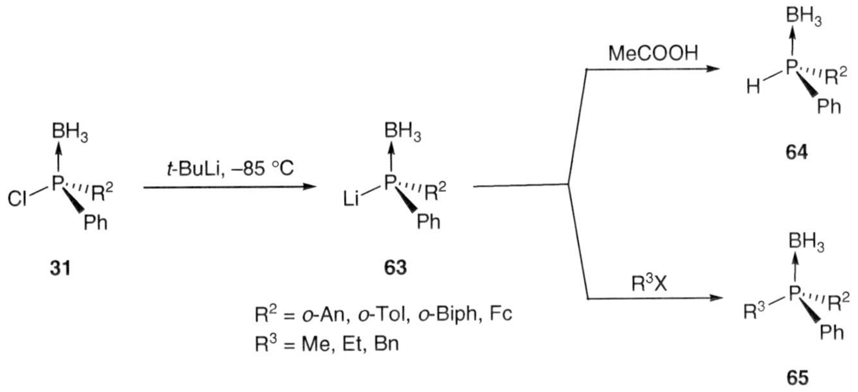

Scheme 4.29 Preparation of phosphide boranes and related derivatives from **31**.

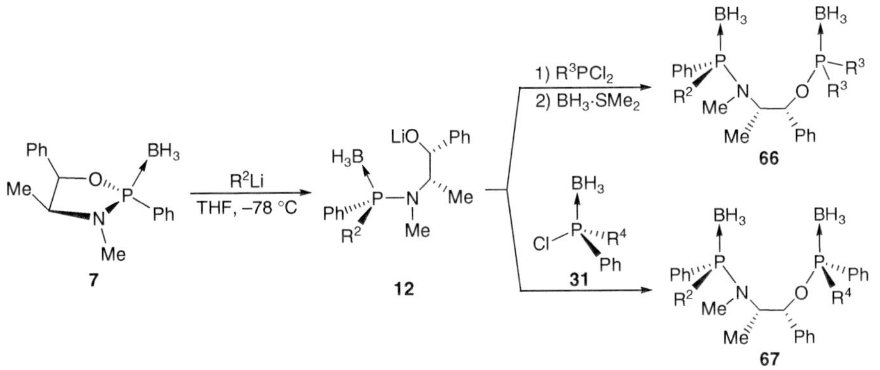

Scheme 4.30 Preparation of *P*-stereogenic AMPP ligands.

feature has been used to prepare a few pairs of diastereomeric compounds (entries 1, 2, 13, 14, 17, 18, 20 and 21).

Kamer and co-workers[47] reported an interesting extension of this chemistry to prepare polymer-supported ligands (Scheme 4.31).

A variety of oxazaphospholidine boranes **7** were successfully reacted with lithiated *p*-bromo-functionalised polystyrene **68**, yielding resins whose ^{31}P NMR chemical shifts indicated the formation of polymer-supported aminophosphines **69** by comparison with the shifts of the molecular analogues. Phosphination of the hydroxyl group and deboronation by standard procedures afforded the desired polymer-supported AMPP ligands **70**. Elemental analysis and gel-phase ^{31}P NMR spectroscopy indicated that resins **70** were formed in high yields and purities. An advantage of resins is that excess reagents can be used since purification is simply performed by successive washings. The downside of this procedure is that only very clean reactions are useful because impurities, if

Table 4.6   *P*-stereogenic AMPP boranes prepared to date.

Entry	R^2	R^3	R^4	Yield (%)	References
1[a]	Me	Ph	–	77	12
2[b]	Me	Ph	–	79	11, 27
3[a]	Me	*p*-An	–	51	12
4[a]	Me	3,5-$(CF_3)_2C_6H_3$	–	92	12
5[a]	Me	2,2′-biphenol	–	66	12
6[b]	Me	–	Me	18	11, 27
7[a]	*n*-Bu	Ph	–	66	12
8[a]	*n*-Bu	*p*-An	–	53	12
9[a]	*n*-Bu	3,5-$(CF_3)_2C_6H_3$	–	78	12
10[b]	*t*-Bu	Ph	–	64	11, 27
11[b]	Cy	Ph	–	38	27
12[b,c]	Ph	Ph	–	63	11, 27
13	Ph	–	*o*-An	73	11
14[b,d]	Ph	–	*o*-An	44–55	11, 27
15[b]	*o*-An	Ph	–	62	11, 27
16[b]	*o*-An	Cy	–	67	11, 27
17	*o*-An	–	*o*-An	40	11, 27
18[b]	*o*-An	–	*o*-An	65	27
19[b]	*o*-MEMC$_6$H$_4$[e]	Ph	–	66	27
20[b]	1-naphthyl	Ph	–	62	11, 27
21	1-naphthyl	Ph	–	79	12
22[b]	2-naphthyl	Ph	–	78	11, 27
23	Ferrocenyl	Ph	–	62	27

[a]Combined yield of the free ligand.
[b]Precursor **7** prepared from (+)-ephedrine.
[c]EPHOS ligand; its phosphorus atoms are not stereogenic.
[d]For this compound both enantiomers of **31** have been used.
[e]MEM = 2-[(2-methoxy)ethoxy]methoxy.

grafted on the resin, are usually impossible to remove. This was the case when a supported version of **7** was employed as starting material and therefore the route in Scheme 4.31 was followed.[47]

4.4.2   Deprotonation of Methylphosphine Boranes

Many *P*-stereogenic phosphines bearing a methyl group have been prepared by the Jugé–Stephan method. The presence of the borane protective group at the phosphorus atom confers a relative acidity to the methyl group making it susceptible to deprotonation by strong bases, usually *s*-BuLi. The corresponding carbanions are configurationally stable and have been used to prepare a variety of mono- and bidentate *P*-stereogenic phosphorus ligands (Scheme 4.32).

From α-lithiated carbanions **72**, formylated (**73**), carboxylated (**74**), silylated (**75**) and aminated (**76**) *P*-stereogenic phosphine boranes have been prepared without loss of optical purity. Some of them can be further functionalised to obtain other interesting compounds. Furthermore, oxidative coupling with Cu(II) salts affords C_2 diphosphine boranes of the DiPAMP family (**77**) and

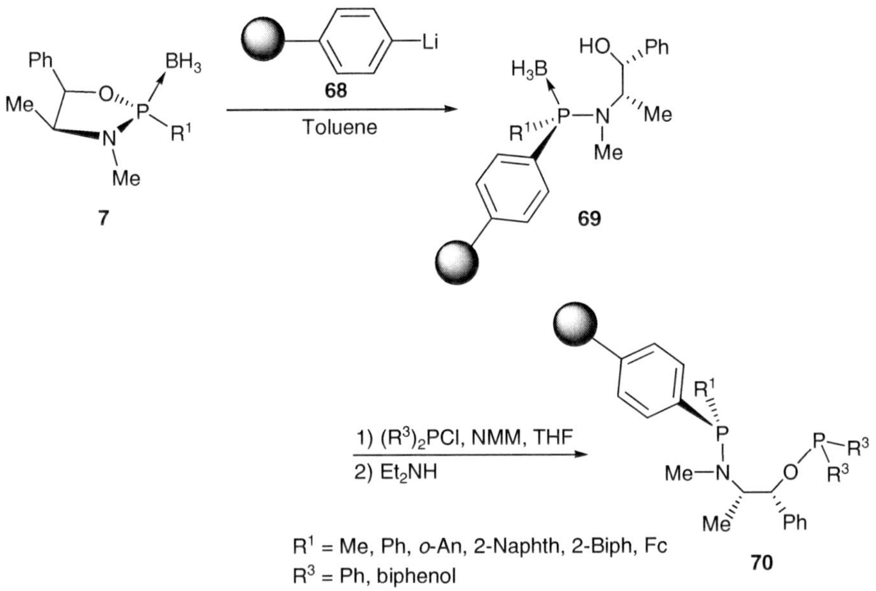

R^1 = Me, Ph, *o*-An, 2-Naphth, 2-Biph, Fc
R^3 = Ph, biphenol

Scheme 4.31   Synthesis of polystyrene-bound aminophosphine phosphinites.

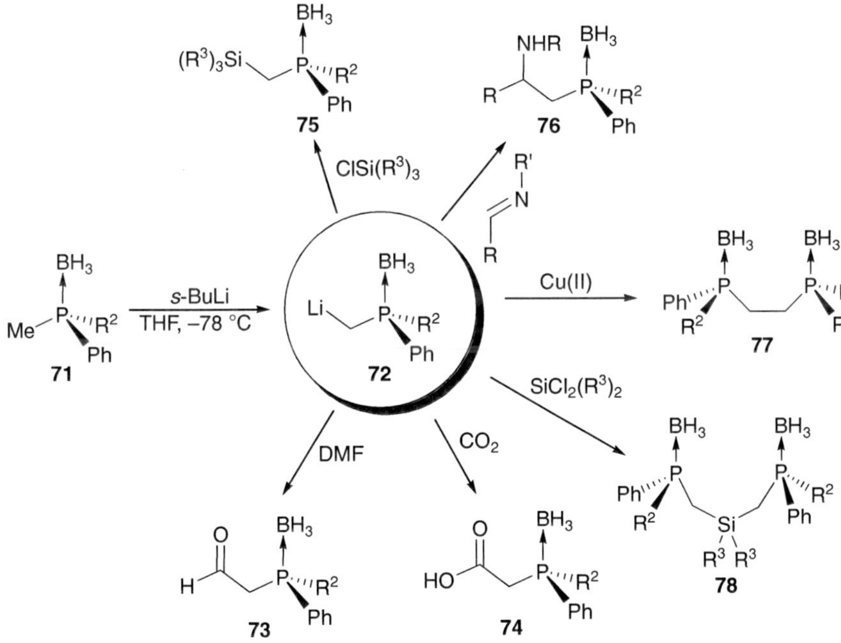

Scheme 4.32   Versatility of α-lithiated methylphosphine boranes.

double nucleophilic substitution on dichlorosilanes gives access to **78**, which is another class of C_2 diphosphine boranes. The preparation of all these compounds is detailed in the next sections.

4.4.2.1 Monophosphines and Bidentate C_1 Ligands

Johansson, Kann and Andersson[48] transformed (*S*)-PAMP·BH_3 into several β-aminophosphine boranes *via* the sequence formylation–reductive amination (Scheme 4.33).

The carbanion derived from (*S*)-PAMP·BH_3 can be easily quenched at low temperature with DMF affording the formylated phosphine borane **73** in 85% yield. This compound, although rather unstable, when freshly prepared reacted smoothly with amines and subsequently sodium triacetoxyborohydride under microwave irradiation (μW) to afford the desired β-aminophosphine boranes **79** in 84–96% yields and in optically pure form.

Andrieu, Jugé and co-workers[49] prepared similar compounds by another route, depicted in Scheme 4.34.

The carbanion derived from (*S*)-PAMP·BH_3 reacted with *N*-benzilideneaniline to afford the β-aminophosphine boranes **76** as an epimeric mixture in 76% yield. The major diastereomer (shown in Scheme 4.34 after deprotection)

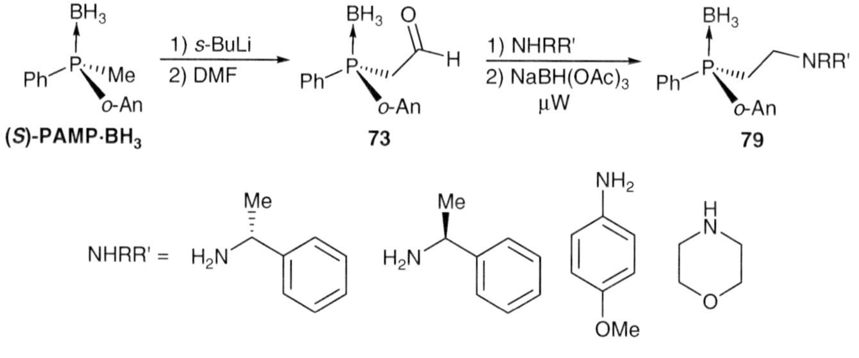

Scheme 4.33 Synthesis of β-aminophosphine boranes by formylation and subsequent reductive amination.

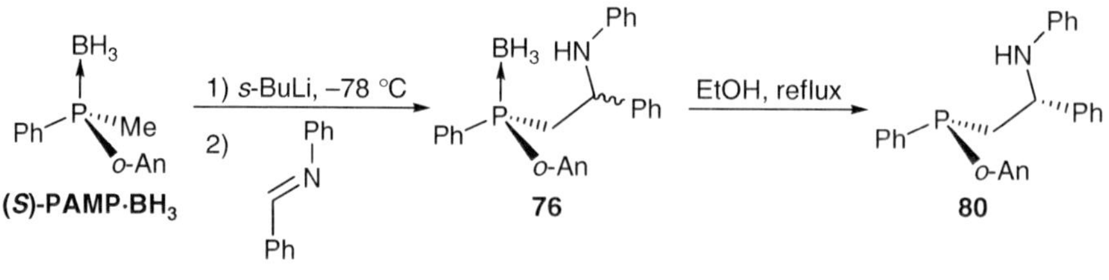

Scheme 4.34 Synthesis of β-aminophosphine boranes by carbanionic attack on *N*-benzilideneaniline.

can be isolated in 92% de after several recrystallisations, albeit in low yield (30%). Interestingly, this compound can be enantioselectively deboronated by simply stirring it in refluxing ethanol to afford **80**.

Hii and co-workers[50] prepared *P*-stereogenic diarylphosphinocarboxylic acids as valuable precursors towards C_1 amido- and aminodiphosphine ligands (Scheme 4.35).

Lithiated **71** (Ar = *o*-An, *o*-Tol and 1-Naphth) reacted with carbon dioxide affording the phosphinocarboxylic acid boranes **74** in 55–60% yields. Their deprotection using TFA, with retention of configuration, and coupling with (*S*)-(−)-2-(diphenylphosphino)methylpyrrolidine furnished amidodiphosphines **81** (48–51% yield). Finally, carbonyl reduction with borane (and concomitant *N*- and *P*-protection) followed by a deboronation step gave access to P,N ligands **82**.

Pizzano and co-workers[51] also used the carboxylic acid derived from (*S*)-PAMP·BH₃ to prepare a *P*-stereogenic phosphine phosphite ligand (Scheme 4.36).

Reduction of the acid to the alcohol and standard deboronation with DABCO proceed smoothly, giving the β-hydroxyphosphine **84** in 86% yield. Condensation of **84** with an enantiopure biphenolphosphorochloridite leads to the desired ligand **85** in 63% yield as a single diastereomer, featuring both a stereogenic phosphorus atom and a stereogenic axis in the hindered biaryl group.

Another type of phosphine phosphite ligand, with a longer alkyl bridge, was reported by van Leeuwen and co-workers (Scheme 4.37).[52]

Metallated phosphine boranes **71** were used for regioselective ring opening of propene and styrene oxides to yield phosphino alcohols **86**, which were condensed with bisphenol phosphochloridite to yield compounds **87**.

Chlorosilanes are also very reactive towards the α-metallated phosphine boranes leading to α-silylated phosphine boranes (Scheme 4.38).[16,32,53]

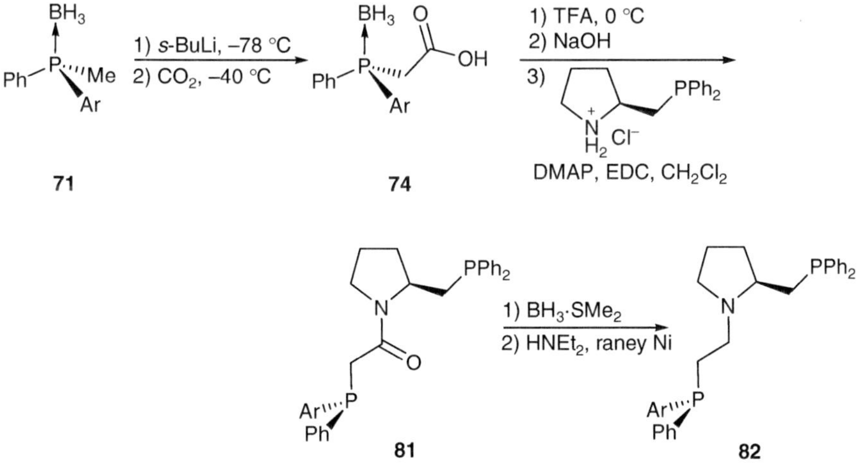

Scheme 4.35 Preparation of *P*-stereogenic phosphinocarboxylic acids, amido- and aminodiphosphines.

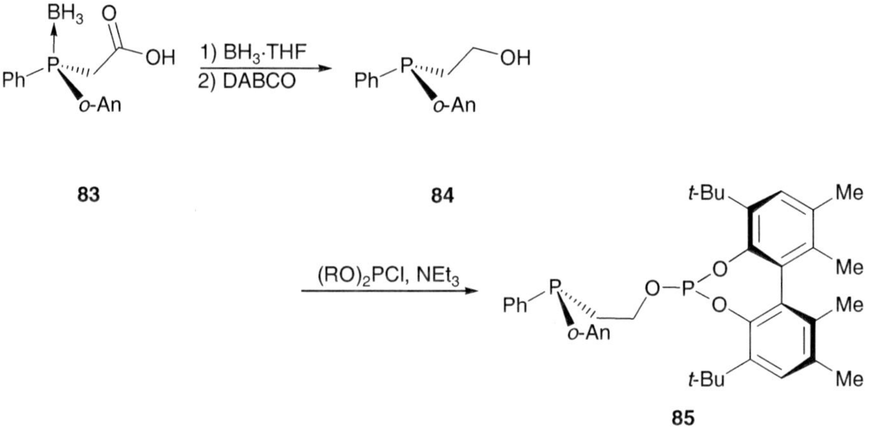

83

84

85

Scheme 4.36 An example of *P*-stereogenic phosphine phosphite ligand.

Ar = *o*-An, 1-Naphth

71

R = Me, Ph

86

87

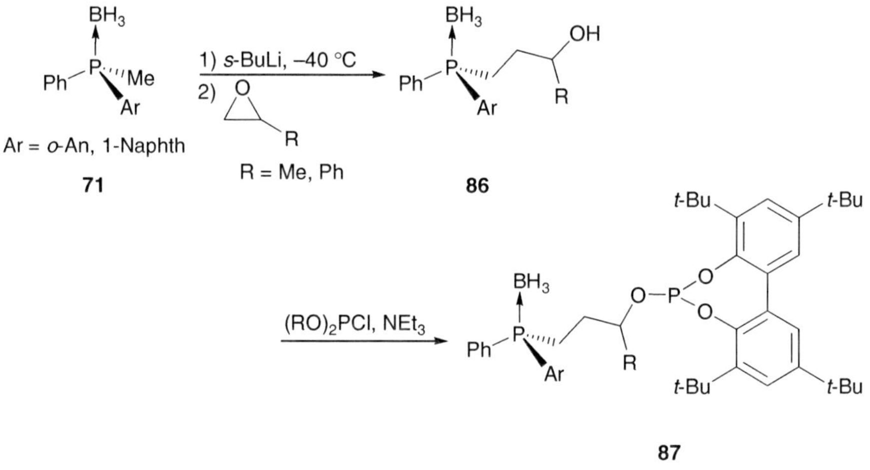

Scheme 4.37 Preparation of phosphine phosphite ligand **87**.

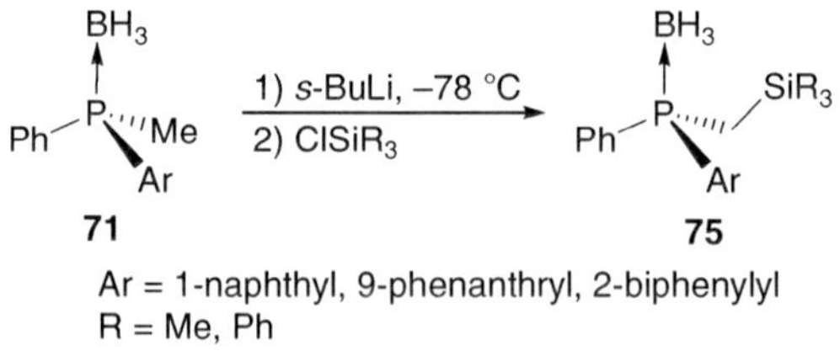

71

75

Ar = 1-naphthyl, 9-phenanthryl, 2-biphenylyl
R = Me, Ph

Scheme 4.38 Preparation of α-silylated phosphine boranes.

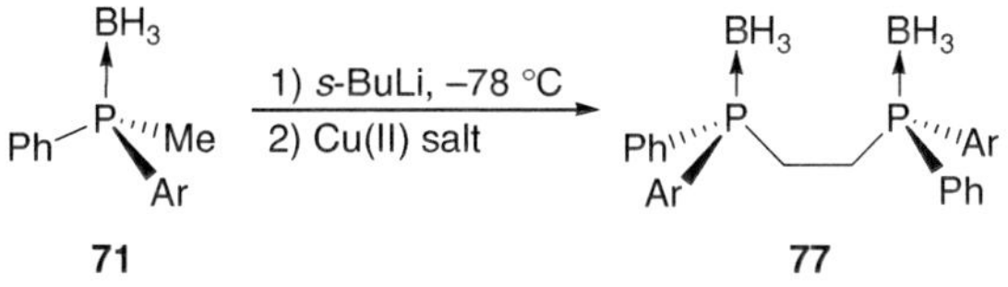

Scheme 4.39 Preparation of DiPAMP analogues by Cu-promoted oxidative coupling.

Phosphine boranes **75** can be obtained pure after acidic work-up in moderate yields even when the electrophile is the very bulky chlorotriphenylsilane.[32] Remarkably, the same reaction has been used to prepare C_2 diphosphines with a central $-SiR_2-$ moiety and also the first dendrimers containing *P*-stereogenic phosphines, compounds discussed in the next sections.

4.4.2.2 C_2-Diphosphines

The Jugé–Stephan methodology has been used to prepare many C_2 diphosphine boranes bearing at least one aryl group at each phosphorus atom. In particular, it is probably the best method to prepare not only DiPAMP itself, but also a large family of analogues with different substituents at the phosphorus atoms and/or different bridges, which have found application in many catalytic processes.

DiPAMP diborane and its closest analogues have an ethylene group bridging the two phosphorus atoms. They can be conveniently prepared by Cu-promoted oxidative dimerisation of the α-carbanions derived from methylphosphine boranes (Scheme 4.39).

This reaction requires the presence of stoichiometric amounts of Cu(II) salts and is thought to proceed though a radical intermediate,[54] implying that the rates of formation of the homochiral isomers of **77** and the *meso* form are very similar. This comes with a bonus: even if **71** is not completely enantiopure, the optical purity of the desired *l* isomer of **77** increases after separation of the *meso* diastereomer. For example,[54] when (*S*)-**71** is enriched to 89% ee the calculated product distribution for **77** [(*S*,*S*):*meso*:(*R*,*R*)] is 296:34:1, meaning that after the *meso* isomer has been removed **77** is obtained in 99.3% ee. The diphosphine boranes **77** prepared by this method are listed in Table 4.7.

In all the entries but the first two, the phosphorus atoms contain two aryl groups. Some crowded diphosphine boranes have also been reported (entries 2, 6, 10, 14 and 16–21). In general, good yields for the coupling reaction are observed, even in the latter cases. The diphosphine boranes are obtained as enantiomerically pure white solids or oils, which can be easily deboronated by amines. The diphosphinite borane of entry 1[38] also deserves some comment because another product could have been formed (Scheme 4.40).

Reaction of **88** with *s*-BuLi/CuCl$_2$ provides **89** in good yields, but the substitution product **90** could have been formed, following the reactivity of Section 4.3.3. The balance between basicity and steric hindrance of *s*-BuLi probably makes it ideal to deprotonate the methyl group but not nucleophilic enough to attack the phosphorus centre to form **90**.

Table 4.7 Diphosphine boranes with an ethylene bridge prepared by the Jugé–Stephan method.

Entry	Diphosphine borane	Yield (%)	References
1	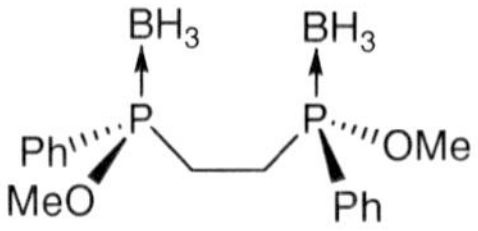 BH_3, BH_3; Ph, P, OMe; MeO, P, Ph	88	38
2	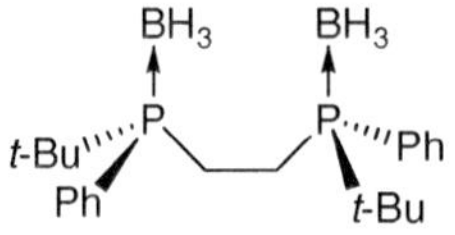 BH_3, BH_3; *t*-Bu, P, Ph; Ph, P, *t*-Bu	44	17
3	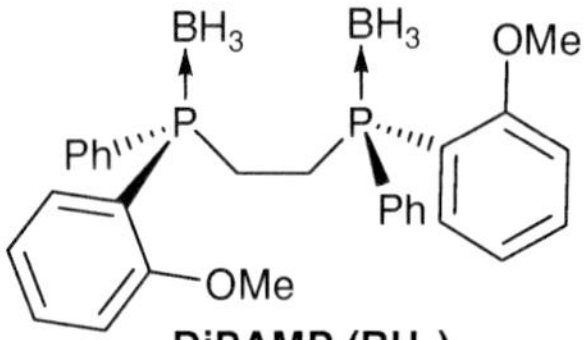 BH_3, BH_3, OMe; Ph, P, P, Ph; OMe; **DiPAMP·(BH$_3$)$_2$**	–	6
4	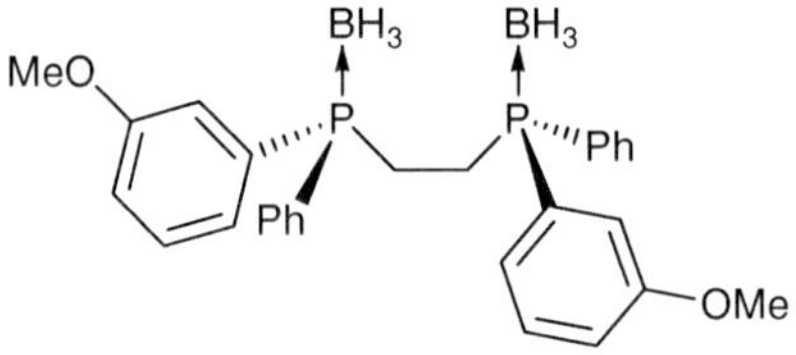 BH_3, BH_3; MeO, P, P, Ph; Ph, OMe	70	23
5	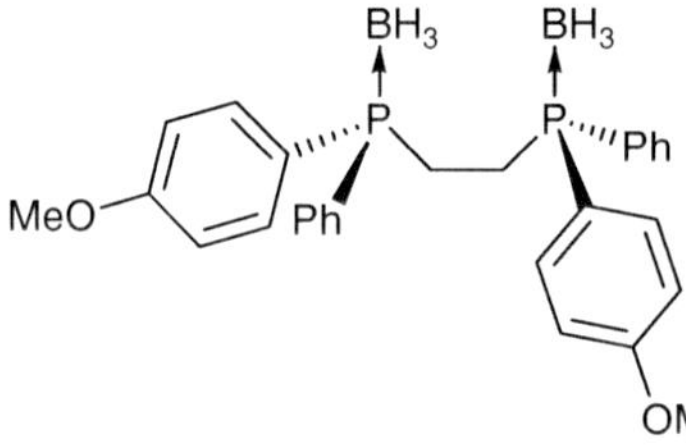 BH_3, BH_3; MeO, P, P, Ph; Ph, OMe	67	23
6	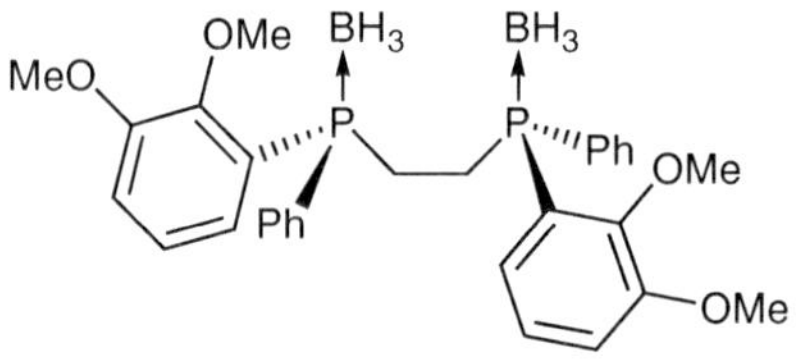 OMe, BH_3, BH_3; MeO, P, P, Ph, OMe; Ph, OMe	63	24
7	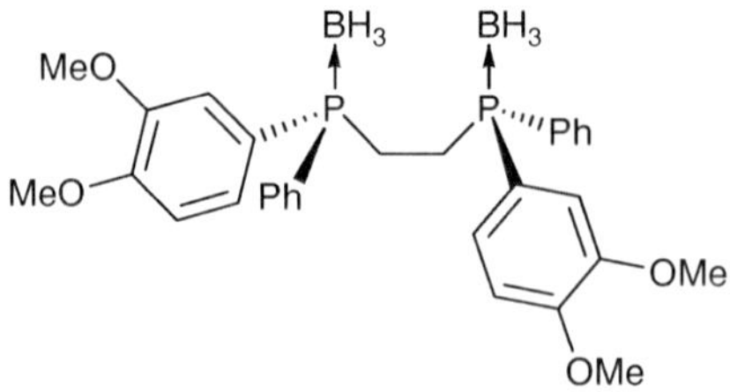 BH_3, BH_3; MeO, P, P, Ph; MeO, Ph, OMe, OMe	56	23

Table 4.7   (*Continued*).

Entry	Diphosphine borane	Yield (%)	References
8		63	23
9		56	24
10		77	24
11		84	23
12		90	24
13		64	24
14		65	24

Table 4.7   (*Continued*).

Entry	Diphosphine borane	Yield (%)	References
15		61	24
16	**4MeBigFUS·(BH₃)₂**	80	24
17	***i*-Pr-SMS-Phos·(BH₃)₂**	97	55
18		93	25
19		66	25
20		94	25

Table 4.7 (*Continued*).

Entry	Diphosphine borane	Yield (%)	References
21	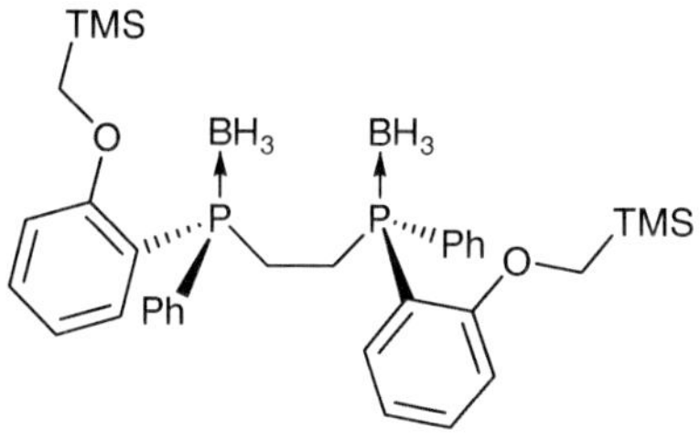	98	25
22	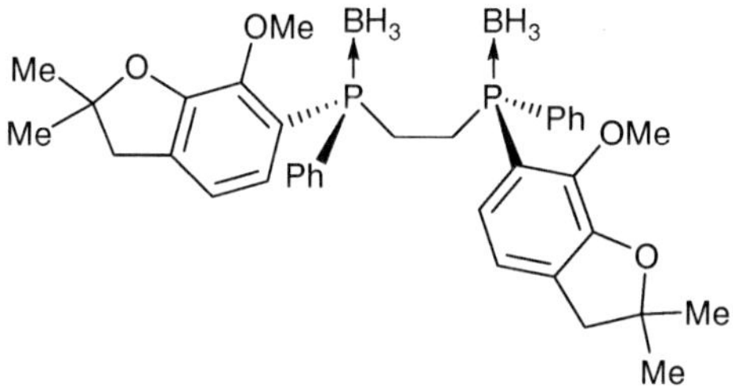	78	24
23	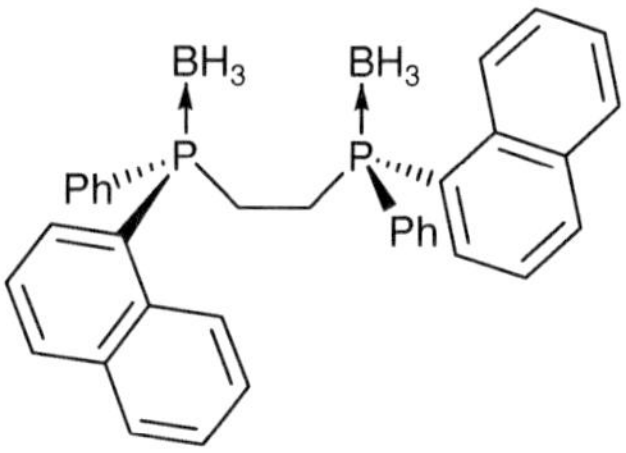 α-bnpe·(BH$_3$)$_2$	66	56
24	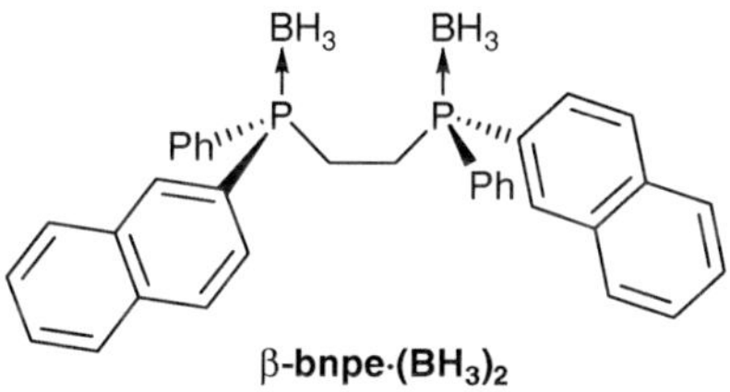 β-bnpe·(BH$_3$)$_2$	–	6
25	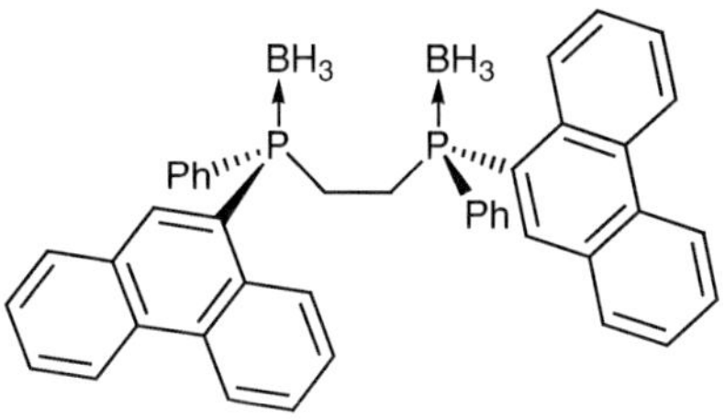	79	16

Table 4.7 (*Continued*).

Entry	Diphosphine borane	Yield (%)	References
26		76	24
27		76	24

Scheme 4.40 Dimerisation *vs.* nucleophilic substitution of **88**.

Nucleophilic displacement of both methoxy groups of compound **89** has been used by Zupancic, Mohar and Stephan[24] as an alternative route to diphosphine boranes (Scheme 4.41).

This substitution reaction is low yielding (33–42%) but it occurs with inversion of configuration at the phosphorus atoms and therefore leads to the other enantiomer of **77** than the traditional oxidative coupling with Cu(II).

Zhang and co-workers[57,58] reported the preparation of the two C_2-diphosphine boranes **92** by nucleophilic attack of the carbanion derived from (*R*)-PAMP to *bis*(benzylic) reagents **91** (Scheme 4.42).

Both compounds **92** were obtained in high yields. Here the success of the synthesis is probably due to the choice of **91** as electrophile, since benzylic compounds are known to be very good substrates for nucleophilic substitutions.

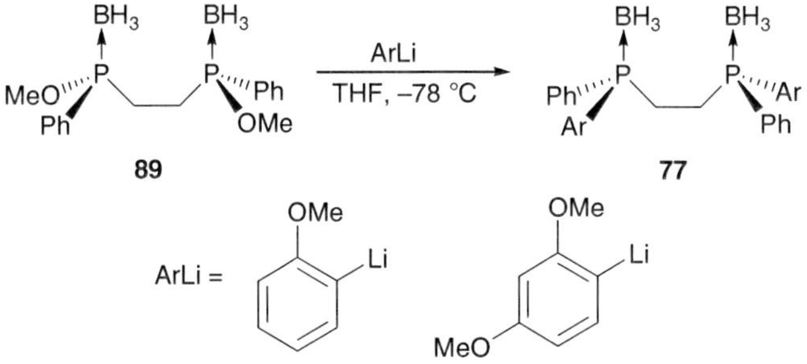

Scheme 4.41 Alternative procedure towards DiPAMP · $(BH_3)_2$ and one of its analogues.

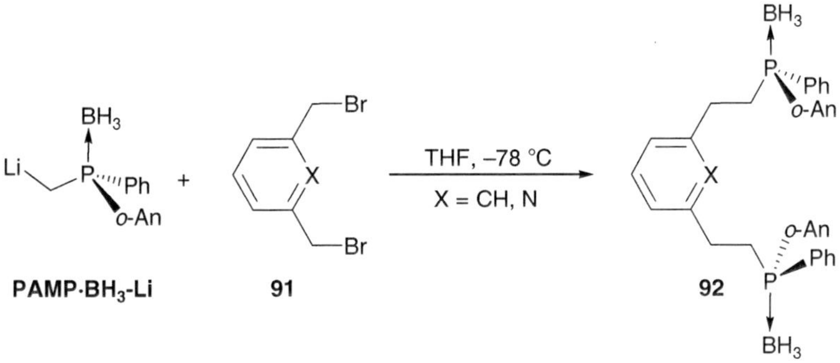

Scheme 4.42 Preparation a diphosphine borane by double nucleophilic substitution.

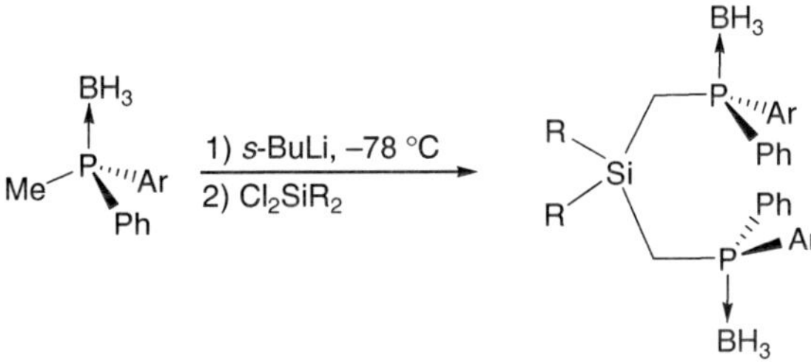

Scheme 4.43 Preparation of diphosphine boranes with 2-silapropyl bridge.

Another set of diphosphine boranes has been prepared by double nucleophilic attack on dichlorosilanes by carbanions derived from methylphosphine boranes (Scheme 4.43 and Table 4.8).

Table 4.8 reveals that the diphosphine boranes are obtained as optically pure compounds in high yields. The compound in entry 3 is noteworthy, because it is prepared in good yield by double nucleophilic substitution on a relatively encumbered silicon atom. Only in the preparation of the diphosphine borane of entry 4 some monosubstituted product was observed[60] although it could be removed by column chromatography.

Table 4.8 Diphosphine boranes with 2-silapropyl bridges.

Entry	Diphosphine borane	Yield (%)	References
1	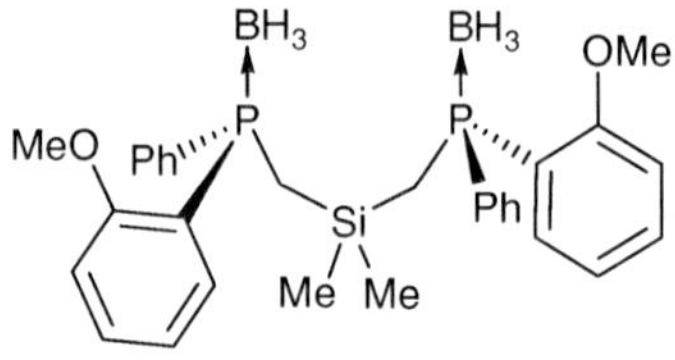 (*S,S*)-DiPAMPSi·(BH₃)₂	85	38
2	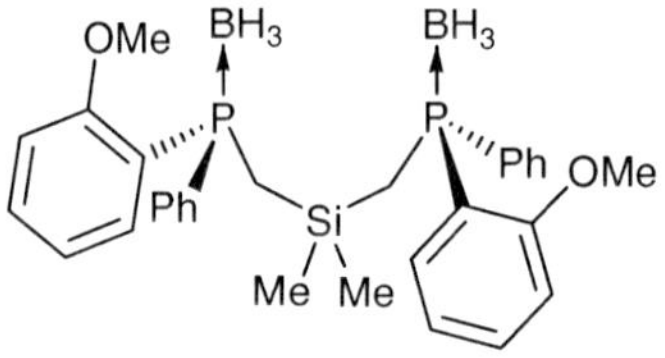 (*R,R*)-DiPAMPSi·(BH₃)₂	76	59
3	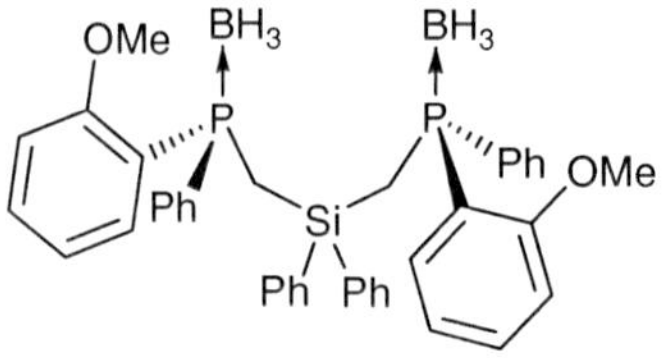	80	38
4[a]	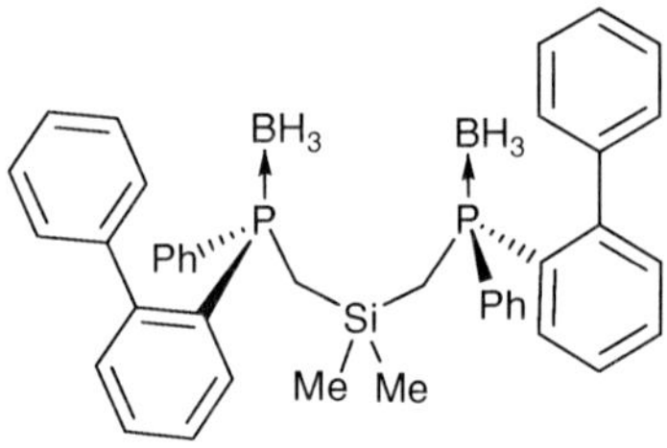	—	60
5	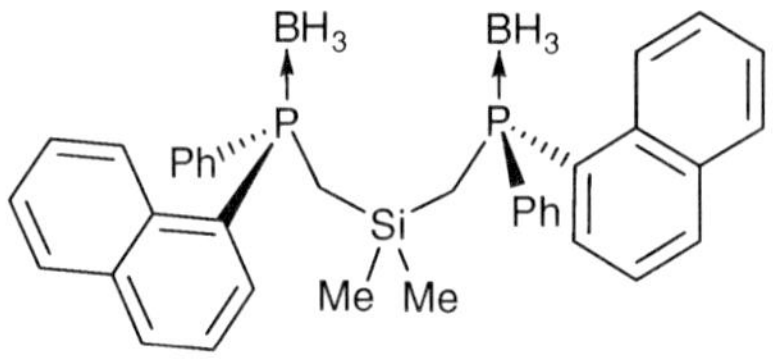	98	28
6[b]	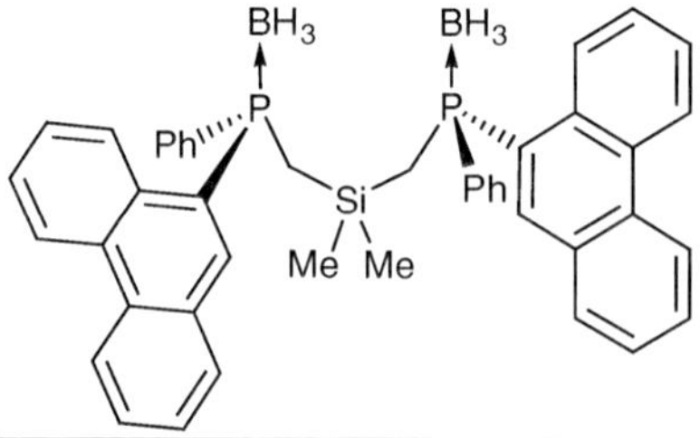	95	16

[a]The yield of the boronated phosphine has not been reported, but is 56% for the free phosphine (contaminated with the methylphosphine borane precursor).

[b]Contaminated with the methylphosphine borane precursor.

4.4.2.3　*P-Stereogenic Dendrimers*

The high reactivity of chlorosilanes towards α-carbanions derived from methylphosphine boranes and the availability of several carbosilane dendrimers possessing Si–Cl bonds made the preparation of the first dendrimers bearing *P*-stereogenic phosphines possible.

The first were reported by Muller, Rossell and co-workers[53] and contain monophosphine groups at the periphery of the dendrimers (Scheme 4.44).

The reaction of an excess of phosphine borane carbanions with **93** yielded a crude product whose ^{1}H NMR spectrum proved the total consumption of **93**. After deprotection of the borane groups in neat morpholine and purification by column chromatography, **94** were obtained as pure solids in 41–56% yields. With a similar procedure **95** could be obtained (15% yield), although in this

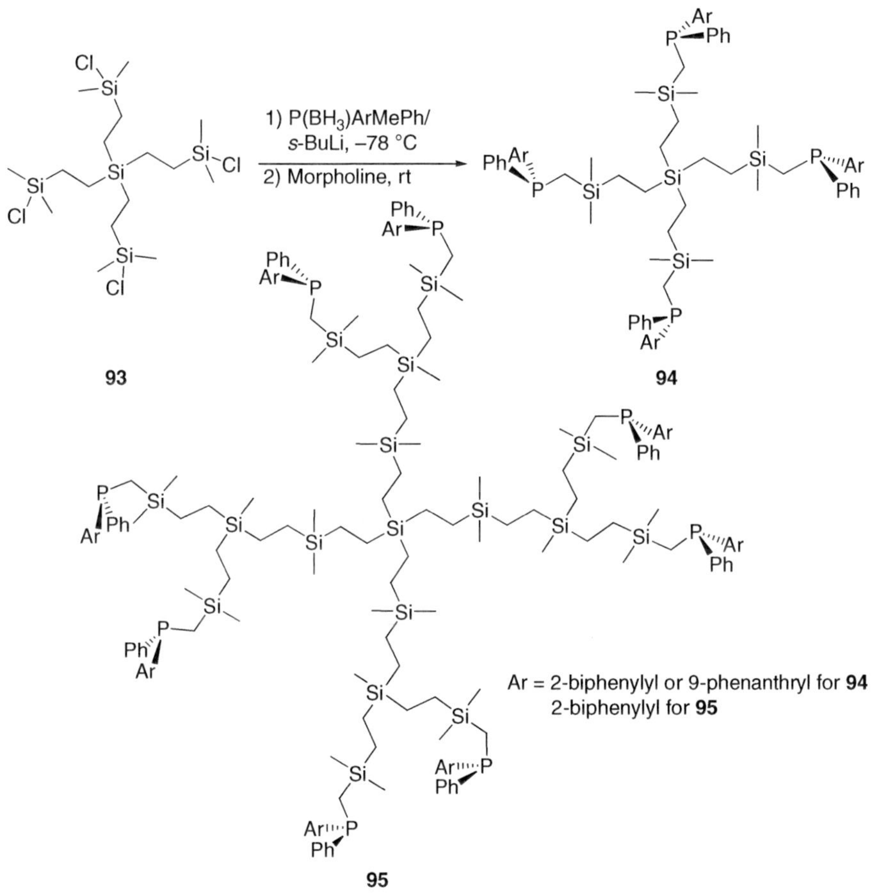

Scheme 4.44　Preparation of the first dendrimers functionalised with *P*-stereogenic phosphines.

case for Ar = 9-phenanthryl only partial functionalisation of the dendrimer branches could be accomplished probably due to the rigidity and steric bulk of the phenanthryl unit.[53]

A similar method was employed by the same group to prepare carbosilane dendrimers bearing diphosphine units on the periphery (Scheme 4.45).[60]

Here the dichloro carbosilane dendrimer **96** afforded the expected product **97** in 29% yield, although in this case it was contaminated with some monophosphine, which could not be removed by column chromatography. The preparation of the highly functionalised dendrimer **98** was also attempted but incomplete phosphination of all the arms was observed, probably due to the high steric congestion in the molecule.[60]

Further work from the same group[61] has lead to the preparation of carbosilane dendrons containing a single phosphine unit at the focal point (Scheme 4.46).

The reaction of triflate **99** with a metallated methylphosphine borane followed by deboronation with morpholine yielded, after column purification, dendron **100** in 31% yield. An analogous procedure was used to prepare **101** in similar yield.[61] Both dendrons can be considered as phosphines bearing an extremely bulky carbosilane dendrimer group.

4.4.3   Other Extensions

Starting from a vinyl-substituted precursor, Brown and Ramsden[20] reported an example of a C_1-symmetric diphosphine prepared by the Jugé–Stephan methodology (Scheme 4.47).

Ring opening of **7** with vinyllithium, followed by hydrophosphination of the vinyl group and borane protection afforded **102** in 89% yield. Acidic methanolysis and reaction of the obtained phosphinite borane with *o*-anisyllithium produced the desired compound **103** in 35% yield over the last two steps. Compound **103** can be easily deboronated with diethylamine.[20]

Gouverneur and co-workers[18,19] used several phosphine boranes and phosphinite boranes bearing alkenyl groups prepared by the Jugé–Stephan method in metathesis reactions (see Chapter 6, Section 6.2.6), but they were prepared as racemates intentionally.

4.5   Other Methodologies

4.5.1   Variations of the Jugé–Stephan Method

Before the Jugé–Stephan method was established and widely used other similar approaches, also based on the use of ephedrine as a chiral auxiliary, were studied. Although they have been completely eclipsed by the 'modern' strategy, they are noteworthy as they give an interesting perspective on how the current method was originally developed.

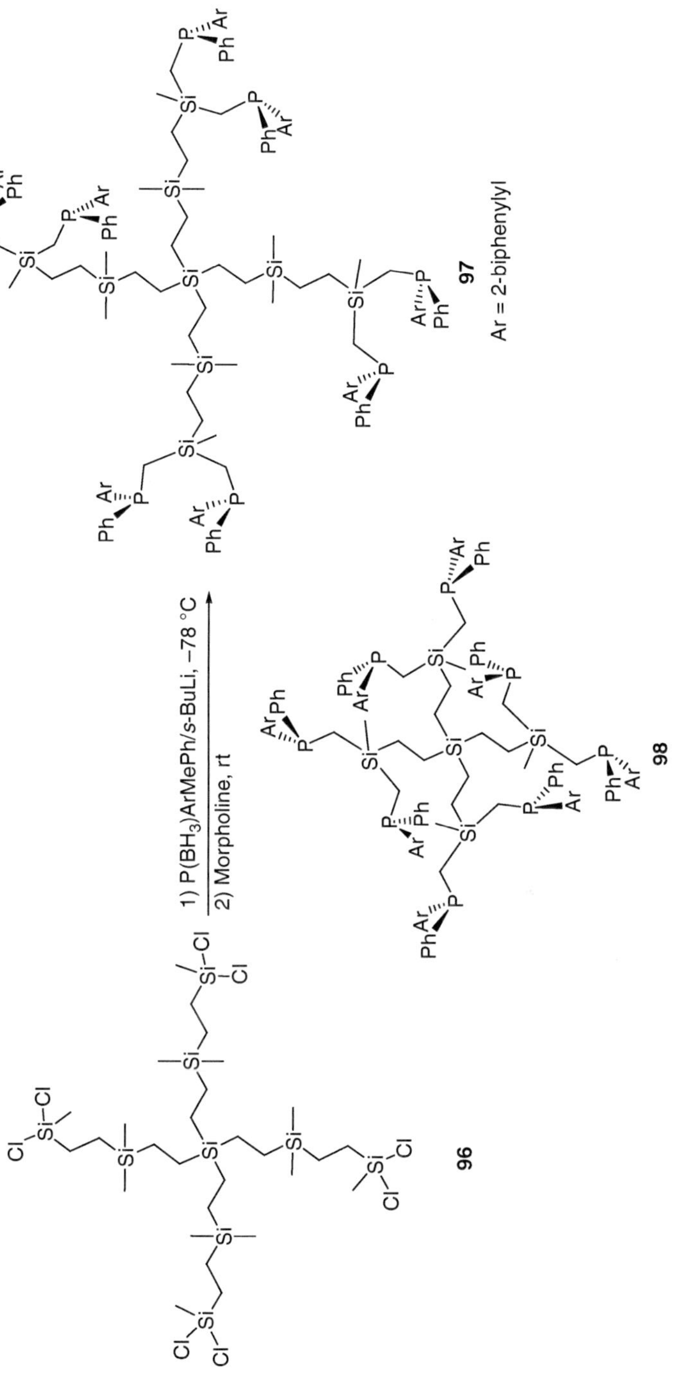

Ar = 2-biphenylyl

97

96

98

1) P(BH$_3$)ArMePh/s-BuLi, −78 °C
2) Morpholine, rt

Scheme 4.45 Preparation of the first dendrimers functionalised with *P*-stereogenic diphosphines.

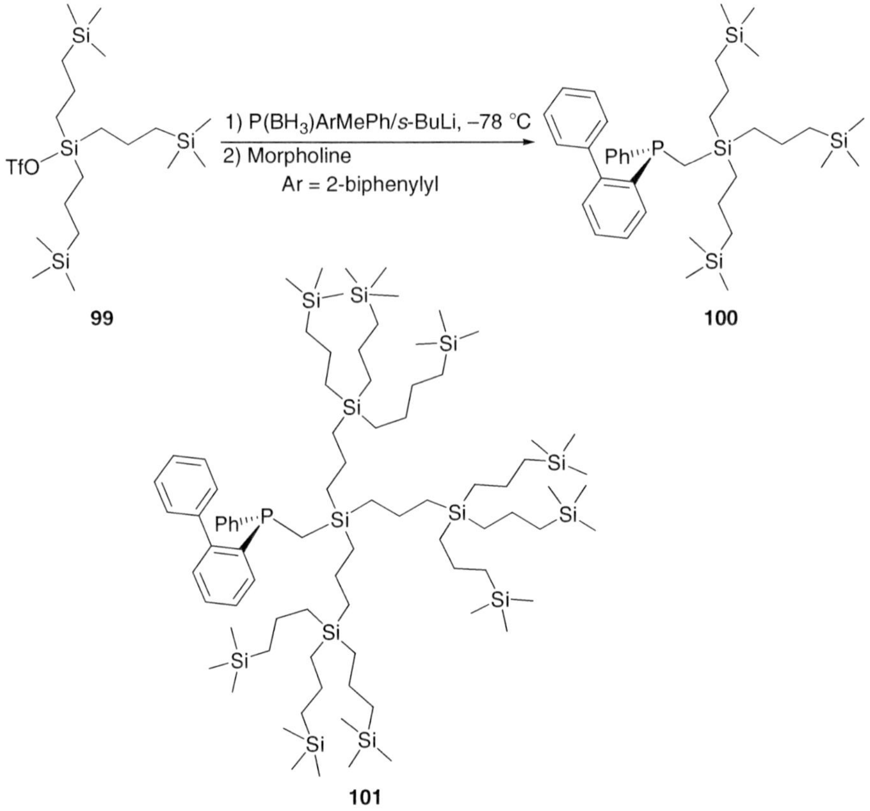

Scheme 4.46 Preparation of dendrons functionalised with a *P*-stereogenic phosphine moiety.

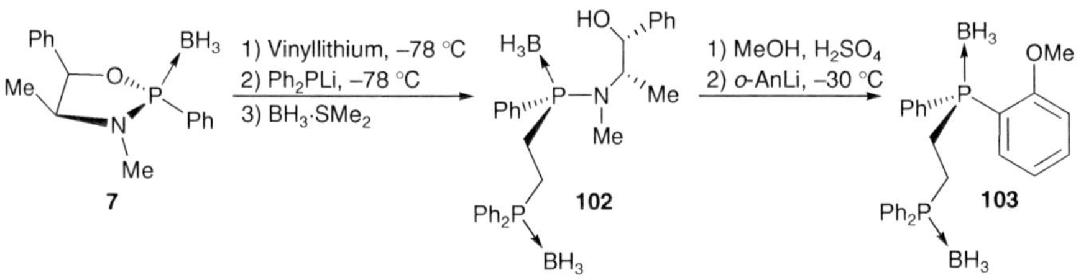

Scheme 4.47 Preparation of a C_1 diphosphine borane.

4.5.1.1 *Pentavalent Oxazaphospholidines Derived from Ephedrine*

Inch and co-workers[62] reported in 1979 the preparation of *P*-stereogenic phosphinothioic acids starting from 1,3,2-oxazaphospholidine sulfides derived from (–)-ephedrine (Scheme 4.48).

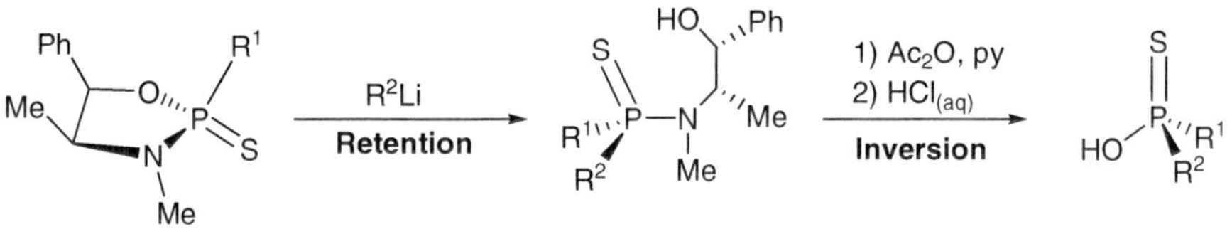

R¹, R² = Me, Et, *n*-Bu, Ph; with R¹ ≠ R²

Scheme 4.48   Preparation of *P*-stereogenic phosphinothioic acids.

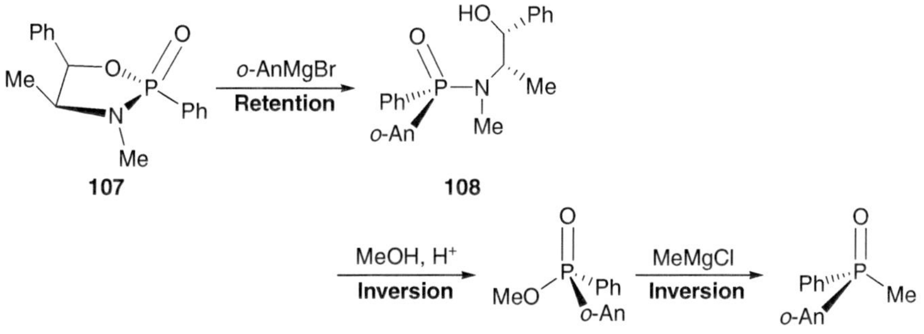

Scheme 4.49   Synthesis of (*R*)-PAMPO *via* the oxide version of the Jugé–Stephan method.

Reaction of optically pure **104** with organolithiums to form **105** occurs with preponderant P–O bond cleavage and retention of configuration (de > 90%), exactly like in the 'modern' Jugé–Stephan method, but at that time those observations were unexpected. The retention of configuration in the ring opening of **104** was explained by the sequence: formation of a pentacoordinated intermediate–pseudorotation–apical elimination. Acidolysis of **105** produced the desired phosphinothioic acids **106** in ee values better than 90%. This step is also directly comparable to the acidic methanolysis in the modern method.

The 'oxide' version of the Jugé–Stephan method was reported the same year as the borane version by Brown and co-workers.[63] Unsurprisingly, the first compound to be prepared was the oxide of PAMP, usually known as PAMPO (Scheme 4.49).

In contrast to the borane route, it was found that the more oxophilic Grignard reagents gave cleaner reactions than organolithiums. Reaction of **107** with *o*-AnMgBr at −30 °C in THF produces **108** in 70% yield, with retention of configuration and exclusive P–O bond cleavage. Acidic methanolysis (MeOH, dry HCl) of **108** afforded **109** in 93% yield with an ee better than 96%. Finally, methoxy displacement by methylmagnesium chloride furnishes PAMPO in almost quantitative yield with an ee of 94%. The last displacement was also efficiently performed with EtMgCl and *p*-AnMgBr but not with

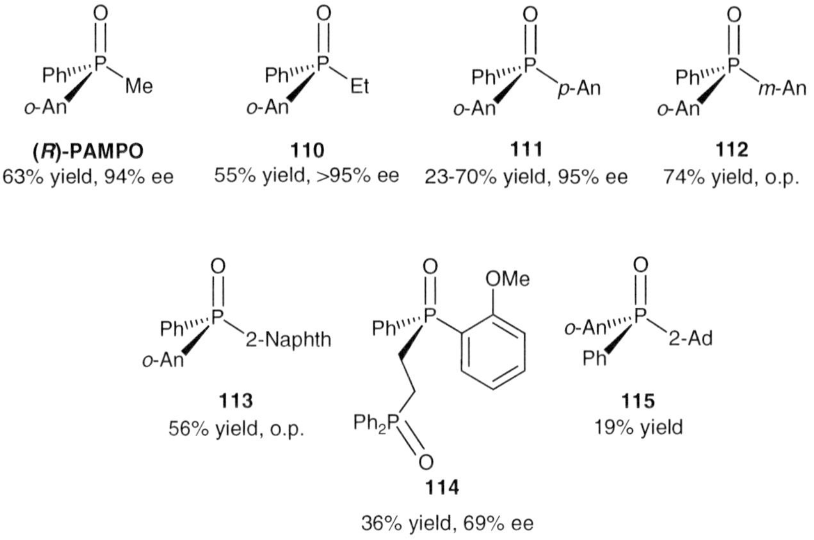

(*R*)-PAMPO
63% yield, 94% ee

110
55% yield, >95% ee

111
23-70% yield, 95% ee

112
74% yield, o.p.

113
56% yield, o.p.

115
19% yield

114
36% yield, 69% ee

Figure 4.3   *P*-stereogenic phosphine oxides prepared with ephedrine methodology.

vinylmagnesium bromide, probably due to the high sensitivity of the product to further nucleophilic attack.[20,63]

There are a few reports following this methodology to prepare *P*-stereogenic phosphine oxides and some were reduced to the free phosphines with silanes.[20,64]

Compounds **110–114** were obtained in similar yields and enantioselectivities to PAMPO (Figure 4.3). Compound **114** is interesting since it represents the dioxide analogue of compound **103** prepared by the Jugé–Stephan method (see Section 4.4.3) and therefore allows for a direct comparison between both procedures. Although the yields are similar, the optical purity of **103** is much higher and can be brought to 100% by a single recrystallisation. Therefore, the phosphine borane methodology is clearly superior in the preparation of *P*-stereogenic phosphines, at least in this case.[20] On the other hand, the extremely bulky 2-adamantyl group instead of Ph in **107** makes substitution reactions exceedingly slow (the first step with *o*-AnMgBr took up to 10 days!) and only proceeded with low stereoselectivity.[31]

4.5.1.2   *Trivalent Oxazaphospholidines Derived from Ephedrine*

There is one report by Buono and co-workers[4] examining the nucleophilic attack on the unprotected oxazaphospholidine **116** (Scheme 4.50).

Ring opening of **116** with *t*-BuLi and subsequent acidolysis with *p*-toluene-sulfonic acid (PTSA) afforded the secondary phosphine oxide **117** in 40% overall yield but with an interesting 85% ee. This reaction is noteworthy because the reader will probably remember that the acidolysis (either with MeOH/H$_2$SO$_4$ or with dry HCl) of *t*-Bu substituted aminophosphine boranes does not occur (see Section 4.3.2).

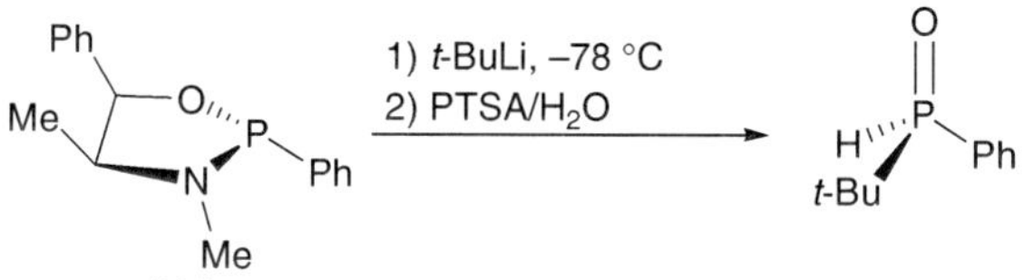

Scheme 4.50 Ring opening and acidolysis of unprotected oxazaphospholidine **116**.

Compound **116** is also reactive toward electrophiles, a feature that has been exploited to prepare enantioenriched *P*-stereogenic tertiary phosphine oxides (Scheme 4.51).[3]

Reaction of unprotected oxazaphospholidine **116** with alkyl halides afforded, after Michaelis–Arbuzov rearrangement, phosphonamides **119** in 85–90% yields and in 40–85% de with predominant retention of configuration at the phosphorus atom. After recrystallisation these compounds could be obtained in optically pure form. The acidic methanolysis of **119** yielded phosphinates **120**, which were treated with Grignard reagents to afford the tertiary phosphine oxides **121**, as seen before. This method is less stereoselective than the 'oxide' version (Section 4.5.1.1) or the Jugé–Stephan method. The loss of stereochemistry has been attributed to the first step in the Michaelis–Arbuzov rearrangement, namely quaternisation of **116** to form the cyclic phosphonium salts **118** and **118'**.[65] ^{31}P NMR spectroscopy revealed that the stereoselectivity in the formation of this salt matches the stereoselectivity observed for **119**. The formation of **118** (which leads to retention of configuration) although preferred, is relatively impaired because the electrophilic attack is on the more hindered side of the oxazaphospholidine and as a consequence some epimerisation occurs, forming **118'**. The exact ratio **118/118'** and therefore the optical purity of the derived products depends on the nature of the electrophilic reagent as well as the experimental conditions of the reaction.

4.5.2 Other Chiral Auxiliaries

The Jugé–Stephan method and its variations reveal that ephedrine is an extraordinary chiral auxiliary in *P*-stereogenic chemistry, highlighted by the simplicity of its structure. In spite of this, there are no reasons to assume that other chiral amino alcohols or, more generally, other heterobifunctional chiral auxiliaries can not be equally good or even outperform ephedrine. In the literature, there are a few early reports on synthetic sequences resembling the Jugé–Stephan method but with other auxiliaries, which are briefly described here.

4.5.2.1 Chiral Amino Alcohols

There is one example reported by Koizumi and co-workers[5] where phosphinates were prepared from bicyclic oxazaphospholidine oxides or sulfides by the sequential nucleophilic substitution and acidic methanolysis (Scheme 4.52).

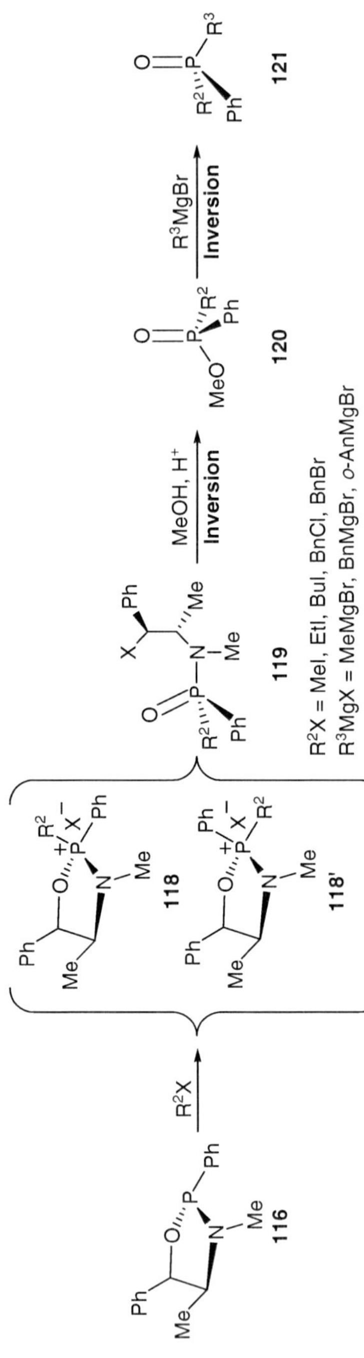

Scheme 4.51 The 'Michaelis–Arbuzov' version of the Jugé–Stephan method.

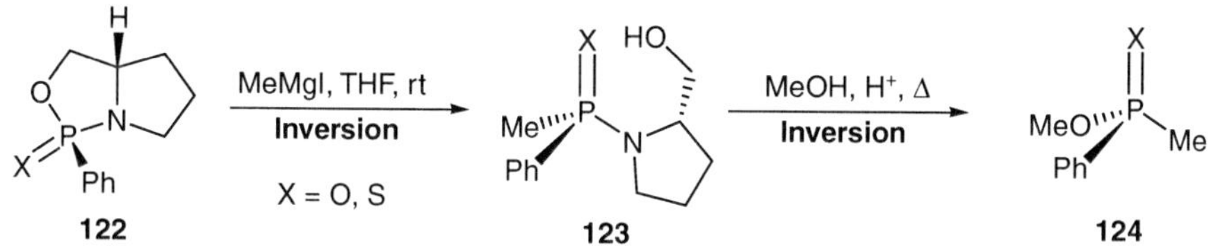

Scheme 4.52 Preparation of a phosphinate from a bicyclic oxazaphospholidine oxide/sulfide.

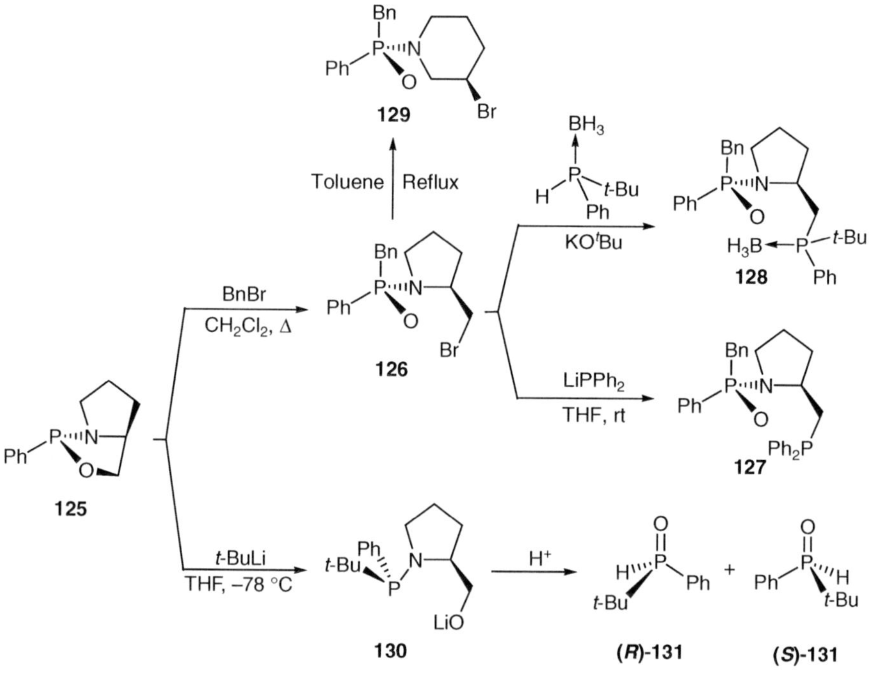

Scheme 4.53 Different transformations of oxazaphospholidine **125**.

Methylmagnesium iodide reacted with **122** (X = O) with P–O bond fission, to afford **123** in quantitative yield. Acidic methanolysis of this compound then produced (***R*_P)-124** in 92% ee. That implies that the first step occurred with inversion of configuration at the phosphorus atom, in contrast with oxazaphospholidine oxides derived from ephedrine. Similar inversion was observed for X = S and for other Grignard reagents. This example emphasises once more the difficulties of drawing general conclusions about the stereochemical outcome of the substitutions at the phosphorus atom.

Oxazaphospholidines derived from (*S*)-prolinol have also been used to prepare some *P*-stereogenic compounds (Scheme 4.53).

Buono and co-workers[66] reacted the compound **125** with benzyl bromide (and other alkyl halides)[67] to produce, after Michaelis–Arbuzov rearrangement, phosphine oxide **126** with retention of configuration both at the phosphorus and at the α carbon atom of the proline ring. Interestingly, it was observed that **126** is thermally unstable and rearranges to the methylpyrrolidine phosphine oxide **129** with retention of configuration at the phosphorus atom. Compound **126** was reacted with lithium diphenylphosphide affording phosphine-phosphine oxide ligand **127** in 64% overall yield.[66] The reaction with racemic *t*-butylphenylphosphide borane afforded **128** as a 2:1 mixture of epimers, which could be separated by column chromatography and the individual diastereomers deboronated with DABCO.[67]

Much more recently,[4] compound **125** has been used to prepare the secondary phosphine oxide **131**. Nucleophilic attack of *t*-BuLi to **125** proceeds by P–O bond splitting, as expected, to afford **130** with retention of configuration at the P atom. An acidic work-up of **130** provides the secondary phosphine oxide **131**. Interestingly, the stereochemistry of the acidolysis step depends on the acid employed. With strong acids (sulfuric, triflic and *p*-toluenesulfonic) it occurs with preponderant retention of configuration (giving **(R)-131** in up to 91% ee) whereas with weaker acids (formic, acetic and *p*-nitrobenzoic) inversion of configuration is observed and the *S* enantiomer of **131** is predominantly produced in up to 85% ee. It has to be noted that in this case the presence of a *t*-Bu group does not block the cleavage of the chiral auxiliary.

Martens and co-workers[68] used several oxazaphospholidine boranes to prepare methylphosphinite boranes (Scheme 4.54).

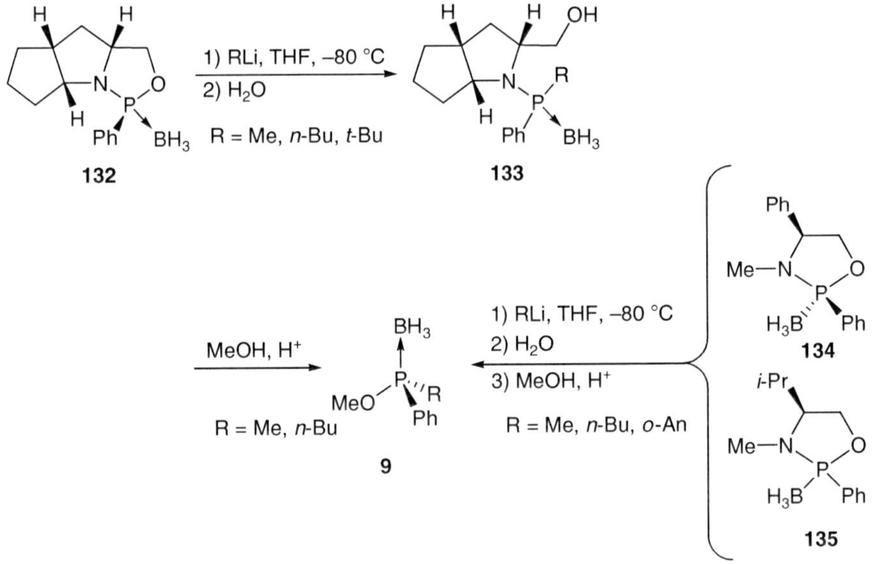

Scheme 4.54 Use of several oxazazaphospholidine boranes to prepare phosphinite boranes.

The reaction of tricyclic compound **132** with alkyllithium reagents afforded the ring opened product **133** as a single diastereomer (although the absolute configurations at the P atom were not established) in 45–56% yields. Their acidic methanolysis furnished the expected methylphosphinite boranes **9** (with *S* absolute configuration) in good yields and optical purities (92% ee for R = Me and 77% for R = *n*-Bu) although it did not occur for R = *t*-Bu. The same study was carried out with compounds **134** and **135** (for the latter, the absolute configuration at the P atom was unknown), yielding **9** with ee values in the range of 76–100%. It has to be mentioned that in both cases the crude ring-opened products were not diastereomerically pure and had to be purified by column chromatography prior to the methanolysis step. The presence of only one stereogenic carbon in the amino alcohol can account for this reduced diastereoselectivity in comparison to ephedrine.

4.5.2.2 Other Heterobifunctional Chiral Auxiliaries

Corey's[69] route to DiPAMP is based on nucleophilic substitutions on the optically pure thiosphosphonate **136**, prepared from a camphor derivative (Scheme 4.55 and Chapter 3, Section 3.4.1).

Nucleophilic substitution on **136** with *o*-AnLi occurred with exclusive P–S bond cleavage and afforded, after sulfur protection, thiophosphinate ester **137** in high yield with clean retention of configuration. An excess of MeLi (7 equivalents) and subsequent quenching with methanol provided the PAMPS as a single enantiomer in 80% yield. The boron trifluoride-etherate step was necessary to cleave the TBS ether and release the chiral auxiliary, which was recovered in high yields. From PAMPS, DiPAMP is accessible by known routes involving DiPAMP disulfide or PAMP · BH$_3$ without affecting the optical purity of the phosphorus atom. With this method another diphosphine of the DiPAMP family (with *t*-Bu groups instead of *o*-An) was also prepared in high optical purity.

This route is often cited in the literature along with the Jugé–Stephan method as an example of a procedure to prepare *P*-stereogenic phosphines *via* enantioselective synthesis. In spite of that, it has unfortunately not been applied to prepare other ligands, so its potential remains unknown. The commercial availability of ephedrine but not of the thio alcohol required to prepare **136** is a likely explanation of the lack of development of this method.

Jugé and co-workers[70] studied the enantioselective synthesis of organophosphorus compounds from a dioxaphospholidine borane (**138**) derived from (*S*)-1,1-diphenylpropane-1,2-diol (**142**) (Scheme 4.56).

Compound **138** reacted with organolithium reagents at very low temperatures (−120 to −78 °C) to afford ring-opened phosphinite boranes **139** in good yields. The reaction was found to proceed regioselectively, with cleavage of the P–O bond next to the carbon bearing the two phenyl groups. Two epimers of **139** were formed with different configurations at the P atom, in up to 86% de. In the case of R^1 = Me one of the epimers was obtained pure after recrystallisation. Reaction, in THF, between a second organolithium and **139**

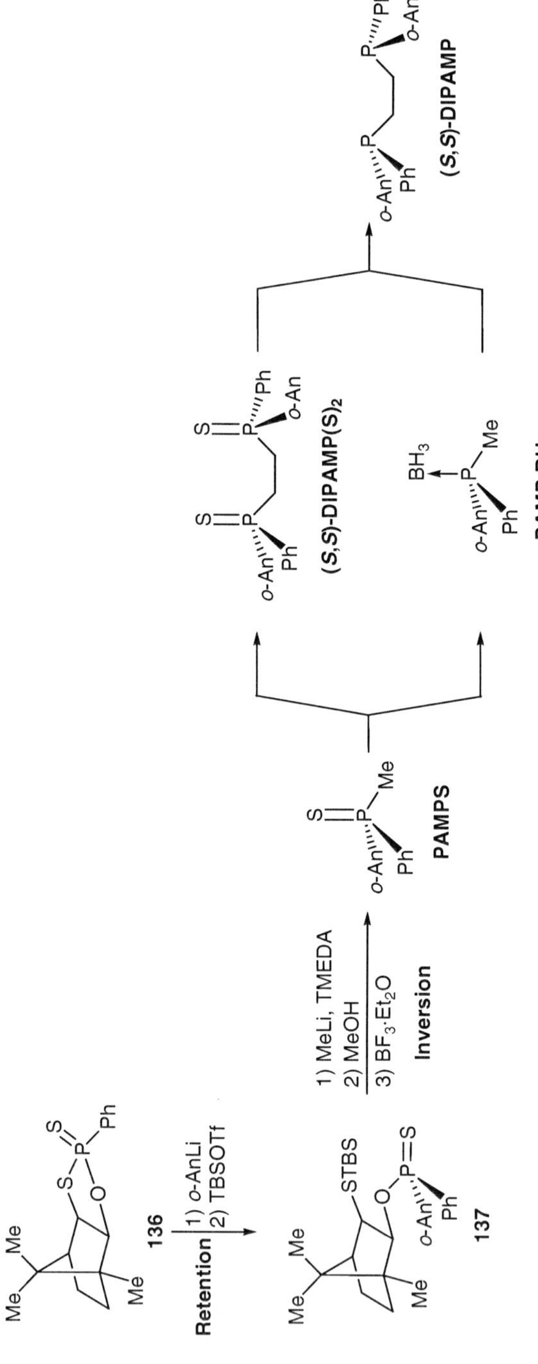

Scheme 4.55 Corey's route to DiPAMP.

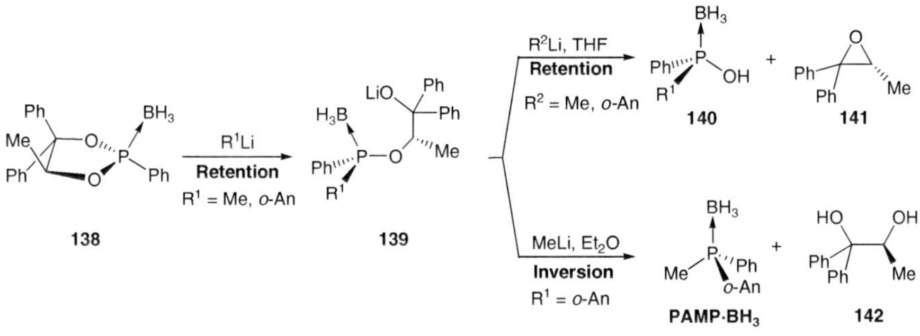

Scheme 4.56 Reactivity of dioxaphospholidine borane **138**. Only one isomer of compounds **139** and **140** is shown.

decomposed it to phosphinous acid borane **140** and epoxide **141**, the latter with the stereogenic carbon atom inverted. In contrast, methyllithium reacted with **139** ($R^1 = o$-An) in diethyl ether through substitution, leading to PAMP·BH$_3$ in 43% yield and 70% ee and recovery of the chiral diol **142**.

The regio- and stereochemistry of the reactions was explained by addition–elimination mechanisms involving pentacoordinated intermediates.

The comparison between this section and a previous one (Section 4.3) about the Jugé–Stephan method shows that ephedrine has totally overshadowed other auxiliaries in the preparation of *P*-stereogenic ligands. However, the Jugé–Stephan method has its limitations and therefore the study of other bifunctional auxiliaries is a rather unexplored area, which awaits further development.

4.6 Conclusion

The Jugé–Stephan method and its variants represent one of the very few relatively general methods to prepare *P*-stereogenic compounds not involving (tedious) separations of diastereomers. It is very well suited to prepare many types of P(III) and P(V) compounds and in particular mono- and diphosphines. Superior ligands of the DiPAMP family, others incorporating one or more ferrocenyl units or dendrimers functionalised with *P*-stereogenic units have become therefore accessible in multi-gram amounts from cheap starting materials.

In spite of all the research carried out, the full potential of the methodology has yet to be completely uncovered. For example, almost all the ligands incorporate a phenyl group in the structure because the phenyloxazapho-spholidine borane is used as starting material. There is no reason to think, however, that other (aryl) groups can not be used and indeed a few ligands devoid of unsubstituted phenyl groups have already appeared. The chlorophosphine boranes, although more difficult to handle, have also shown their high reactivity towards *soft* nucleophiles (alcohols, amines, thiols, ...), which represents a doorway towards a huge number of *P*-stereogenic compounds. The recently discovered metallation with *t*-BuLi is also an interesting advance, which awaits further development.

Despite its versatility, the ephedrine methodology clearly suffers from some limitations. Exceedingly bulky phosphines (bearing mesityl or 9-anthryl groups for example) can only be prepared in low yields if at all. This is because the method is based on nucleophilic substitutions at the phosphorus atom, which become progressively more difficult with increased steric shielding. Another limitation is that all the phosphines prepared bear at least one aryl group (almost exclusively phenyl) and are therefore very basic, trialkylphosphines are not accessible *via* the ephedrine method. This limitation has been partially overcome by another complementary and extremely important method: the enantioselective deprotonation of dimethylphosphine derivatives described in detail in the next chapter.

References

1. S. Jugé, *Phosphorus, Sulfur, and Silicon*, 2008, **183**, 233.
2. C. Darcel, J. Uziel and S. Jugé, in *Phosphorus Ligands in Asymmetric Catalysis: Synthesis and Applications*, A. Börner, (ed.), Wiley-VCH, Weinheim, 2008, p. 1211.
3. S. Jugé and J. P. Genêt, *Tetrahedron Lett.*, 1989, **30**, 2783.
4. A. Leyris, D. Nuel, L. Giordano, M. Achard and G. Buono, *Tetrahedron Lett.*, 2005, **46**, 8677.
5. T. Koizumi, R. Yanada, H. Takagi, H. Hirai and E. Yoshii, *Tetrahedron Lett.*, 1981, **22**, 571.
6. S. Jugé, M. Stephan, J. A. Laffitte and J. P. Genêt, *Tetrahedron Lett.*, 1990, **31**, 6357.
7. S. Jugé, M. Stephan, R. Merdès, J. P. Genêt and S. Halut-Desportes, *J. Chem. Soc., Chem. Commun.*, 1993, 531.
8. A. J. Rippert, A. Linden and H. Hansen, *Helv. Chim. Acta*, 2000, **83**, 311.
9. M. Stephan, D. Sterk, B. Modec and B. Mohar, *J. Org. Chem.*, 2007, **72**, 8010.
10. H. Yang, N. Lugan and R. Mathieu, *Organometallics*, 1997, **16**, 2089.
11. D. Moulin, C. Darcel and S. Jugé, *Tetrahedron: Asymmetry*, 1999, **10**, 4729.
12. R. Ewalds, E. B. Eggeling, A. C. Hewat, P. C. J. Kamer, P. W. N. M. van Leeuwen and D. Vogt, *Chem. Eur. J.*, 2000, **6**, 1496.
13. E. B. Kaloun, R. Merdès, J. P. Genêt, J. Uziel and S. Jugé, *J. Organomet. Chem.*, 1997, **529**, 455.
14. C. Bauduin, D. Moulin, E. B. Kaloun, C. Darcel and S. Jugé, *J. Org. Chem.*, 2003, **68**, 4293.
15. E. A. Colby and T. F. Jamison, *J. Org. Chem.*, 2003, **68**, 156.
16. A. Grabulosa, PhD Thesis, Universitat de Barcelona, Barcelona, 2005.
17. F. Maienza, F. Spindler, M. Thommen, B. Pugin, C. Malan and A. Mezzetti, *J. Org. Chem.*, 2002, **67**, 5239.
18. M. Schuman, M. Trevitt, A. Redd and V. Gouverneur, *Angew. Chem. Int. Ed.*, 2000, **39**, 2491.

19. K. S. Dunne, S. E. Lee and V. Gouverneur, *J. Organomet. Chem.*, 2006, **691**, 5246.
20. J. A. Ramsden, J. M. Brown, M. B. Hursthouse and A. I. Karalulov, *Tetrahedron: Asymmetry*, 1994, **5**, 2033.
21. H. Yang, M. Alvarez-Gressier, N. Lugan and R. Mathieu, *Organometallics*, 1997, **16**, 1401.
22. U. Nettekoven, P. C. J. Kamer, P. W. N. M. van Leeuwen, M. Widhalm, A. L. Spek and M. Lutz, *J. Org. Chem.*, 1999, **64**, 3996.
23. B. Zupancic, B. Mohar and M. Stephan, *Tetrahedron Lett.*, 2009, **50**, 7382.
24. B. Zupancic, B. Mohar and M. Stephan, *Adv. Synth. Catal.*, 2008, **350**, 2024.
25. B. Zupancic, B. Mohar and M. Stephan, *Org. Lett.*, 2010, **12**, 1296.
26. M. B. Tollefson, J. J. Li and P. Beak, *J. Am. Chem. Soc.*, 1996, **118**, 9052.
27. C. Darcel, D. Moulin, J. Henry, M. Lagrelette, P. Richard, P. D. Harvey and S. Jugé, *Eur. J. Org. Chem.*, 2007, 2078.
28. R. M. Stoop, A. Mezzetti and F. Spindler, *Organometallics*, 1998, **17**, 668.
29. U. Nettekoven, M. Widhalm, H. Kalchhauser, P. C. J. Kamer, P. W. N. M. van Leeuwen, M. Lutz and A. L. Spek, *J. Org. Chem.*, 2001, **66**, 759.
30. U. Nettekoven, P. C. J. Kamer and P. W. N. M. van Leeuwen, *Organometallics*, 2000, **19**, 4596.
31. J. M. Brown and J. C. P. Laing, *J. Organomet. Chem.*, 1997, **529**, 435.
32. A. Grabulosa, G. Muller, J. I. Ordinas, A. Mezzetti, M. A. Maestro, M. Font-Bardia and X. Solans, *Organometallics*, 2005, **24**, 4961.
33. J. Omelanczuk, *J. Chem. Soc. Chem. Commun.*, 1992, 1718.
34. W. Perlikowska, M. Gouygou, J. C. Daran, G. Balavoine and M. Mikolajczyk, *Tetrahedron Lett.*, 2001, **42**, 7841.
35. A. Rauk, L. C. Allen and K. Mislow, *Angew. Chem. Int. Ed.*, 1970, **9**, 400.
36. S. Humbel, C. Bertrand, C. Darcel, C. Bauduin and S. Jugé, *Inorg. Chem.*, 2003, **42**, 420.
37. D. Moulin, S. Bago, C. Bauduin, C. Darcel and S. Jugé, *Tetrahedron: Asymmetry*, 2000, **11**, 3939.
38. C. Darcel, E. B. Kaloun, R. Merdès, D. Moulin, N. Riegel, S. Thorimbert, J. P. Genêt and S. Jugé, *J. Organomet. Chem.*, 2001, **624**, 333.
39. U. Nettekoven, M. Widhalm, P. C. J. Kamer, P. W. N. M. van Leeuwen, K. Mereiter, M. Lutz and A. L. Spek, *Organometallics*, 2000, **19**, 2299.
40. F. Maienza, M. Wörle, P. Steffanut, A. Mezzetti and F. Spindler, *Organometallics*, 1999, **18**, 1041.
41. U. Nettekoven, M. Widhalm, P. C. J. Kamer and P. W. N. M. van Leeuwen, *Tetrahedron: Asymmetry*, 1997, **8**, 3185.
42. L. S. von Chrzanowski, M. Lutz and A. L. Spek, *Acta Cryst.*, 2007, **E63**, m2098.
43. D. Vinci, N. Mateus, X. Wu, F. Hancock, A. Steiner and J. Xiao, *Org. Lett.*, 2006, **8**, 215.
44. F. Agbossou, J. Carpentier, F. Hapiot, I. Suisse and A. Mortreux, *Coord. Chem. Rev.*, 1998, **178–180**, 1615.
45. F. Agbossou-Niedecorn and I. Suisse, *Coord. Chem. Rev.*, 2003, **242**, 145.

46. F. Agbossou-Niedecorn, in *Phosphorus Ligands in Asymmetric Catalysis: Synthesis and Applications*, A. Börner, (ed.), Wiley-VCH, Weinheim, 2008, p. 477.

47. R. den Heeten, B. H. G. Swennenhuis, P. W. N. M. van Leeuwen, J. G. de Vries and P. C. J. Kamer, *Angew. Chem. Int. Ed.*, 2008, **47**, 6602.

48. M. J. Johansson, K. H. O. Andersson and N. Kann, *J. Org. Chem.*, 2008, **73**, 4458.

49. J. M. Camus, J. Andrieu, P. Richard, R. Poli, C. Darcel and S. Jugé, *Tetrahedron: Asymmetry*, 2004, **15**, 2061.

50. H. Lam, P. N. Horton, M. B. Hursthouse, D. J. Aldous and K. K. Hii, *Tetrahedron Lett.*, 2005, **46**, 8145.

51. S. Vargas, M. Rubio, A. Suárez, D. del Río, E. Álvarez and A. Pizzano, *Organometallics*, 2006, **25**, 961.

52. S. Deerenberg, P. C. J. Kamer and P. W. N. M. van Leeuwen, *Organometallics*, 2000, **19**, 2065.

53. L. Rodríguez, O. Rossell, M. Seco, A. Grabulosa, G. Muller and M. Rocamora, *Organometallics*, 2006, **25**, 1368.

54. T. Imamoto, T. Hoshiki, T. Onozawa, T. Kusumoto and K. Sato, *J. Am. Chem. Soc.*, 1990, **112**, 5244.

55. M. Stephan, D. Sterk and B. Mohar, *Adv. Synth. Catal.*, 2009, **351**, 2779.

56. R. M. Stoop, C. Bauer, P. Setz, M. Wörle, T. Y. H. Wong and A. Mezzetti, *Organometallics*, 1999, **18**, 5691.

57. G. Zhu, M. Terry and X. Zhang, *Tetrahedron Lett.*, 1996, **37**, 4475.

58. G. Zhu, M. Terry and X. Zhang, *J. Organomet. Chem.*, 1997, **547**, 97.

59. F. Maienza, F. Santoro, F. Spindler, C. Malan and A. Mezzetti, *Tetrahedron: Asymmetry*, 2002, **13**, 1817.

60. L. I. Rodríguez, O. Rossell, M. Seco and G. Muller, *J. Organomet. Chem.*, 2009, **694**, 1938.

61. L. Rodríguez, O. Rossell, M. Seco and G. Muller, *Organometallics*, 2008, **27**, 1328.

62. C. R. Hall, T. D. Inch and I. W. Lawston, *Tetrahedron Lett.*, 1979, **29**, 2729.

63. J. M. Brown, J. V. Carey and M. J. H. Russell, *Tetrahedron*, 1990, **46**, 4877.

64. J. V. Carey, M. D. Barker, J. M. Brown and M. J. H. Russell, *J. Chem. Soc. Perkin Trans.*, 1993, **1**, 831.

65. S. Jugé, M. Wakselman, M. Stephan and J. P. Genêt, *Tetrahedron Lett.*, 1990, **31**, 4443.

66. B. Faure, A. Archavlis and G. Buono, *J. Chem. Soc. Chem. Commun.*, 1989, 805.

67. B. Faure, O. Pardigon and G. Buono, *Tetrahedron*, 1997, **53**, 11577.

68. V. Peper, K. Stingl, H. Thumler, W. Saak, D. Haase, S. Pohl, S. Jugé and J. Martens, *Liebigs Ann.*, 1995, 2123.

69. E. J. Corey, Z. Chen and G. J. Tanoury, *J. Am. Chem. Soc.*, 1993, **115**, 11000.

70. J. Uziel, M. Stephan, E. B. Kaloun, J. P. Genêt and S. Jugé, *Bull. Soc. Chim. Fr.*, 1997, **134**, 379.

CHAPTER 5

P-Stereogenic Phosphines Prepared by Enantioselective Deprotonation

5.1 Introduction

The methods described in this chapter are based on the enantioselective deprotonation of alkyl groups in – usually achiral – phosphine derivatives, which provide highly enantioenriched α-carbanions that are versatile precursors of a variety of mono- and diphosphines. This method was developed by combining two known facts: firstly, that the methyl group in methylphosphines and their oxides, boranes and sulfides can be deprotonated with strong bases[1] and secondly common organolithium reagents (*n*-, *s*- or *t*-BuLi) can exert enantioselective deprotonations in prochiral substrates in the presence of certain chiral auxiliaries.[2,3]

In 1989 White, Raston and co-workers[4] put into practice these ideas using the chiral diamine (–)-sparteine ((–)-sp, Scheme 5.1).

The (–)-enantiomer of sparteine is obtained from natural sources as a commercially available and relatively inexpensive oil. The metallation of **1** with *n*-BuLi in hexane, in the presence of (–)-sp, produced the α-carbanion **2** which was subsequently quenched affording the known phosphine oxide **3**[5,6] in approximately 14% ee. Although it has been suggested that the low stereoselectivity is due to the disruption of the (–)-sp/BuLi complex by complexation of the oxygen atom to the lithium cation,[7] the same reaction with unprotected dimethylphenylphosphine provides completely racemic products. Due to the low enantioselectivity, this method was initially ignored in phosphorus chemistry until a seminal paper of Evans and co-workers[8] in which they described similar reactions with phosphine boranes and sulfides. In this case, the enantioselectivities were much higher and the method could be successfully applied

RSC Catalysis Series No. 7
P-Stereogenic Ligands in Enantioselective Catalysis
By Arnald Grabulosa
© Arnald Grabulosa 2011
Published by the Royal Society of Chemistry, www.rsc.org

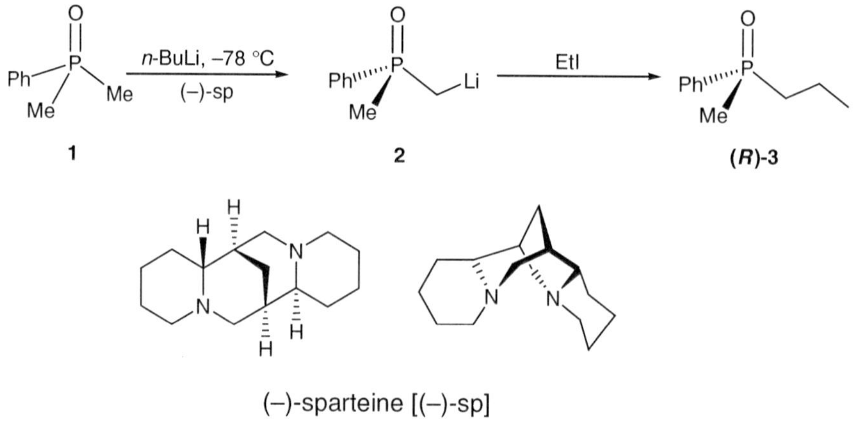

(−)-sparteine [(−)-sp]

Scheme 5.1 First enantioselective deprotonation–substitution sequence in a dimethylphosphorus compound and two views of the structure of (−)-sparteine.

to the synthesis of several C_2-diphosphines, as discussed in detail in this chapter. Other groups have expanded this chemistry enormously to access a range of compounds in high optical purity (Scheme 5.2).[7]

This method is not only restricted to dimethylphosphine boranes and sulfides. It has been also applied to cyclic diphosphine derivatives by deprotonation of one of the enantiotopic methylene groups α to the phosphorus atom (Scheme 5.3). In this case, a second stereogenic centre is created and therefore it is necessary to control also the diastereoselectivity of the reactions.

This chapter describes the preparation of the compounds depicted in Schemes 5.2 and 5.3 by the enantioselective deprotonation protocol. It has become an extremely important procedure to prepare some interesting families of ligands including many electron-rich bulky diphosphines, thus complementing the Jugé–Stephan method.

5.2 Deprotonation of Dimethylphosphine Derivatives

5.2.1 Dimethylphosphine Boranes

5.2.1.1 *Monophosphines*

In 1995 Evans and co-workers[8] reported the first synthetically useful enantioselective deprotonation–electrophilic quenching of dimethylphosphine boranes (Scheme 5.4).

The deprotonation with *s*-BuLi of diethyl ether solutions of several aryldimethylphosphine boranes **4** in the presence of (−)-sp generated the corresponding carbanions **5**, which were quenched with benzophenone to afford derivatives **6** in good yield and optical purity (Table 5.1).

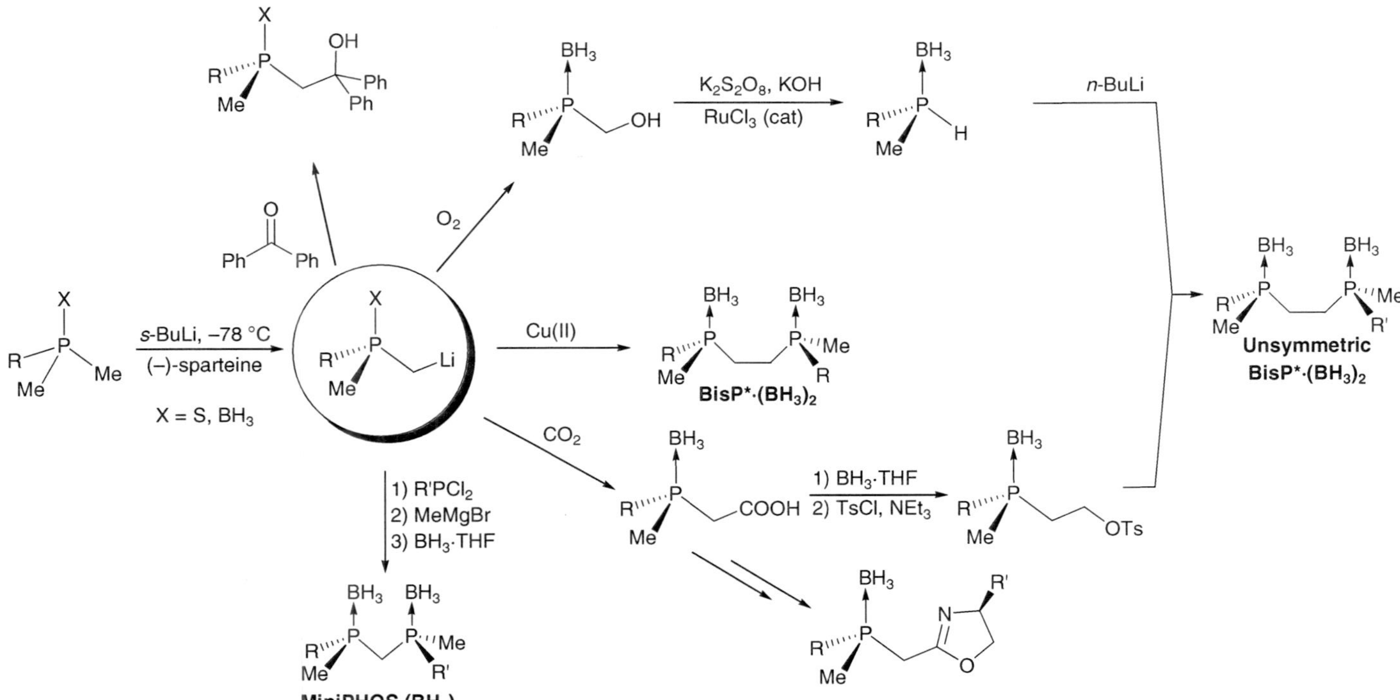

Scheme 5.2 Overview of compounds obtainable by enantioselective deprotonation of dimethylphosphine derivatives.

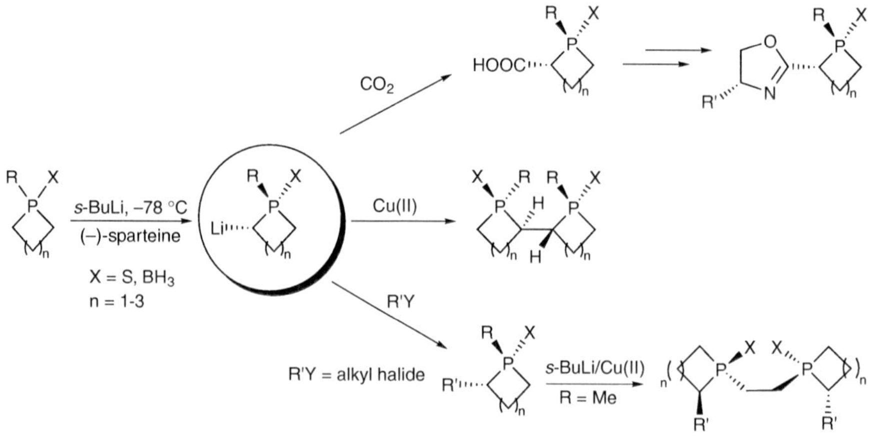

Scheme 5.3 Overview of compounds obtainable by enantioselective deprotonation of cyclic phosphine derivatives.

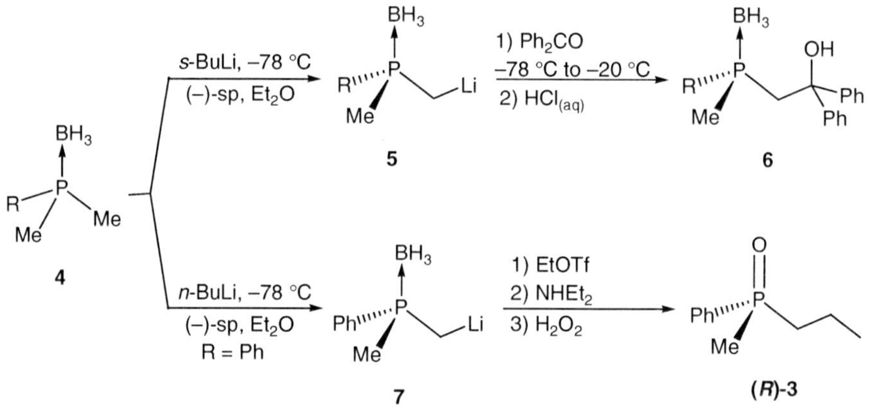

Scheme 5.4 Enantioselective deprotonation–electrophilic quenching of aryldimethylphosphine boranes.

Table 5.1 Enantioselective deprotonation of aryldimethylphosphine boranes (Scheme 5.4).

Entry	R	Yield (%)	ee (%)
1	Ph	88	79
2	o-An	81	83
3	o-Tol	84	87
4	1-naphthyl	86	82

The absolute configuration of the major enantiomer of **6**, *S*, was determined by chemical correlation thanks to the preparation of the known phosphine oxide **3**,[5,6] and was assumed to be the same for the rest of phosphine boranes **6**.

Livinghouse and co-workers[9,10] reported a variant of the method in order to improve the optical purity of the products (Scheme 5.5).

The alkoxide group in **6** was acylated with trimethylacetyl (pivaloyl) chloride to furnish adducts **8** in very good yields. These products were found to be highly crystalline, allowing their optical purity to be increased to >99% ee by a simple recrystallisation. Once optically pure, they were reduced by lithium naphthalenide and protonated with methanol affording the optically pure secondary phosphine boranes **9** (Table 5.2).

In short, the sequence enantioselective lithiation–electrophilic trapping–reductive elimination represents an enantioselective demethylation of **4** to produce optically pure **9**. As seen through this book, secondary phosphine boranes are versatile synthons in *P*-stereogenic chemistry. In this case, they were alkylated with 2-(chloromethyl)benzothiophene providing phosphine boranes **10** in good yields, which were used in HPLC analysis to evaluate the optical purity of **9**.

The enantioselective deprotonation protocol has been expanded to alkyldimethylphosphine boranes. Kann and co-workers[11] subjected *t*-butyl- and

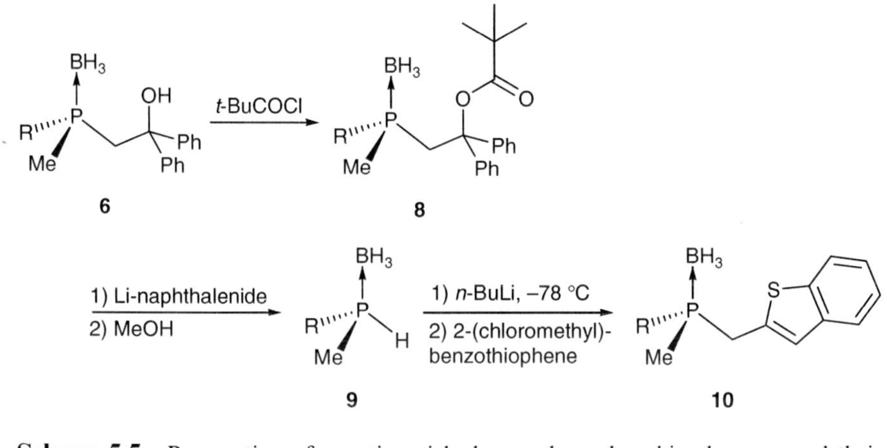

Scheme 5.5 Preparation of enantioenriched secondary phosphine boranes and their alkylation.

Table 5.2 Secondary phosphine boranes **9** (Scheme 5.5).

Entry	R	Yield (%)	ee (%)
1	Phenyl	94	>99
2	o-(i-Pr)C$_6$H$_4$	81	>98
3	o-An	88	>99
4	o-Tol	71	>99

cyclohexyldimethylphosphine borane to the reactions of Scheme 5.4, resulting in 76% yield and 70% ee respectively for **6**. Imamoto and co-workers[12] oxidised the α-carbanions with molecular oxygen to afford highly enantioenriched alkyl(hydroxymethyl)methylphosphine boranes **11** (Scheme 5.6 and Table 5.3).

In all cases the products were obtained in good yields and enantioselectivities. It was found that the latter increased with the steric hindrance of R, reaching the highest value for the very bulky 1-adamantyl group (entry 3). In this case, a single recrystallisation afforded a virtually enantiopure product. Unfortunately, the same procedure could not be applied to the other compounds because of their oily nature.

Hydroxymethylphosphine boranes **11** were found to be useful intermediates for further transformations (Scheme 5.7).

Following usual methods, **11** could be easily converted to the phosphine/phosphite diboranes **12**[12] and to compounds **13**,[12,13] which bear a leaving group

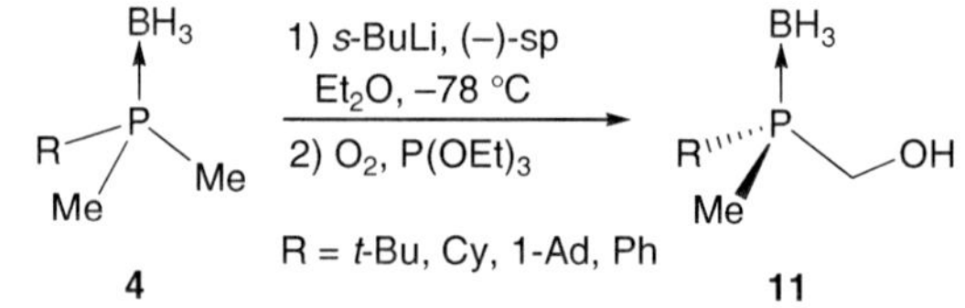

Scheme 5.6   Preparation of enantioenriched hydroxymethylphosphine boranes.

Table 5.3   Hydroxymethylphosphine boranes (Scheme 5.6).

Entry	R	Yield (%)	ee (%)
1	*t*-butyl	73	91
2	Cyclohexyl	70	75
3	1-adamantyl	74	93 (crude) 99 (recr.)
4	Phenyl	67	81

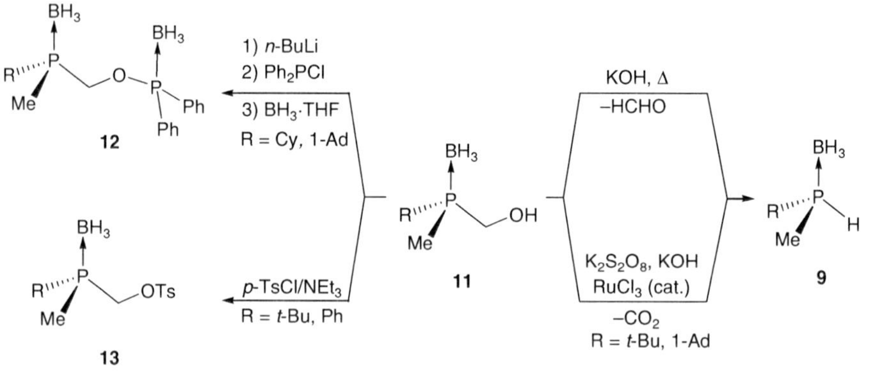

Scheme 5.7   Transformations of hydroxymethylphosphine boranes **11**.

in the α position. Both transformations proceeded in good yields and did not affect the optical purity of the phosphorus centre. More interestingly, the one-carbon degradation of the hydroxymethyl unit of **11** (R = 1-Ad, 99% ee) was investigated as a route to optically enriched secondary phosphine boranes **9**. Deformylation with excess of KOH in toluene required high temperatures (>80 °C) to proceed to acceptable conversions but at the expense of optical purity of **9** (5–38% ee). In contrast, oxidative degradation met with more success. Typical Cr(VI) oxidative reagents provided **9** in low yields but in high optical purity, proving that the transformation was enantioselective. Finally, Ru-catalysed oxidation using potassium peroxodisulfate afforded **9** (R = 1-adamantyl) in 91% yield and 99% ee. For R = *t*-Bu the reactions proceed equally well, but in this case the enantiomeric excess of starting material **11** was only 91% (Table 5.3). In order to increase the optical purity of the final product, the hydroxyl group in **11** was benzylated (BnCl, pyridine) to afford the corresponding ether as a solid that was recrystallised twice to give the enantiopure compound.[14,15] After hydrolysis (KOH, EtOH), the same oxidative degradation was followed to furnish the optically pure compound **9** in 80% yield. A limitation of the oxidative degradation method is that it cannot be applied to the phenyl precursor **14** because another product was formed (Scheme 5.8).

It was found that **14** was over-oxidised to phosphinous acid borane **15**, probably due deprotonation of the formed secondary phosphine borane and immediate oxidation of the phosphide anion. Interestingly, the analysis of the optical purity of **16**, prepared by methylation of **16**, showed that the reactions were enantioselective.

The optically pure secondary phosphine boranes are easily deprotonated, giving extremely nucleophilic phosphide borane anions, which preserve the absolute configuration at the P atom at low temperatures. This fact prompted Danjo, Imamoto and co-workers[16] to couple the phosphide boranes derived from **9** with aryl cation equivalents. In particular, they found that fluoroarene tricarbonylchromium complexes such as **17** were excellent substrates for the S$_N$Ar process due to the high electron deficiency of the arene ring and significant electrophilicity of the *ipso* carbon (Scheme 5.9).

The reactions proceed smoothly in THF at room temperature with *n*-BuLi as a base, although an excess of phosphine borane (up to 3.3 equivalents) was required to obtain high yields of **18**. Further optimisation was needed to ensure good enantioselectivity. It was found that the reaction had to be carried out at

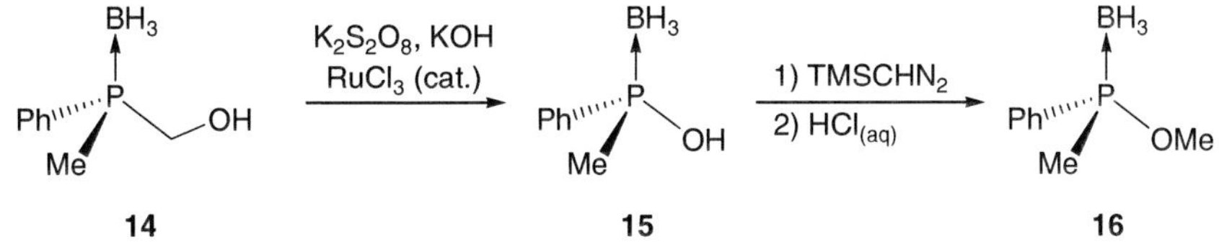

Scheme 5.8 Over-oxidation of **14** to **15** and methylation of the latter.

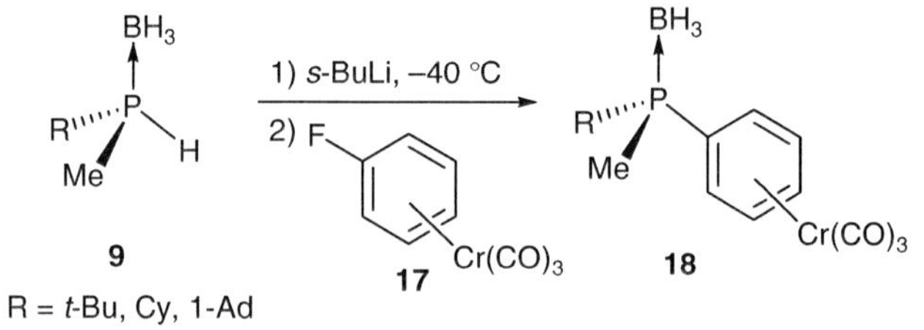

Scheme 5.9 S$_N$Ar between **9** and fluorobenzene tricarbonylchromium.

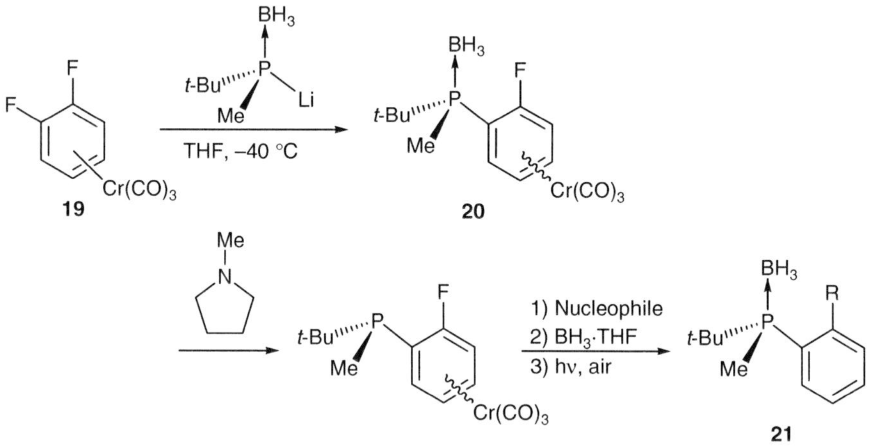

Scheme 5.10 Preparation of *ortho*-substituted phosphine boranes.

low temperature ($-40\,^{\circ}$C or $-78\,^{\circ}$C) and with *s*-BuLi instead of *n*-BuLi to avoid partial racemisation of the phosphide borane. Under these conditions, the optical purity of **18** was virtually the same as the precursor **9**, and the high yields were maintained (81–93%). The same reactions were unsuccessful with bromo- and chlorobenzene tricarbonylchromium complexes, in accordance with the known fact that the small and electronegative fluorine atom accelerates the first step in the addition–elimination mechanism of aromatic nucleophilic substitution.

The same method was applied to *o*-difluorobenzene tricarbonylchromium with the aim of obtaining *ortho*-substituted monophosphines and diphosphines (Scheme 5.10). The first nucleophilic attack provided the desired mono-substituted compound (**20**) in 97% yield, but no traces of diphosphine borane could be detected, even using excess of phosphide borane. Compound **20** possesses a second stereogenic element (a plane) and it was found to be present as a mixture of separable diastereomers in 60% de. When this compound was reacted with MeMgBr it was found that the large steric hindrance of the phosphine borane group inhibited the second nucleophilic attack and therefore the deprotection of the phosphine was required (Scheme 5.10).

Once the phosphorus atom was deprotected, the reaction was successful and a range of phosphine boranes could be prepared after reboronation of the phosphorus atom and photochemical decomplexation of the tricarbonylchromium unit (Table 5.4).

Grignard and organolithium reagents were competent nucleophiles for the reaction (entries 1–7) but less basic thiolate and selenide nucleophiles were also effective (entries 8–10). The cyano and azido derivatives (entries 11 and 12 respectively), prepared from the trimethylsilyl precursors, were further functionalised yielding *o*-heterocyclic derivatives (Scheme 5.11).

Table 5.4 Compounds **21** (Scheme 5.10) reported to date.

Entry	Nucleophile	Yield (%)
1	MeMgBr	90
2	*n*-BuMgBr	75
3	*n*-BuLi	22 (60[a])
4	CyMgBr	41
5	CyLi	69
6	PhMgBr	87
7	PhLi	88
8	PhSLi	40
9	*t*-BuSLi	55
10	PhSeLi	52
11[b]	TMSCN	91
12[b]	TMSN$_3$	68

[a]TMEDA (2 equiv.) was added.
[b]Isopropanol (2 equiv.) was added.

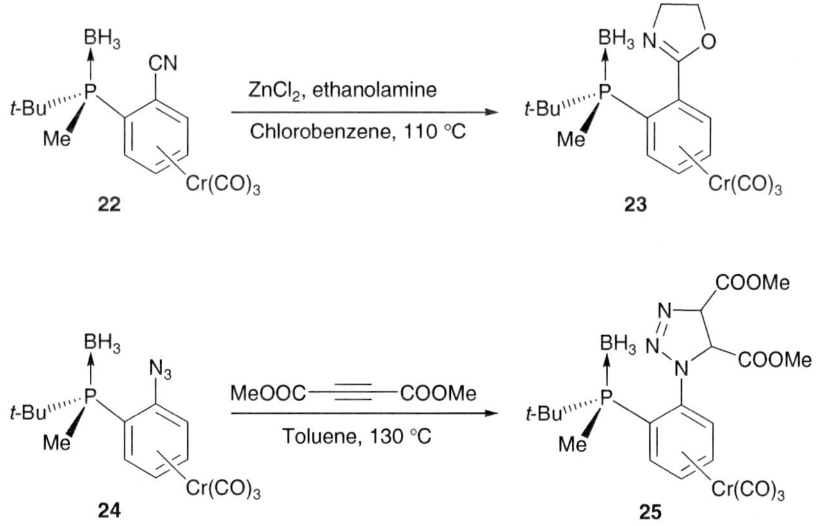

Scheme 5.11 Preparation of *o*-heterocyclic phosphine boranes.

Compounds **23** (42% yield) and **25** (66% yield) are potential P/N hybrid ligands with an oxazoline and triazine ring respectively. It has to be remarked that both the borane and the tricarbonylchromium moieties remained untouched under the rather harsh conditions required to form the heterocycle.

The chemistry of the secondary phosphine boranes has been expanded in a recent contribution of Imamoto's group,[17] in which they disclosed a procedure to convert the P–H into a P–halogen bond stereoselectively (Scheme 5.12).

Enantiomerically pure phosphine borane **26** was deprotonated at low temperature and halogenated with 1,2-dibromo- or 1,2-diiodoethane to furnish compounds **27**, which could be isolated as crystalline solids in good yields. In analogy to chlorophosphine boranes (see Section 4.3.2.2), they are not configurationally stable and racemise slowly at room temperature. Therefore, compounds **27** were not isolated but used *in situ* with several carbanionic reagents to afford the corresponding tertiary phosphine borane with inversion (**28**) or retention of configuration (**28′**) (Table 5.5).

Entries 1–5 constitute the first examples of stereoselective substitutions of halogens by alkynyl groups in *P*-stereogenic compounds. Grignard reagents were also effective in the substitution, albeit in low or moderate yield (entries 6 and 8). Contrary to expectations, the bromide was found to be a more suitable leaving group than the iodine in some cases (compare entry 1 with entry 2 and

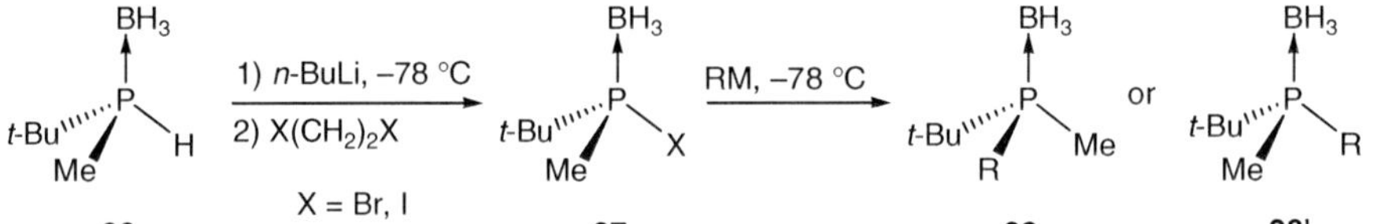

Scheme 5.12   Preparation and reactivity of halophosphine boranes.

Table 5.5   Substitution reactions in bromo- or iodophosphine boranes **27** (Scheme 5.12).

Entry	X	RM	Yield (%)	ee (%)
1	Br	PhCCLi	83	97 (inv.)
2	I	PhCCLi	12	32 (inv.)
3	Br	*t*-BuCCLi	82	98 (inv.)
4	Br	TMSCCLi	81	99 (inv.)
5	Br	(*i*-Pr)$_3$SiCCLi	86	98 (inv.)
6	Br	PhMgBr	28	83 (inv.)
7	I	PhMgBr	0[a]	–
8	Br	*o*-AnMgBr	68	93 (inv.)
9	Br	BuLi	63	91 (ret.)
10	I	BuLi	79	89 (ret.)
11	Br	BnMgCl	94	96 (ret.)
12	I	BnMgCl	92	93 (ret.)

[a]**(S)-26** was recovered in high yield.

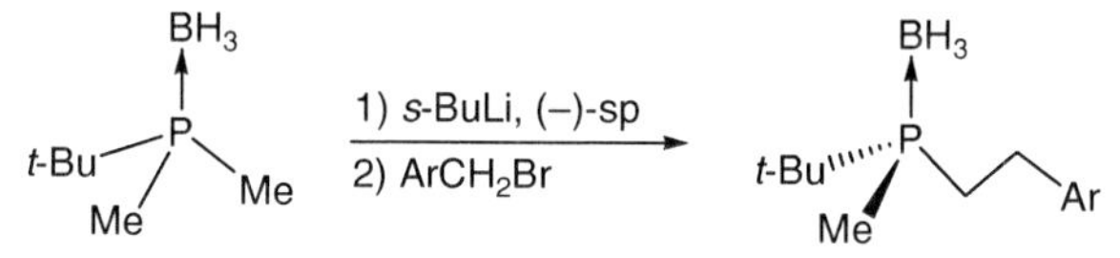

Scheme 5.13 Preparation of phenethylphosphine boranes.

entry 6 with entry 7). In entries 1–8 the reactions occur with inversion of configuration, as expected for a S_N2 process. Unexpectedly, *n*-BuLi and benzylmagnesium chloride afforded products with retention of configuration (entries 9–12). This outcome can be explained by the sequence metal–halogen exchange followed by reaction of the phosphide anion with the alkyl halide formed in the metallation step. This hypothesis is supported by the fact that treatment of **27** (X = Br) first with *t*-BuLi followed by bromide afforded a product with retention of configuration in 99% ee (see end of Section 4.3.3.2).

Reaction of enantioenriched carbanions with benzylic halides allowed Muller and co-workers[18] the access to a group of phenethylphosphine boranes (Scheme 5.13 and Table 5.6).

The yields were low to moderate although with a single recrystallisation the products were found to be optically pure by HPLC.

Reaction of the enantioenriched carbanions with carbon dioxide was originally employed by Ohashi and Imamoto[19,20] and more recently by others[21–23] to access *P*-stereogenic α-carboxyphosphine boranes and some reduced derivatives (Scheme 5.14).

The desired acids **31** were obtained by simply bubbling CO_2 through a solution of the deprotonated phosphine boranes **4** and could be quantitatively reduced to β-hydroxyphosphines **32**. The hydroxyl group of **32** was then mesylated or tosylated in good yields to afford compounds **33**, which were employed to prepare other compounds like P/N and P/S bidentate ligands (see later in this section) and unsymmetric BisP* ligands (Section 5.2.1.3). The yields of compounds **31–33** are listed in Table 5.7.

All compounds were obtained in good yields and in all cases they were highly enantioenriched ($>87\%$ ee), proving that none of the reactions caused any loss of optical purity.

Imamoto and co-workers[24] used carboxylic acids **34** to prepare of a family of *P*-stereogenic phosphine/oxazoline ligands (Scheme 5.15).

Here the coupling system EDC/HOBt was used to form the amide bond between **34** and the desired amino alcohol to afford **35**. The mesylation of the hydroxyl group prompted the cyclisation providing **36**, which could be easily deboronated with tetrafluoroboric acid and used in Pd-catalysed allylic substitution reactions. The compounds reported are listed in Table 5.8.

The desired compounds were prepared in good yields and excellent optical purity. The stereogenic centres at the phosphorus and in the oxazoline ring could be independently tuned with several substituents or eliminated (entries 3

Table 5.6 Yields of the reported phenethylphosphine boranes **30** (Scheme 5.13).

Entry	$ArCH_2Br$	Yield (%)[a]
1	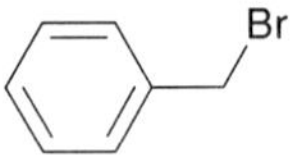	67
2	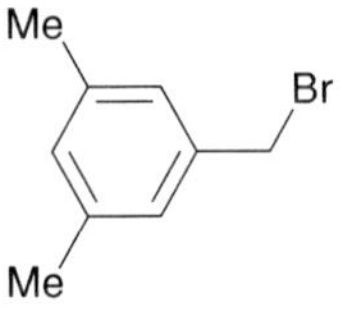	33
3	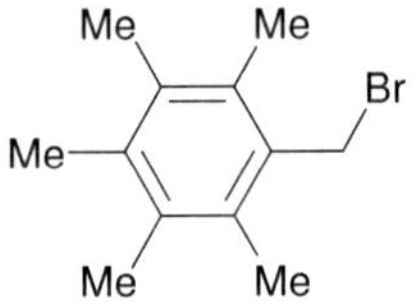	37
4	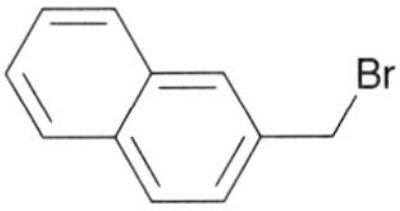	31
5	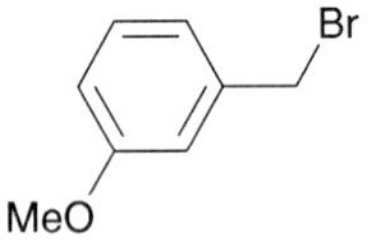	49
6	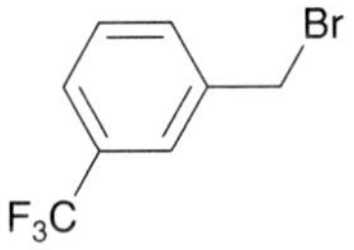	44

[a]Final yields of **30** after a single recrystallisation.

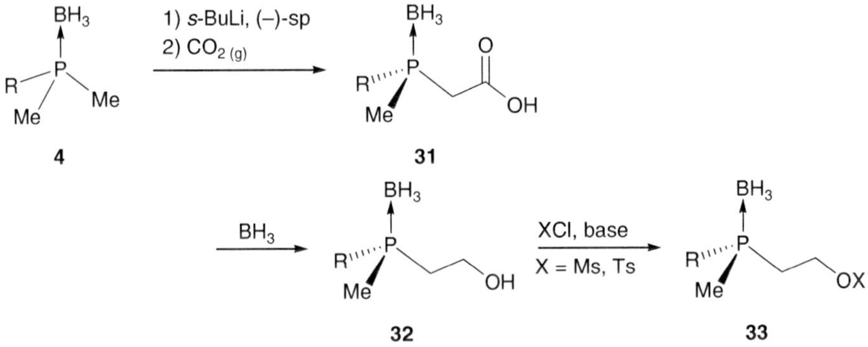

Scheme 5.14 Preparation of boronated α-carboxyphosphine boranes and some derivatives.

Table 5.7 Reported yields (%) for compounds **31**–**33**.

Entry	R	31	32	33	References
1	*t*-Bu	70	95–quant.	70–94	19–21
2	Cy	60	88–quant.	64–96	19,21
3	1-Ad	70	98–quant.	45–97	19–21
4	Ph	–	71 (87% ee)	NP[a]	23
5	Mes	90	80	65[b]	21
6	Fc	–	80	84	21,24

[a]Not prepared.
[b]The mesylate or tosylate derivatives could not be prepared, instead the bromide was obtained by reaction of **32** with CBr_4 and triphenylphosphine.[21]

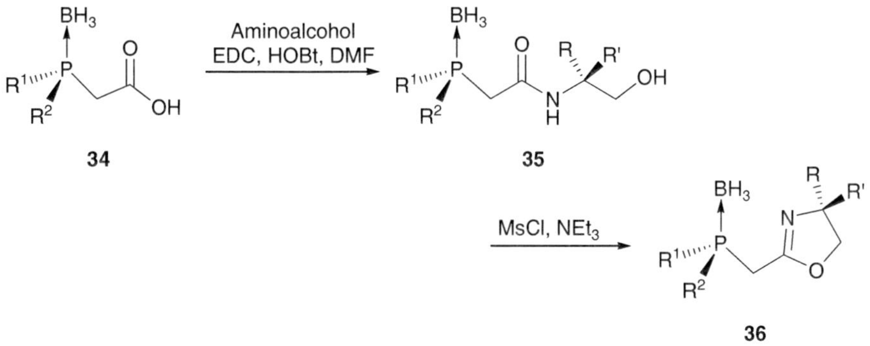

Scheme 5.15 Preparation of *P*-stereogenic phosphine/oxazoline boranes.

Table 5.8 *P*-stereogenic phosphine/oxazoline boranes **36** (Scheme 5.15).

Entry	R^1	R^2	R	R'
1	*t*-Bu	Me	H	*i*-Pr
2	*t*-Bu	Me	*i*-Pr	H
3[a]	Cy	Cy	H	*i*-Pr
4	*t*-Bu	Me	Me	Me
5	*t*-Bu	Me	H	*t*-Bu
6	1-Ad	Me	H	*t*-Bu
7	Ph	Me	H	*t*-Bu
8	Fc	Me	H	*t*-Bu
9[b]	*t*-Bu	Ph	H	H

[a]In this compound the phosphorus atom is not stereogenic.
[b]See text.

and 4) to study the effect of the different stereogenic elements in the outcome of the catalysis. The compound of entry 9 has a methyl group at the carbon next to the oxygen and was prepared by another method (see Chapter 2, Section 2.2.4) by Reek and co-workers.[25]

Further work by the same group[13] also applied the deprotonation/electrophilic quenching method for an efficient synthesis of a family of *P*-stereogenic phosphine/sulfide hybrid ligands (Scheme 5.16).

Treatment of tosylates **37** or **38** with *in situ* prepared sodium thiolates at room temperature afforded phosphine/sulfide boranes **39** in high yields (Table 5.9). Alternatively, to avoid the preparation of **37**, compounds **39** with $n = 1$ and $R^2 = Ph$ could also be prepared directly from **4** by enantioselective deprotonation and treatment with diphenyldisulfide. The compounds reported are listed in Table 5.9.

As reflected in Table 5.9, reactions proceed smoothly and all the compounds could be obtained in very high yields except the one containing the electron withdrawing *p*-nitrophenyl substituent at the sulfur atom (entry 8).

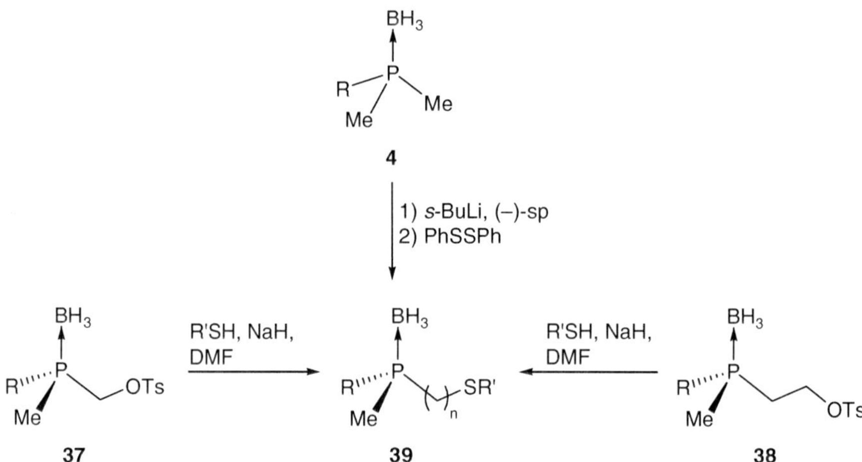

Scheme 5.16   Different methods to prepare *P*-stereogenic phosphine/sulfide boranes.

Table 5.9   *P*-stereogenic phosphine/sulfide boranes **39** of Scheme 5.16.

Entry	R	R'	n	Yield (%)
1	*t*-Bu	*t*-Bu	1	98
2	Ph	*t*-Bu	1	85
3	*t*-Bu	Ph	1	97
4	*t*-Bu	Ph	2	96
5	*t*-Bu	*o*-Tol	1	96
6	*t*-Bu	*p*-Tol	1	89
7	*t*-Bu	2-Naphth	1	98
8	*t*-Bu	$p\text{-NO}_2\text{C}_6\text{H}_4$	1	26
9	*t*-Bu	Cy	1	92
10	*t*-Bu	Bn	1	93

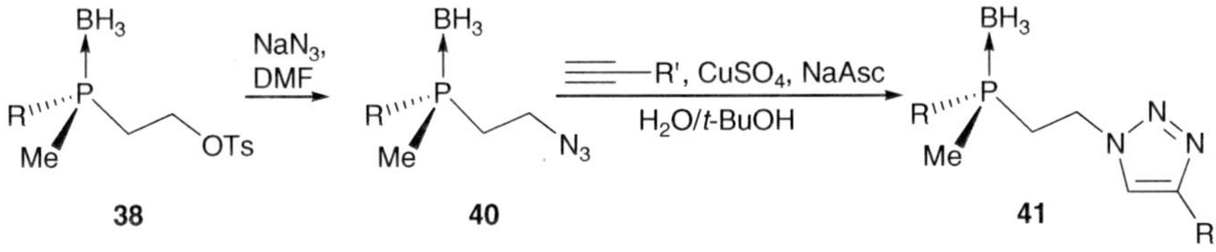

Scheme 5.17 Cu-catalysed cycloaddition between alkynes and azides **40**.

More recently, Kann and co-workers[21] transformed tosylates **38** into azides which were employed in the well-known Cu-catalysed version of the Huisgen 1,3-dipolar cycloaddition between alkynes and azides (Scheme 5.17).

Products **41** contain a coordinating triazole ring and therefore once the phosphorus is deprotected they are potential P/N bidentate ligands. This synthesis is highly modular, especially in the triazole ring given the wide variety of terminal alkynes available. Indeed, they coined the name 'ChiraClick' for the new family of ligands **41**. This is the first report on a library of *P*-stereogenic phosphines prepared through a combinatorial approach. The alkynes and precursors **40** gave access to 19 ligands, whose yields are displayed in Table 5.10.

All the yields were from good to excellent, with the exception of the compound bearing both *t*-butyl and 1-adamantyl groups, probably due to steric reasons. It has to be mentioned that a quick filtration of the reaction mixture through a short pad of silica gel was in general sufficient to afford pure crystalline materials.

Further work of the same group[22] was focused on the obtention of *P*-stereogenic β-aminophosphine boranes. To this end, they employed α-formylated and α-carboxylated phosphine boranes (Scheme 5.18).

The stereoselective formylation of one of the methyl groups was easily achieved by deprotonation followed by quenching with DMF. With this procedure, compounds **42** were obtained in 92 and 82% yields for $R^1 = Ph$ and Fc respectively and in 81% ee in both cases. The formylated phosphine boranes **42** were then subjected to reductive amination (see Chapter 4, Section 4.4.2.1). After purification, β-aminophosphine boranes **44** were obtained in 71–90% yields. For the *t*-butyl derivative, this procedure caused the decomposition of the aldehyde **42** during column chromatography purification leading to very low yields of isolated product. Therefore, an alternative route was developed. The α-carboxyphosphine borane **31** was prepared using CO_2 and converted to α-amidophosphine boranes **43** using typical coupling agents (see also Scheme 5.15) in good yields (71–92%). A final reduction step with borane furnished the desired compounds, also in good yields (71–95%).

5.2.1.2 C_2-Diphosphines

Enantioselective deprotonation followed by oxidative coupling with Cu(II) salts has been employed to prepare C_2-symmetric 1,2-*bis*(methylphosphine borane)ethanes, described in the following paragraphs and in Section 5.2.1.3.

Table 5.10 Yields for the reported 'ChiraClick' phosphine boranes **41** (Scheme 5.17).

Yield (%)

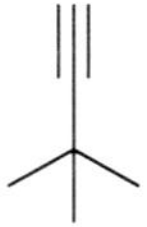

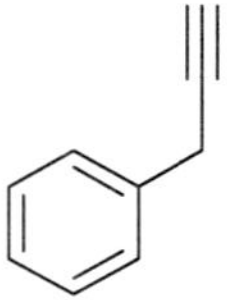

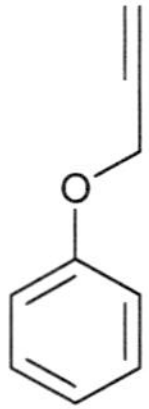

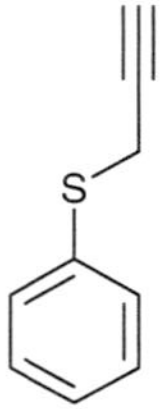

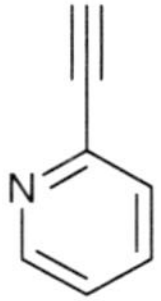

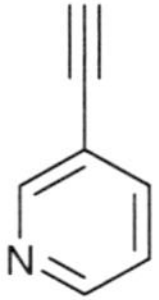

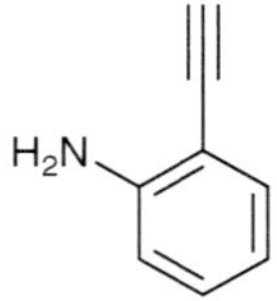

Substrate	tBu alkyne	benzyl alkyne	O-phenyl propargyl	S-phenyl propargyl	2-pyridyl alkyne	3-pyridyl alkyne	2-aminophenyl alkyne
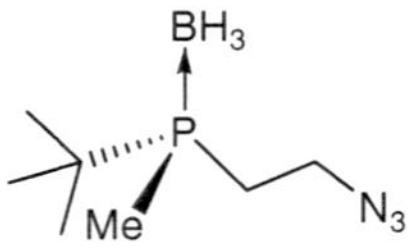 tBu phosphine borane	77	95	95	95	92	96	90
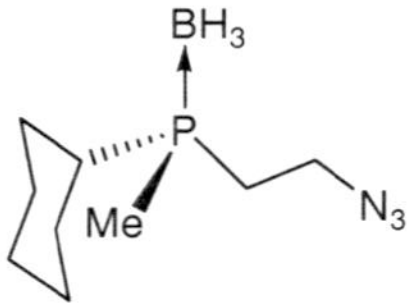 cyclohexyl phosphine borane	86	90	90	NP	95	NP	NP
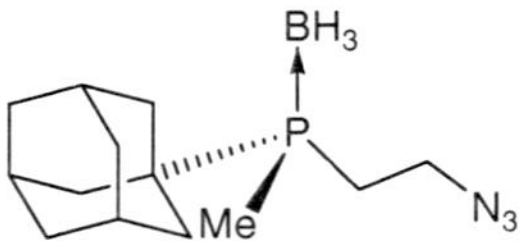 adamantyl phosphine borane	20	76	73	NP	95	NP	NP
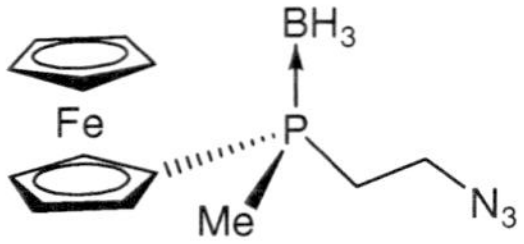 ferrocenyl phosphine borane	89	92	96	NP	98	NP	NP

NP: not prepared.

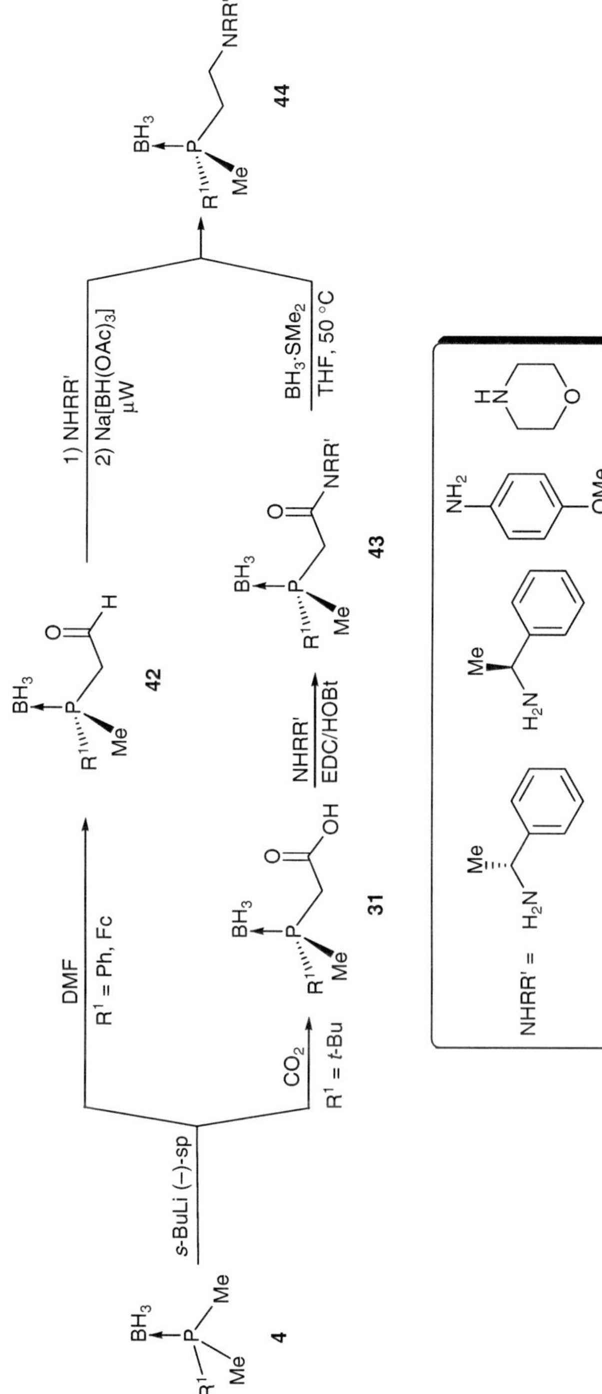

Scheme 5.18 Two possible routes to prepare β-aminophosphine boranes.

Evans and co-workers[8] described the preparation of 1,2-*bis*(arylmethyl-phosphine borane)ethanes by oxidative coupling of anions derived from aryl-dimethylphosphine boranes with Cu(II) pivalate (Scheme 5.19).

As the carbanions were not obtained optically pure (Table 5.1), the major **(S,S)-45** products were invariably accompanied by some **meso-45**. The *meso* compounds could be removed by column chromatography and the optical purity of **(S,S)-45** increased by recrystallisation. A few other reports have also used Cu(II) salts and precursors **4** with an aryl substituent. The compounds of this kind reported to date are listed in Table 5.11.

Starting from precursors **4** with unsubstituted or *o*-substituted aryl groups, the yields are moderate and the enantioselectivities very high (entries 1–4 and 8). Entries 5 and 7 show that the introduction of the *o,o'*-disubstituted groups causes a severe drop to the optical purity and therefore this method does not seem suitable to prepare such compounds in optically pure form. Diphosphine boranes **45** were deprotected with diethylamine, although they were found to be configurationally unstable in chloroform.[8]

The enantioselective deprotonation/double nucleophilic attack on dichlor-osilanes (see Chapter 4, Section 4.4.2.2) has also been used (Scheme 5.20).

Evans and co-workers[8] prepared crude **46** (R = *o*-An) with 77% de. A single recrystallisation eliminated efficiently the *meso* diastereomer, yielding enantio-pure **46**. This diphosphine was easily deprotected with diethylamine although it

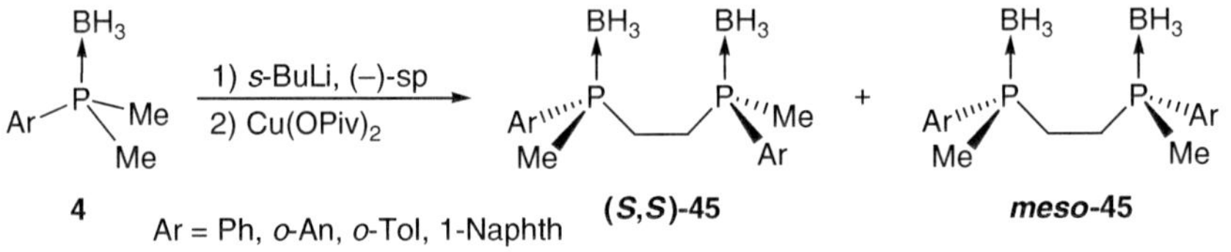

Scheme 5.19 Synthesis of 1,2-*bis*(arylmethylphosphine borane)ethanes.

Table 5.11 Reported 1,2-*bis*(arylmethylphosphine borane)ethanes **45** (Scheme 5.19).

Entry	Ar	Yield (%) 45[a]	de (%) (l)-45[b]	Yield (%) (S,S)-45	ee (%) (S,S)-45	References
1	Ph	65	66	54	98	8
2	*o*-An	76	71	65	99	8
3	*o*-Tol	65	72	56	99	8
4	*o*-CF$_3$C$_6$H$_4$	62	77	55	–	10
5	Mes	62	100	62	37	26
6	1-naphthyl	55	86	51	96	8
7	9-anthryl	70	72	39	18	26
8	Ferrocenyl	43	90	33	o.p.	27

[a]Combined yield of all isomers of **45**.
[b](l) = **(R,R)-45** + **(S,S)-45**.

was found to be configurationally unstable in chloroform solution. A few other phosphine boranes have been prepared with this method (Table 5.12).

Also in these syntheses the enantiomeric purity of the recrystallised products is very high. However, the reaction seems sensitive to the steric bulk of the R groups at the silicon atom as suggested by the decreasing yields going from entry 2 to 4.

Imamoto and co-workers[17] used a few optically pure alkynyl(*t*-butyl)-methylphosphine boranes **47**, obtained by enantioselective deprotonation, to prepare the corresponding 1,2-diphosphinoethanes **48** (Scheme 5.21).

The corresponding compounds **48** were obtained in 73–78% yields by the standard method in optically pure form. The compound with $R = Si(i\text{-Pr})_3$ could be easily desilylated with TBAF to furnish **49**, which was methylated to afford **50**.

Several diphosphine boranes have been prepared by taking advantage of the availability of a few optically pure secondary phosphine boranes (Section 5.2.1.1). After deprotonation, the phosphide boranes were easily alkylated by *bis*(benzylic) substrates (Scheme 5.22).

Gridnev Higashi and Imamoto[28] prepared **51** in 27% yield after recrystallisation as an optically pure compound. Livinghouse and co-workers[10] followed a similar procedure to obtain the potentially tridentate ligand precursors **52** and **53**.

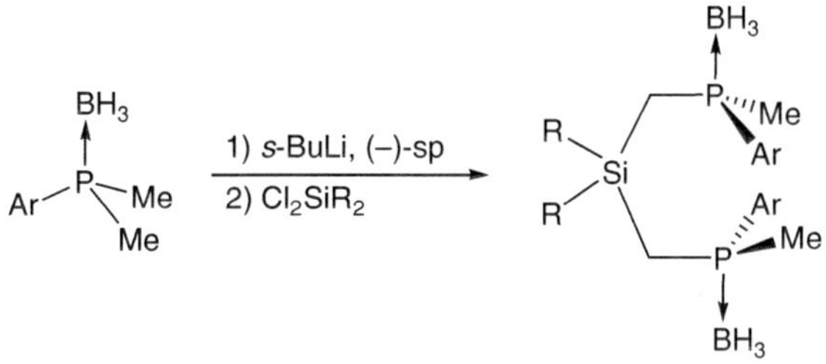

Scheme 5.20 Preparation of diphosphine boranes **46**.

Table 5.12 Diphosphine boranes **46** (Scheme 5.20).

Entry	Ar	R	Yield (%) 46^a	de (%) $(l)\text{-}46^b$	Yield (%) (R,R)-46	ee (%) (R,R)-46	References
1	*o*-An	Me	–	78	73	>99	8
2	*o*-CF$_3$C$_6$H$_4$	Me	69	66	57	o.p.	10
3	*o*-CF$_3$C$_6$H$_4$	*i*-Pr	64	78	57	o.p.	10
4	*o*-CF$_3$C$_6$H$_4$	Cy	49	46	36	o.p.	10

^{a}Combined yield of all isomers of **46**.
^{b}$(l) = (R,R)\text{-}46 + (S,S)\text{-}46$.

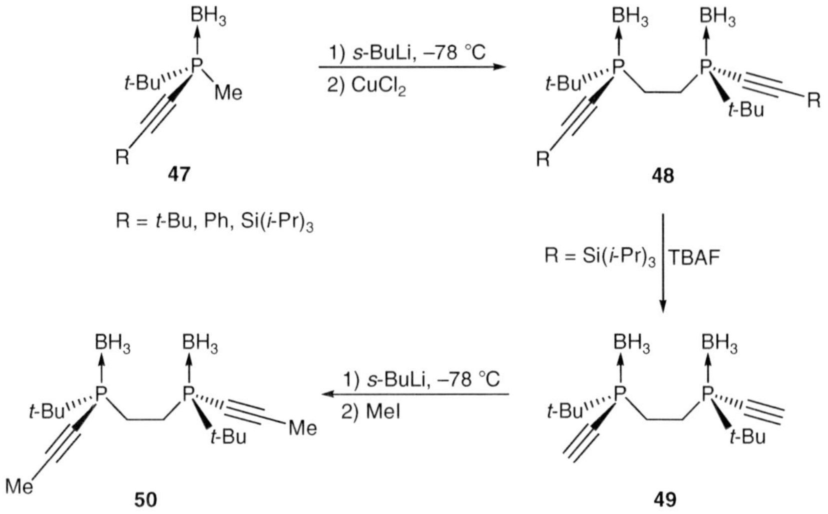

Scheme 5.21   Synthesis of 1,2-*bis*(alkynyl(*t*-butyl)phosphine borane)ethanes.

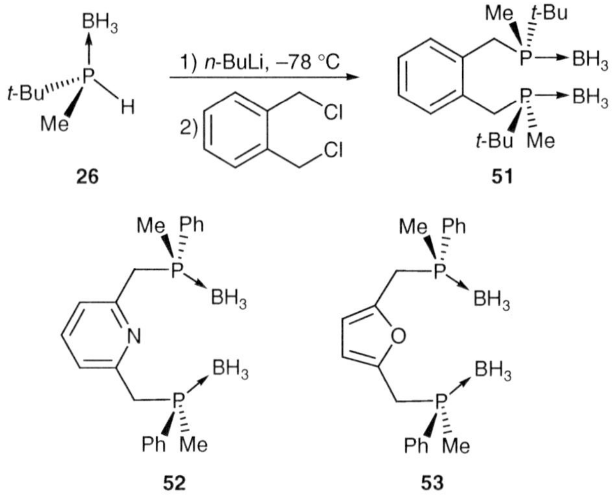

Scheme 5.22   Synthesis of diphosphine boranes by double nucleophilic substitution.

The phosphide borane anions are also reactive towards haloarenes, giving substitution products *via* S$_N$Ar. This fact was used by Imamoto and co-workers[14,15] to prepare two ligands known as QuinoxP* (Scheme 5.23).

These two ligands were easily prepared by reaction of the enantiomerically pure phosphide borane anions with 2,3-dichloroquinoxaline followed by deboronation with TMEDA, in 75–80% combined yield. Remarkably, none of the ligands were either oxidised or epimerised on standing in air for months.

This is probably due to the significant electron-withdrawing character of the quinoxaline backbone, which reduces the electron density at the phosphorus atoms. Indeed, *t*-Bu-QuinoxP* is nowadays one of the very few commercially available *P*-stereogenic ligands.[29]

Further work by the same group[14,30] aimed to obtain *o*-diphosphino-benzenes (**55**) by double nucleophilic aromatic substitutions on *o*-difluoro-benzene tricarbonylchromium complexes (Scheme 5.24).

Even with an excess of *tert*-butylmethylphosphide, the desired compound **55** was not obtained and only **20** could be isolated, probably due to both steric and electronic reasons. Accordingly, the X-ray structure of **20** shows effective shielding of the *ipso* carbon of the remaining fluorine atom by the phosphine borane group. Deboronated **20** (Scheme 5.10) did not react with the phosphide either. Unexpectedly, when the reaction mixture containing **19** and the phosphide treated with acid at low temperature the main product turned out to be the *para*-substituted diphosphine borane **54**.[30] Compound **54** arises from nucleophilic attack at the *meta* position to the fluoro group; therefore the

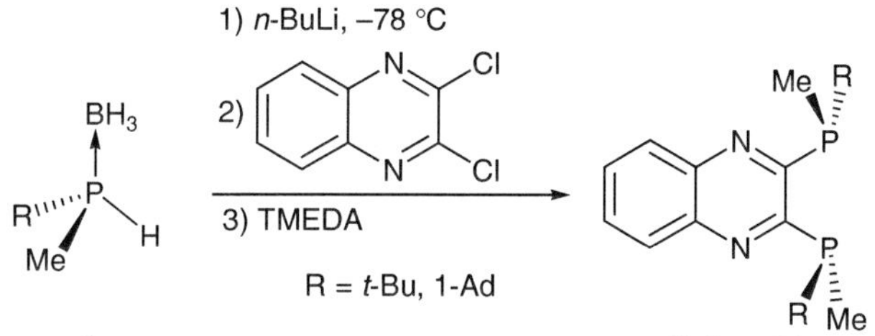

Scheme 5.23 Preparation of QuinoxP* ligands.

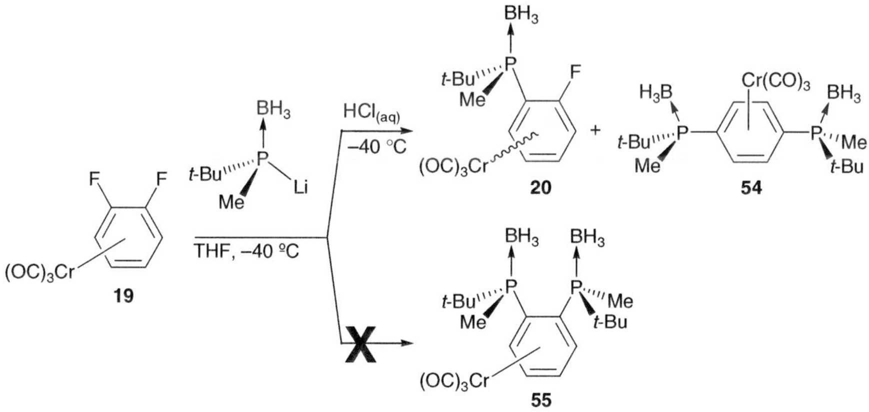

Scheme 5.24 Synthesis of mono- and bidentate *meta-tele*-substituted phosphino-benzene boranes.

process can be named as a *meta-tele*-S_NAr. The attack at this position is advantageous because the stabilisation of the Meisenheimer intermediate by the phosphine borane group. Experiments either with deuterated substrate **19** or with deuterated acid revealed that the *meta-tele* substitution proceeds *via* nucleophilic attack of the phosphide to the *para* position to the phosphine borane group to form a Meisenheimer-type intermediate. Then two processes take place: the protonation of the fluoro group by the acid and a 1,5-hydrogen shift of the Meisenheimer intermediate. Whichever comes first, both lead to **54** after elimination of HF.[30]

After some optimisation, a family of 1,4-diphosphine borane benzenes was obtained after photochemical dechromination (Scheme 5.25 and Table 5.13).

With this method, one C_2 diphosphine borane was prepared (entry 1) along with several C_1 analogues bearing one (entries 2, 4, 5 and 7–12) or two (entries 3 and 6) stereogenic phosphorus atoms. The yields were moderate to good, except when the nucleophile was sterically hindered (entries 2 and 5).

The same group[31] persisted in the preparation of 1,2-diphosphinobenzenes and finally met success with a clever use of *bis*(phosphine)boronium salts. In these compounds, two secondary phosphines are connected through a BH_2 bridge. They are prepared by the reactions depicted in Scheme 5.26.

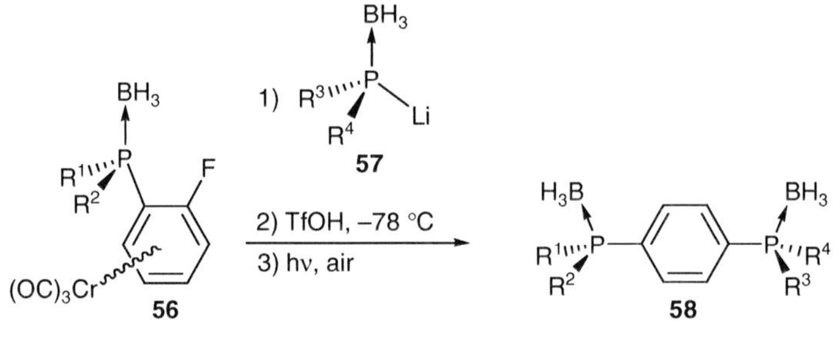

Scheme 5.25   Preparation of 1,4-diphosphine borane benzenes.

Table 5.13   1,4-diphosphine borane benzenes **58** (Scheme 5.25).

Entry	R^1	R^2	R^3	R^4	Yield (%)
1	*t*-Bu	Me	*t*-Bu	Me	73
2	*t*-Bu	Me	Cy	Cy	11
3	*t*-Bu	Me	Cy	Me	70
4	*t*-Bu	Me	*n*-Bu	*n*-Bu	84
5	*t*-Bu	Me	*t*-Bu	*t*-Bu	28
6	*t*-Bu	Me	1-Ad	Me	59
7	Cy	Cy	*t*-Bu	Me	86
8	*n*-Bu	*n*-Bu	*t*-Bu	Me	70
9	*n*-Bu	*n*-Bu	*t*-Bu	Me	79
10	*i*-Pr	*i*-Pr	*t*-Bu	Me	48
11	*t*-Bu	*t*-Bu	*t*-Bu	Me	62
12	Ph	Ph	*t*-Bu	Me	83

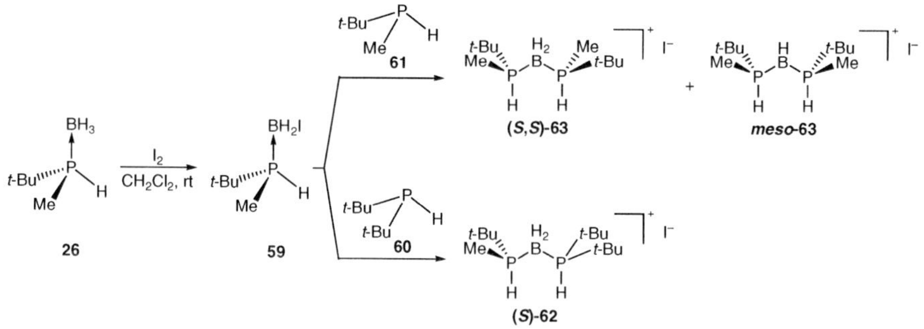

Scheme 5.26 Preparation of *P*-stereogenic *bis*(phosphine)boronium iodides.

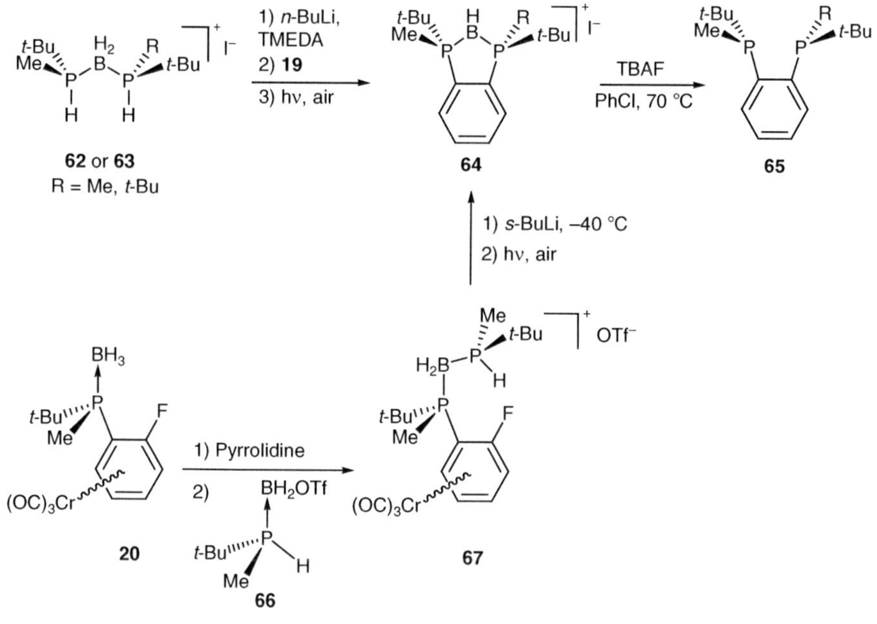

Scheme 5.27 Preparation of *P*-stereogenic 1,2-diphosphinobenzenes.

Monoiodination of **26** with 0.5 equivalents of iodine furnished **59**, which was reacted with **60** to afford enantiomerically pure **(S)-62** in 62% yield. The same reaction with racemic **61** afforded **63** as an equimolar mixture of the *meso* and (*S,S*) isomers. Luckily, the chiral isomer was obtained by recrystallisation from THF in 96% de.

In compounds **62** and **63**, the diphosphino borane groups are ideally preorganised to be installed in contiguous position into the aromatic ring (Scheme 5.27).

Deprotonation of **62** or **63** with two equivalents of *n*-BuLi and double S$_N$Ar on **19** afforded optically pure diphosphinobenzeneboronium salts **64** after photochemical dechromination in 87% and 78% yields for R = Me and *t*-Bu respectively. The BH_2 linkage proved to be resistant to usual deboronation reagents such as tetrafluoroboric acid or *N*-methylpyrrolidine probably due to the stability of its bicyclic structure. It could only be cleaved using TBAF at relatively high temperature, to release the free diphosphines **65** without loss of optical purity. Interestingly, compound **64** with R = Me could be obtained in 58% yield also in stepwise manner from **20** *via* intermediate **67**.

Throughout this book the borane group constantly appears but 'only' as a protective group that has to be removed when the free phosphorus compound is required. In the last syntheses, however, the reader will have noticed the different usage of the borane moiety as a 'modifiable' group able to play an active role in the chemistry involved enabling, in this case, the preparation of the crowded diphosphines **65**, difficult to obtain by other ways.

5.2.1.3   *BisP* Diphosphines*

The diphosphines discussed in this section are of the same type as those seen in the beginning of Section 5.2.1.2, the only difference being that they do not bear any aromatic substituent, *i.e.* they are 1,2-*bis*(alkylmethylphosphino)ethanes. They have been proved to be superb in hydrogenation reactions (Chapter 7) and other catalytic transformations (Chapter 8) and form one of the most important types of 'new generation' *P*-stereogenic ligands. The name BisP* was coined when their synthesis was first reported by Imamoto and co-workers.[32] Their preparation consists of the enantioselective deprotonation protocol followed by Cu(II) oxidative coupling (Scheme 5.28 and Table 5.14).

The desired phosphine boranes were prepared from phosphorus trichloride in a few steps using inexpensive materials. Although the diastereoselectivities were not very high, particularly in the phosphine of entry 5, the *meso*-**68** compounds could be efficiently separated by conventional methods (chromatography or recrystallisation) affording chemically pure C_2 compounds. Their optical purities could be easily increased by recrystallisation because the *meso* isomers were found to be more soluble than the C_2 counterparts with the exception of the compound in entry 1.[33] It has to be said that **68** could not be deboronated by the classic treatment with amines and the acidic method had to be used (see Chapter 1, Section 1.3.2).[35]

All the ligands bear the smallest alkyl group, methyl, and a bulky alkyl group. Upon coordination to transition metals, they form five-membered

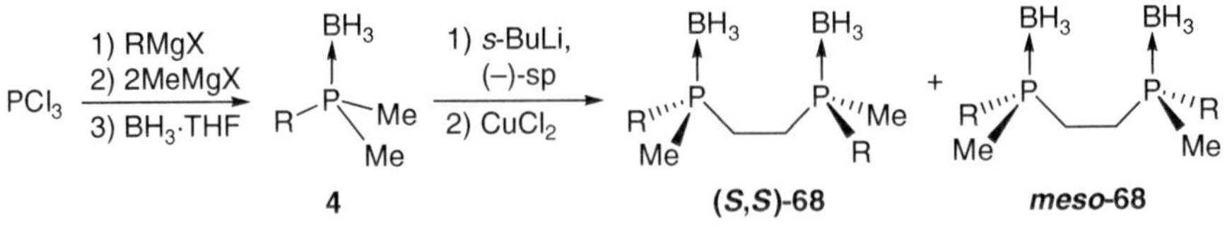

Scheme 5.28   Preparation of BisP* diphosphine boranes from PCl_3.

C_2-symmetric rings with a strong steric dissymmetry, which proved to be exceptionally beneficial for catalytic performance, especially in hydrogenation reactions. The downside is that due to their trialkyl nature, the ligands are very electron-donating and therefore extremely sensitive to oxidation.

The practical unavailability of (+)-sparteine means that the standard enantioselective deprotonation protocol can not be used to prepare (R,R)-BisP* diphosphines (see, however, Section 5.4.1). Some work has been performed to overcome this limitation. Imamoto and co-workers[36] were able to prepare both enantiomers of t-Bu-BisP* · (BH$_3$)$_2$ by using both enantiomers of *tert*-butylmethylphosphine borane. Unfortunately, the preparation of the latter compounds involved a preparative HPLC separation step and therefore the strategy cannot be easily adapted for large-scale synthesis.

Another possibility is the conversion of (S,S)-t-Bu-BisP* · (BH$_3$)$_2$ into (R,R)-t-Bu-BisP* · (BH$_3$)$_2$, displayed in Scheme 5.29.

Deboronation of (S,S)-t-Bu-BisP* · (BH$_3$)$_2$ and consequent oxidation easily provides (R,R)-t-Bu-BisP* · (O)$_2$.[37] Both steps occur with retention of configuration; the change in the descriptors of the absolute configuration merely agrees with the priority order in the CIP rules. This oxide was then stereoselectively reduced and reboronated with a procedure known to provide inverted products (see Chapter 1, Section 1.3.1).[38] Although the desired compound was obtained in optically pure form, the yield was disappointingly low (24%).

A third possibility was disclosed by Crépy and Imamoto[39] and it is based on the *bis*(secondary) phosphine borane **71** (Scheme 5.30).

This compound was prepared from oxidative coupling of deprotonated **69** to afford **70** (50% yield) and double one-carbon degradation of the latter (Section 5.2.1.1). Compound **71** was obtained in around 80% yield before recrystallisation. Although **69** was not completely enantiopure (91% ee), **71** was obtained

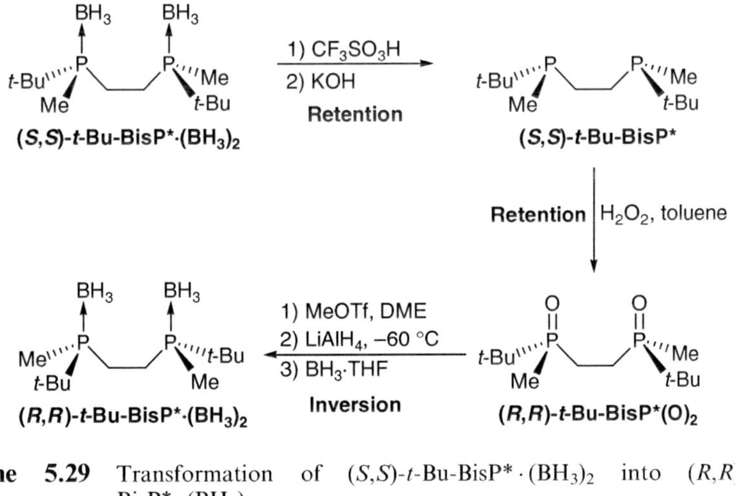

Scheme 5.29 Transformation of (S,S)-t-Bu-BisP* · (BH$_3$)$_2$ into (R,R)-t-Bu-BisP* · (BH$_3$)$_2$.

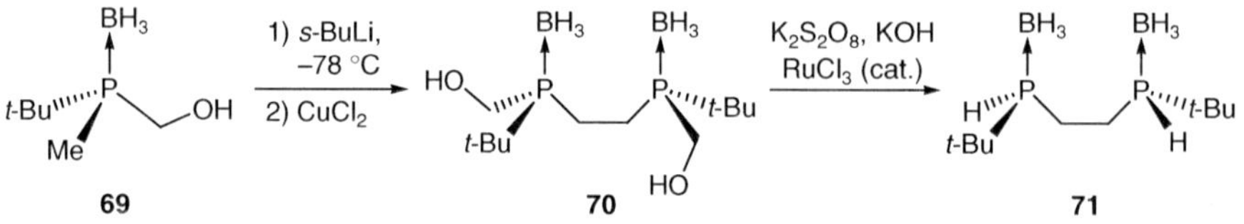

Scheme 5.30   Synthesis of 1,2-*bis*(*tert*-butylphosphinoborane)ethane **71**.

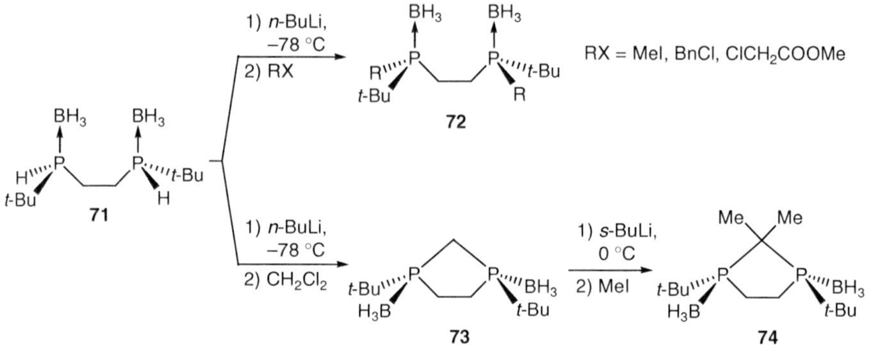

Scheme 5.31   Reactivity of deprotonated **71** towards several electrophiles.

in optically pure form even before recrystallisation. This fact has been attributed to the destabilisation of ***meso*-70** due to coulombic repulsion of the deprotonated alcoholate groups during the synthesis. Compound **71** was found to be a suitable precursor to several diphosphine boranes (Scheme 5.31).

Deprotonation and electrophilic quenching with MeI afforded (*R*,*R*)-*t*-Bu-BisP* · $(BH_3)_2$ in quantitative yield as an optically pure product. Other electrophiles also reacted smoothly to yield other diphosphine boranes **72**. The interesting cyclic diphosphine borane **73** was obtained in 35–53% yield when the dianion of **71** was treated with CH_2Cl_2. The methylene group bridging the two phosphorus atoms could also be alkylated to afford **74** in 35% yield.

Another expansion of the BisP* family was the preparation C_1-diphosphines with different alkyl groups on each phosphorus atom (unsymmetric BisP* ligands), disclosed by Ohashi and Imamoto.[19,20,40] Unsymmetric BisP* ligands were prepared by coupling two of the precursors discussed in Section 5.2.1.1: the lithiated secondary phosphine boranes **9** and the mesylates or tosylates **33** (Scheme 5.32).

The corresponding diphosphine boranes **75** were obtained in excellent optical purity (>97% de) with only minor amounts of other diastereomers. The compounds prepared are listed in Table 5.15.

Entries 3, 6 and 8 show that combination of 1-adamantyl and cyclohexyl or 1-adamantyl and *tert*-butyl groups provide quantitative yield of the coupling product. In contrast, *tert*-butyl and isopropyl or cyclohexyl groups are only badly coupled together (entries 1 and 2), due to the formation of undesired (and unknown) by-products.[19,20] This procedure is also suitable for the preparation of symmetric BisP* ligands[19] and indeed the coupling step compares favourably to the classic oxidative coupling with Cu(II) salts (compare entry 10 in Table 5.15 and entry 4 in Table 5.14), although the latter procedure is preferred because it involves a much shorter synthetic route.

The method discussed here has been adapted to the preparation of C_1-diphosphines **79** bearing a single *P*-stereogenic atom (Scheme 5.33).[19,20]

The nucleophilic attack of dicyclohexyl- or diphenylphosphine to **77** provided the desired compounds but only in 35% yield for R = Ph and in trace amounts for R = Cy. The alternative route starting from **79** was most successful, providing **78** with 61%, 73% and 60% yields for R = Me, Cy and Ph respectively.

5.2.1.4 MiniPHOS Diphosphines

The very good results of BisP* ligands in several catalytic reactions prompted Yamanoi and Imamoto[41] to simplify even further their structure but maintain

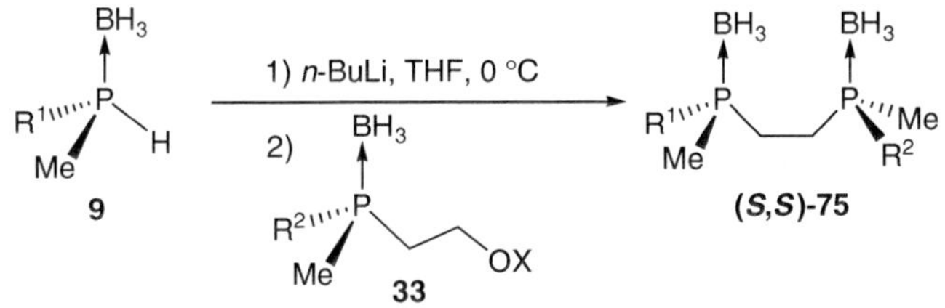

Scheme 5.32 Preparation of unsymmetric BisP* diphosphine boranes (**75**).

Table 5.14 BisP* diphosphine boranes reported to date (Scheme 5.28).

Entry	R	Yield (%) 68^a	de (%) (l)-68^b	Yield (%) (S,S)-68	ee (%) (S,S)-68	References
1	*i*-Pr	–	–	–	80^c	33
2	*t*-Bu	80	68	30–40	o.p.	32–34
3	Et₃C	–	$-^d$	69	99	32,33
4	1-Ad	–	70	33	o.p.	32,33
5	Cyclopentyl	75	22	34	o.p.^{e}	32,33
6	Cy	–	ND^f	55	o.p.	32,33
7	1-MeCy	–	–	41	o.p.	33

^{a}Combined yield of **(l)-68** + **meso-68**.
^{b}**(l)** = **(R,R)-68** + **(S,S)-68**.
^{c}After four recrystallisations.
^{d}A 'small amount' of *meso* isomer is reported.
^{e}The enantiomeric excess of **(S,S)-68** was 75% before recrystallisation.
^{f}Not determined.

Table 5.15   Unsymmetric BisP* phosphine boranes **75** (Scheme 5.32) reported
to date.

Entry	R^1	R^2	Yield (%)	References
1	*t*-Bu	*i*-Pr	19	20
2	*t*-Bu	Cy	10	19,20,40
3	*t*-Bu	1-Ad	quant.	19,20,40
4	Cy	Cyclopentyl	70	20
5	1-Ad	*i*-Pr	62	20
6	1-Ad	*t*-Bu	quant.	19,20,40
7	1-Ad	Cyclopentyl	75	20
8	1-Ad	Cy	quant.	19,20,40
9	1-Ad	Ph	62	20
10[a]	1-Ad	1-Ad	40	19,20

[a]The compound is a 'normal', symmetric BisP* · $(BH_3)_2$.

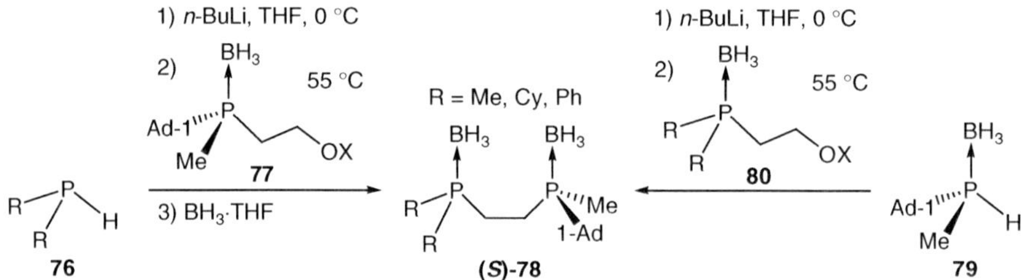

Scheme 5.33   Preparation of unsymmetric BisP* diphosphine boranes with a single
P-stereogenic atom.

the steric dissymmetry at the phosphorus atoms. Therefore they prepared the
methylene-bridged analogues, which were appropriately called MiniPHOS
because they are quite small in comparison with most of the other chiral
diphosphines (Scheme 5.34 and Table 5.16).

1,2-*Bis*(alkylmethylphosphino)methanes **81** were prepared by enantioselective deprotonation followed by treatment of the α-carbanion with one
equivalent of dichlorophosphine and nucleophilic substitution of the remaining
chloro group by methyllithium. The latter two reactions were not stereoselective because an equimolar mixture of **(R,R)-81** and **meso-81** was formed.
Luckily, the C_2 diastereomers were more insoluble than the *meso* and finally
(R,R)-81 could be obtained in optically pure form (Table 5.16).

Although the isolated yields of the diphosphine boranes may seem very low,
it has to be noted that they account for the overall one-pot, four-step transformation from **10** to **(R,R)-81**, implying that each step has an average yield of
50–70%. The phosphine borane of entry 5 is electronically very different from
the others, because it bears phenyl rings at the phosphorus atoms and as a
result is considerably less basic. The interesting MiniPHOS diphosphine borane
of entry 6 was prepared using *t*-BuPCl$_2$ in the second step, proving that the
protocol is also suitable to prepare unsymmetric MiniPHOS ligands.

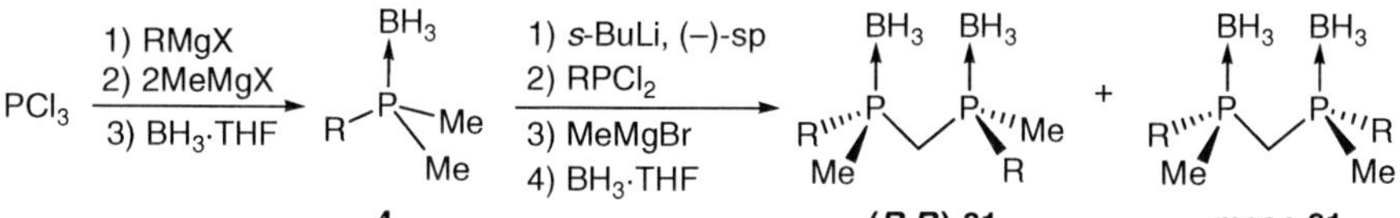

Scheme 5.34 Preparation of MiniPHOS diphosphine boranes (**81**).

Table 5.16 MiniPHOS diphosphine boranes reported to date (Scheme 5.34).

Entry	(R,R)-81	Yield (%)	ee (%)	References
1	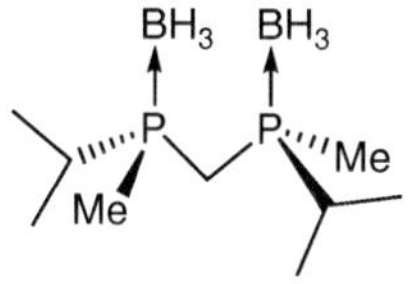	13	o.p.	33,41
2	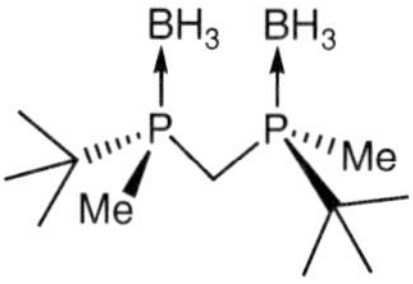	22	o.p.	33,41
3	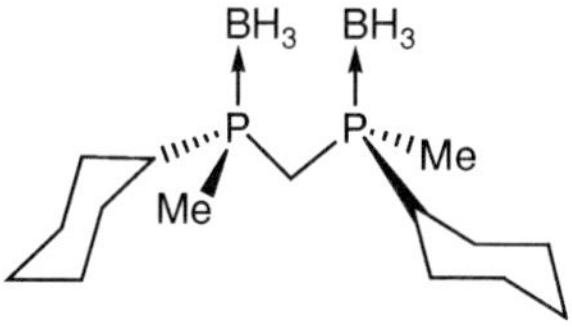	19	o.p.	33,41
4[a]	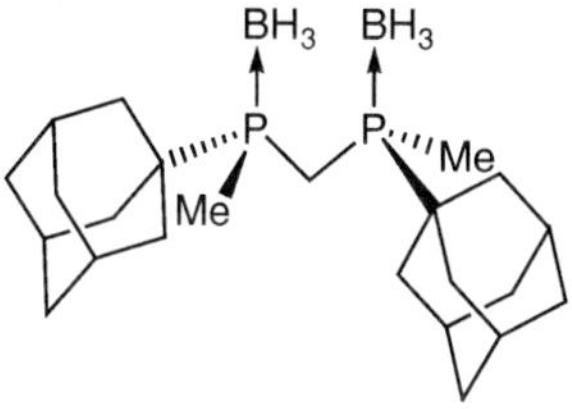	7	o.p.	42
5	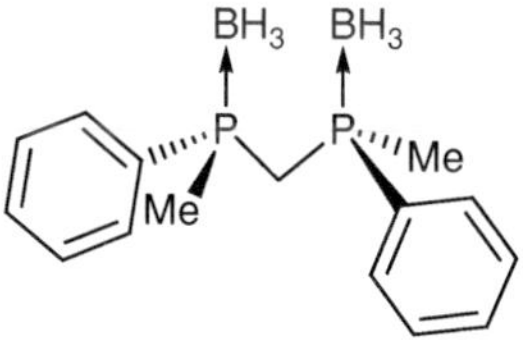	28	o.p.	33,41
6[b]	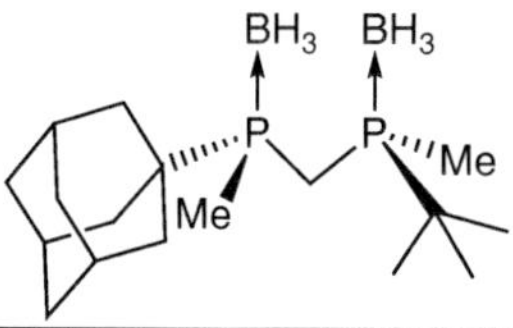	20	o.p.	42

[a]In this case the purification also involved preparative HPLC.
[b]See discussion in text.

5.2.1.5 *P-Stereogenic Dendrimers, Polyphosphines, Polymers and Macrocycles*

The enantioselective deprotonation protocol has been found to be well suited for the preparation of dendrimers, macrocycles, polyphosphines and polymers containing *P*-stereogenic atoms. To date, most of the reported compounds are derived from *t*-Bu-BisP*.

Chujo and co-workers[43] deprotonated both methyl groups in (*S*,*S*)-*t*-Bu-BisP*·(BH₃)₂ with two equivalents of *s*-BuLi and reacted the dianion with benzyl ether electrophilic dendrons to form first and second generation dendrimers with stereogenic phosphorus atoms in the core (Scheme 5.35).

First generation dendrimer **82** was obtained in 64% yield after purification by HPLC to remove the monofunctionalised compound, formed in 18% yield. This dendrimer was also prepared in optically inactive form, starting from *rac*-*t*-Bu-BisP*·(BH₃)₂, in 82% yield. Second generation dendrimer **83** was then prepared similarly in 70% yield after HPLC purification. The diphosphine borane **84** was also prepared in 42% yield to be used as a model.[44] The optically active dendrimers were analysed by CD measurements and it was found that they exhibited the Cotton effect, induced by the stereogenic phosphorus atoms, in the 280 nm π–π* absorption band of the aromatic rings. Unexpectedly, compound **84** did not display any Cotton effect.[43]

Imamoto, Gridnev and co-workers[45] reacted the secondary diphosphine borane **71** (Section 5.2.1.3) with tosylate **85** (Section 5.2.1.1) yielding the expected tetraphosphine borane **86** in 65% yield (Scheme 5.36).

Compound **86** was obtained as a single enantiomer and could be deprotected by standard methods (HBF₄) affording the corresponding free tetraphosphine in 95% yield, which was employed in Rh-catalysed hydrogenation. Its coordination chemistry has been recently investigated by Danjo, Yamaguchi and co-workers[46] who found that the ligand was able to form mono-, di- or tri-nuclear homo- and heterometallic complexes with Rh, Ru and Pd depending on the metallic precursor and the reaction conditions.

The diboronated BisP* diphosphines contain two methyl groups amenable to be deprotonated and oxidatively oligomerised to afford C₂-oligophosphine boranes with an even number of phosphorus atoms. This chemistry has been extensively studied by Chujo and co-workers (Scheme 5.37).[47]

Deprotonation of *t*-Bu-BisP*·(BH₃)₂ with 1.2 equivalents of *s*-BuLi/(–)-sparteine formed a mixture of mono- and dilithiated *t*-Bu-BisP*·(BH₃)₂ whose – intermolecular – oxidative coupling with Cu(II) afforded a mixture of tetraphosphine borane **86**[47,48] and hexaphosphine borane **87**.[47–49] The use of (–)-sparteine may surprise the reader, as the reaction does not involve any enantioselective deprotonation. However, it was found that its presence was beneficial to activate the *s*-BuLi,[50] showing that (–)-sp is not only a chiral inductor, but also modifies the reactivity of the organolithium reagents. After separation by column chromatography, **86** and **87** were obtained in 26% and 14% isolated yields respectively. Tetraphosphine borane **86** was subjected to similar reactions, affording compounds **88** (29% yield) and **90** (8% yield), which contain 8

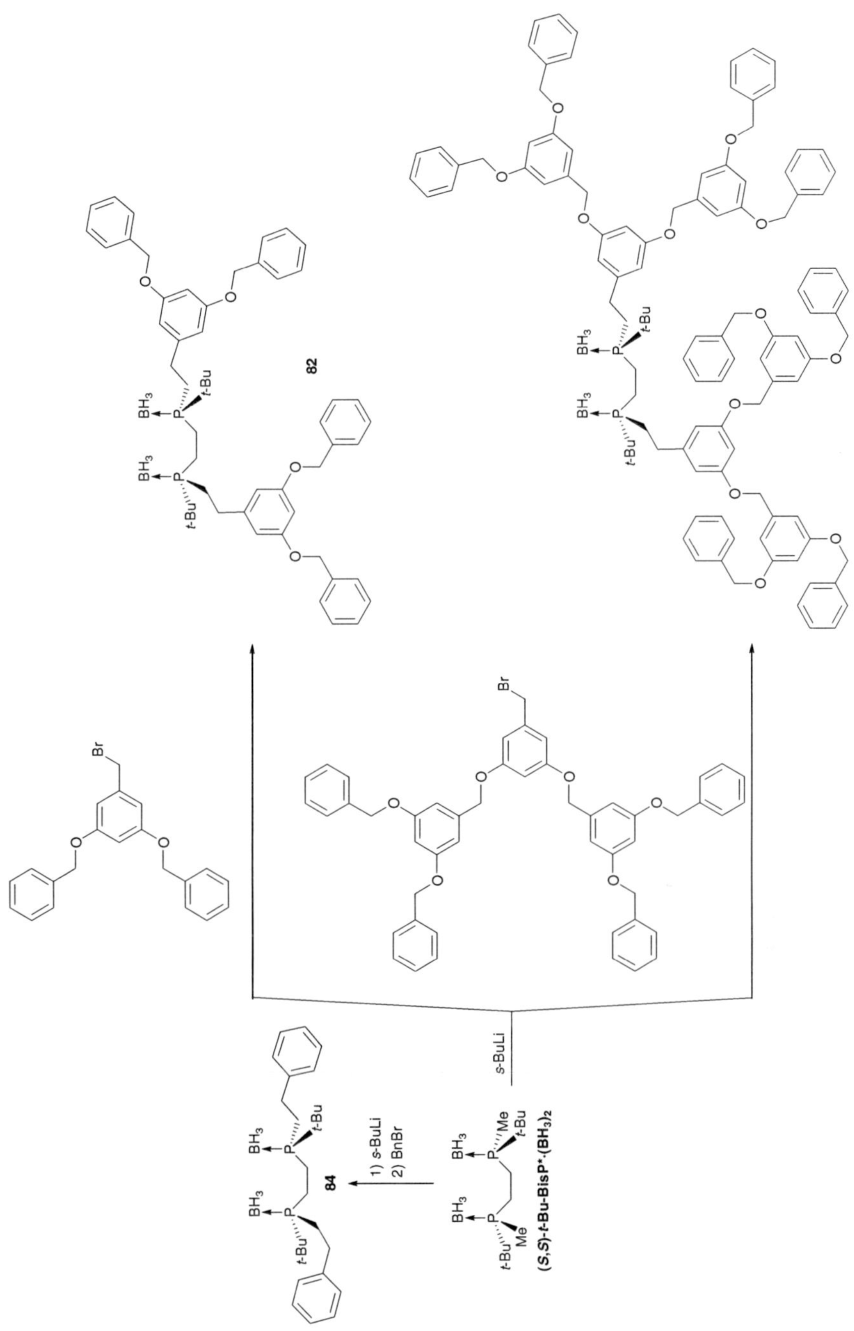

Scheme 5.35 Synthesis of *P*-stereogenic dendrimers from *t*-Bu-BisP* · (BH₃)₂.

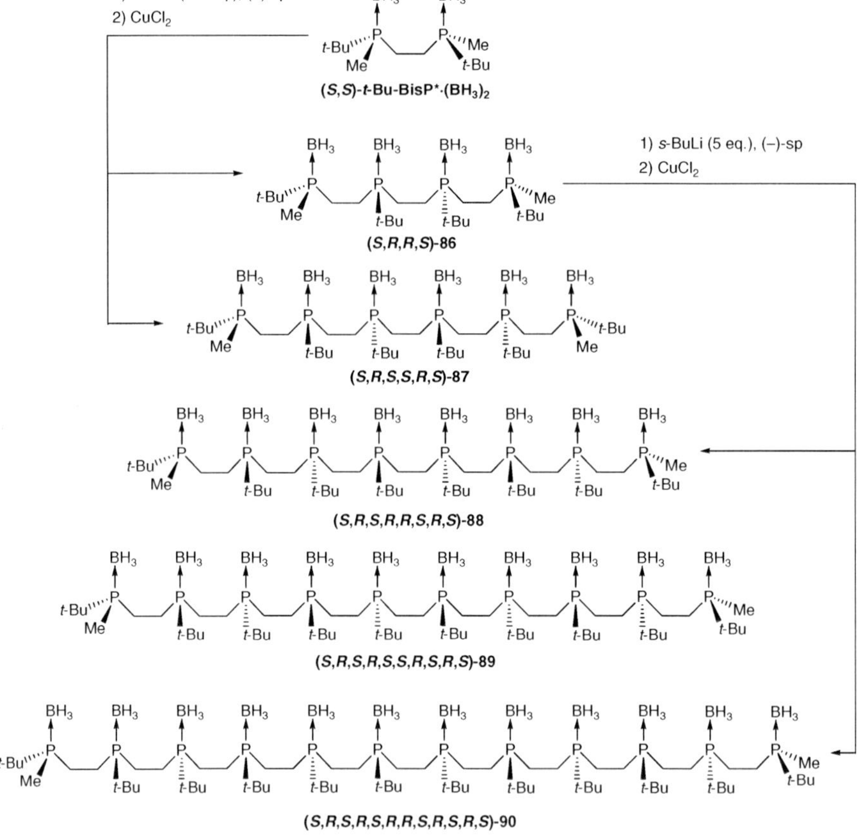

Scheme 5.36   Preparation of tetraphosphine borane **86**.

Scheme 5.37   Preparation of *P*-stereogenic oligophosphine boranes from *t*-Bu-BisP* · (BH₃)₂. Decaphosphine borane **89** has not been reported but is included for the sake of completeness.

and 12 stereogenic phosphorus atoms respectively and could be separated by HPLC.[47] The properties of these polyphosphine boranes were investigated by differential scanning calorimetry (DSC), powder X-ray diffraction and polarimetry at variable temperature. The conclusion is that *t*-Bu-BisP* and **86** behave as 'normal' crystalline small molecules whereas **88** and **90** behave as amorphous polymers.[47] For example, DSC shows that *t*-Bu-BisP* and **86** have sharp melting points (T_m) and lack any glass transition temperature (T_g) while **88** and **90** present a characteristic T_g. Hexaphosphine **87** exhibits an interesting intermediate behaviour between a crystalline small molecule and an amorphous polymer. It presents crystalline and amorphous states and therefore constitutes an example of an amorphous small molecule, a so-called molecular glass.[47,49]

Oligophosphine boranes *t*-Bu-BisP*, **86**, **87** and **90** have been also prepared in optically inactive form, *i.e.* synthesised without the presence of sparteine. Comparison of these diastereomeric mixtures with the corresponding optically active oligophosphine boranes revealed that the latter are more crystalline.[47,49] For example, the optically inactive tetraphosphine borane exhibits DSC thermograms and XRD profiles similar to the optically pure hexaphosphine borane **87**.

Chujo and co-workers[44,51,52] have also described several copolymers containing (*S,S*)-*t*-Bu-BisP* diphosphine borane units (Scheme 5.38).

Reaction of the dianion of (*S,S*)-*t*-Bu-BisP* with 4-iodobenzyl bromide afforded the monomer **91** in 15% yield after separation from the monosubstituted compound.[51] Pd-catalysed Sonogashira coupling polymerisation of **91** with 1,4-diethynylbenzene produced the desired polymer **92** (11% yield) after removal of low-molecular weight species by HPLC.[51]

Direct copolymerisation of the dianion of (*S,S*)-*t*-Bu-BisP* with any of the three regioisomers of α,α′-dibromoxylene or with an azobenzylic dibromide derivative furnished polymers **93** (11–15% yields) and (*E*)-**94** respectively, the latter in a remarkable 62% yield.

Polymers **92** and **93** are stable solids soluble in common organic solvents and therefore could be characterised by UV spectroscopy and multinuclear and multinuclear NMR. CD spectroscopy revealed the presence of Cotton effects around the area of the π-π* transitions of the aromatic rings, except for polymer *o*-**93**.[44,51] The presence of a Cotton effect means that in solution the polymers adopt an optically active conformation, such as a single-handed helix, induced by the stereogenic phosphorus atoms.

Polymer (*E*)-**94** constitutes the first example of stimuli-responsive *P*-stereogenic polymer.[52] That is because it can be isomerised to the *Z* form upon UV irradiation at 300–420 nm and isomerised back to the more stable *E* form by heating to 40 °C. This polymer is also the only *P*-stereogenic polymer that has been deboronated (with trifluoromethanesulfonic acid) and complexed to Pt.[52] Differences in the properties of boronated and platinated polymers **94** were observed. The most remarkable is the presence of Cotton effect in the complex but not in the boronated polymer, indicating that the complexation generates a higher-order chiral structure. Furthermore, the photochemical *E-Z* isomerisation of the complexed polymer was slower than in the (*E*)-**94**, indicating complexation-induced rigidification.

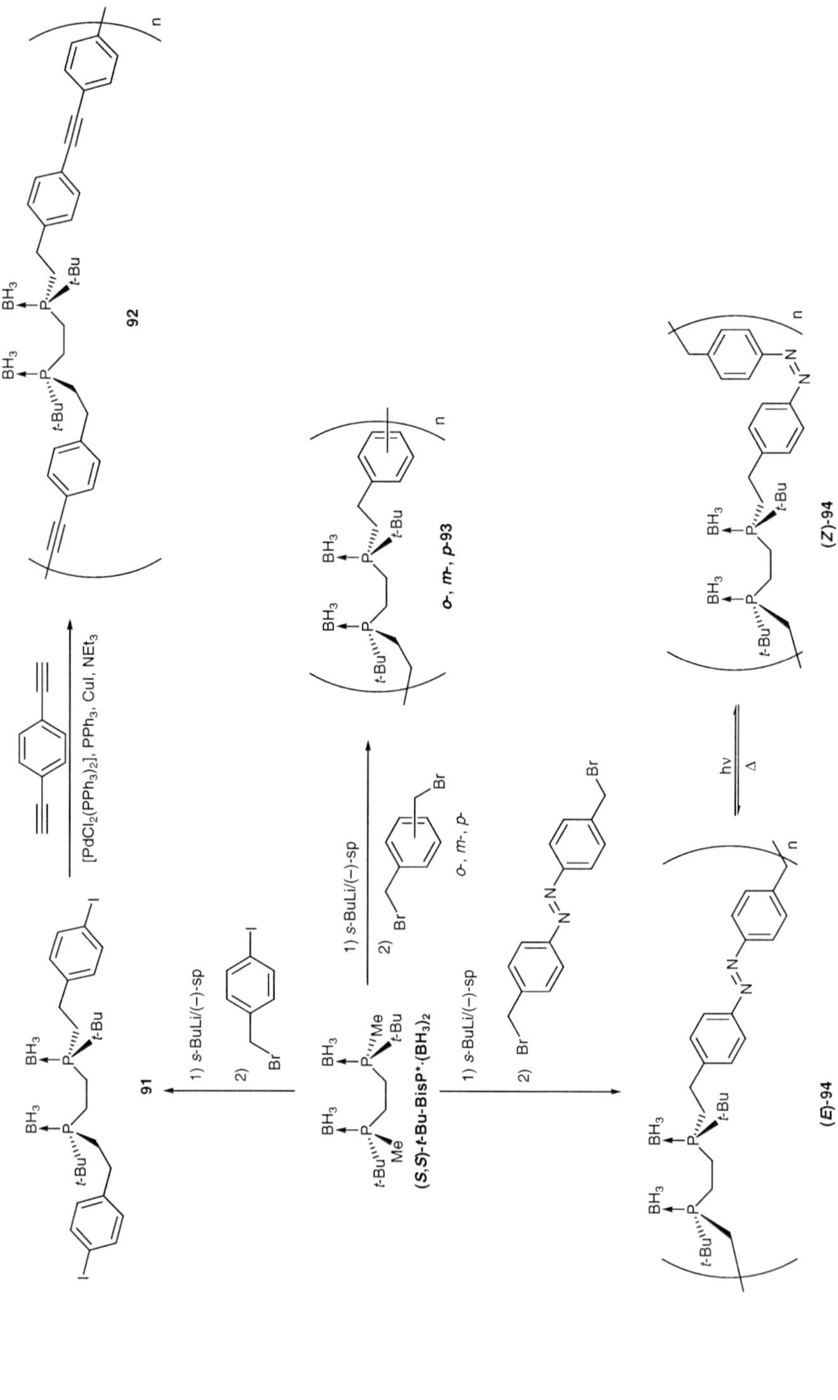

Scheme 5.38   Polymers containing *P*-stereogenic atoms.

An important parameter for any polymer is the average molecular weight and its distribution. For the *P*-stereogenic polymers discussed here, these parameters were obtained by gel permeation chromatography (GPC). The values of the number-average molecular weight (M_n), the weight-average molecular weight (M_W) and the molecular weight distribution (M_W/M_n) are listed in Table 5.17.

Table 5.17 reveals that polymer **92** achieves the highest molecular weight and *o*-**93** the lowest. In the case of polymers **93**, an increase in the molecular weight is observed going from *o*- to *p*-**93** due to the decrease in the steric hindrance.[44] An interesting observation is that the molecular weights of **(E)-94** decrease upon complexation owing to the more folded structure of the platinated polymer (compare entries 5 and 6).[52]

During the preparation of oligophosphine boranes, the principal side reaction is the intramolecular oxidative coupling of the dianions to produce cyclic compounds.[48] These reactions have been studied in more detail and constitute a stereoselective route to several interesting cyclic (but achiral) polyphosphine boranes. The simplest one is the 1,4-diphosphacyclohexane borane (Scheme 5.39).[50]

Table 5.17 Data for the polymers **92–94** (Scheme 5.38).

Entry	Polymer	M_n	M_W	M_W/M_n	Yield (%)	References
1	**92**	6100	10800	1.8	21	51
2	***o*-93**	1000	3200	3.1	11	44
3	***m*-93**	1500	1900	1.3	11	44
4	***p*-93**	1900	2700	1.4	15	44
5	**(E)-94**	3000	3500	1.6	62	52
6	**(E)-94 + Pt**[a]	1900	2400	1.3	quant.	52

[a]The unprotected diphosphine groups were complexed with [PtCl$_2$] units.

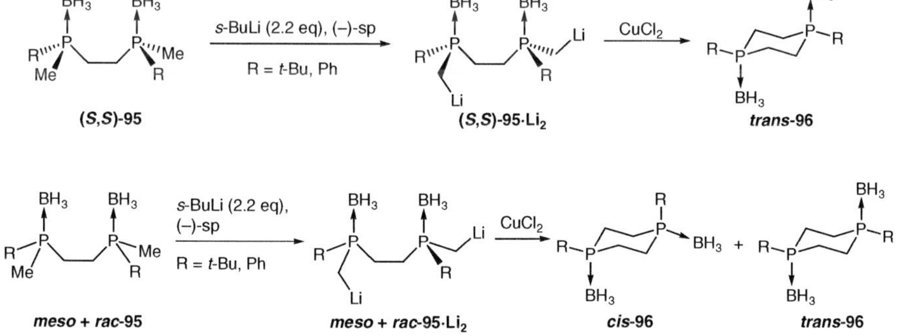

Scheme 5.39 Stereoselective synthesis of *cis*- and *trans*-1,4-diphosphacyclohexane borane.

Treatment of optically active **(*S,S*)-95** with 2.2 equivalents of *s*-BuLi generates the corresponding dilithiated intermediates **(*S,S*)-95·Li$_2$**, which suffer intramolecular coupling affording ***trans*-96** as a single product in 73% and 56% yields for R = *t*-Bu and Ph respectively. X-ray crystal structures of ***trans*-96** confirmed the expected chair conformation with the R groups in equatorial positions and the boranes in axial positions. It has to be pointed out that ***trans*-96** is not chiral and therefore the same result would have been obtained starting from **(*R,R*)-95**. The same synthetic route was applied to optically inactive **95** (prepared without (–)-sparteine), present as a mixture of the *meso* and *rac* diastereomers. The cyclisation of ***rac*-95** generated ***trans*-96** in 16–18% yield whereas ***meso*-95** produced ***cis*-96** in 20% (R = *t*-Bu) and 17% (R = Ph) yields. The *cis* compounds could be separated from the *trans* by column chromatography. Both ***cis*-** and ***trans*-96** were easily deboronated by standard methods in high yields. It has to be mentioned that while *cis*-diphosphacycloalkanes are useful bidentate ligands for transition metals the *trans* counterparts have potential as building blocks in self-assembly processes to produce metal–organic structures such as coordination polymers.[50]

The same method, applied to the tetraphosphine borane **86**, led to the 12-phosphacrown-4 skeleton (Scheme 5.40).

After dilithiation of **86**, the cyclic compound **97** could be isolated in 15% yield.[48] This compound consists of four ethylene units bridging four P(*t*-Bu)(BH$_3$) groups. The *tert*-butyl groups are located alternatively below and above the plane defined by the four phosphorus atoms according to the X-ray structure of **97**.[48]

More recently Morisaki, Chujo and co-workers[23] reported the synthesis of chiral *P*-stereogenic diphosphacrowns containing a diphosphine borane unit. To this end, they first prepared both enantiomers of a β-hydroxydiphosphine borane (Scheme 5.41).

The deprotonation and carboxylation of both methyl groups in **(*S,S*)-98** produced **(*S,S*)-99**, which was not isolated but reduced with borane to afford **(*S,S*)-100** in a combined yield of 42%. For the preparation of the other enantiomer, the β-hydroxymonophosphine borane **(*S*)-101** (87% ee), prepared by the enantioselective deprotonation protocol, was oxidatively coupled to afford directly **(*R,R*)-100** in 36% yield after purification by repeated recrystallisations to remove the *meso* compound and improve the optical purity of the final product.

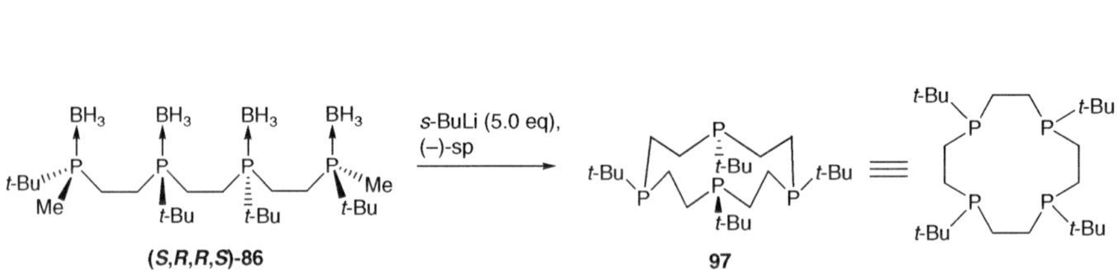

Scheme 5.40 Stereoselective synthesis of 12-phosphacrown-4 borane (**97**). The borane groups on **97** have been omitted for clarity.

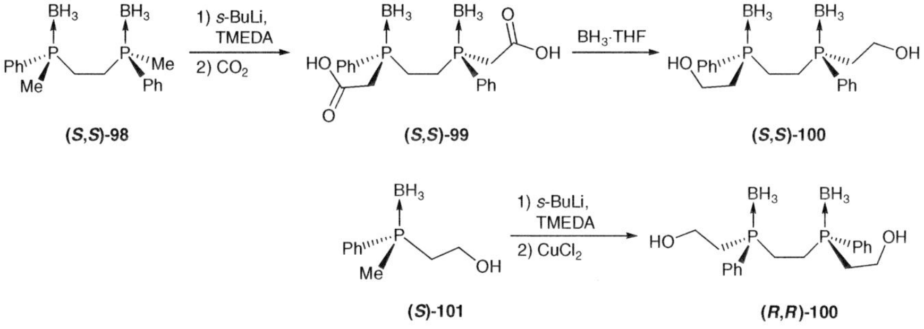

Scheme 5.41 Preparation of both enantiomers of a β-hydroxydiphosphine borane.

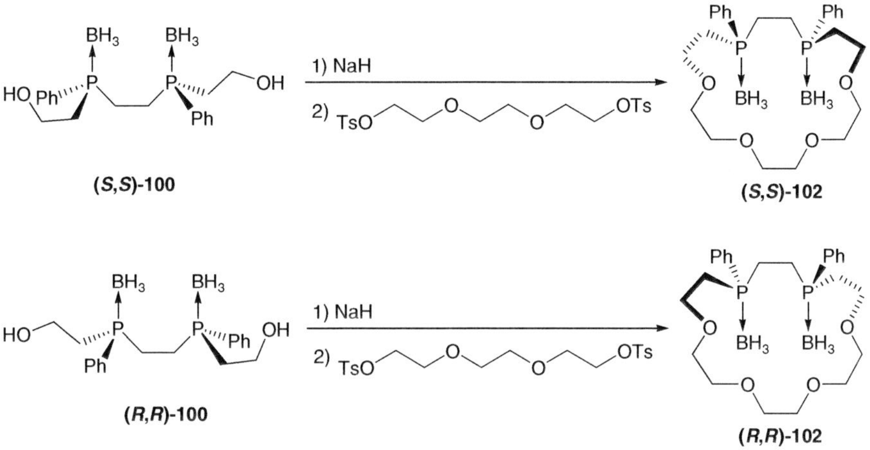

Scheme 5.42 Preparation of both enantiomers of *P*-stereogenic diphosphacrowns **102**.

The diphosphine boranes **100** were then deprotonated with sodium hydride and reacted with triethyleneglycol ditosylate to afford the desired 18-phosphacrown-6 ethers **102** (Scheme 5.42).

Compounds **(S,S)-** and **(R,R)-102** were obtained in 17% and 22% yield as optically pure solids. Both compounds were easily deboronated with DABCO and could be coordinated to Pd(II) and Pt(II) moieties.[23]

5.2.2 Dimethylphosphine Sulfides

The enantioselective deprotonation protocol has also been applied with certain success to dimethylphosphine sulfides (Scheme 5.43).

The main difference between phosphine boranes and phosphine sulfides is that for the latter compounds *n*-BuLi rather than *s*-BuLi is generally used as the

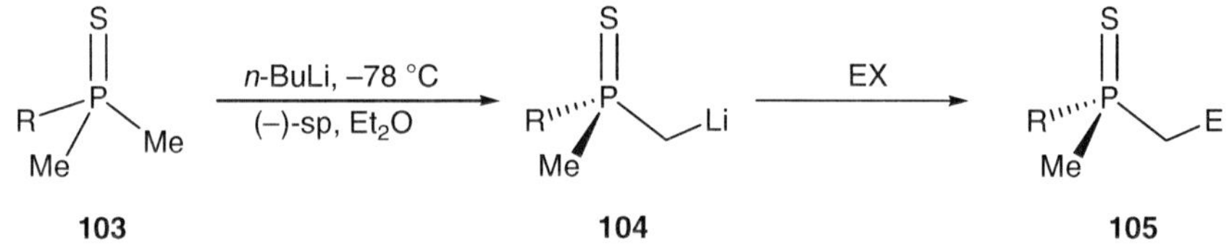

Scheme 5.43 Enantioselective deprotonation–electrophilic quenching of dimethylphosphine sulfides.

deprotonating agent because it achieves higher levels of enantioselection. Both
phosphine boranes and phosphine sulfides afford products with the same sense
of optical induction with the standard protocol with (–)-sparteine.[8] Table 5.18
lists the compounds **105** prepared by the enantioselective deprotonation of
phosphine sulfides.

Comparison between Table 5.18 and Table 5.1 reveals that although phosphine sulfides provide slightly better yields, the enantioselectivities are somewhat lower. It also seems that primary alkyl-lithiums are more enantioselective
than the secondary ones (compare entries 5 and 7 with entry 6).

The lower enantioselectivity achieved by phosphine sulfides and other
practical considerations (like product crystallinity or ease of purification)[8] have
made phosphine boranes the reagents of choice in enantioselective deprotonation. Recently, however, it has been showed that in certain cases phosphine
sulfides outperform the borane counterparts. O'Brien and co-workers[54]
developed a strategy to prepare both enantiomers of phosphine derivatives
using enantioselective deprotonation with (–)-sparteine (Scheme 5.44).

The standard protocol provides **105**, as seen throughout this chapter. The
other enantiomer, **105′**, could in theory be obtained by functionalisation of **103**
with a removable group A to afford **106**. Then a regioselective lithiation into
the less hindered methyl group (a kinetically controlled process) would furnish
107, which should provide **105′** after removal of the group A.

This procedure was applied to the preparation of mono- and diphosphine
sulfides with the opposite configuration to those prepared by the standard
method (Scheme 5.45).[54]

The trimethylsilyl derivative **109** was chosen because its optical purity can be
easily increased by recrystallisation until a virtually enantiopure compound is
obtained in 56% yield. Some screening[54] showed that polyamines such as
PMDETA in combination with *s*-BuLi constitute the optimum conditions in
order to minimise undesired functionalisation of the methylene group. Under
these conditions, **109** was regioselectively lithiated and the carbanion further
elaborated by known methods to afford, after a standard desilylation step,
α-hydroxyphosphine sulfide **110** (55% yield, 94% ee) and β-hydroxyphosphine
sulfide **111** (50% yield over two steps, 96% ee). The method was also adapted
to yield diphosphine sulfides **112** (69% yield, 90% ee) and **113** (46% yield, 99%
ee). The latter compound is indeed the disulfide of (*S*,*S*)-*t*-Bu-MiniPHOS.
During its synthesis it was observed that no *meso* compound was formed,

Table 5.18 Enantioselective deprotonation of dimethylphosphine sulfides (Scheme 5.43).

Entry	105	Yield (%)	ee (%)	Reference
1	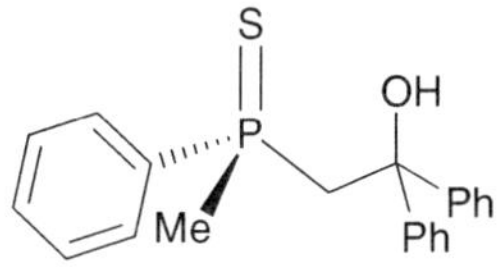	94	78	8
2	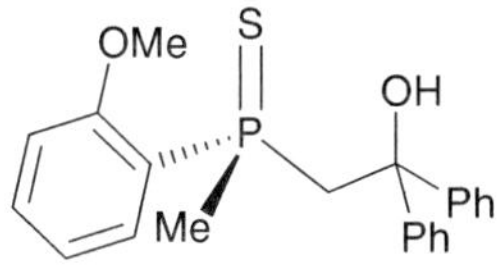	83	76	8
3	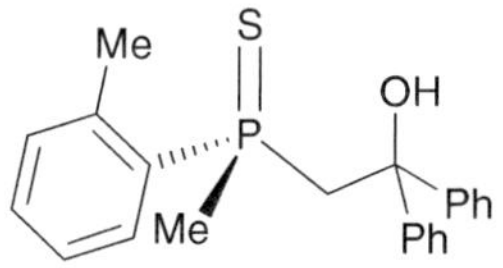	79	79	8
4	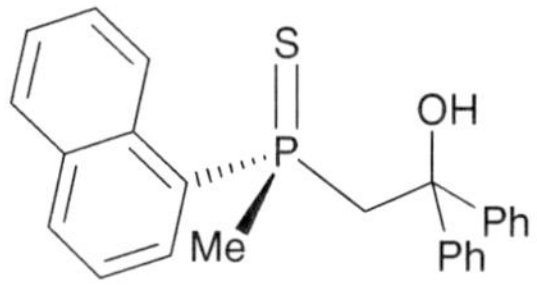	80	60	8
5	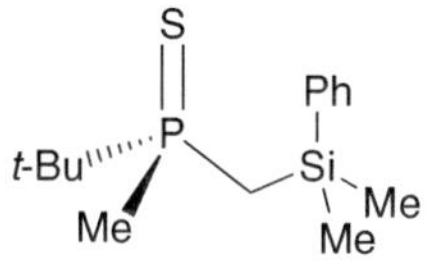	88	76	53,54
6[a]	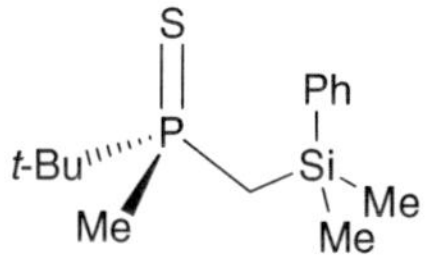	74	68	53,54
7[b]	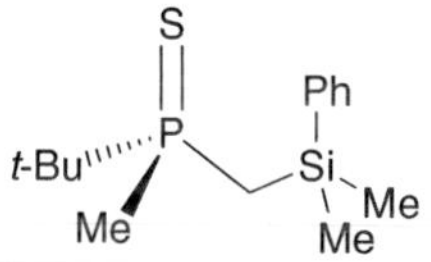	88	76	53,54

[a] s-BuLi as a deprotonating agent.
[b] $(CH_3)_3SiCH_2Li$ as a deprotonating agent.

meaning that the attack of lithiated **109** on t-BuPCl$_2$ to create the second stereogenic phosphorus atom is highly stereoselective, in contrast with the normal preparation *via* phosphine boranes where ∼1:1 diastereoselectivity is observed (Section 5.2.1.4). This highlights the different reactivity between phosphine boranes and phosphine sulfides.

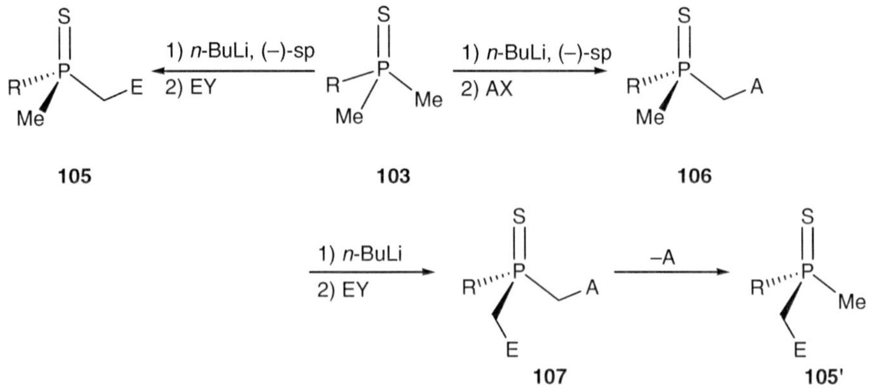

Scheme 5.44   Strategy for preparing both enantiomers of **105** *via* enantioselective deprotonation.

5.3   Deprotonation of Cyclic Phosphine Derivatives

5.3.1   Phosphetane Boranes and Sulfides

There is a single example of enantioselective deprotonation of phosphetane derivatives, reported by Imamoto and co-workers.[55] They prepared a C_2-diphosphine borane (abbreviated as DiSquareP*) by enantioselective deprotonation of 1-*tert*-butylphosphetane borane, **114** (Scheme 5.46).

Enantioselective deprotonation and oxidative coupling provided a mixture of C_2-symmetric and *meso*-DiSquareP* $\cdot$ (BH$_3$)$_2$ in isolated yields of 24% and 31% respectively. Remarkably, the C_2-symmetric compound was present in 95% ee. After recrystallisation, the major enantiomer of the C_2-symmetric compound (shown in Scheme 5.46) was isolated in 15% yield and its absolute configuration unequivocally determined by X-ray diffraction. This compound could be deboronated with HBF$_4$ to yield the highly air-sensitive free DiSquareP*, which was complexed to Rh and the complex tested in hydrogenation reactions.[55] The DiSquareP* ligand resembles *t*-Bu-BisP* but contains two rigid phosphetane rings. The same route depicted in Scheme 5.46 was attempted using phosphine sulfides.[55] Although successful, the yields in the oxidative coupling and in the desulfurisation (with hexachlorodisilane) steps were inferior compared to the phosphine borane route.

5.3.2   Pholpholane Boranes, Sulfides and Oxides

Most of the work in enantioselective deprotonation of cyclic phosphine derivatives has been carried out in phospholane derivatives. Simpkins and co-workers[56,57] studied the stereoselective deprotonation of a readily available triphenylphospholane oxide and prepared several *P*-stereogenic compounds (Scheme 5.47).

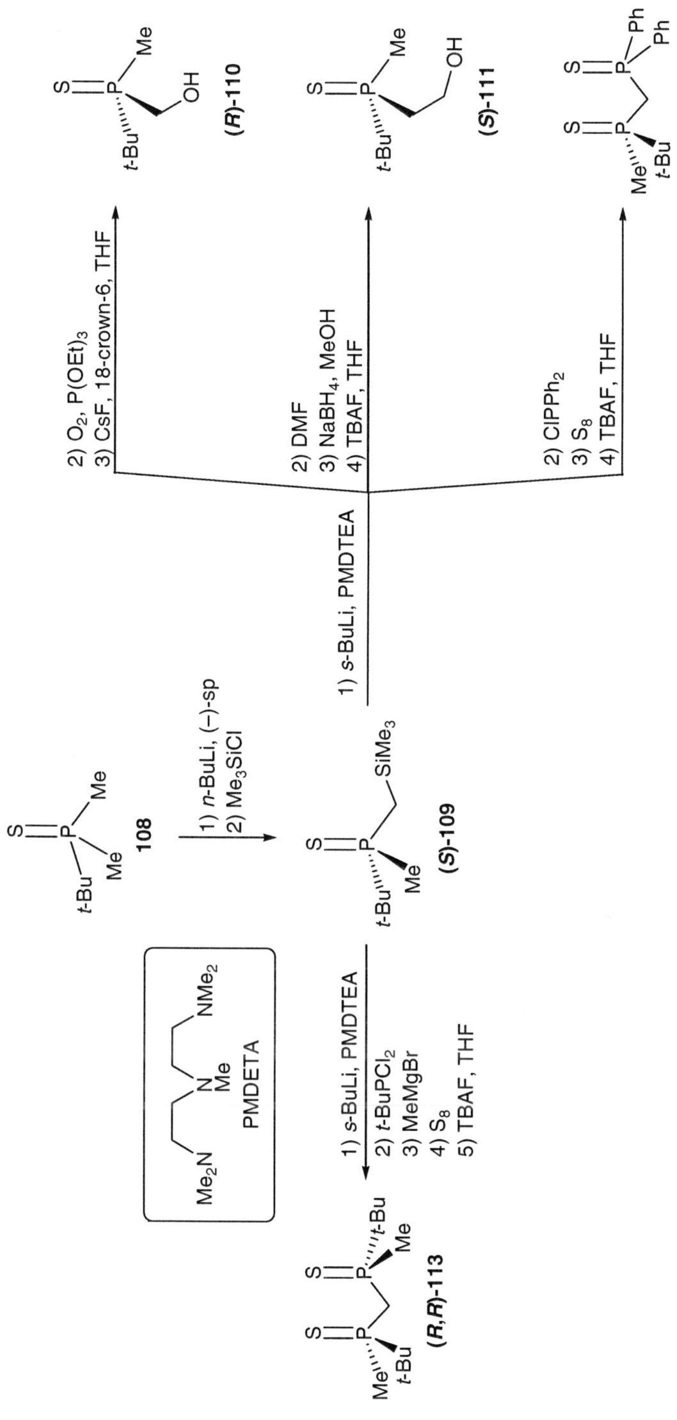

Scheme 5.45 Synthesis of mono- and diphosphine sulfides.

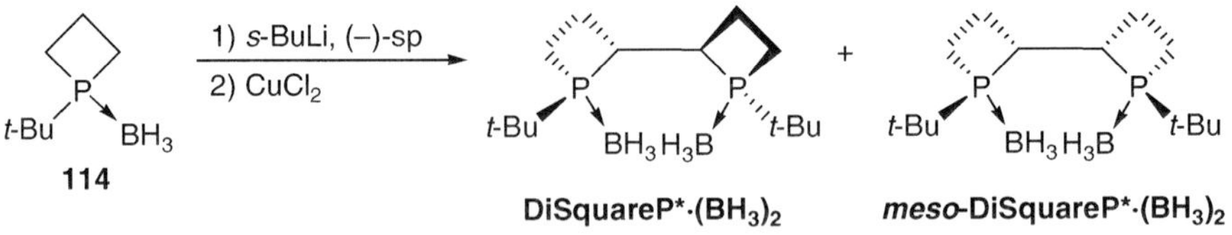

Scheme 5.46 Preparation of DiSquareP* · (BH$_3$)$_2$.

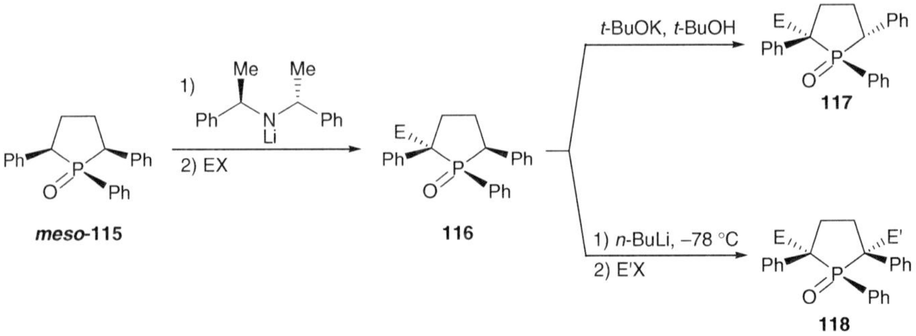

Scheme 5.47 Desymmetrisation of achiral phospholane oxide ***meso*-115**.

Table 5.19 Stereoselective deprotonation of phospholane oxide ***meso*-115** (Scheme 5.47).

Entry	EX	Yield (%)	ee (%)	Reference
1	MeI	87	84	56,57
2	EtI	89	90	56,57
3	BnBr	62	ND	56,57
4	Allyl bromide	72	87	56,57
5	Prenyl bromide	93	88	57
6	PhSSO$_2$Ph	60	82	56,57
7	Ph–CHO	82	92[a]	56,57
8	Cy–CHO	74	80[a]	56,57
9	Acetone	85	86	56,57
10	Picolyl chloride	61	84	57
11	2-chloromethyl-4,4-dimethyl oxazoline	74	82	57

[a]The product is obtained as a mixture of diastereomers.

Reaction of ***meso*-115** with a chiral lithium amide at very low temperature (−100 °C) furnished, after addition of several electrophiles, the chiral products **116** in good yields and optical purities (Table 5.19).

Only the diastereomer maintaining the *syn* arrangement of the three phenyl rings was formed, in comparable yields and rather similar enantioselectivities of around 85% ee irrespective of the electrophile employed. The optical purity of the products could be improved by a single recrystallisation. When a third stereogenic centre is formed (entries 7 and 8) the products results in a mixture of diastereomers although some selectivity was observed in the case of the product of entry 7. The stereoselectivity of the reaction is thought to be due to steric shielding of one face of the substrate by the two fixed phenyl groups because the lithiated phosphine oxides have been shown to be configurationally unstable under the conditions of the reaction.[57]

Some of the enantiopure pholane oxides **116** were used for further transformations. For example, they could be isomerised by treatment with base for several hours in a protic solvent to afford **117** or subjected to a second deprotonation to produce **118** as a single isomer. Attempts to prepare *bis*(phospholanes) by double nucleophilic attack of the anions derived from **116** were unsuccessful. Some of the phospholane oxides were reduced with tri-chlorosilane/pyridine to afford the free phospholane with retention of config-uration.[57] The same strategy was applied to 1-phenylphospholane oxide but low yields of racemic products were obtained. This suggests that substrate **115** is probably a special case due to the formation of a more stable benzylic carbanion.[57]

Ohashi and Imamoto[58] studied the stereoselective deprotonation of 1-*tert*-butylphospholane borane (**119**) in presence of sparteine at − 78 °C (Scheme 5.48).

After quenching the α-carbanion with methyl iodide, the monophosphine borane **120** was obtained in quantitative yield as a single diastereomer but in racemic form. The oxidative coupling with Cu(II) was found to be troublesome and give only 10% of achiral ***meso*-121** and 40% of recovered starting material. The reader will probably have noticed that to form ***rac*-120** and ***meso*-121** the lithiation of **119** occurred exclusively *trans* to the bulky *t*-Bu group but unfortunately not in an enantioselective way. This disappointing result con-trasts with the successful (albeit low-yielding) synthesis of optically pure DiS-quareP* · (BH$_3$)$_2$, discussed in Section 5.3.1, and with the enantioselective deprotonation of 1-*tert*-butylphospholane sulfide, discussed later in this section.

Better results have been achieved in the enantioselective deprotonation of 1-phenylphospholane borane by Kobayashi and co-workers (Scheme 5.49).[59]

In this case the carbanion was quenched with dry ice in order to prepare the α-carboxylated phospholane borane **123**. This compound was obtained as a

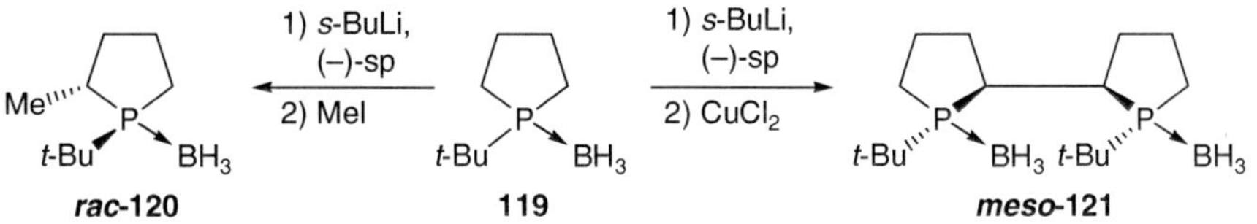

Scheme 5.48 Enantioselective deprotonation of **119**. Only one of the enantiomers for **120** is represented.

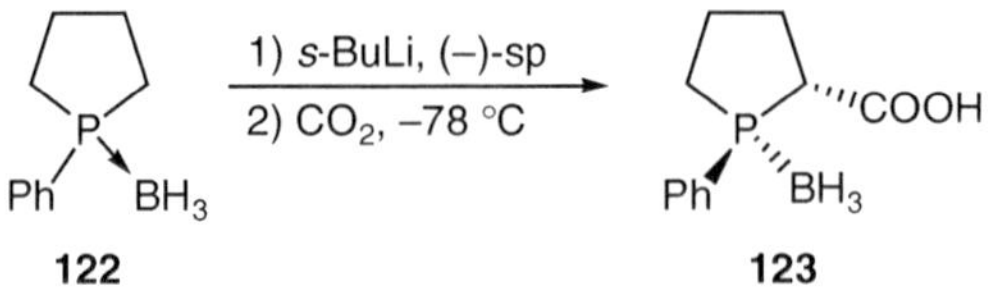

Scheme 5.49 Stereoselective deprotonation/carboxylation of 1-phenylphospholane borane. Only one enantiomer of the *trans* diastereomer of **123** is represented.

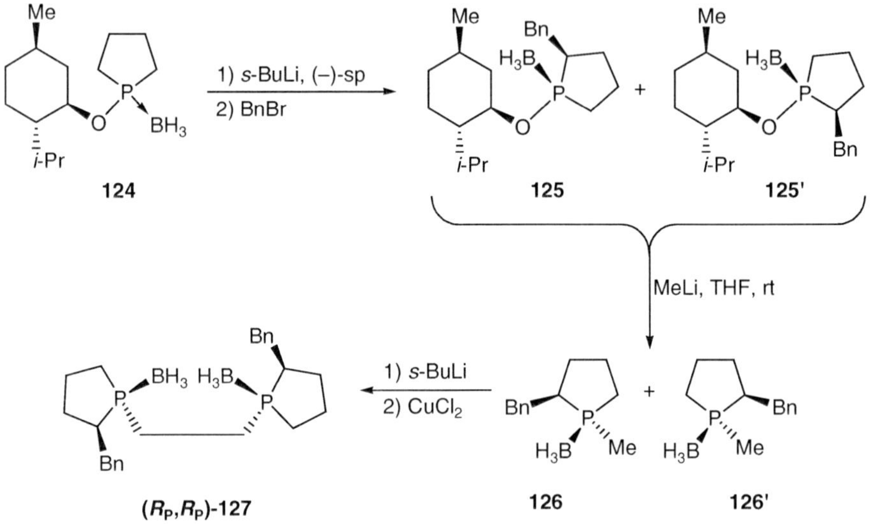

Scheme 5.50 Preparation of the *P*-stereogenic *bis*phospholane borane **127**.

diastereomeric mixture consisting of predominantly the *trans* diastereomer in up to 60% de.[60] The enantioselectivity of ***trans*-123** was found to be rather capricious and depended on the number of equivalents of *s*-BuLi and (–)-sp, the temperature and even on the scale of the synthesis. In a 25 mmol preparation, the best results (55% yield, 60% de, 92% ee) were obtained with equimolar quantities of **122**, *s*-BuLi and (–)-sp at a temperature of 25 °C during the deprotonation. The optical purity of **123** could be increased by recrystallisation but at the expense of the yield whereas the absolute configuration of optically pure **123** was determined by X-ray diffraction analysis of its menthyl ester.[59] In a more recent paper, **123** has been used to prepare a *P,N* ligand containing a tetrahydroisoquinoline skeleton.[61]

The preparation of a *P*-stereogenic 1,2-*bis*phospholane was reported by Hoge[62] in a route involving a diastereoselective deprotonation step in a molecule already containing stereogenic centres (Scheme 5.50).

The diastereoselective lithiation, in the presence of (–)-sp, of one α-carbon of the (–)-menthoxyphospholane borane **124** and benzylation produce **125** and **125′** as an inseparable diastereomeric mixture in 77% yield and 80% de. Without sparteine, an equimolar mixture was obtained in 94% yield. In both cases the alkylation took place exclusively *cis* to the borane group. The synthesis was carried forward using the diastereoenriched mixture. Methyl-lithium cleanly displaced the menthoxy group in **125** to afford **126** and **126′** in 53% yield. The de of the mixture was nearly identical to the starting materials and hence the transformation occurred without loss of optical purity and with complete retention of configuration at the phosphorus atom. Finally, depro-tonation and oxidative coupling with Cu(II) afforded the desired boronated ligand **127**. A recrystallisation step furnished the optically pure (R_P,R_P) product in 41% yield. Its absolute configuration was confirmed by X-ray diffraction analysis. Interestingly, the (S_P,S_P) enantiomer of **127** can be prepared from a chiral alcohol, by a completely different procedure, discussed in Chapter 2, Section 2.4.1.

Zhang and co-workers[63,64] explored the enantioselective metallation of 1-*tert*-butylphospholane sulfide (**128**) to prepare bidentate *P,P* and *P,N* ligands (Scheme 5.51).

The enantioselective deprotonation of **128** and oxidative addition of the carbanion with Cu(II) afforded TangPhos disulfide as a mixture of diaster-eomers where the chiral C_2-symmetric product was the major compound (66% de, 95% ee).[63] A single recrystallisation afforded the desired optically pure TangPhos sulfide in 20% yield. The desulfuration was performed with hexa-chlorodisilane to yield the very air-sensitive Tangphos diphosphine in 88% yield. The synthesis of a family of phospholane-oxazoline ligands[64] started with the preparation of carboxylic acid **129** by the procedure in 40% yield after recrystallisation. The crude product was obtained in relatively low optical

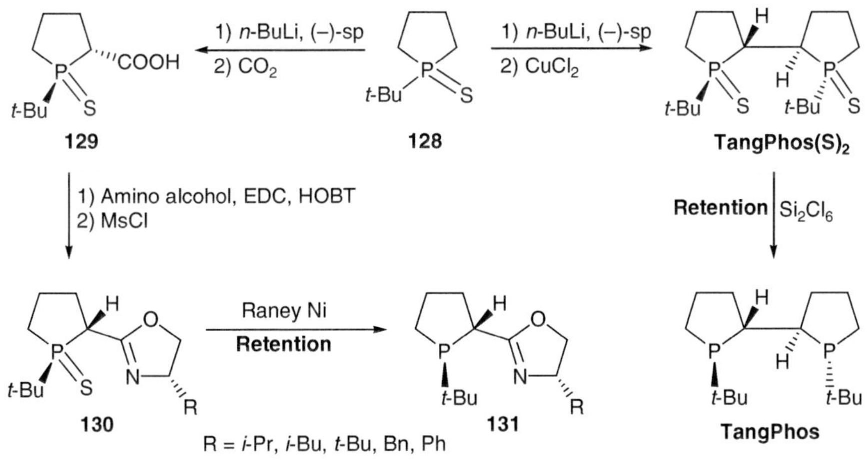

Scheme 5.51 Synthesis of phosphinooxazoline **131** and TangPhos.

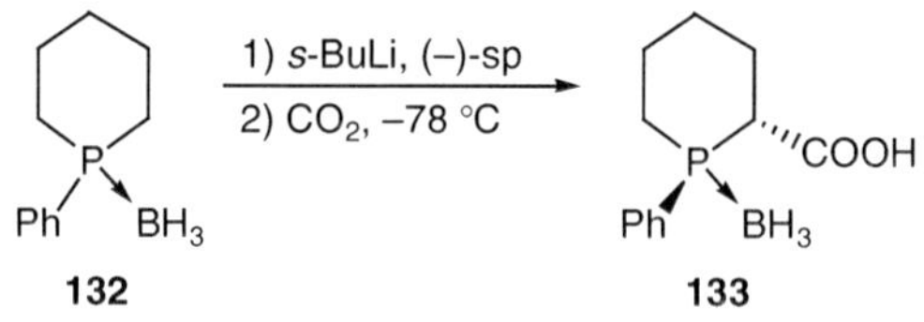

1) *s*-BuLi, (−)-sp
2) CO_2, −78 °C

Ph　BH₃　　**132**　　　　Ph　COOH　BH₃　　**133**

Scheme 5.52　Stereoselective deprotonation/carboxylation of **132**. Only one enantiomer of the *trans* diastereomer of **133** is represented.

purity (72% ee) but the recrystallisation step yielded enantiopure **129**. Its condensation with chiral amino alcohols and oxazoline ring formation was carried out by standard methods producing **130** in 70–80% yield. The final unprotected ligands **131** were obtained by a high-yielding Raney Ni-mediated desulfuration step.

5.3.3　Phosphorinane Boranes

There is a single report of Kobayashi and co-workers[59] describing the enantioselective deprotonation of 1-phenylphosphorinane borane (**132**) (Scheme 5.52).

With this substrate high diastereoselectivities towards the *trans* isomers could be obtained (up to 97% de) with yields and ee values reaching 79 and 90% respectively under optimised conditions. Compared to 1-phenylphospholane borane (**122**) (Scheme 5.49), **132** seems to be a better substrate for enantioselective deprotonation, especially regarding to diastereoselectivity. That could be due to the more 'rigidly stable'[59] chair conformation adopted by the phosphorinane ring in comparison to the five-membered phospholane ring. This is an interesting result because up to now the six-membered phosphorinane ring has been rarely present in ligands for enantioselective catalysis,[65] especially compared to the very common five-membered phospholane counterpart.[66] Therefore, further developments in this area should be seen in the near future.

5.4　Extensions of the Method

5.4.1　Surrogates of (+)-Sparteine

One of the principal limitations of the enantioselective deprotonation protocol with sparteine is that only one of the enantiomers of the phosphine can be prepared, although sometimes this limitation has been circumvented, as in the preparation of (*R*,*R*)-*t*-Bu-BisP* seen in Section 5.2.1.3. That is because (+)-sparteine is only available by multistep synthesis or resolution of the racemic natural product lupanine.[67] Unfortunately, none of these procedures can give multigram quantities of (+)-sparteine easily and cheaply. This has been called the '(+)-sparteine problem'[67] and has been partially solved by the synthesis of (+)-sparteine mimics known as (+)-sparteine surrogates, **134** (Figure 5.1).

The working hypothesis that guided the design of **134** was that because modelled structures show that the D ring in (+)-sparteine is far away from the coordinated lithium atom, so a 'simplified' version without this ring should not significantly alter the chiral environment around the lithium atom.[68] Therefore, several surrogates **134** were prepared from (−)-cytisine[9] and tested in enantioselective deprotonation of phosphine boranes and sulfides (Scheme 5.53 and Table 5.20).

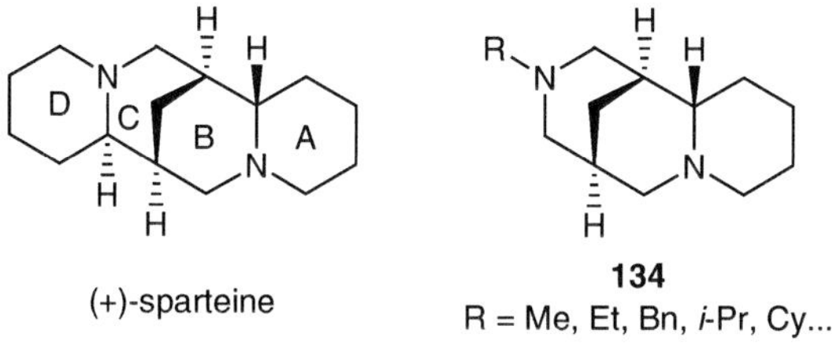

Figure 5.1 Chemical structures of (+)-sparteine and some of its surrogates.

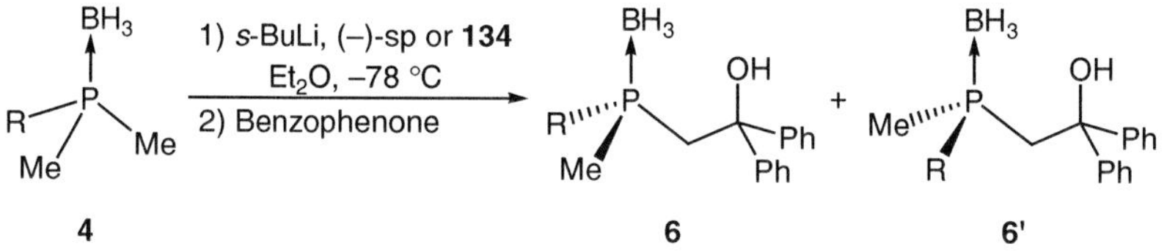

Scheme 5.53 Enantioselective deprotonation–electrophilic quenching of **4**.

Table 5.20 Some results in enantioselective deprotonation of **4** (Scheme 5.53).

Entry	R in 4	Amine or R in 134	Conf.[a]	ee (%)	Yield (%)	References
1	Ph	(−)-sp	S	78	87	11,69
2	Ph	Me	R	67	89	11,69
3	Ph	i-Pr	R	63	82	11,69
4	Ph	Bn	–	0	16	69
5	Cy	(−)-sp	S	70	67	11,69
6	Cy	Me	R	74	77	11,69
7	Cy	i-Pr	R	62	86	11,69
8	t-Bu	(−)-sp	S	76	83	11,69
9	t-Bu	Me	R	92	78	11,69,70
10	t-Bu	Et	R	90	82	69
11	t-Bu	i-Pr	R	75	72	11,69
12	t-Bu	Bn	R	50	35	69

[a]Absolute configuration of the major enantiomer of **6**.

Table 5.20 shows that this approach allows the synthesis of the other enantiomer to that obtained with (–)-sp. Furthermore, in some cases the surrogate considerably outperformed (+)-sparteine (compare entries 6 with 5 or 10 and 11 with 8). A competition experiment with equimolar amounts of (–)-sp and **Me-134** in the deprotonation of **4** (R = *t*-Bu) with *s*-BuLi showed that the product was obtained with a 50% ee with the sense of induction of (+)-sparteine, confirming that **Me-134** is more reactive than sparteine.[71] The choice of the R group in **134** is crucial. For example, benzyl gives disappointing results probably due to deprotonation at the benzylic position (entries 4 and 12).[69] Up to now, the best surrogate seems to be the diamine **Me-134**[72] and this has been used to prepare the enantiomers of some of the phosphine boranes and sulfides discussed earlier in this section.[53,71,73,74]

5.4.2 Catalytic Enantioselective Deprotonation

It is well known that certain diamines increase the reactivity of organolithium compounds by complexation to the Li atom. Consequently, with the appropriate chiral diamines it should be possible to perform catalytic asymmetric deprotonations. In the original report of Evans and co-workers[8] it is mentioned that the enantioselection can be maintained with only 0.7 equivalents of (–)-sparteine, with no further details. These ideas have been recently exploited for *t*-butyldimethylphosphine borane and the analogous sulfide with sparteine and its surrogates. It can be understood with the catalytic cycle of Scheme 5.54, reported by O'Brien and co-workers.[53,73]

In Scheme 5.54, sparteine (or a surrogate) is present in a catalytic amount forming RLi · (–)-sp, which in step I enantioselectively deprotonates **135** yielding **(*S*)-136** · (–)-sp. At this point, an incoming equivalent of the organolithium could displace sparteine to form **(*S*)-136 · L** and regenerate the **RLi · (–)-sp** (step II). Electrophilic quenching of either **(*S*)-136 · (–)-sp** or **(*S*)-136** would produce the desired optically pure product. The presence of a stoichiometric amount of an achiral 'sacrificial' ligand may be needed for the ligand exchange in the second step of the cycle. The terms *one-ligand* and *two-ligand catalysis* are usually employed when the stoichiometric achiral ligand is absent or present respectively.[71,74] The success of this approach depends on an efficient ligand exchange in the second step but also on a faster catalytic deprotonation of **135** by **RLi · (–)-sp** compared to the background stoichiometric deprotonation by RLi · L, which yields *rac*-**136**.[73]

In the first experiments the two-ligand catalysis approach was followed: *t*-butyldimethylphosphine borane, bispidine **137** and substoichiometric amounts (0.2 equiv.) of (–)-sp or **Me-134**.[73] Later it has been found that the absence of the stoichiometric ligand (one-ligand catalysis) also lead to satisfactory results.[74] For example, an application of the catalytic enantioselective deprotonation can be seen in Scheme 5.55 (Table 5.21).

Here both enantiomers of *t*-Bu-BisP*(BH₃)₂ are prepared in almost enantiopure form with substoichiometric amounts of chiral inductor. A severe

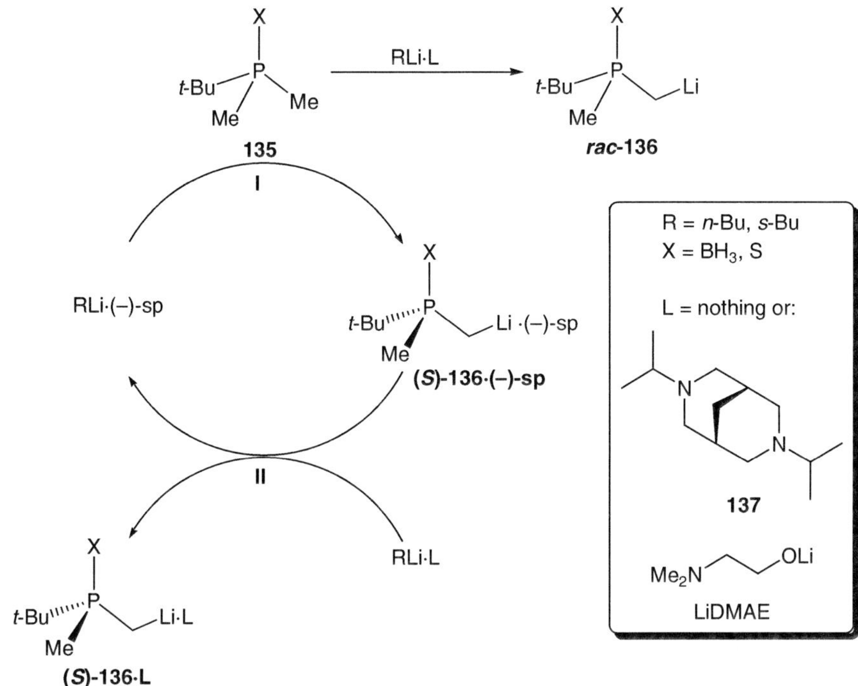

Scheme 5.54 Catalytic enantioselective deprotonation.

Scheme 5.55 Preparation of both enantiomers of *t*-Bu-BisP*(BH$_3$)$_2$ by enantioselective deprotonation.

increase of the amount of the *meso* compound is observed when lowering too much the load of chiral diamine (entries 3 and 6). In spite of that, the result of entry 6 is remarkable because just 10% of (–)-sparteine surrogate **Me-134** is enough to produce a respectable amount of enantiopure product.

Table 5.21   Results in the catalytic reactions of Scheme 5.55.

Entry	Diamine, equiv.	Yield (1) (%)	Yield meso (%)	ee (1)	References
1	(–)-sp, 1.2	67 (40[a])	13	o.p.	32,33
2	(–)-sp, 0.5	61	15	98	74
3	(–)-sp, 0.2	48	27	98	74
4	**Me-134**, 0.5	59	15	98	74
5	**Me-134**, 0.2	56	17	98	74
6	**Me-134**, 0.1	45	25	98	74

[a]Final yield after purification.

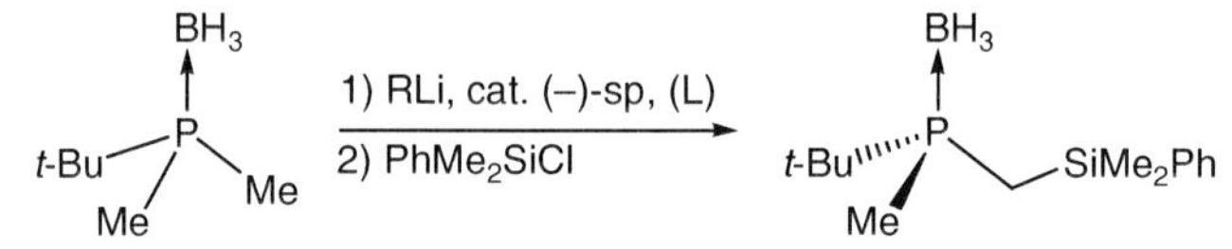

Scheme 5.56   One- or two-ligand catalytic enantioselective deprotonation of **29**.

A broad comparison between one- and two-ligand catalysis for the catalytic enantioselective deprotonation of phosphine borane **29** with either *n*-BuLi or *s*-BuLi has been recently carried out by O'Brien and co-workers (Scheme 5.56).[71] In the two-ligand catalysis they employed bispidine **137**, LiDMAE or *N*-methylmorpholine.

They found that although *n*-BuLi can be used in the stoichiometric reaction (76% yield, 78% ee for **138**), alone it does not lithiate **29**. Unfortunately, the one-ligand catalysis (0.2 equiv. of (–)-sp) is not possible at $-78\,°\mathrm{C}$ since no turnover of (–)-sp is observed. Increasing the temperature to $-50\,°\mathrm{C}$, however, allowed the isolation of **138** in 50% yield and 62% ee. In the presence of stoichiometric amounts of LiDMAE, the reaction proceeded at $-78\,°\mathrm{C}$ leading to very similar results. With *s*-BuLi, two-ligand catalysis with LiDMAE or bispidine **137** worked slightly better than one-ligand catalysis, also with similar yields and enantioselectivities. They concluded that for the system studied using one-ligand catalysis with *n*-BuLi at $-50\,°\mathrm{C}$ or two-ligand catalysis with *s*-BuLi/LiDMAE or *s*-BuLi/**137** at $-78\,°\mathrm{C}$ is approximately equivalent.

Further work from the same group has focused on the comparison between one-ligand catalytic deprotonation with (–)-sp of *t*-butyldimethylphosphine borane and sulfide with *n*- and *s*-butyllithium.[53] After a series of experiments similar to those of Scheme 5.56, they arrived at the conclusions summarised in Table 5.22.

The somewhat frustrating situation is as follows: *n*-BuLi does not lithiate **29** in the absence of ligands (so no background reaction) but it can not be used in the catalysis reaction because step II in Scheme 5.54 does not occur. In all the other combinations, there is catalytic turnover but also background deprotonation, which decreases the enantioselectivity. Consequently, the only way

Table 5.22 Differences between **29** and **108** in enantioselective catalytic deprotonation.

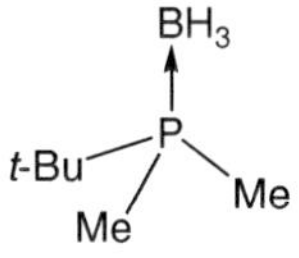

	29		**108**	
	n-BuLi	*s-BuLi*	*n-BuLi*	*s-BuLi*
Direct deprotonation of **29** or **108**?	No	Yes	Yes	Yes
Turnover in Scheme 5.54?	No	Yes	Yes	Yes

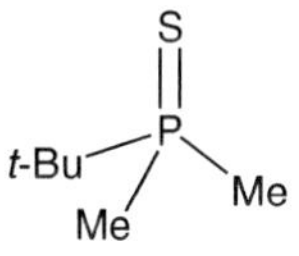

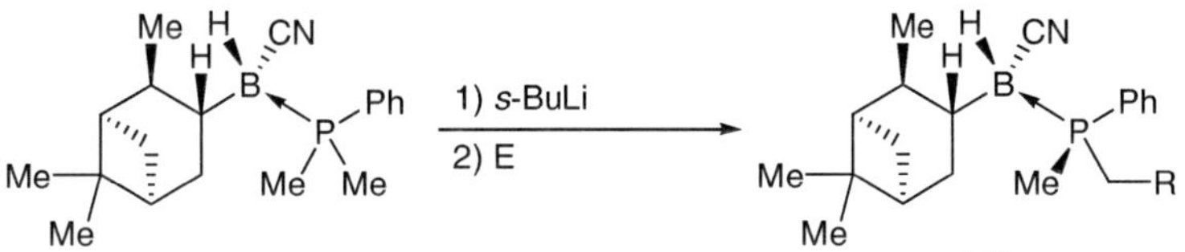

Scheme 5.57 Diastereoselective alkylation of **139**.

forward is the optimisation of the conditions to minimise the background deprotonation. They found that high yields and enantioselectivities could be obtained with phosphine sulfide **108** and *n*-BuLi. For example, with 0.2 equivalents of (–)-sp, they obtained 82% yield and 66% ee for the sulfide analogue to **138**. Interestingly, they found that toluene was a better solvent, which allowed the use of only 5% of (–)-sparteine, generating the product whose yield and optical purity (88%, 70% ee) was close the stoichiometric result (90%, 78% ee). Under these optimised conditions, they were able to prepare both enantiomers of the diborane analogues of diphosphine sulfide **112**, seen in Section 5.2.2.

5.4.3 Deprotonation of Other Substrates

There a few examples of enantioselective deprotonation of other types of substrates. Le Gall, Mioskowski and co-workers[75] reported the deprotonation–alkylation of **139**, a substrate bearing a chiral borane (Scheme 5.57).

The optically pure borane was prepared in four steps starting from (*R*)-α-pinene and $BH_3 \cdot SMe_2$. The alkylated products **140** were obtained with yields in the range 77–92%. The best ee was 74%, obtained in the reaction with acetophenone. The functionalised borane could be cleaved by a conventional treatment with DABCO.

Ortho-lithiations on phenyl rings have also been used for the desymmetrisation of P(O)Ph$_2$ groups. The first example, reported by Ortiz and co-workers,[76] was the diastereoselective desymmetrisation of a diphenylphosphinamide (Scheme 5.58).

The metallation of **141** with an excess of *n*-BuLi produced a dianion whose electrophilic quenching afforded *ortho*-functionalised phosphinamides **142**. The disastereomeric excess of **142** varied from 33 to 66%, the best result being with E = Ph$_3$SiCl. In all the cases, the epimers could be separated by column chromatography.

The same group extended this chemistry to enantioselective desymmetrisations of achiral substrates with the system *n*-BuLi/(–)-sparteine (Scheme 5.59).[77]

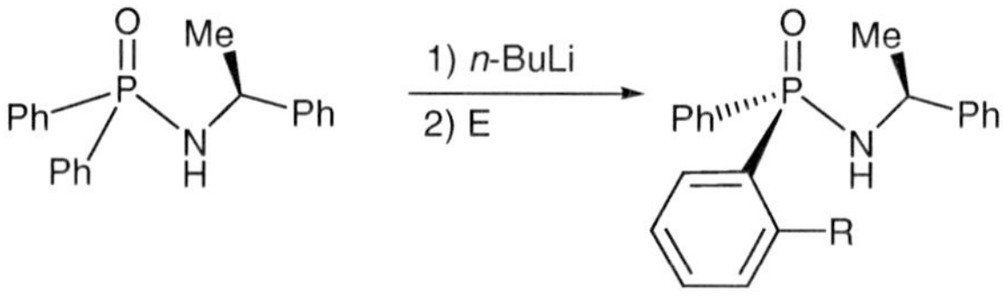

Scheme 5.58 Diastereoselective desymmetrisation of a **141**. Only one epimer of **142** is shown.

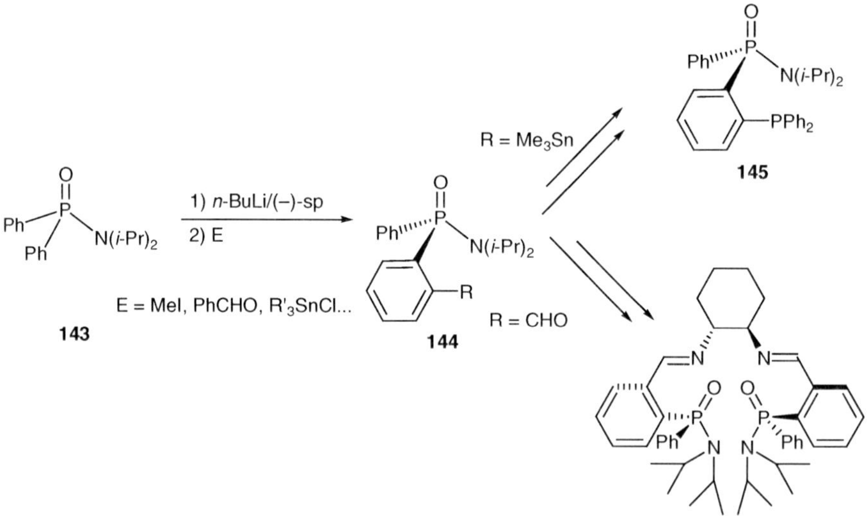

Scheme 5.59 Enantioselective desymmetrisation of **143**. Only one enantiomer of **144** is shown.

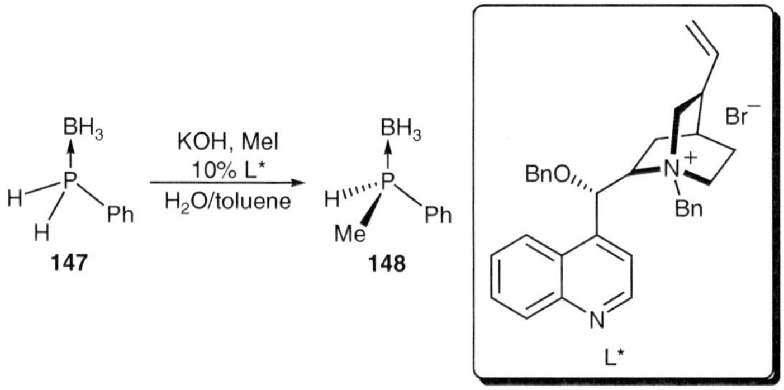

Scheme 5.60   Enantioselective methylation of **147**.

The best results were obtained by treating **143** with the preformed *n*-BuLi/(−)-sp complex in toluene at − 90 °C followed by quenching with E at this temperature. Good yields and ee values in the 45–63% range were obtained. In some cases, optically pure products could be obtained after recrystallisation. The metallation did not take place in the absence of sparteine whereas *t*-BuLi (a stronger metallating agent) gave **144** as racemates. Further elaboration of **144** led in a few steps to optically pure phosphine-phosphine oxide **145** and to the 'salen'-type ligand **146**.

All the methods discussed in this chapter are based in the deprotonation of C–H bonds by organolithiums. There is a single example of deprotonation of a P–H bond by the much milder base KOH in a biphasic solution containing a chiral auxiliary (Scheme 5.60).

Monoalkylation of **147** was achieved using one equivalent of potassium hydroxide and 10% of L*, which served as chiral phase-transfer catalyst. Unfortunately, **148** was obtained in 17% ee only.

5.5   Conclusions

The enantioselective deprotonation protocol is both a short and elegant route to enantiopure *P*-stereogenic phosphines. Along with the Jugé–Stephan method, discussed in Chapter 4, it constitutes the main modern method to prepare families of *P*-stereogenic compounds. More importantly, the scopes of the two methods are complementary. The enantioselective deprotonation protocol is particularly well suited to prepare very electron-rich diphosphines such as BisP*, MiniPHOS and TangPhos, which demonstrated superior performance in catalysis, especially in hydrogenation, despite their simple structure. The method is particularly useful to access many ligands bearing both a *tert*-butyl (or 1-adamantyl) and methyl groups at the phosphorus atom. This feature has proved to be an excellent chiral induction factor in homogeneous catalysis.

Although the desymmetrisation of *tert*-butyldimethylphosphine borane and the analogous sulfide has been widely used, the application of the same chemistry to other substrates, especially cyclic phosphines, is still relatively undeveloped and further advances are likely to appear.

The use of (+)-sparteine surrogates and the catalytic version of the protocol (now using *substoichiometric* amounts of chiral diamine) are very exciting extensions that undoubtedly will see further progress, for example in order to decrease the amounts of chiral inductor required for the reaction to work efficiently. Furthermore, the generalisation of this catalytic variant of the method to the synthesis of more phosphines remains to be seen.

Finally, the method relies almost exclusively[78] on sparteine or its surrogates, making the exploration of other chiral inductors a worthy effort.

References

1. C. A. Maryanoff, B. E. Maryanoff, R. Tang and K. Mislow, *J. Am. Chem. Soc.*, 1973, **95**, 5839.
2. P. Beak, A. Basu, D. J. Gallagher, Y. S. Park and S. Thayumanavan, *Acc. Chem. Res.*, 1996, **29**, 552.
3. P. Beak, D. R. Anderson, M. D. Curtis, J. M. Laumer, D. J. Pippel and G. A. Weisenburger, *Acc. Chem. Res.*, 2000, **33**, 715.
4. L. T. Byrne, L. M. Engelbardt, G. E. Jacobsen, W. Leung, R. I. Papasergio, C. L. Raston, B. W. Skelton, P. Twiss and A. H. White, *J. Chem. Soc. Dalton Trans.*, 1989, **105**.
5. O. Korpiun and K. Mislow, *J. Am. Chem. Soc.*, 1967, **89**, 4784.
6. K. M. Pietrusiewicz, M. Zablocka and J. Monkiewicz, *J. Org. Chem.*, 1984, **49**, 1522.
7. T. Imamoto, in *Phosphorus Ligands in Asymmetric Catalysis: Synthesis and Applications*, ed. A. Börner, Wiley-VCH, Weinheim, 2008, p. 1201.
8. A. R. Muci, K. R. Campos and D. A. Evans, *J. Am. Chem. Soc.*, 1995, **117**, 9075.
9. B. Wolfe and T. Livinghouse, *J. Org. Chem.*, 2001, **66**, 1514.
10. H. Heath, B. Wolfe, T. Livinghouse and S. K. Bae, *Synthesis*, 2001, 2341.
11. M. J. Johansson, L. O. Schwartz, M. Amedjkouh and N. Kann, *Eur. J. Org. Chem.*, 2004, 1894.
12. K. Nagata, S. Matsukawa and T. Imamoto, *J. Org. Chem.*, 2000, **65**, 4185.
13. H. Sugama, H. Saito, H. Danjo and T. Imamoto, *Synthesis*, 2001, **15**, 2348.
14. T. Imamoto, K. Sugito and K. Yoshida, *J. Am. Chem. Soc.*, 2005, **127**, 11934.
15. T. Imamoto, A. Kumada and K. Yoshida, *Chem. Lett.*, 2007, **36**, 500.
16. K. Katagiri, H. Danjo, K. Yamaguchi and T. Imamoto, *Tetrahedron*, 2005, **61**, 4701.
17. T. Imamoto, Y. Saitoh, A. Koide, T. Ogura and K. Yoshida, *Angew. Chem. Int. Ed.*, 2007, **46**, 8636.

18. R. Aznar, G. Muller, D. Sainz, M. Font-Bardia and X. Solans, *Organometallics*, 2008, **27**, 1967.
19. A. Ohashi and T. Imamoto, *Tetrahedron Lett.*, 2001, **42**, 1099.
20. A. Ohashi, S. Kikuchi, M. Yasutake and T. Imamoto, *Eur. J. Org. Chem.*, 2002, 2535.
21. F. Dolhem, M. J. Johansson, T. Antonsson and N. Kann, *J. Comb. Chem.*, 2007, **9**, 477.
22. M. J. Johansson, K. H. O. Andersson and N. Kann, *J. Org. Chem.*, 2008, **73**, 4458.
23. Y. Morisaki, H. Imoto, K. Tsurui and Y. Chujo, *Org. Lett.*, 2009, **11**, 2241.
24. H. Danjo, M. Higuchi, M. Yada and T. Imamoto, *Tetrahedron Lett.*, 2004, **45**, 603.
25. P. E. Goudriaan, X. Jang, M. Kuil, R. Lemmens, P. W. N. M. van Leeuwen and J. N. H. Reek, *Eur. J. Org. Chem.*, 2008, 6079.
26. F. Maienza, F. Spindler, M. Thommen, B. Pugin, C. Malan and A. Mezzetti, *J. Org. Chem.*, 2002, **67**, 5239.
27. N. Oohara, K. Katagiri and T. Imamoto, *Tetrahedron: Asymmetry*, 2003, **14**, 2171.
28. I. D. Gridnev, N. Higashi and T. Imamoto, *Organometallics*, 2001, **20**, 4542.
29. T. Imamoto, M. Nishimura, A. Koide and K. Yoshida, *J. Org. Chem.*, 2007, **72**, 7413.
30. Y. Yamamoto, H. Danjo, K. Yamaguchi and T. Imamoto, *J. Organomet. Chem.*, 2008, **693**, 3546.
31. Y. Yamamoto, T. Koizumi, K. Katagiri, Y. Furuya, H. Danjo, T. Imamoto and K. Yamaguchi, *Org. Lett.*, 2006, **8**, 6103.
32. T. Imamoto, J. Watanabe, Y. Wada, H. Masuda, H. Yamada, H. Tsuruta, S. Matsukawa and K. Yamaguchi, *J. Am. Chem. Soc.*, 1998, **120**, 1635.
33. I. D. Gridnev, Y. Yamamoi, N. Higashi, H. Tsuruta, M. Yasutake and T. Imamoto, *Adv. Synth. Catal.*, 2001, **343**, 118.
34. K. V. L. Crépy, T. Imamoto, G. Seidel and A. Fürstner, *Org. Synth.*, 2005, **82**, 22.
35. L. McKinstry and T. Livinghouse, *Tetrahedron Lett.*, 1994, **35**, 9319.
36. T. Miura, H. Yamada, S. Kikuchi and T. Imamoto, *J. Org. Chem.*, 2000, **65**, 1877.
37. S. Matsukawa, H. Sugama and T. Imamoto, *Tetrahedron Lett.*, 2000, **41**, 6461.
38. T. Imamoto, S. Kikuchi, T. Miura and Y. Wada, *Org. Lett.*, 2001, **3**, 87.
39. K. V. L. Crépy and T. Imamoto, *Tetrahedron Lett.*, 2002, **43**, 7735.
40. A. Ohashi and T. Imamoto, *Org. Lett.*, 2001, **3**, 373.
41. Y. Yamanoi and T. Imamoto, *J. Org. Chem.*, 1999, **64**, 2988.
42. S. Taira, K. V. L. Crépy and T. Imamoto, *Chirality*, 2002, **14**, 386.
43. Y. Ouchi, Y. Morisaki and Y. Chujo, *Polym. Bull.*, 2007, **59**, 339.
44. Y. Morisaki, Y. Ouchi, K. Tsurui and Y. Chujo, *J. Polym. Sci. Part A: Polym. Chem.*, 2007, **45**, 866.

45. T. Imamoto, K. Yashio, K. V. L. Crépy, K. Katagiri, H. Takahashi, M. Kouchi and I. D. Gridnev, *Organometallics*, 2006, **25**, 908.
46. K. Yashio, M. Kawahata, H. Danjo, K. Yamaguchi, M. Nakamura and T. Imamoto, *J. Organomet. Chem.*, 2009, **694**, 97.
47. Y. Morisaki, Y. Ouchi, K. Naka and Y. Chujo, *Chem. Asian J.*, 2007, **2**, 1166.
48. Y. Morisaki, Y. Ouchi, T. Fukui, K. Naka and Y. Chujo, *Tetrahedron Lett.*, 2005, **46**, 7011.
49. Y. Morisaki, Y. Ouchi, K. Naka and Y. Chujo, *Tetrahedron Lett.*, 2007, **48**, 1451.
50. Y. Morisaki, H. Imoto, Y. Ouchi, Y. Nagata and Y. Chujo, *Org. Lett.*, 2008, **10**, 1489.
51. Y. Morisaki, Y. Ouchi, K. Tsurui and Y. Chujo, *Polym. Bull.*, 2007, **58**, 665.
52. Y. Ouchi, Y. Morisaki, T. Ogoshi and Y. Chujo, *Chem. Asian J.*, 2007, **2**, 397.
53. J. J. Gammon, S. J. Canipa, P. O'Brien, B. Kelly and S. Taylor, *Chem. Commun.*, 2008, 3750.
54. J. J. Gammon, P. O'Brien and B. Kelly, *Org. Lett.*, 2009, **11**, 5022.
55. T. Imamoto, N. Oohara and H. Takahashi, *Synthesis*, 2004, 1353.
56. S. C. Hume and N. S. Simpkins, *J. Org. Chem.*, 1998, **63**, 912.
57. A. J. Blake, S. C. Hume, W. Li and N. S. Simpkins, *Tetrahedron*, 2002, **58**, 4589.
58. A. Ohashi and T. Imamoto, *Acta Cryst. Sect. C*, 2000, **C56**, 723.
59. S. Kobayashi, N. Shiraishi, W. Lam and K. Manabe, *Tetrahedron Lett.*, 2001, **42**, 7303.
60. X. Sun, K. Manabe, W. Lam, N. Shiraishi, J. Kobayashi, M. Shiro, H. Utsumi and S. Kobayashi, *Chem. Eur. J.*, 2005, **11**, 361.
61. X. Sun, M. Koizumi, K. Manabe and S. Kobayashi, *Adv. Synth. Catal.*, 2005, **347**, 1893.
62. G. Hoge, *J. Am. Chem. Soc.*, 2003, **125**, 10219.
63. W. Tang and X. Zhang, *Angew. Chem. Int. Ed.*, 2002, **41**, 1612.
64. W. Tang, W. Wang and X. Zhang, *Angew. Chem. Int. Ed.*, 2003, **42**, 943.
65. J. Holz, M. Gensow, O. Zayas and A. Börner, *Curr. Org. Chem.*, 2007, **11**, 61.
66. T. P. Clark and C. R. Landis, *Tetrahedron: Asymmetry*, 2004, **15**, 2123.
67. P. O'Brien, *Chem. Commun.*, 2008, 655.
68. M. J. Dearden, C. R. Firkin, J. R. Hermet and P. O'Brien, *J. Am. Chem. Soc.*, 2002, **124**, 11870.
69. M. J. Johansson, L. Schwartz, M. Amedjkouh and N. Kann, *Tetrahedron: Asymmetry*, 2004, **15**, 3531.
70. A. J. Dixon, M. J. McGrath and P. O'Brien, *Org. Synth.*, 2006, **83**, 141.
71. S. J. Canipa, P. O'Brien and S. Taylor, *Tetrahedron: Asymmetry*, 2009, **20**, 2407.
72. M. J. Dearden, M. J. McGrath and D. F. O'Brien, *J. Org. Chem.*, 2004, **69**, 5789.

73. M. J. McGrath and P. O'Brien, *J. Am. Chem. Soc.*, 2005, **127**, 16378.
74. C. Genet, S. J. Canipa, D. F. O'Brien and S. Taylor, *J. Am. Chem. Soc.*, 2006, **128**, 9336.
75. P. Vedrenne, V. Le Guen, L. Toupet, T. Le Gall and C. Mioskowski, *J. Am. Chem. Soc.*, 1999, **121**, 1090.
76. I. Fernandez, P. O. Burgos, G. R. Gomez, C. Bled, S. Garcia-Granda and F. L. Ortiz, *Synlett*, 2007, 611.
77. C. Popovici, P. Oña-Burgos, I. Fernandez, L. Roces, S. Garcia-Granda, M. J. Iglesias and F. L. Ortiz, *Org. Lett.*, 2010, **12**, 428.
78. D. Stead, P. O'Brien and A. Sanderson, *Org. Lett.*, 2008, **10**, 1409.

Preparation by Enantioselective Catalysis

6.1 Introduction

The synthetic procedures described in the preceding chapters constitute the main methods of preparation of enantiomerically pure *P*-stereogenic compounds. They are based on resolution of racemates or enantioselective synthesis, requiring stoichiometric amounts of chiral auxiliaries, such as ephedrine or sparteine, or expensive preparative HPLC columns. A much more efficient way would be the use of enantioselective catalysis starting from achiral or racemic reagents. Such an approach has scarcely been used in the preparation of *P*-stereogenic compounds until the last decade,[1] when transition metal complexes were found to catalyse a variety of reactions leading to optically enriched phosphines and derivatives.[2–4] These reactions are noteworthy in the sense that the products, often unprotected phosphines, are good ligands themselves and therefore may compete with the other phosphorus ligands usually present in the catalyst. That is not necessarily bad because coordination of the obtained ligand to the metal can have the positive effect of 'chirality breeding', *i.e.* the obtained phosphine catalysing its own formation.

In addition to metal-catalysed reactions, it has been found that enzymatic catalysis is also an effective tool for some transformations leading to some types of enantioenriched *P*-stereogenic compounds.

In this chapter, catalytic methods for ligand synthesis are described in detail. In spite of that, the enantioselective deprotonation of *tert*-butyldimethylphosphine borane with a catalytic amount of (−)-sparteine or a (+)-sparteine surrogate, reported by O'Brien and co-workers,[5,6] is included for convenience in Chapter 5, Section 5.4.2, following the discussion on the general strategy of desymmetrisation by enantioselective deprotonation. The coverage of Section 6.2 is mainly limited to systems in which the chiral catalyst acts in the step where the

RSC Catalysis Series No. 7
P-Stereogenic Ligands in Enantioselective Catalysis
By Arnald Grabulosa

Published by the Royal Society of Chemistry, www.rsc.org

formation of the *P*-stereogenic compound takes place, excluding other metal-catalysed transformations.[7] In Section 6.3 the coverage is limited to kinetic resolutions and desymmetrisations through esterification, saponification and nitrile hydrolysis catalysed by enzymes.

6.2 Metal-Catalysed Synthesis of *P*-Stereogenic Compounds

6.2.1 Metal Phosphido Complexes

Activation of an electrophile by coordination to an electron-poor metal centre (a Lewis acid) and subsequent attack by an external nucleophile is extremely common in organic chemistry. The opposite scenario – the activation of the nucleophile by coordination to an electron-rich metallic centre and electrophilic attack – has been much less pursued. This chemistry can be studied with late transition metals phosphido complexes [$L_nM–PR_2$], known to have highly nucleophilic phosphorus atoms.[8–10]

Primary and secondary phosphines can oxidatively add to low-valent transition metal complexes to form phosphido complexes. In contrast to phosphines, metal phosphido complexes are known to undergo fast pyramidal inversion, often on the NMR timescale. Inversion barriers for some platinum complexes, determined by NMR spectroscopy, range from 42 to 67 kJ mol^{-1}.[11] As a consequence, phosphido complexes containing other chiral ligands are mixtures of interconverting diastereomers. Reaction of these complexes with electrophiles yields tertiary phosphines and derivatives (Scheme 6.1).

The interplay between the rates of phosphorus inversion (which interconverts **2** and **2'**) and the rates of electrophilic attack to each diastereomer of **2** will dictate the enantiomeric excess of **3**.

This is the basis of several methods for metal-catalysed preparation of *P*-stereogenic phosphines and derivatives (Scheme 6.2).

The three reactions are likely to have phosphido complexes as intermediates. Pt-, Pd- and Ln-catalysed hydrophosphination of activated alkenes and Pd-catalysed phosphination of aryl halides (a cross-coupling reaction) have been known for some time whereas Pt and Ru-catalysed alkylation of secondary phosphines are more recent.

Two metal-catalysed reactions not based on phosphido complexes are also discussed in this chapter: Rh-catalysed desymmetrisation of prochiral dialkynyl-phosphine oxides and Ru- or Mo-catalysed metathesis.

6.2.2 Hydrophosphination of Alkenes and Alkynes

Hydrophosphination reactions,[12] *i.e.* reactions where a P–H bond is added across a multiple bond (usually of an activated alkene) can proceed uncatalysed[13–16] or catalysed by transition metal complexes. With suitable substrates,

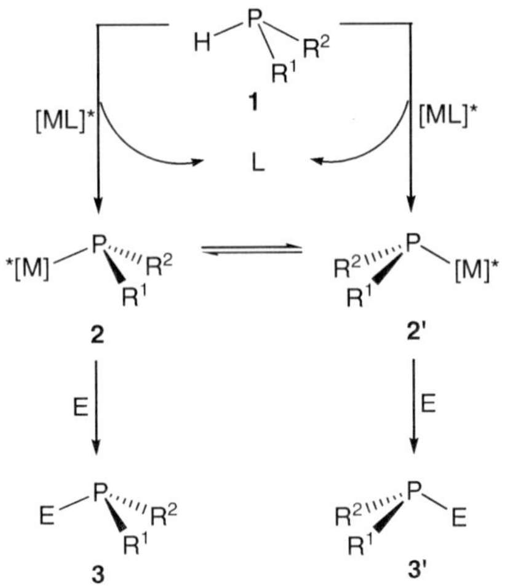

Scheme 6.1   Metal-catalysed synthesis of *P*-stereogenic phosphines and derivatives *via* phosphido complexes.

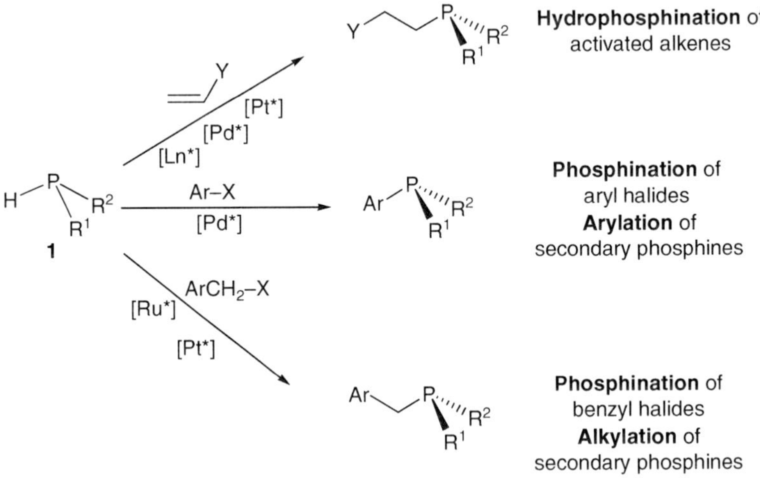

Scheme 6.2   Metal-catalysed enantioselective synthesis of *P*-stereogenic phosphines.

stereogenic phosphorus and/or carbon atoms are generated, which are susceptible to enantiocontrol by chiral catalysts (Scheme 6.3).

More than 20 years ago, Pringle and co-workers[17] showed that hydrophosphination is a suitable reaction to prepare useful phosphine ligands for homogeneous catalysis by using platinum metal complexes. More recently, this reaction has been extended to the preparation of *P*-stereogenic phosphines and derivatives.

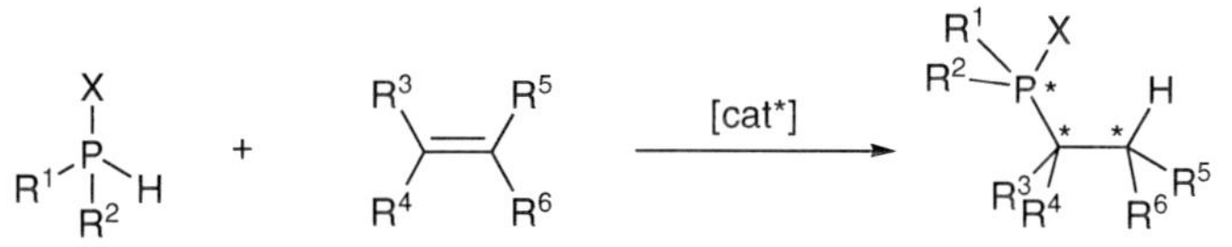

Scheme 6.3   Catalytic hydrophosphination of alkenes (X = lone pair, BH_3 or oxygen).

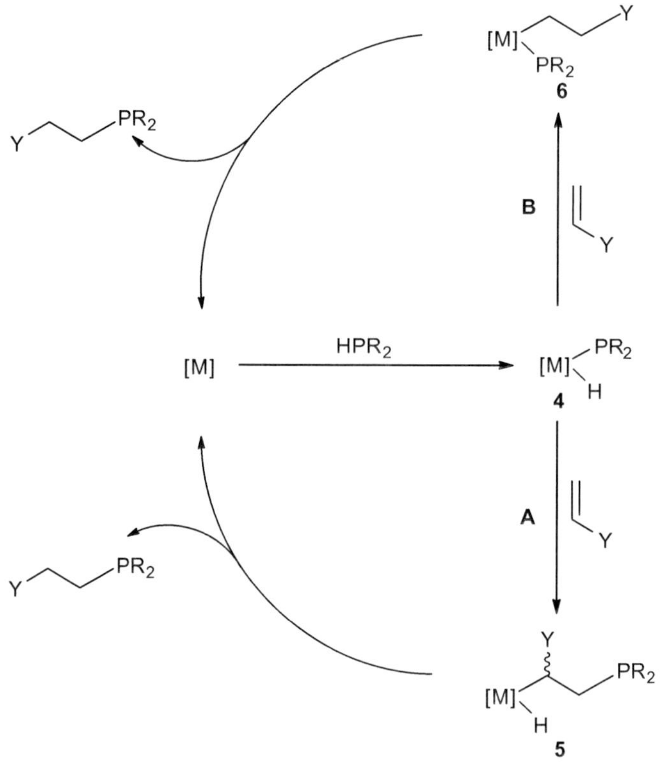

Scheme 6.4   Two possible pathways for metal-catalysed hydrophosphination (only one regioisomer is shown).

6.2.2.1   Pt-Catalysed Hydrophosphination

Mechanistically, it is accepted that hydrophosphination involves oxidative addition of the P–H bond to a low-valent metal centre, often a Pt(0) complex, followed by alkene insertion and reductive elimination to yield the product. Reports on metal-catalysed addition reactions to alkenes show that two different pathways are possible (Scheme 6.4).[12,18] Path **A** involves alkene insertion into the M–P bond to form complexes **5** followed by C–H reductive elimination whereas in path **B** the alkene inserts into the M–H bond to form complexes **6** followed by P–C reductive elimination.

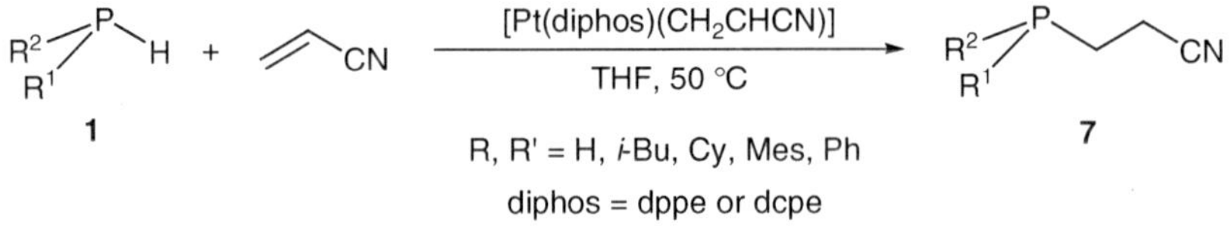

$$\text{1} \quad + \quad \text{CN} \quad \xrightarrow[\text{THF, 50 °C}]{[Pt(diphos)(CH_2CHCN)]} \quad \text{7}$$

R, R' = H, *i*-Bu, Cy, Mes, Ph

diphos = dppe or dcpe

Scheme 6.5   Platinum-catalysed acrylonitrile hydrophosphination.

Table 6.1   Results of Pt-catalysed hydrophosphination of acrylonitrile (Scheme 6.5).

		Selectivity (%)	
Phosphine 1[a]	Product 7	dppe	dcpe
PH$_2$Ph	P(ce)$_2$Ph	68	29
PH$_2$Mes*	PH(ce)Mes*	>95	>95
PHPh$_2$	P(ce)Ph$_2$	95	88
PHMesPh	P(ce)MesPh	>95	>95
PHCyPh	P(ce)CyPh	82	45
PH(*i*-Bu)Ph	P(*i*-Bu)(ce)Ph	82	57
PHMePh	P(ce)MePh	88	55

[a]Conditions: 10% catalyst, 1.5:1 acrylonitrile to phosphine ratio (3:1 for PH$_2$Ph); ce = 2-cyanoethyl, CH$_2$CH$_2$CN; Mes* = 'supermesityl' = 2,4,6-tris(*tert*-butyl)phenyl.

Detailed studies with Pt(diphosphine) complexes in acrylonitrile hydrophosphination by Glueck and co-workers[18,19] allowed the spectroscopic observation and/or isolation of all the species in Scheme 6.4 for both pathways. A range of secondary and primary phosphines **1** was successfully added to acrylonitrile, yielding cyanoethyl phosphines **7**, as shown in Scheme 6.5.

For the systems studied it was concluded that the reaction proceeds *via* acrylonitrile insertion into the Pt–P bond followed by alkene-induced C–H reductive elimination (path **A**, Scheme 6.4). Some of the results with these systems are summarised in Table 6.1.

As expected, the reaction rate is faster for smaller phosphines and in many cases other phosphine products are formed, identified by ^{31}P NMR spectroscopy, especially with dcpe as the ligand in the catalyst. Hydrophosphination also occurs for most of the substrates in the absence of catalyst but more slowly than in the catalysed process.

Most of the resulting phosphines listed in Table 6.1 are *P*-stereogenic but as the catalysts were achiral, they were obtained as racemates. The replacement of the achiral ligands by typical chiral diphosphines (Chiraphos, BINAP) was studied by the same group.[20] ^{31}P NMR analysis showed that cationic complexes [Pt(diphos*)(Me)(PHR^{1}R^2)]$^+$ exist as a mixture of diastereomers but for neutral phosphido analogues, [Pt(diphos*)(Me)(PR^{1}R^2)], only a single set of resonances could be detected even at low temperature. These observations are inconclusive since they are consistent either with the presence of only one

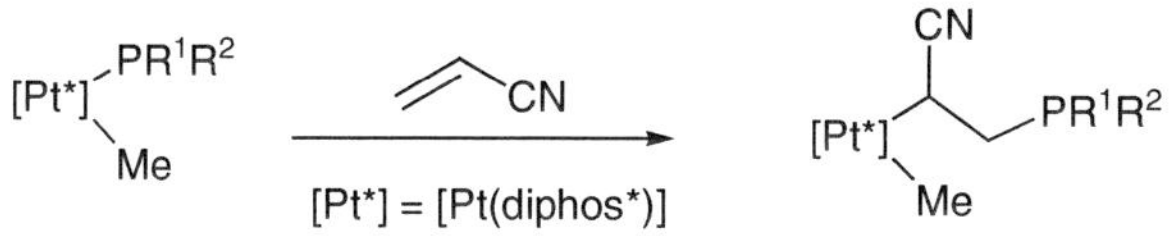

Scheme 6.6 Acrylonitrile insertion in chiral platinum phosphido complexes.

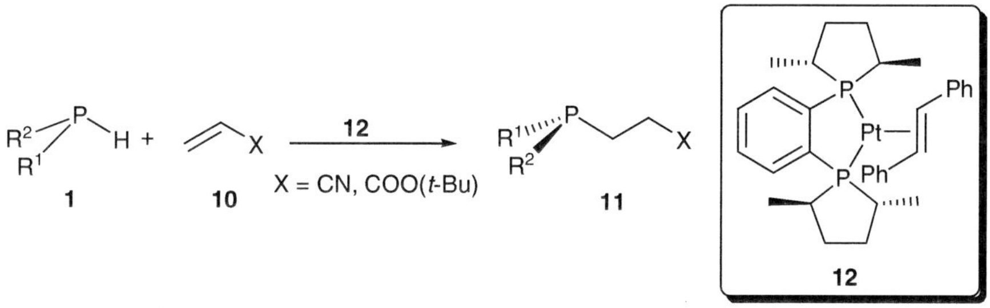

Scheme 6.7 Pt-catalysed enantioselective hydrophosphination.

diastereomer in solution or with a mixture of rapidly interconverting phosphido complexes. The reactivity of these complexes (protonation and acrylonitrile insertion), however, can be more readily explained with the second hypothesis. An important finding was that chiral Pt(II) phosphido complexes undergo diastereoselective acrylonitrile insertion (Scheme 6.6).

A mixture of all four possible stereoisomers of **9** was detected by ^{31}P NMR, which is not consistent with the existence of **8** as single diastereomer and suggests it to be a mixture of two of them rapidly interconverting *via* phosphorus inversion.

The diastereoselective insertion of acrylonitrile into the Pt–P bond should lead to enantioenriched *P*-stereogenic phosphines after reductive elimination in hydrido complexes **5** (Scheme 6.4). This idea was published one year later by the same group[21] and is summarised in Scheme 6.7.

It was reasoned that a successful hydrophosphination catalyst requires a tightly bound chiral ligand that will not be displaced by the substrates or products. Building on previous work,[22] Pt(0) complex [Pt((*R*,*R*)-Me-DuPhos)(*trans*-stilbene)] (**12**), was chosen because it has shorter Pt–P bond distances than other analogues, consistent with tighter binding.

The mechanism of Scheme 6.8 was proposed. Pt(0) complex **12** reacts with the secondary phosphine PH(Is)Ph (Is = 'isityl' = 2,4,6-*tris*(isopropyl)phenyl) to give the phosphido hydride **14** after displacement of **13**.[20,23] This compound also showed a single set of resonances in the ^{31}P NMR spectra, consistent with rapid inversion on the NMR time scale, although the crystal structure **14** revealed that it was present as a single diastereomer. Acrylonitrile inserts diastereoselectively into the Pt–P bond of **14** to yield a mixture of all four possible

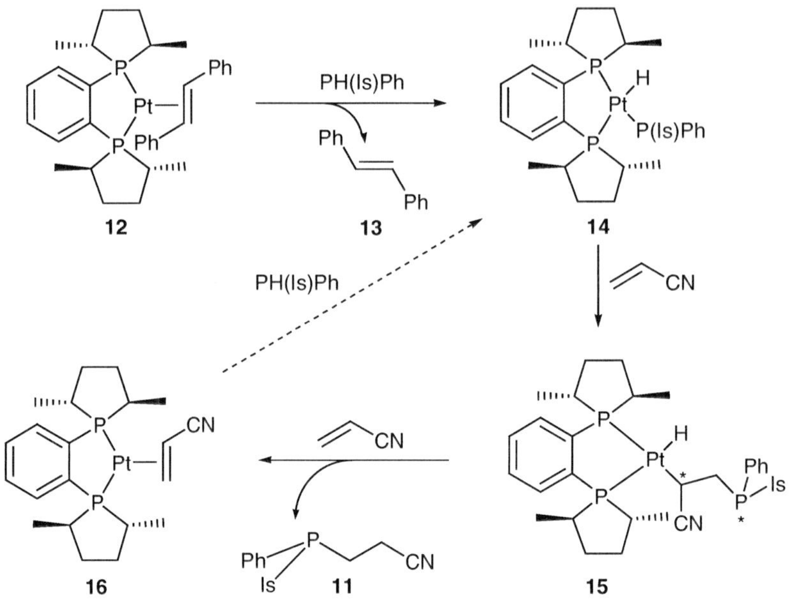

Scheme 6.8 Mechanism of Pt-catalysed enantioselective hydrophosphination.

isomers of **15** in 1:7:9:20 ratio at $-20\,^{\circ}$C, determined by ^{31}P NMR. Warming this mixture to room temperature provokes a reductive elimination to yield the expected phosphine P(ce)(Is)Ph in 63–71% ee together with the new complex **16**.

It is clear that if the starting secondary phosphine can oxidatively add to complex **16**, Scheme 6.8 would represent a catalytic cycle. With this idea in mind, a range of phosphines and activated alkenes were tested with a catalytic amount of complex **12**.[21] The results are listed in Table 6.2. It should be noted that the reaction also occurs in some cases in the absence of catalyst, albeit more slowly.

The comparison of entry 1 and the stoichiometric steps described before for the phosphine PH(Is)Ph shows that the catalytic reaction is very slow and the ee is much lower. Reactions with less bulky and more nucleophilic phosphines (entries 2–6) were faster but less selective. This data shows that complex **16** is rather inert towards P–H oxidative addition probably because acrylonitrile is more strongly bound to Pt than *trans*-stilbene and therefore more difficult to be displaced by the phosphine. To promote oxidative addition a bulky alkene, *tert*-butyl acrylate, was used in entries 7–9, which resulted in enhanced rates and slightly improved enantioselectivities, though still very modest.

^{31}P NMR spectra revealed the presence of other phosphorus products whose full characterisation has given mechanistic insights into Pt-catalysed hydrophosphination. A closer look at the products formed during the reaction in entry 8 of Table 6.2 showed that complex **12** catalyses the addition of PH

Table 6.2 Results of Pt(Me-DuPhos)-catalysed enantioselective hydrophosphination.

Entry[a]	Phosphine	Alkene 10	Product 11	TOF[b]	Selectivity (%)	ee (%)
1	PH(Is)Ph	CN	IsPhP—CN	$<1\,\mathrm{d}^{-1}$	>90	17
2	PHMesPh	CN	MesPhP—CN	$1\,\mathrm{d}^{-1}$	80	13
3	PH(o-An)Ph	CN	(o-An)PhP—CN	$7\,\mathrm{d}^{-1}$	80	5
4	PHCyPh	CN	CyPhP—CN	$10\,\mathrm{d}^{-1}$	60	–[c]
5	PH(i-Bu)Ph	CN	(i-Bu)PhP—CN	$10\,\mathrm{d}^{-1}$	60	5
6	PHMePh	CN	MePhP—CN	$10\,\mathrm{d}^{-1}$	70	5
7	PHMesPh	COO-t-Bu	MesPhP—COO-t-Bu	$5\,\mathrm{d}^{-1}$	100	0
8	PH(i-Bu)Ph	COO-t-Bu	(i-Bu)PhP—COO-t-Bu	$1\,\mathrm{min}^{-1}$	95	20
9	PHMePh	COO-t-Bu	MePhP—COO-t-Bu **17**	$1\,\mathrm{min}^{-1}$	95	22

[a]Conditions: 5% complex **12**, 0.24 mmol phosphine and 0.25 mmol alkene in 0.5 ml of THF.
[b]TOF: turnover frequency = equivalents of secondary phosphine converted per equivalent of catalyst per unit of time.
[c]Not determined due to overlap of NMR peaks.

(*i*-Bu)Ph to *tert*-butyl acrylate to give the expected hydrophosphination product **17** along with some minor by-products containing more than one inserted acrylate, as shown in Scheme 6.9.[24]

It was found that the amount of by-products increased by lowering the temperature and with the presence of an excess of acrylate. To account for the formation of the expected hydrophosphination product as well as these by-products the mechanism of the reaction was re-examined (Scheme 6.10).

The previously accepted pathway consisted of P–H oxidative addition to Pt(0) to form **19** followed by coordination and insertion of the alkene in the Pt–P bond to form **20** and a final reductive elimination to furnish the product and regenerate the catalyst. Another possibility is the nucleophilic attack of phosphido complex **19** to the alkene ('Michael addition' mechanism, as in anionic polymerisation) to generate the zwitterionic intermediate **21**. This complex can yield the hydrophosphination product **11** *via* one of two complementary pathways. Carbanion attack at the cationic platinum hydride (*i.e.* intramolecular hydrogen transfer) would yield the final phosphine complexed to Pt(0) that would be displaced by an equivalent of PHR^1R^2 to furnish, after oxidative addition, starting complex **19**. Alternatively, the anionic carbon atom in **21** could attack the platinum centre directly, forming the cyclic intermediate **22**. From here Pt–P bond dissociation would generate **20**, which would furnish the product after reductive elimination.

The Michael addition mechanism offers an easy explanation for by-product formation if zwitterion **21** attacks a second molecule of alkene instead of the proton transfer in Scheme 6.10. In this case (Scheme 6.11), the newly generated zwitterion **21** can yield a by-product with two alkene fragments or attack another alkene to eventually produce by-products derived from the insertion of three or more alkenes (**23**).

To test the validity of this mechanism, it was reasoned that a weak acid (*t*-BuOH or water) should quench the zwitterion **21** and suppress or at least decrease the formation of by-products. This is indeed the case, although the addition of more alkene increases the quantity of by-products, even in the presence of *t*-BuOH.[24] It should be noted that the presence of these protic additives is not innocent, since it also increases the reaction rates and affects the enantioselectivity. For example adding 20 equivalents of *t*-BuOH to the reaction of PH(Is)Me with *tert*-butyl acrylate halves the time for the completion of the reaction (from 5 to 2 days) and doubles the enantiomeric excess from 28% to 56%. The latter enantioselectivity is the best obtained to date with the systems discussed in this section. More evidence for the Michael addition mechanism came from trapping intermediate **21** with electrophiles other than a proton.[25] Scheme 6.12 shows that performing the hydrophosphination reaction in the presence of benzaldehyde produced some of the three-component coupling product **25**.

Reaction of zwitterion **21** with the aldehyde produces another zwitterion, **24**, which after proton transfer and loss of the platinum moiety generates the product **25**. This mechanism is very similar to the Morita–Baylis–Hillman reaction of classic organic chemistry, which is the condensation of an aldehyde

Scheme 6.9 Pt-catalysed hydrophosphination of *tert*-butyl acrylate with PHPh(*i*-Bu).

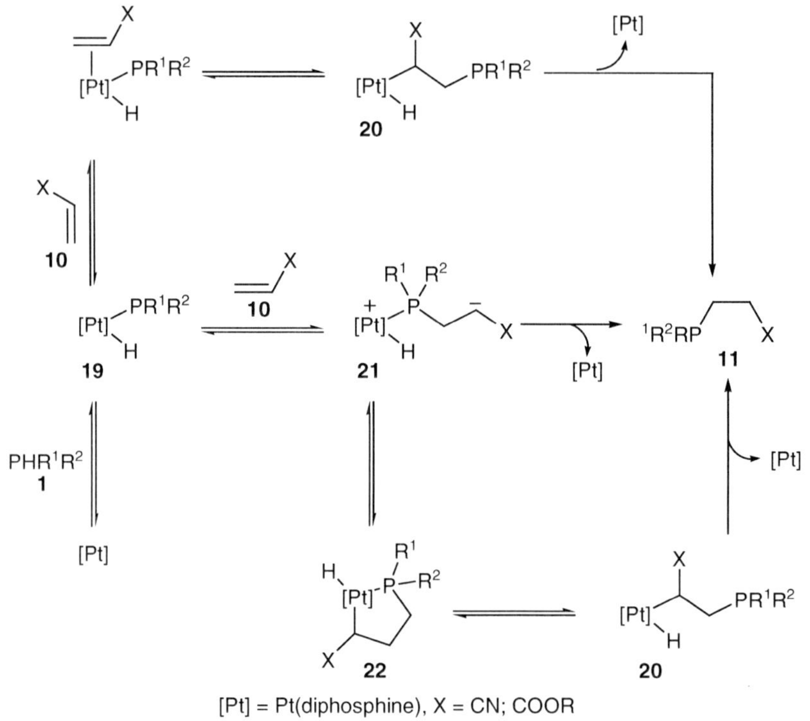

[Pt] = Pt(diphosphine), X = CN; COOR

Scheme 6.10 Possible mechanism of Pt-catalysed hydrophosphination of activated alkenes.

[Pt] = Pt(diphosphine), X = CN; COOR

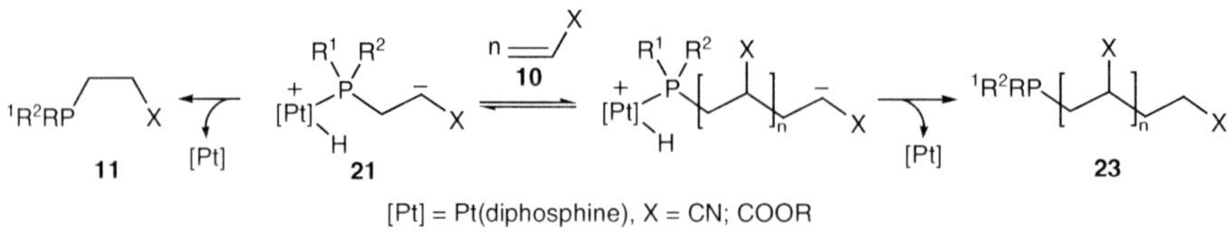

Scheme 6.11 Mechanism for by-product formation in Pt-catalysed hydrophosphination of activated alkenes.

and a conjugated alkene in the presence of an amine or a tertiary phosphine.[26] Hydroxyphosphines **25** were produced as unequal mixtures of diastereomers, although neither of them was enantiomerically enriched. As shown in Scheme 6.12, this reaction competes with normal hydrophosphination and hence the ratio **25**/[$R^1R^2PCH_2CH_2X$] reflects the selectivity of zwitterion **21** for internal (H) or external (PhCHO) electrophiles. When the alkene employed was *tert*-butyl acrylate, a ratio of approximately 1:1 was found and this increased, as expected, on addition of more benzaldehyde.

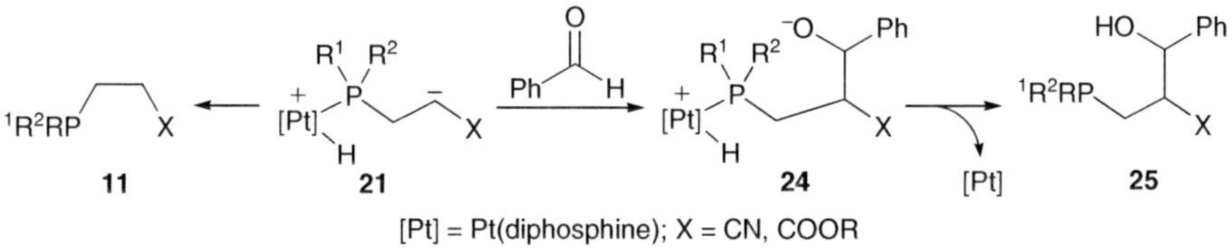

Scheme 6.12 Morita–Baylis–Hillman reaction with platinum complexes.

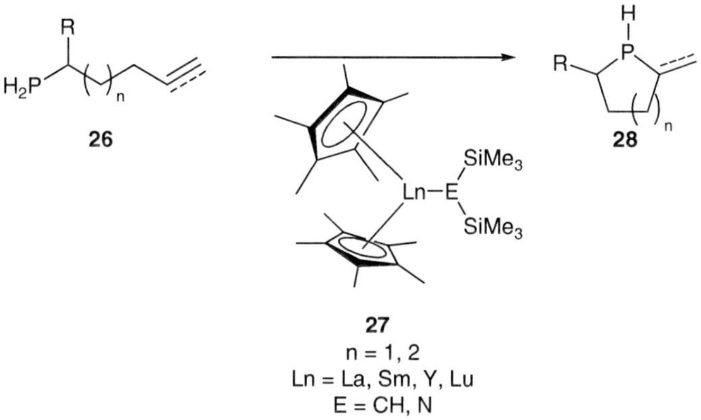

Scheme 6.13 Ln-catalysed hydrophosphination of alkenes and alkynes.

6.2.2.2 Ln-Catalysed Alkene Hydrophosphination

Platinum catalysts of the preceding section activate only the nucleophile (a secondary phosphine), which then reacts with the electrophile (an activated alkene). There are, however, interesting examples of lanthanide complexes activating both the nucleophile and the electrophile towards intramolecular diastereoselective hydrophosphination/cyclisations. In 2000 Marks and co-workers[27] found that certain lanthanide complexes catalyse the intramolecular hydrophosphination of alkenes and alkynes (Scheme 6.13).

With around 2% of **27** present, the reactions proceed until the starting material **26** was consumed, providing phospholanes and phosphorinanes **28**. The reactions were performed at low temperatures and in the absence of light to avoid non-catalysed intramolecular P–H addition to furnish the endocyclic ring product. Attempts were also made to cyclise alkenes with $n = 3$ but the reactions were sluggish and the substrates underwent uncatalysed oligomerisation and decomposition.[28]

Although no enantioselectivity could be expected, several diastereomers, distinguishable by ^{31}P NMR spectroscopy, were observed. Figure 6.1 represents the diastereomers for $R = Me$.

It has to be noted that the phosphorus atoms in **30a** and **32a** are not stereogenic centres but *chirotopic, i.e.* 'a centre in a chiral environment' as defined

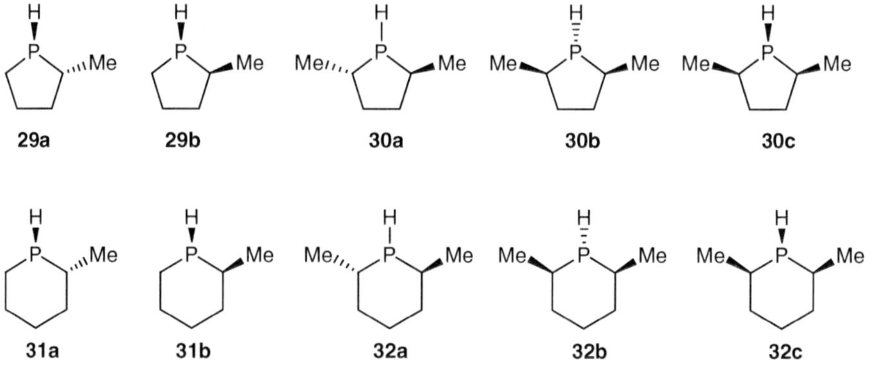

29a　　**29b**　　**30a**　　**30b**　　**30c**

31a　　**31b**　　**32a**　　**32b**　　**32c**

Figure 6.1　Different diastereomers for pholpholanes and phosphorinanes. In chiral molecules, only one of the enantiomers is shown.

Table 6.3　Results of Ln-catalysed hydrophosphination.

Entry	Product	Ln	Distribution (%)[a]		
			a	b	c
1	**29**	La	28	52	–
		Sm	34	60	–
		Y	29	54	–
		Lu	20	33	–
2	**30**	La	12	30	trace
		Sm	72	17	trace
		Y	35	53	trace
		Lu	26	64	trace
3[b]	**30**	Sm	44	46	3
		Y	64	12	19
		Lu	76	8	8
4	**31**	Sm	89	11	–
		Y	89	9	–
5	**32**	Sm	12	66	2
		Y	11	77	12

[a]Conditions: reactions at 25 °C; non-catalytic endocyclic ring formation accounts for the remainder in the product mixture.
[b]Addition of ~10 equiv. of *n*-propylamine.

by Mislow.[29] The reader should also note that compounds **30b**, **30c**, **32b** and **32c** are *meso* and, therefore, not chiral in spite of possessing stereogenic phosphorus and carbon atoms. The diastereomeric ratios are listed in Table 6.3.

From Table 6.3 (entries 1, 4 and 5) it is clear that the ring closure diastereoselectivity heavily depends on the substrate whereas the catalyst has a secondary role. That is not true for the formation of phospholanes **30**, whose isomeric distribution shows a large dependence on the lanthanide ion employed.

With the aim of developing chiral organolanthanide catalysts for enantio-selective transformations, the same group modified the *ansa*-lanthanocene framework in complexes **27** with a chiral group.[30] The new catalytic precursors consist of C_1-symmetric octahydrofluorenyl complexes bearing an enantio-merically pure (−)-menthyl moiety in the lower Cp ring. Figure 6.2 depicts their structure.

It was hoped that this stereodemanding, electron-donating octahydro-fluorenyl, coupled to the smaller Cp-(−)-menthol group would lead to increased stereoselection in catalysis.

Diastereomerically pure catalysts **33** were obtained after recrystallisation and tested in intramolecular hydrophosphination of alkenes to yield phospholanes **35** (Scheme 6.14).

Their performance in hydrophosphination is summarised in Table 6.4.

It was found that the highest turnover frequency corresponds to Sm^{3+} and decreases in parallel with the lanthanide ionic radius. All the catalysts screened display high selectivities for the formation of products with relative *trans* configurations (**29a** and **30a** in Figure 6.1). Despite the potential enantios-electivity of the catalysts, no attempts to quantify the ratios or separate indi-vidual enantiomers were reported.

Organolanthanide complexes differ from late d-block transition metal complexes in several aspects. They are electrophilic, kinetically labile and lack conventional oxidative addition/reductive elimination pathways in their reactions. They have alternative mechanisms to perform catalytic transforma-tions and are being increasingly used in homogeneous catalysis. The hydro-phosphination reaction was proposed to proceed through the cycle depicted in

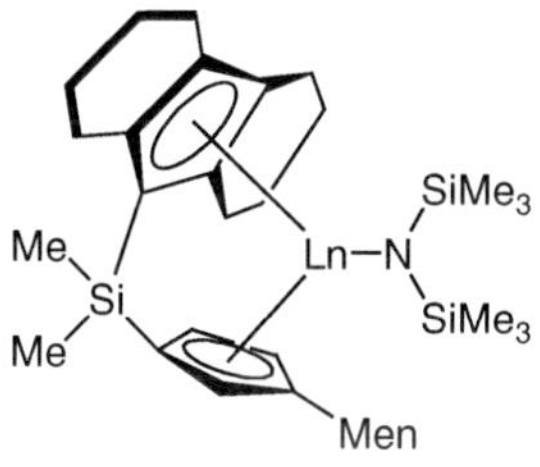

Figure 6.2 Chiral organolanthanide complexes used in hydrophosphination.

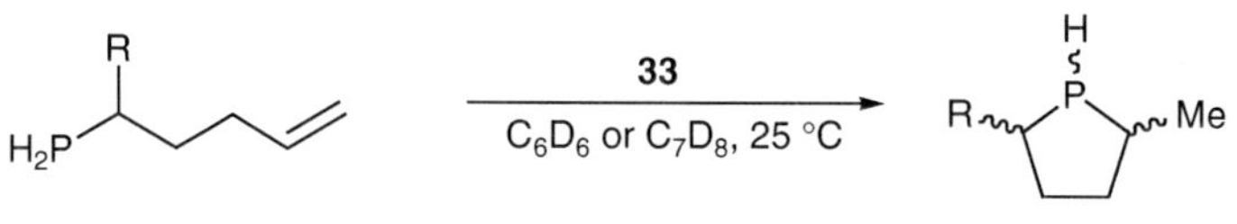

Scheme 6.14 Preparation of phospholanes *via* alkene hydrophosphination.

Table 6.4 Results of enantioselective Ln-catalysed hydrophosphination (Scheme 6.14).

Entry[a]	*R*	*Ln*	*Product 35*[b] *(%)*	*de (%)*[c]
1	H	Sm	81	88
2	H	Y	75	78
3	H	Lu	78	83
4	Me	Sm	82	91
5	Me	Sm	86	96
6	Me	Y	68	77
7	Me	Lu	72	90

[a]Conditions: 0.7–3% [Ln*], C_6D_6 or C_7D_8, initiated with $\sim$10 equiv. of *n*-propylamine (see text) and monitored by ^{1}H and ^{31}P NMR.
[b]Sum of all the isomers of the product.
[c]Determined by ^{31}P NMR.

Scheme 6.15. More recently this cycle has been studied by DFT analysis in analogous systems.[31]

After cleavage of the Ln–N bond of **33** by protonation, the phosphinoalkene **34** coordinates to the metallic centre to give the active catalyst **36**, which undergoes insertion of the double bond into the Ln–P bond producing intermediate **37**. An incoming molecule of the substrate induces protonolysis of the newly formed Ln–C bond producing the product **35** and regenerating the catalyst. This is thought to be the rate-limiting step of the cycle.[31] It was found that in phosphinoalkenes the formation of catalyst **36** is extremely slow even at elevated temperatures. However, the addition of $\sim$10 equivalents of *n*-propylamine effectively removed $HN(TMS)_2$ from **33** yielding a much more active catalyst.

6.2.2.3   *Pd-Catalysed Alkyne Hydrophosphination*

Han, Tanaka and co-workers[32] described the hydrophosphinylation of alkynes by the optically pure phosphinate **38** in the presence of a Pd(II) complex as catalytic precursor (Scheme 6.16).

Very good yields (65–96%) of phosphinates **40** were obtained with a variety of terminal alkynes, including acetylene. X-ray crystallography demonstrated the reaction proceeds with retention of configuration at the phosphorus atom.

Gaumont and co-workers[33,34] reported the Pd-catalysed hydrophosphination of terminal alkynes with racemic secondary phosphine boranes. The general reaction is depicted in Scheme 6.17 and is expected in principle to afford two different types of products: **42** (*anti*-Markovnikov addition) and **43** (Markovnikov addition).

In a initial study it was found that microwave thermal activation (70–80 °C) of a solution of **41a** or **41b** in neat 1-octyne (**39**, R = *n*-C_6H_{13}) for less than 1 hour yielded products **42a,b** exclusively with an appreciable stereoselectivity (Z/E ratios of 80/20 and 70/30 respectively).[33] No trace of regioisomer **43** was found. Isolated yields were 82% for **42a** and 49% for **42b**, reflecting the lower reactivity of **41b**, presumably due to the steric hindrance of the *t*-Bu group.

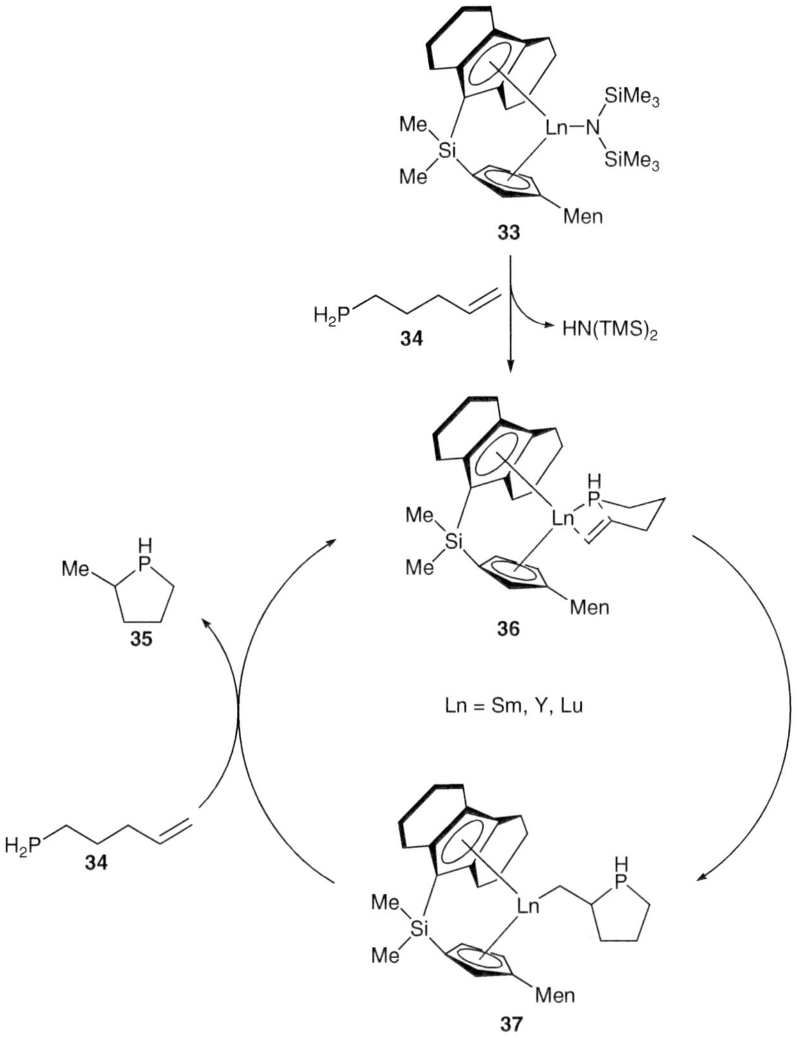

Scheme 6.15 Proposed mechanism of Ln-catalysed cyclisation of phosphinoalkenes.

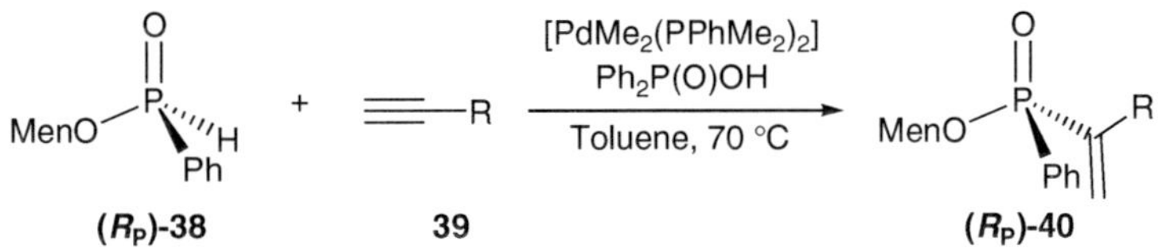

Scheme 6.16 Stereoselective Pd-catalysed hydrophosphinylation of alkynes.

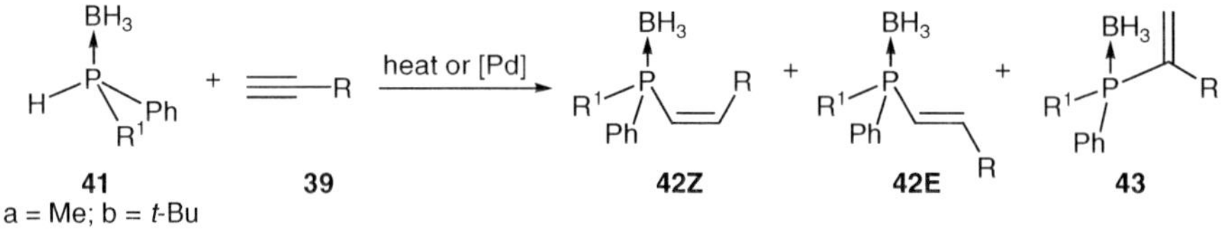

Scheme 6.17 Product distribution in hydrophosphination of terminal alkynes.

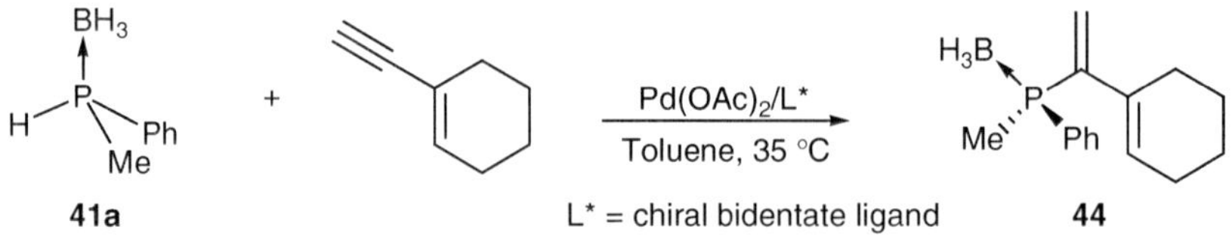

Scheme 6.18 Enantioselective hydrophosphination of 1-ethynylcyclohexene.

The same reaction was performed in the presence of a source of Pd and a bidentate phosphine in toluene at 50 °C. Methylphenylphosphine borane (**41a**) was reacted with 1-octyne in the presence of palladium acetate (5%) and dppp (10%) for 35 minutes. Interestingly, the Markovnikov product **43a** is obtained in 85% yield, with no trace of **42**, meaning that in this case the phosphorus attacks the internal carbon of 1-octyne. The same reaction using bdpp (2,4-*bis*(diphenylphosphino)pentane) and phenylacetylene (**39**, R = Ph) also yields regioisomer of **43** with 53% yield. Several experiments with diphenylphosphine borane under different reaction conditions (palladium source, temperature, alkyne) showed that the reaction does not occur in the absence of palladium (0) and that hydrophosphination of 1-ethynylcyclohexene (**42**) undergoes selective hydrophosphination at the triple bond, with the alkeneic bond untouched.

The same study was performed with chiral enantiopure ligands to investigate the enantioselective version of the reaction, giving vinylic *P*-stereogenic phosphine boranes (Scheme 6.18).[34]

The reaction was performed under kinetic resolution conditions (with a ratio **41a**:alkyne of 1:0.5) for a variety of diphosphines and other bidentate ligands with different stereogenic elements (Table 6.5).

The results show important ligand effects in both activity and enantioselectivity. Reasonable conversions (limited to 50% due to the kinetic resolution conditions used) are only achieved with Josiphos, the phosphino oxazoline in entry 3, BDPP and Tol-BINAP and total inactivity was observed with the thiazoline of entry 4. Interestingly, there is a good correlation between the donating ability of the ligand employed and the enantioselectivities. Electron poor *bis*(arylphosphines) (entries 5–9) give very low ee regardless of their stereogenic elements whereas P–N ligands in entries 2 and 3 give low but measurable ee. The mixed alkyl-aryl diphosphine Josiphos in entry 1 gives better enantioselectivity but the best results are obtained with the electron-rich phosphines of the DuPhos family (entries 10 and 11). Using Me-DuPhos as a ligand, several parameters were varied in order to improve these results, such as the solvent, the

Table 6.5 Performance of several ligands in the reaction of Scheme 6.18.

Entry[a]	*L**	*Conversion*[b] *(%)*	*ee (%)*
1	(*R,R*)-Josiphos	30	20
2	(*R,S*)-PPFA	<10	15
3	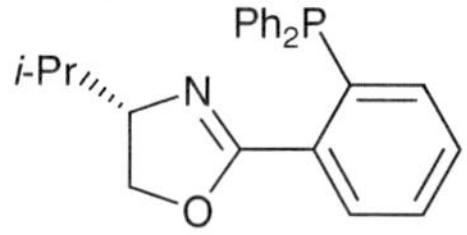	30	10
4	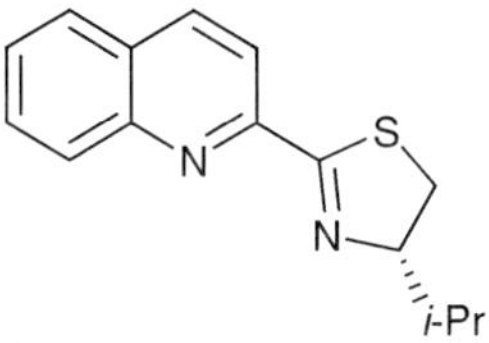	0	–
5	(*R,R*)-BDPP	30	<10
6	(*S,S*)-Chiraphos	<10	<10
7	(*R,R*)-Norphos	<10	<10
8	(*S*)-PHANEPHOS	<10	<10
9	(*S*)-Tol-BINAP	20	<10
10	(*R,R*)-Me-DuPhos	12	30
11	(*R,R*)-*i*-Pr-DuPhos	8	30

[a]Conditions: ratio **41**a/alkyne/Pd(OAc)$_2$/L* of 100/50/5/7.5 for 17 h.
[b]Maximum conversion 50%.

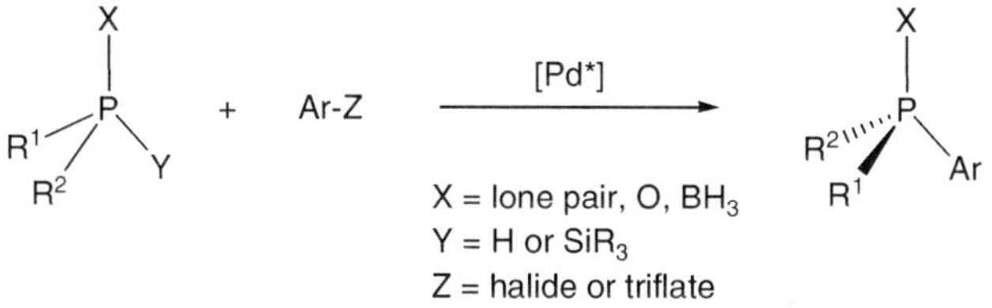

Scheme 6.19 Cross-coupling of an aryl halide or triflate with a secondary phosphine or derivative.

Pd/L* ratio and the source of the metal, but with little success. It was found that increasing the temperature to 50 °C tripled the conversion from 12 to 35% and, unexpectedly, slightly improved the enantioselectivity (from 30 to 42%). Further increase of the temperature is detrimental for both activity and enantioselectivity. This is probably due to deprotection of the phosphine borane **41a** to the free phosphine, which can compete with L* in palladium complexation.

6.2.3 Phosphination of Aryl Halides

The most intensely studied metal-catalysed reaction producing *P*-stereogenic phosphines is the Pd-catalysed phosphination of aryl halides with secondary phosphines and derivatives, a cross-coupling reaction (Scheme 6.19).

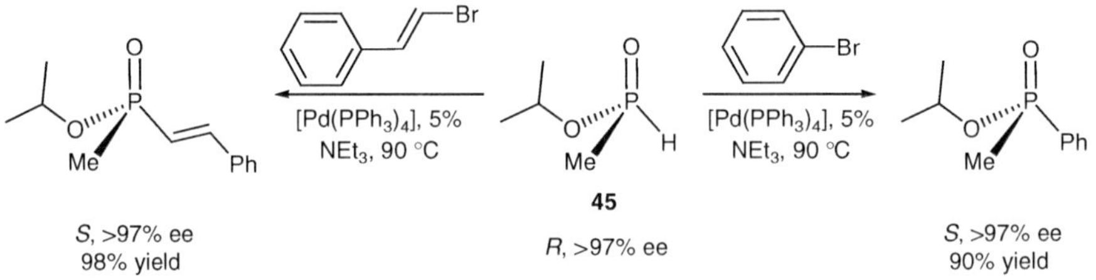

Scheme 6.20   Examples of Pd-catalysed alkenylation and arylation of **45**. Apparent inversions in the absolute configurations are due to the Cahn–Ingold–Prelog priority rules.

When enantiomerically pure phosphines or derivatives and achiral palladium catalysts are employed, the stereochemistry of the P–C bond formation depends on the protective group X but also on the reaction conditions. Xu and co-workers[35–37] reported that isopropyl methylphosphinate, **45**, couples with aryl and vinyl halides with clean retention of configuration at the phosphorus atom (Scheme 6.20).

With phosphine boranes, a more complicated scenario was discovered by Imamoto and co-workers.[38,39] Stereochemistry of the Pd-catalysed P–C bond formation ranges from almost complete retention to inversion, depending on the solvent and base used. A dramatic example of the effect of the solvent in the arylation of phosphinite borane **46** is shown in Scheme 6.21.

More recently, Livinghouse and co-workers[40] studied the effect of Cu(I) on Pd-catalysed arylation of secondary phosphine boranes. Apart from a marked beneficial effect on the coupling efficiency, they discovered that Cu(I) is able to suppress the base-mediated racemisation of the phosphide borane intermediate, to generate a configurationally stable metallophosphide **47** (Scheme 6.22). Consequently, tertiary phosphine boranes **48** were obtained with retention of configuration, in good yields and excellent enantioselectivities.

The last examples demonstrate that secondary phosphines and derivatives are versatile precursors for the synthesis of tertiary phosphines. However, their preparation in optically pure form requires stoichiometric quantities of chiral reagents. For this reason more recently the work has shifted towards racemic secondary phosphines and derivatives and chiral enantiopure metal catalysts.

6.2.3.1   *Pd-Catalysed Arylation of Secondary Phosphines*

Glueck and co-workers[41,42] studied in detail the Pd-catalysed enantio-selective coupling of bulky racemic secondary phosphines with phenyl halides (Scheme 6.23).

Secondary phosphines **49** were chosen with the hope that steric dissymmetry between the aromatic and methyl groups would provide better enantioselectivity in **51**. Me-DuPhos palladium complex **50** was selected for the first set of experiments.[41] The rather unusual sodium trimethylsilanoate was chosen based

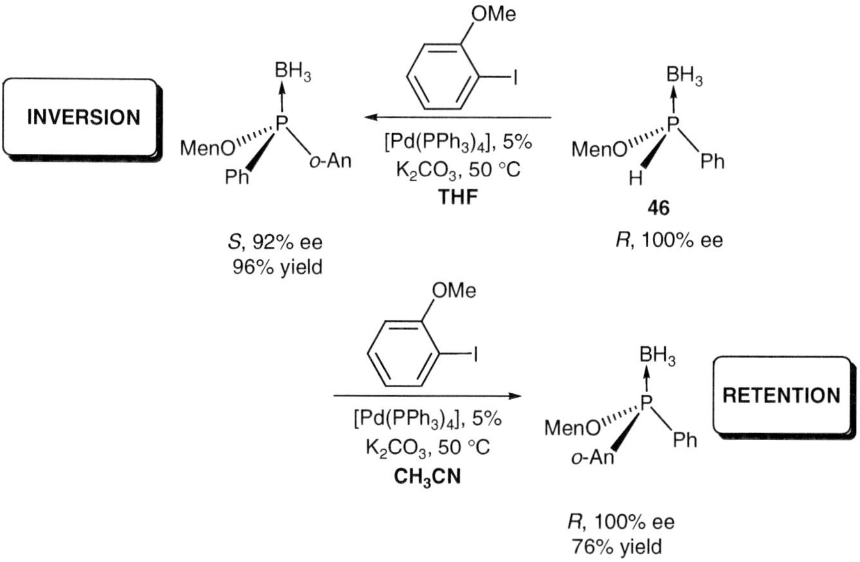

OMe
I
INVERSION
BH_3
MenO
P
Ph
o-An
[Pd(PPh$_3$)$_4$], 5%
K_2CO_3, 50 °C
THF
BH_3
MenO
P
H
Ph
46
S, 92% ee
96% yield
R, 100% ee
OMe
I
[Pd(PPh$_3$)$_4$], 5%
K_2CO_3, 50 °C
CH_3CN
BH_3
MenO
P
o-An
Ph
RETENTION
R, 100% ee
76% yield

Scheme 6.21 Solvent effect on the stereochemistry of Pd-catalysed cross-coupling reaction.

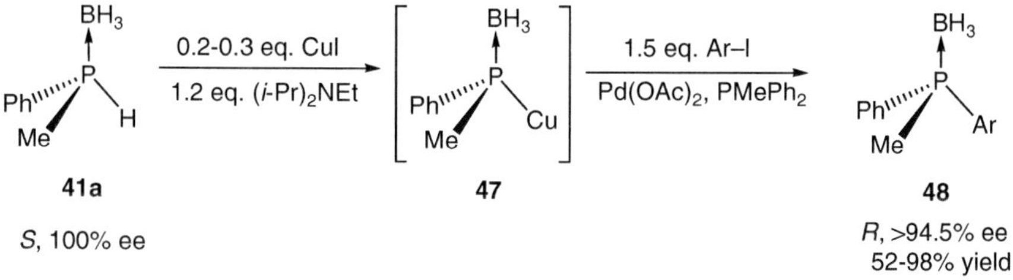

BH_3
Ph
P
Me
H
41a
S, 100% ee
0.2-0.3 eq. CuI
1.2 eq. (i-Pr)$_2$NEt
BH_3
Ph
P
Me
Cu
47
1.5 eq. Ar–I
Pd(OAc)$_2$, PMePh$_2$
BH_3
Ph
P
Me
Ar
48
R, >94.5% ee
52-98% yield

Scheme 6.22 Pd(0)-Cu(I) co-catalysed arylation of **41a**.

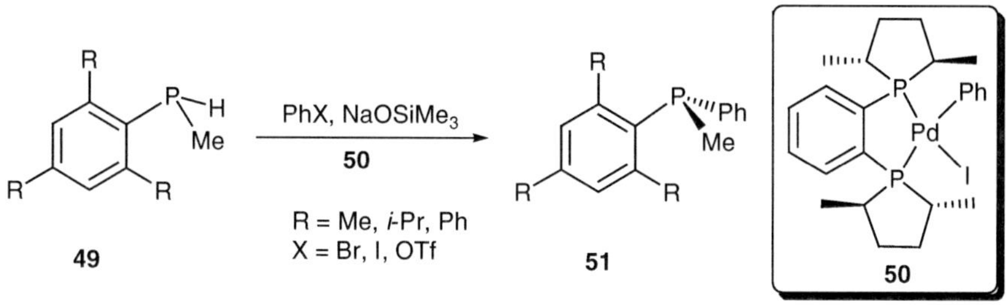

R
P
H
Me
R
R
49
PhX, NaOSiMe$_3$
50
R = Me, i-Pr, Ph
X = Br, I, OTf
R
P
Ph
Me
R
R
51
P
Ph
Pd
I
P
50

Scheme 6.23 Pd-catalysed enantioselective phosphination.

Table 6.6 Results of Pd-catalysed enantioselective synthesis of P(Is)MePh (Scheme 6.23).

Entry	PhI equiv.	Cat. loading (%)	NaSiOMe₃ equiv.	Solvent	T (°C)	Yield[a] (%)	ee[b] (%)
1	2	5	1	toluene	22	71	73
2	1.05	3	1	toluene	22	93	72
3	1.07	2.5	1	toluene	22	88	73
4	1.1	5	1	pentane	22	90	78
5	2	7	1	THF	22	69	66
6	2	7	1	MeCN	22	60	58
7	1.1	5	1	DMSO	22	78	55
8	1	5	1	toluene	4	82	82
9	2	5	1	toluene	50	60	42
10[c]	1.05	5	1	toluene	22	91	70
11[d]	1.1	2.5	2	toluene	22	90	73
12[d]	1.1	2.5	3	toluene	22	89	75
13[d]	1.1	2.5	10	toluene	22	96	74
14[d,e]	1.1	2.5	1	toluene	22	83	75

[a]Determined by ¹H NMR or isolated yield after column chromatography.
[b]Determined by ¹H NMR after complexation to enantiopure $[Pd((S)\text{-}Me_2NCH(Me)C_6H_4)(\mu\text{-}Cl)]_2$.
[c]Reaction carried out with 500 mg of **49** (R = *i*-Pr).
[d]Catalytic precursor: [Pd((*R,R*)-Me-DuPhos)(*trans*-stilbene)].
[e]Base: Na[N(SiMe₃)₂].

on the work of Gaumont *et al.*[43] for being a mild base unable to deprotonate the free phosphine **49** until it is coordinated to palladium.

Firstly, they looked at the precursor with the 'isityl' group in **49**, screening the influence of several parameters (Table 6.6) in order to find the optimal conditions.

It was found that an excess of base or phenyl iodide is not required (entries 1 and 2 and 10–14) and that the catalyst loading can be lowered to 2.5% without significant loss of enantioselectivity. Apolar solvents (toluene and pentane) are preferred, as more polar ones slightly reduce enantioselectivity (entries 5–7). As expected, lower temperatures increase enantioselectivities (compare entries 1, 8 and 9).

Several other palladium complexes [Pd(L*)(Ph)(I)] and [Pd(L*)(*trans*-stilbene)] with different chiral diphosphine ligands L* were also employed as catalyst precursors. The yield and enantioselectivity of the product are strongly dependent on the diphosphine. It was found that the DuPhos series (Me/Et/*i*-Pr-DuPhos) produces results similar to those listed in Table 6.6. In contrast, changing to the more flexible (*R,R*)-Me-BPE gives a very active catalytic system but with a disappointing 18% ee. With DuXantphos a very slow system results, with the formation of by-products. Ferrocenyl phosphines (Me-FerroLANE, Et- FerroTANE, Josiphos, BoPhoz) were also employed but were found to be inferior to DuPhos. Finally, with several Ni(Me-DuPhos) catalyst precursors, reactions do not reach completion and formation of by-products is observed.

Once the conditions were optimised, the substrate scope for the reaction was investigated (Table 6.7).

Table 6.7 Substrate screening for Pd-catalysed arylation of **49** (Scheme 6.23).

Entry[a]	Substrate 49	ArX	Time (h)	Yield[b] (%)	ee[c] (%)
1[d]	PH(Is)Me	PhBr	1	53	38
2[d]	PH(Is)Me	PhOTf	1	70	50
3	PH(Is)Me	PhI	1	97	76
4	PH(Is)Me	p-PhOC$_6$H$_4$I	1	89	88
5	PH(Is)Me	p-MeOC$_6$H$_4$I	6	96	82
6	PH(Is)Me	p-NH$_2$OC$_6$H$_4$I	1	88	79
7	PH(Is)Me	p-MeC$_6$H$_4$I	22	98	78
8	PH(Is)Me	p-CF$_3$OC$_6$H$_4$I	2	96	45
9	PH(Is)Me	p-IC$_6$H$_4$I	2	68	40
10[e]	PH(Is)Me	p-NO$_2$OC$_6$H$_4$I	1	73	12
11[f]	PH(Is)Me	p-IC$_6$H$_4$I	2	91	33[g]
12	PH(Is)Me	8-QuinolylI	6	75	22
13	PH(Is)Me	3,5-Me$_2$C$_6$H$_3$I	456	93	27
14	PH(Is)Me	o-C$_5$H$_4$NI	6	98	15
15	PH(Is)Me	1-NaphthI	5	98	12
16	PHMe(Mes)	PhI	4	59	nd[h]
17	PHMe(Mes)	p-MeOC$_6$H$_4$I	48	90	56
18	PHMe(Mes)	p-CF$_3$OC$_6$H$_4$I	48	75	7
19	PHMe(Mes)	PhI	24	91	−59[i]
20	PHMe(Phes)	p-MeOC$_6$H$_4$I	1	81	−65[i]
21	PHMe(Phes)	p-PhOC$_6$H$_4$I	24	84	−82[i]

[a]Catalyst precursor: 5% of [Pd((*R,R*)-Me-DuPhos)(*trans*-stilbene)], reactions performed at room temperature in toluene, Phes = 2,4,6-Ph$_3$C$_6$H$_4$.
[b]Isolated yield after column chromatography.
[c]Determined by ^{1}H or ^{31}P NMR after complexation to enantiopure [Pd((*S*)-Me$_2$NCH(Me)C$_6$H$_4$)(μ-Cl)]$_2$.
[d]Catalyst precursor: **50**.
[e]Catalyst precursor: [Pd((*R,R*)-Me-DuPhos)(p-NO$_2$C$_6$H$_4$)(I)].
[f]2 equiv. PH(Is)Me, 2 equiv. NaSiOMe$_3$.
[g]Diastereomeric excess.
[h]Not determined.
[i]The negative sign indicates that NMR chemical shifts trends for the Pd adducts of the products were opposite of those observed for PAr(Is)Me and PArMe(Mes).

Entries 1–3 in Table 6.7 demonstrate that phenyl bromide and triflate can also be used but give lower yields and selectivities than with phenyl iodide. All the other entries show that in general coupling of secondary phosphines with aryl iodides takes place with high yields but very variable enantioselectivity. Results with p-substituted aryl iodides (where steric effects are minimal) show that donating substituents increase the enantioselectivity. A plot of ee *vs* Hammet substituent constants, σ$_P$, gives a reasonable correlation. Comparing the ee obtained for the different secondary phosphines with one aryl iodide reveals a steric effect of the phosphines on the enantioselectivity (compare entries 3, 16 and 19; entries 5, 17 and 20; entries 8 and 18; entries 4 and 21). The phosphines with a Mes group show lower ee than those with Is or Phes, probably because of the smaller size of the former. Unfortunately, the even bulkier substrate PHMeMes* was found to be unreactive.

In entries 9 and 11, p-diiodobenzene is coupled to one or two phosphine moieties respectively. In entry 13 the reaction is extremely sluggish; it seems that the double *meta* substitution of the substrate slows the reaction. A further limitation was found with *o*-substituted substrates: whereas *o*-iodopyridine (entry 14) and *o*-iodomethylbenzene react smoothly but with low enantioselectivity, the catalytic turnover is very slow with other *o*-substituted substrates (*o*-ClC$_6$H$_4$I and *o*-OMeC$_6$H$_4$I) and the reaction do not occur at all with mesityl iodide. Changing the catalyst to [Pd(Et-FerroTANE)] (see Scheme 6.26) allowed the coupling of *o*-substituted aryl iodides and iodoferrocene but the products were practically racemic.

Studying each stoichiometric step in the coupling of PH(Is)Me with PhI to produce the phosphine P(Is)MePh provided information required to propose a mechanism for the reaction, depicted in Scheme 6.24.

Treatment of catalytic precursor **50** with secondary phosphine **49** reversibly gives the cationic complex **53**, as a mixture of the two expected diastereomers. The equilibrium favours neutral complex **50**. NaSiOMe$_3$ reacts neither with **49** nor with complex **50**, but it deprotonates **53** to yield the key intermediate **54**. This complex was found to be present in a very unequal proportion of diastereomers (*ca.* 40:1). It undergoes reductive elimination to yield the arylated phosphine **51**, although the exact reaction path depends on the conditions. Under catalytic conditions, excess of substrate **49** produces complex **52** as a mixture of the four diastereomers, which reacts with phenyl iodide to regenerate the catalyst. Phenyl iodide was found to react also with **54**, promoting reductive elimination of **51** to directly form the catalyst **50**. Finally, in the absence of a trapping compound, **54** gives the unstable tricoordinated complex **55**, which decomposes to form product **51** and [Pd((R,R)-Me-DuPhos)$_2$], hence deactivating the catalyst. Detailed kinetic studies on the reductive elimination showed that interconversion of diastereomers of **54** (*via* P-inversion) is indeed much faster than P–C bond formation by reductive elimination. In addition, the establishment of the absolute configurations of product **51** (*S*) and intermediate **54** (R_P) suggested that the major diastereomer of **54** forms the major enantiomer of **51**. However, it was also discovered that the minor diastereomer of **54**, (S_P), undergoes reductive elimination about three times faster than the major one, reducing the overall enantioselectivity.

The same group developed an intramolecular version of Pd-catalysed phosphination to prepare *P*-stereogenic benzophospholanes (Scheme 6.25).[44]

The rate and selectivity are strongly dependent on the precursor and the diphosphine used in the catalyst. For R = Ph, phenylbenzophospholane **57** is formed slowly but with good yields and moderate enantioselectivity when phospholane-based ligands were used (up to 66% ee for *i*-Pr-DuPhos). A racemic product is obtained with ligands with a ferrocene backbone. For R = Cy, the reaction is much more rapid giving very good product yields (up to 95% with Et-FerroTANE) in minutes, although as a racemate regardless of the ligand. In this case, displacement of the chiral phosphine from Pd was observed, probably due to the high basicity of the obtained phospholane. Around 50% yield and very little diastereoselectivity is observed when R = (−)-Men is used.

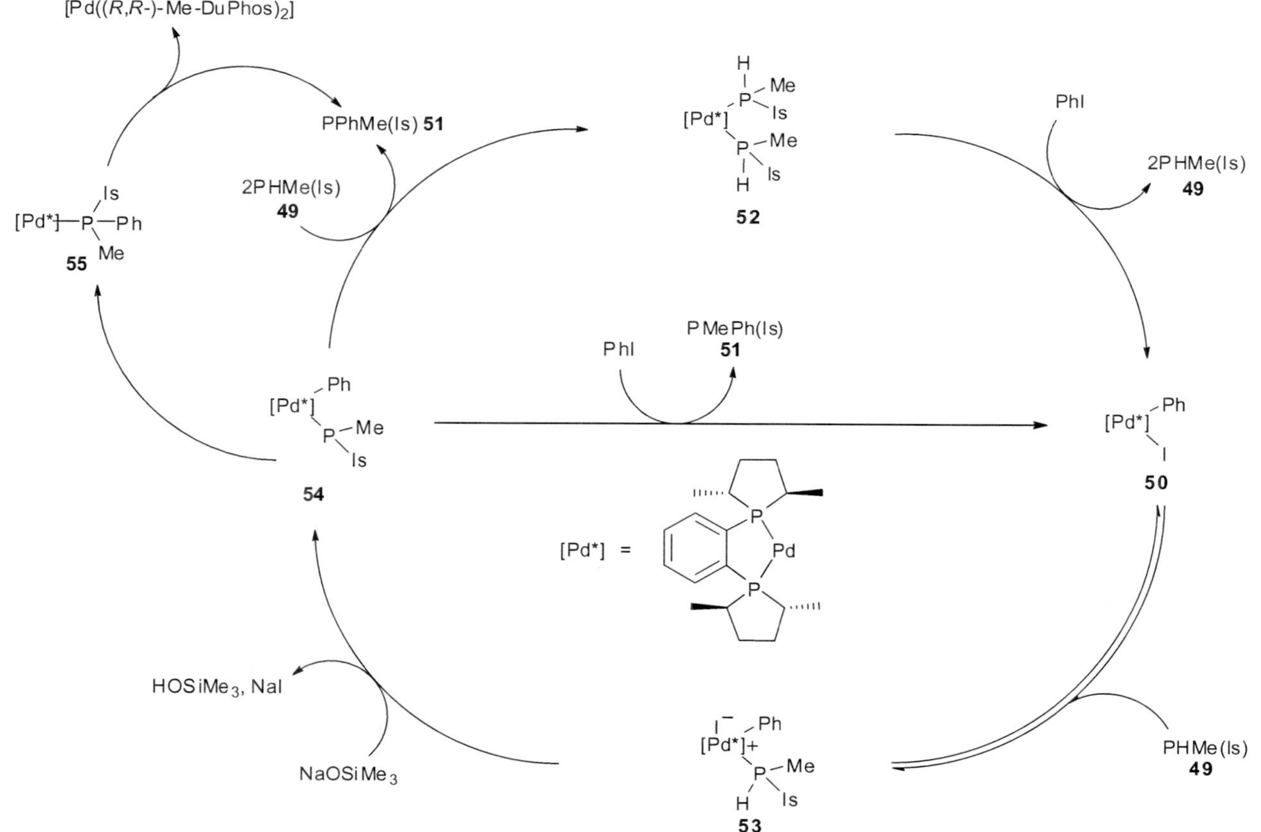

Scheme 6.24 Proposed mechanism for Pd-catalysed arylation of secondary phosphines.

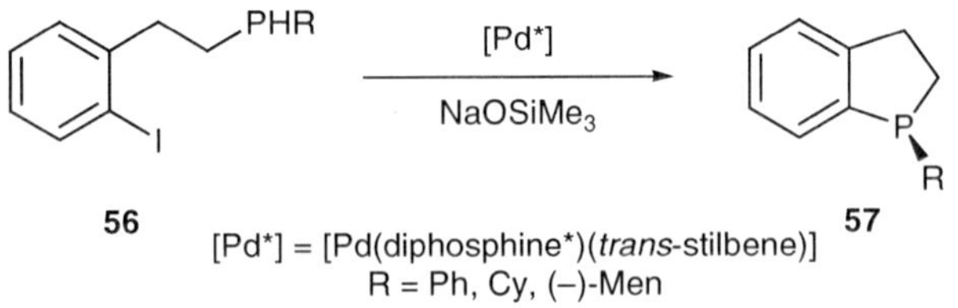

Scheme 6.25 Intramolecular Pd-catalysed phosphination.

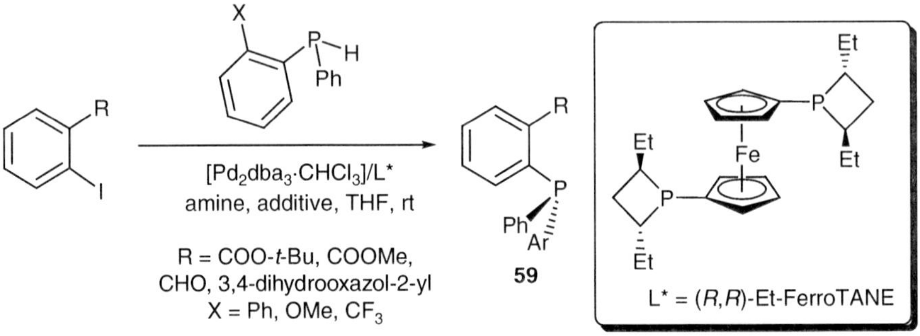

Scheme 6.26 Pd-catalysed phosphination of *o*-substituted aryl iodides.

In Chapter 2, Section 2.4.4 it is mentioned that Helmchen and co-workers[45] developed series of *P*-stereogenic phosphino oxazoline ligands (PHOX) with a stereogenic phosphorus atom by either nucleophilic or electrophilic substitutions at the appropriate phosphorus precursors. However, only major diastereomers are accessible and the method was not applicable to achiral substrates. To overcome this limitation, their attention was turned to Pd-catalysed phosphination of *o*-substituted aryl iodides (Scheme 6.26).[46]

They developed a system where the catalyst is generated *in situ* from Pd(0) and a chiral diphosphine. After screening several diphosphines, the best enantioselectivity of product **59** was obtained with the (*R,R*)-Et-FerroTANE ligand and the rest of the experiments were therefore carried out with this ligand. The results are listed in Table 6.8.

In this case, the postulated intermediate $[PdL^*(PHR^1R^2)]^+I^-$ (like **53** in Scheme 6.24) bears two aryl groups at the P atom and tertiary amines such as NEt_3 are sufficient to deprotonate it, promoting the reaction.

The most striking feature is probably the effect of additives on the reaction outcome. The optimal rate and selectivity are obtained with LiBr; replacing Li with ammonium lowers the enantioselectivity (entry 8). However, the effects of the amine and LiBr are not additive (compare entries 1 and 3 with entries 6 and 9). Phosphines with an electron donor or an electron-withdrawing substituent give low and high enantioselectivities respectively (entries 10 and 11). Interestingly, entry 15 shows that in this case the enantioselectivity decreases with the addition

Table 6.8 Results of Pd-catalysed phosphination of *o*-substituted aryl iodides (Scheme 6.26).

Entry[a]	R	X	Amine	Additive	Yield[b] (%)	ee[c] (%)
1[d]	COO-*t*-Bu	Ph	NEt$_3$	–	63	71
2[d]	COO-*t*-Bu	Ph	NBnMe$_2$	–	45	77
3	COO-*t*-Bu	Ph	*N*-MePip[d]	–	76	84
4	COO-*t*-Bu	Ph	NEt$_3$	LiF	76	66
5	COO-*t*-Bu	Ph	NEt$_3$	LiCl	66	86
6	COO-*t*-Bu	Ph	NEt$_3$	LiBr	76	90
7	COO-*t*-Bu	Ph	NEt$_3$	LiI	58	87
8	COO-*t*-Bu	Ph	NEt$_3$	NBu$_4$Br	76	25
9	COO-*t*-Bu	Ph	*N*-MePip	LiBr	43	86
10	COO-*t*-Bu	OMe	NEt$_3$	LiBr	79	40[c]
11	COO-*t*-Bu	CF$_3$	NEt$_3$	LiBr	39	93
12	COOMe	Ph	NEt$_3$	LiBr	69	85
13	CHO	Ph	NEt$_3$	LiBr	71	63
14	oxazoline	Ph	NEt$_3$	–	45	90[c]
15	oxazoline	Ph	NEt$_3$	LiBr	58	68[c]
16	oxazoline (*S*)-*i*-Pr	Ph	NEt$_3$	–	53:14[e]	4:1[f]
17[g]	oxazoline (*S*)-*i*-Pr	Ph	NEt$_3$	–	5:66[e]	1:13[f]

[a]Reactions carried out in THF at room temperature, 1% catalyst.
[b]Isolated yield.
[c]The major enantiomer had *S* absolute configuration except in entry 11, which was *R* (the apparent inverion is only due to the Cahn–Ingold–Prelog priority rules) and in entries 10, 14 and 15, the absolute configurations were not reported.
[d]*N*-methylpiperidine.
[e](S_C,S_P):(S_C,R_P) isolated yields.
[f](S_C,S_P):(S_C,R_P) diastereomeric ratio.
[g](*S*,*S*)-Et-FerroTANE as catalyst.

of LiBr. The reader will notice that in entries 16 and 17 the enantiopure oxazoline ring allows the study of double asymmetric induction. The results demonstrate that catalyst control dominates over substrate control and it is therefore possible to prepare both diastereomers of **59** by changing only the configuration of the FerroTANE ligand. It seems clear then that entries 16 and 17 represent the mismatched and matched cases, respectively.

Glueck and co-workers[47] described a case of 'chirality breeding' in the synthesis of menthylmethylphenylphosphine *via* Pd-catalysed phosphination of iodobenzene (Scheme 6.27).

The starting material PHMe((–)-menthyl) was obtained as a 1:1.3 mixture of R_P and S_P diastereomers. The phosphination of phenyl iodide was carried out in the presence of a catalytic amount of Pd(0) precursors ([Pd(Me-DuPhos)(*trans*-stilbene)], [Pd(*i*-Pr-DuPhos)(*trans*-stilbene)], [Pd(dba)$_2$], [Pd(PPh$_3$)$_4$]) and Pd(II)

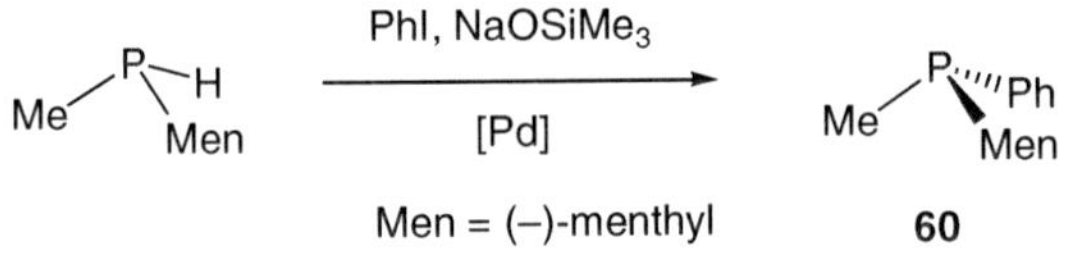

Men = (–)-menthyl   **60**

Scheme 6.27 Pd-catalysed diastereoselective arylation of menthylmethylphosphine.

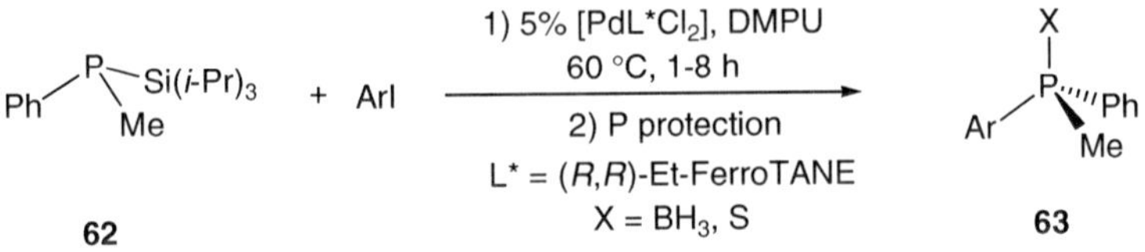

1) 5% [PdL*Cl$_2$], DMPU
60 °C, 1-8 h
2) P protection
L* = (*R,R*)-Et-FerroTANE
X = BH$_3$, S

62   **63**

Scheme 6.28 Pd-catalysed arylation of silylphosphines.

([Pd(PPh$_3$)$_2$(I)(Ph)], [Pd(dcpe)(I)(Ph)] and Pd(OAc)$_2$). The expected product **60** is obtained in 40–90% yield and in 42–91% diastereomeric excess. In all cases, the reaction is selective towards the S_P diastereomer, even when achiral catalysts are employed or when the absolute configuration of the *i*-Pr-DuPhos ligand is changed. This clearly shows that diastereoselectivity is primarily governed by the substrate. Monitoring the reaction by ^{31}P NMR indicated that the catalyst precursors are modified during the reaction and that several intermediates are formed. Some of these intermediates contained the coordinated phosphine **60** or starting PHMe((–)-menthyl). The complexes formed after displacement of the phosphines in the original catalyst precursors were prepared independently. The three diastereomers of [Pd(I)(Ph)(**60**)$_2$] were prepared and the (S_P,S_P) could be isolated diastereomerically pure. Interestingly, it was found that [Pd(I)(Ph) ((S_P,S_P)-**60**)$_2$] (**61**) catalyses the formation of the S_P diastereomer of **60**. This constitutes an example of formal 'chirality breeding', in which the diastereomerically enriched tertiary phosphine **60** acts as a ligand in the catalyst of its own production (*i.e.* *P*-stereogenic phosphines catalysing the formation of *P*-stereogenic phosphines!). In spite of that, although complex **61** acts as a catalyst precursor and is reformed at the end of the reaction, ligand substitution occurs giving a complex mixture of intermediates.

6.2.3.2 *Pd-Catalysed Arylation of Silylphosphines*

More than 20 years ago Stille and Tunney[48] reported a Pd-catalysed coupling of achiral silylphosphines with aryl iodides. Bergman, Toste and Chan[49] recently described a modification of this reaction as a new route to prepare *P*-stereogenic phosphines by enantioselective arylation of tertiary racemic silylphosphines (Scheme 6.28).

The enantioselectivity was optimised for the coupling of *m*-iodoanisole with silylphosphine **62**. The best conditions were 5% of Et-FerroTANE catalyst in

N,*N*'-dimethylpropylene urea (DMPU) as solvent. After borane protection, **63** (Ar = *m*-anisole, X = BH$_3$) was obtained in 91% yield and 56% ee. Then, following the work of Helmchen and co-workers[46] (see previous section), the attention was turned to *o*-iodoarenes (Table 6.9).

Entries 1–10 show that the system is quite tolerant to a range of coordinating substituents, yielding the products in good yields and in modest to excellent enantioselectivities. The amine group (entry 3) is less satisfactory whereas the *o*-benzoates and *o*-benzamides are excellent substrates, although the length of the linker with the amide group proved crucial (entry 10). An exceptional enantioselectivity was found with 2-iodo-*N*,*N*-diisopropylbenzamide (entry 9) and further experiments with *o*-iodo-*N*,*N*-diisopropylbenzamides were therefore carried out (entries 11–22). It was found that changing the electronics either *para* to the iodide or to the benzamide group has little effect on yields and ee (entries 11–15). More electron-rich iodides (entries 16 and 17) react slower but still furnish the products in excellent enantioselectivities. It is remarkable that good results are obtained with a variety of heteroaryl iodides, provided they contain the *N*,*N*-diisopropylamido group (enties 18–22).

Finally, the coupling with 2-iodo-*N*,*N*-diisopropylbenzamide was extended to other phosphines (Scheme 6.29 and Table 6.10).

Silylphosphine **64** containing an electron-poor difluorophenyl group causes a dramatic decrease in the enantioselectivity (entry 1), in contrast to changes in the alkyl group which are well tolerated (entries 2–4).

The mechanism for the reaction (Scheme 6.30) can be described as 'phospha-Stille': oxidative addition of the aryl iodide to the Pd(0) catalyst, transmetallation with the silylphosphine to give the phosphido complex **66** and reductive elimination to form the P–C bond, which furnishes the product **67** and regenerates the catalyst.

Complex **66** is supposed to be present as a mixture of two diastereomers (epimers). Rapid interconversion between them relative to reductive elimination enables the dynamic kinetic resolution to take place. The striking performance of 2-iodo-*N*,*N*-diisopropylbenzamides was not mechanistically investigated but coordination of the amide carbonyl group to the Pd centre, forming a five-membered palladacycle, was suggested.

6.2.3.3 Pd-Catalysed Arylation of Secondary Phosphine Boranes

As seen in previous chapters, phosphine boranes are indispensable intermediates in the synthesis of many types of *P*-stereogenic compounds. Unsurprisingly, racemic secondary phosphine boranes have also found application as substrates in enantioselective Pd-catalysed cross-coupling reactions. Compared to free secondary phosphines or silylphosphines, phosphine boranes have the advantage that they cannot inhibit the catalyst by displacement of its phosphine ligands. Their air-stability is also advantageous when carrying out their catalytic syntheses and when isolating and purifying the products.

Table 6.9 Results of Pd-catalysed arylation of silylphosphines with *o*-iodoarenes (Scheme 6.28).

Entry[a]	ArI	R	Yield[b] (%)	ee (%)
1	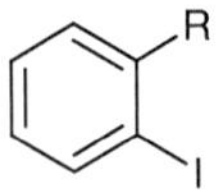	OMe	74	55 (*R*)
2		SMe	73	63 (*R*)
3		NMe$_2$	77	43
4	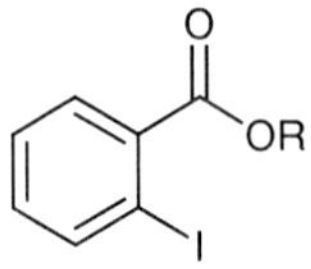	*t*-Bu	83	78
5		2,6-Me$_2$Ph	76	82
6		2,6-(*i*-Pr)$_2$Ph	77	75
7	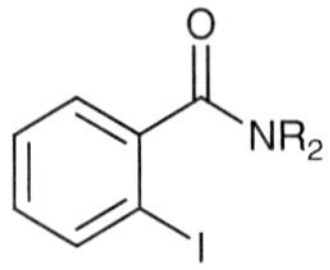	Me(OMe)	65	81
8		Et	63	79
9		*i*-Pr	53	98
10	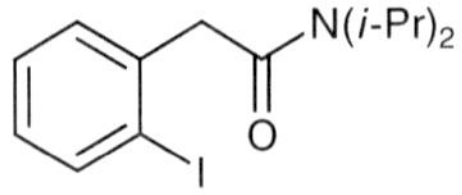	–	82	32
11	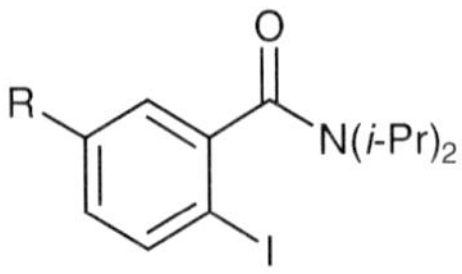	Me	61	97
12		Cl	83	97
13	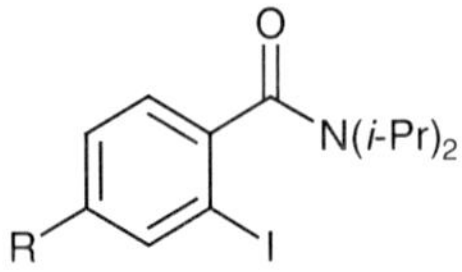	OMe	55	97
14		CF$_3$	79	97
15		Cl	80	95
16	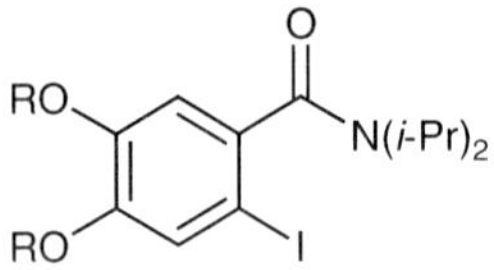	Me	67	97
17		–CH$_2$–	66	97
18	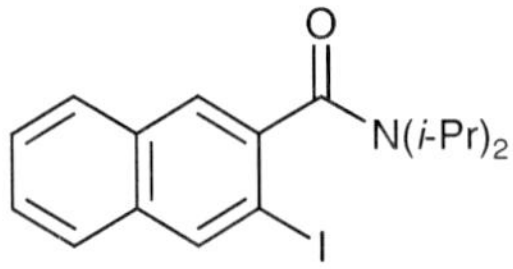	–	89	93
19	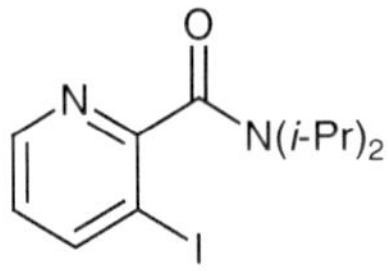	–	75	94

Table 6.9 (*Continued*).

Entry[a]	ArI	R	Yield[b] (%)	ee (%)
20	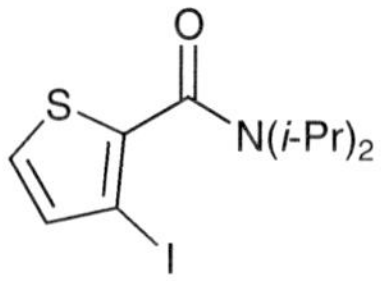	–	44	92
21	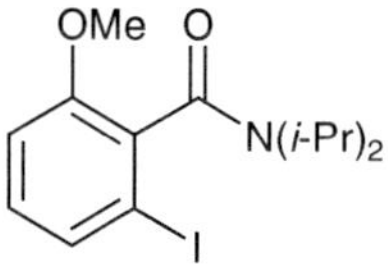	–	52	93
22	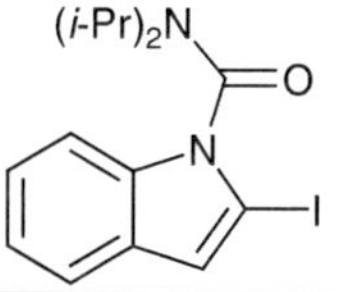	–	62	86

[a]Reaction conditions as in Scheme 6.28, reaction time: 1 h for entries 1–6; 2.5 h for entries 7–15, 18, 19, 21 and 22; 4.5 h for entries 16 and 17 and 8 h for entry 20.
[b]Isolated yields after borane or sulfide protection.

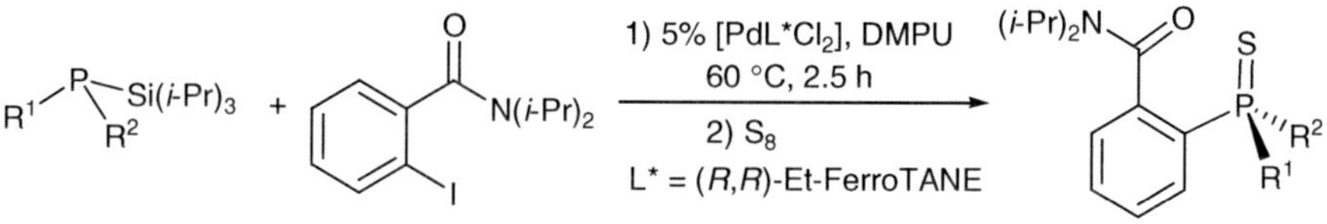

Scheme 6.29 Pd-catalysed arylation with 2-iodo-*N*,*N*-diisopropylbenzamide.

Table 6.10 Results of Pd-catalysed arylation with 2-iodo-*N*,*N*-diisopropyl-benzamide.

Entry[a]	R^1	R^2	Yield[b] (%)	ee (%)
1	$3,5\text{-}F_2\text{-}C_6H_3$	Me	72	46
2	Ph	Bn	76	92
3	Ph	 OMe	69	92
4	Ph	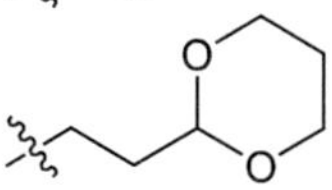	89	93

[a]Reaction conditions as in Scheme 6.29, 2 h of reaction time.
[b]Isolated yields of **65** after sulfide protection.

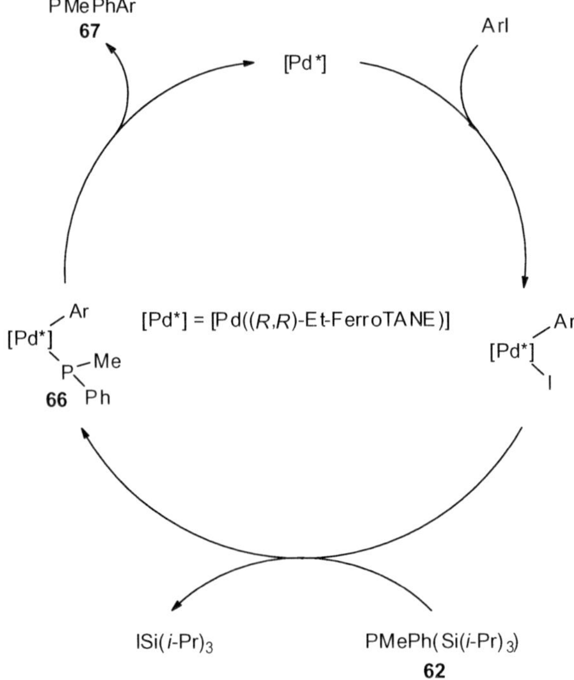

Scheme 6.30   Proposed mechanism for Pd-catalysed arylation of silylphosphines.

Scheme 6.31   Pd-catalysed cross-coupling of *o*-AnI with **41a**.

Glueck's group[50,51] investigated the Pd-catalysed enantioselective synthesis of PAMP·BH_3 (Scheme 6.31).

A range of enantiomerically pure diphosphines (Chiraphos, Me-DuPhos, Tol-BINAP and *t*-Bu-Josiphos) and bases (BuLi, NaOSiMe$_3$, *i*-Pr$_2$NEt, K_2CO_3, Cs_2CO_3, KOH, pentamethylpiperidine) were tested. Catalytic turnover was only observed for Tol-BINAP and *t*-Bu-Josiphos. Under the best conditions (THF, 40 °C, 4% (*R,S*)-*t*-Bu-Josiphos Pd complex, proton sponge), PAMP·BH_3 was obtained in a 76% yield but in only 10% ee and after 10 days. It was found that (*R,S*)-*t*-Bu-Josiphos favours the formation of the *R* enantiomer of PAMP·BH_3. The proposed mechanism is shown in Scheme 6.32.

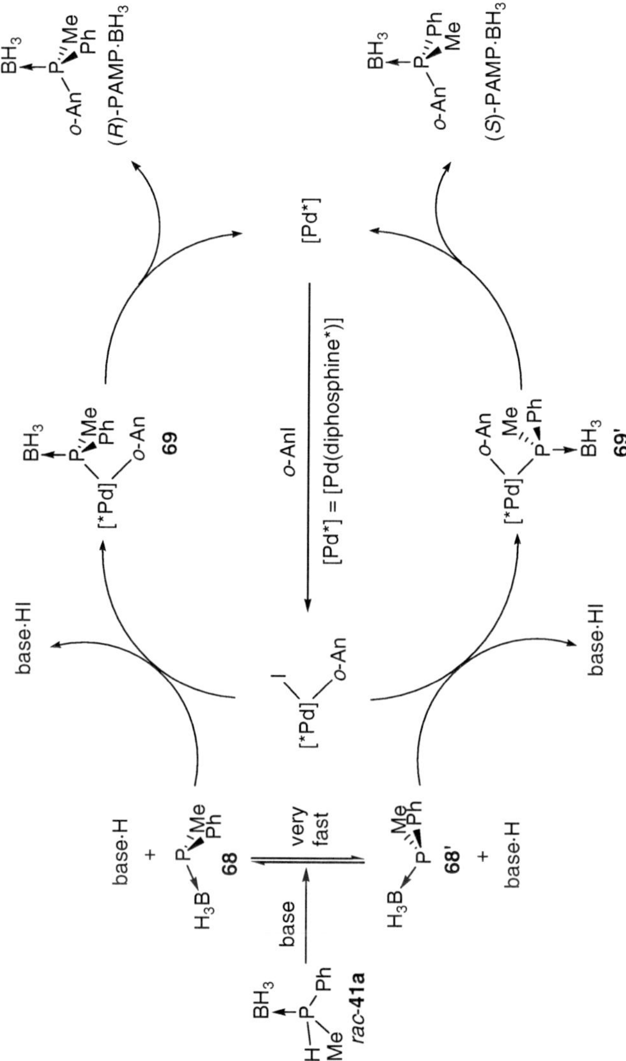

Scheme 6.32 Proposed mechanism for Pd-catalysed cross-coupling of **41a** with *o*-AnI.

Deprotonation of **41a** gives phosphide boranes **68** and **68′**, which are expected to interconvert very quickly. If the rates of nucleophilic attack of **68** and **68′** to the Pd catalyst are different (but slower than the interconversion between them), a dynamic kinetic resolution takes place to afford enantioenriched PAMP·BH$_3$ after reductive elimination. Stoichiometric experiments were conducted to validate the proposed catalytic cycle. With Chiraphos as chiral phosphine and racemic **41a**, complexes **69** and **69′** were prepared, separated and characterised, including their X-ray crystal structures. The latter complexes were also prepared from enantiomerically pure **41a**, in high diastereoselectivity. Determination of the absolute configurations at the phosphorus atoms in complexes **69** and **69′** and in the resulting PAMP·BH$_3$ confirmed two long-held assumptions in organometallic chemistry: Pd–P bond formation (transmetallation) and P–C bond formation (reductive elimination) proceed with retention of configuration.[50] Although the rates of both processes differ for each diastereomer, the rates of transmetallation are not strikingly different and comparable with inversion rates of phosphide borane anions. That probably accounts for the low enantioselectivity of the process.

An intramolecular version of the reaction has been reported by the same group[44] with the boronated version of **56** (Scheme 6.25, Section 6.2.3.1, with Me-BPE or Me-DuPhos as ligands and a proton sponge instead of NaSiOMe$_3$) to avoid displacement of the diphosphine in the catalyst. For R = Cy and (–)-Men no cyclisation was observed but for R = Ph the expected reaction occurred to yield the benzophospholanes borane in 60-89% yield with ee up to 70% with Me-DuPhos.

Following previous work on the achiral version,[43] Pican and Gaumont[52] used racemic secondary phosphine boranes devoid of methyl groups (which have a similar size to BH$_3$) with the aim of increasing steric discrimination between the substituents at the phosphorus and attain better enantioselectivities in the cross-coupling reaction (Scheme 6.33).

With **41b** (R = t-Bu) the reaction was monitored by ^{31}P NMR spectroscopy and stopped before it reached 50% conversion, to maximise the enantioselectivity in the kinetic resolution. The reaction is sluggish at room temperature but runs smoothly at 40 °C. Several typical bidentate ligands were tested and the results are listed in Table 6.11.

Under these reaction conditions, several ligands give racemic products (entries 1–5), including the DuPhos ligand, which had been successfully employed in similar reactions. In contrast, ligands of entries 6–9 gave more

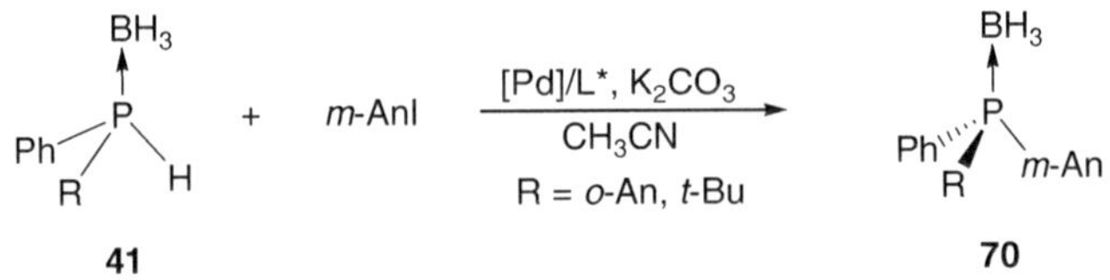

Scheme 6.33   Pd-catalysed cross-coupling of *m*-AnI with secondary phosphine boranes.

Table 6.11 Results of Pd-catalysed cross coupling of **41b** with *m*-AnI.

Entry[a]	Ligand	Yield (%)	ee (%)
1	(R)-BINAP	38	0
2	(S)-PHANEPHOS	47	0
3	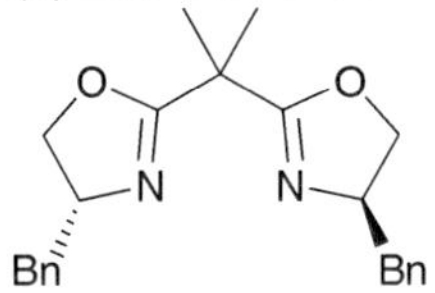 [bis-oxazoline ligand, Bn / Bn]	5	0
4	(R,S)-PPFA	32	0
5	(R,R)-Me-DuPhos	25	0
6	(S,S)-BDPP	20	17
7	(R,R)-Et-FerroTANE	22	12
8	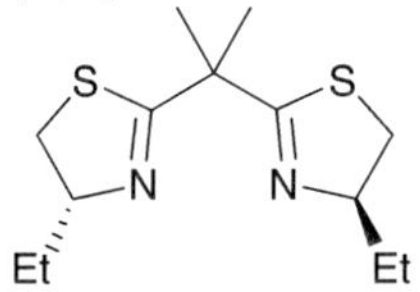 [bis-thiazoline ligand, Et / Et]	40	13
9	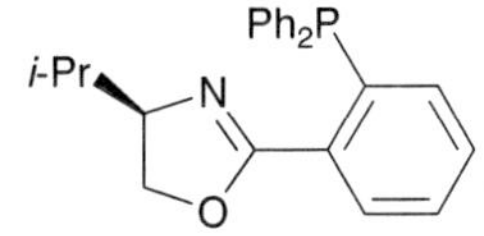 [*i*-Pr-PHOX ligand, Ph$_2$P]	22	27

[a]Reaction conditions: 5% Pd(OAc)$_2$, 2 equiv. K$_2$CO$_3$, MeCN, 40 °C.

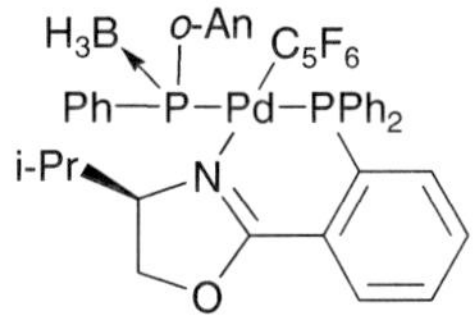

71
−70 °C, 50:50 dr
+55 °C, 23:77 dr

Figure 6.3 Transmetallated complex **71**.

promising results, especially *i*-Pr-PHOX (entry 9). By lowering the temperature to 20 °C the ee jumps to 45%, but lower temperatures could not be used because the reaction becomes too slow. With the other phosphine borane **41** (R = *o*-An), **70** was obtained as a racemate regardless of the conditions.

The mechanism was investigated with the preparation of the transmetallated complex **71** (Figure 6.3, analogous to **69** in Scheme 6.32) bearing the pentafluorophenyl group to slow down the reductive elimination step.

The diastereomeric ratio changes upon heating until thermodynamic equilibration is reached. The major diastereomer, with the absolute configuration S_P, was separated and characterised crystallographically. It seems that, at least

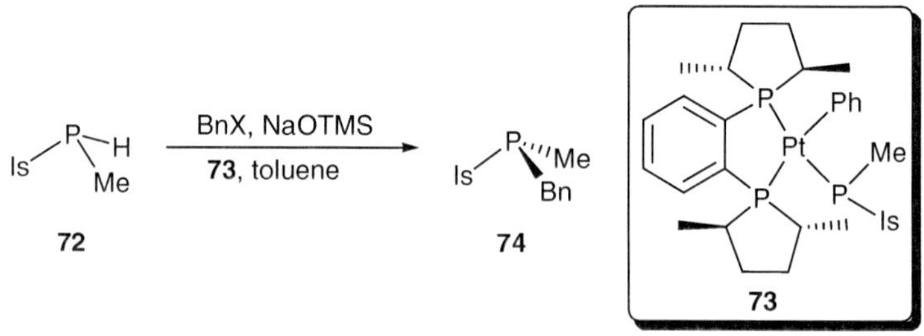

Scheme 6.34 Pt-catalysed enantioselective alkylation of secondary phosphines.

for the diaryl derivatives, the enantiodiscrimination is under thermodynamic control.

6.2.4 Alkylation of Secondary Phosphines

6.2.4.1 *Pt-Catalysed Alkylation of Secondary Phosphines*

Platinum complexes have been applied by Glueck and co-workers[53] in catalytic enantioselective alkylation of racemic secondary phosphines. The basic studies were conducted using the reaction depicted in Scheme 6.34.

This reaction is only plausible if phosphido complex **73** is the most nucleophilic species present in the mixture. Therefore, the base must not be too strong to deprotonate **72**, otherwise the generated phosphide anion would quickly react with the benzyl halide to give racemic **74**. Sodium trimethylsilanoate appeared to be a good choice. When benzyl bromide was used as the electrophile in the presence of 5% of **73**, the tertiary phosphine **74** was obtained in quantitative yield and with 77% ee. Interestingly, with benzyl chloride the reaction proceeds equally well but is extremely slow: it took one month to reach completion compared to 3 h with benzyl bromide. These remarkable results were extended to a variety of benzylic substrates and secondary phosphines (Table 6.12). In these reactions instead of using the rather sensitive complexes **73** or analogues, the air-stable precursor [Pt((R,R)-Me-DuPhos)(Cl)(Ph)] was employed.

Good yields and moderate to good enantioselectivities are obtained for most substrates. *Ortho*-substituted benzyl bromides (entries 1 and 2) and anthracenylmethyl chloride (entry 3) are well tolerated, as well as a range of secondary phosphines (entries 4–10). Diastereomeric mixtures (containing racemic C_2 and the achiral *meso* isomers) of *bis*(benzylic) substrates are also phosphinated (entries 11–14). The diastereoselectivities towards the chiral C_2 diphosphines are moderate to low. In contrast, enantioselectivities are exceedingly good (entries 11 and 12). If the catalyst favours one absolute configuration at the phosphorus atom, for example R_P, a monoalkylated intermediate with S_P configuration is much more likely to yield the *meso* compound (S_P,R_P) than the chiral (S_P,S_P).

Table 6.12 Results of Pt-catalysed enantioselective benzylation of secondary phosphines.

Entry[a]	Electrophile	Product	Yield[b] (%)	ee (%)
1	2-iodobenzyl bromide	2-iodobenzyl PMeIs	88	55
2	2-cyanobenzyl bromide	2-cyanobenzyl PMeIs	86	50
3[c]	9-(chloromethyl)anthracene	9-(CH₂PMeIs)anthracene	77	66
4	BnBr	PBnMe(Phes)	86	81
5	BnBr	PBnMe(Mes)	86	69
6	BnBr	PBnMePh	84	35
7	BnBr	PBnMe(Men)	87	56% de
8	BnBr	P(*o*-An)BnPh	85	9
9	BnBr	PBnCyPh	93	48
10	BnBr	PBn(*t*-Bu)Ph	90	42
11[d]	BnBr	PhBnP~~~PBnPh	81	47% de / 91% ee
12	BnBr	PhBnP~~~PBnPh	87	59% de / 93% ee
13[c]	1,3-bis(bromomethyl)benzene	1,3-bis(CH₂PMeIs)benzene	86	55% de / 69% ee
14[c]	2,6-bis(bromomethyl)pyridine	2,6-bis(CH₂PMeIs)pyridine	90	17% de / 72% ee

[a]Reaction conditions: 5% [Pt((R,R)-Me-DuPhos)(Ph)(Cl)], NaOTMS, THF, room temperature.
[b]Isolated yields after chromatography.
[c]Complex **73** as catalyst precursor, toluene as solvent.
[d]Temperature -15 to $-5\,°C$ and 2.5% of catalyst.

These promising results on diphosphine synthesis prompted the same group to perform a substrate and catalyst screening for the alkylation of *bis*(secondary) phosphines **75** (Scheme 6.35).[54]

A ^{31}P NMR monitoring method was developed to evaluate the rate and diastereoselectivity of the reaction quickly and efficiently. This is possible because in the products the ^{31}P NMR corresponding to C_2-**76** and *meso*-**76** isomers are well separated.

The reaction of **75b** with benzyl bromide was used to optimise the catalyst. After screening Pt complexes containing several typical chiral diphosphines, the best catalysts were again found to be [Pt(DuPhos)(Cl)(Ph)]. A variety of benzylic, naphthylmethyl and anthracenylmethyl halides were screened as alkylating agents. Low to moderate diastereoselectivies (0–60% de to C_2-**76**) were observed and in one case (**75b**, Ar′ = *o*-trifluoromethylbenzyl) both enantiomers of compound C_2-**76** were isolated enantiomerically pure on a gram scale.

Some trends were observed in the catalysis. The reactivity of **75** follows the order **c** > **b** > **d** ≈ **a**, while the diastereoselectivity towards C_2-**76** is similar, **c** > **b** ≈ **d** > **a**. No direct structure/selectivity relationship is evident among the benzyl halides, although as expected bromides react faster than chlorides.

A requirement for the NMR monitoring method is that all the species must be soluble. That is not the case in the reaction of anthracenyl methyl chloride with **75b**. In this case an anomalously high diastereoselectivity (97% de towards C_2-**76**) was found, which could not be reproduced when the reaction (30% yield, 70% ee) was scaled up. This can be explained by the insolubility of *meso*-**76**, which precipitates during the catalysis.

Further work of the same group[55] studied the diastereo- and enantioselectivity of the reaction with the *bis*(secondary) phosphine **77** with 2-(bromomethyl)naphthalene as alkylating agent (Scheme 6.36).

Good diastereoselectivity towards *meso*-**78** under the standard conditions was found. To clarify the origin of this selectivity, the monoalkylated diphosphine **79** derived from **1** was prepared and used as a substrate in the catalytic reaction. Once again, good selectivity towards *meso*-**78** (66% de) was observed. A quantitative analysis was made, which is depicted in Scheme 6.37.

The absolute configuration of the favoured enantiomer of **78** is unknown, but is assumed to be (*R*,*R*), since the same catalyst favours the *R* absolute configuration of **74** (Scheme 6.34). The scheme shows that there is substrate control with negative cooperativity, *i.e.* in **78** the absolute configuration of the tertiary phosphorus atom favours the alkylation of the second centre with opposite configuration, leading to the preferred formation of *meso*-**78**.

A detailed account on the mechanism of the reaction has been reported after studying the individual stoichiometric steps (Scheme 6.38).[56]

The nucleophilic phosphorus atom in complex **73** attacks benzyl bromide to give a mixture of phosphine **74** and neutral bromo complex **80**. In more polar solvents, this mixture is in equilibrium with cationic phosphine complex **81**. The silanoate base reacts with the electrophilic complex **81** to displace the product **74** and generate the silanoate complex **82**, which can also be formed directly from

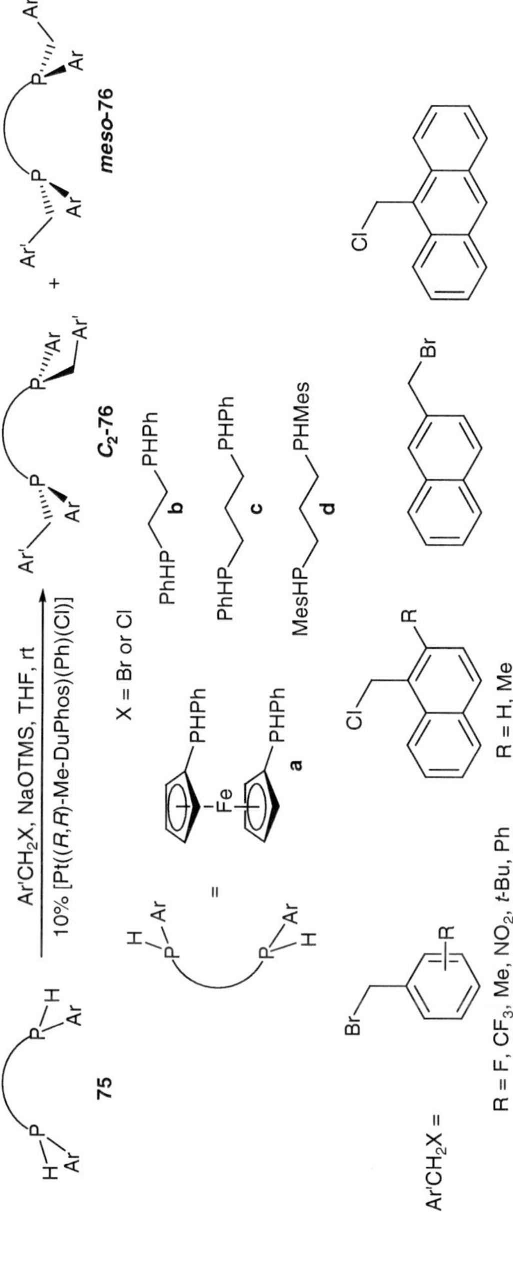

Scheme 6.35 Pt-catalysed alkylation of *bis*(secondary) phosphines.

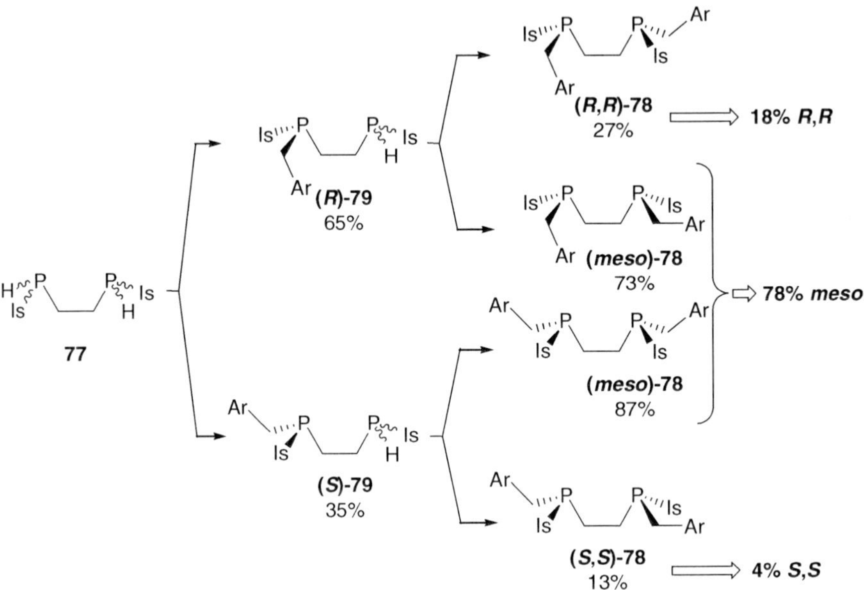

Scheme 6.36   Pt-catalysed alkylation of **77** with 2-(bromomethyl)naphthalene.

Scheme 6.37   Stereoselectivity of the reaction of Scheme 6.36.

80. The platinum silanoate **82** reacts smoothly with secondary phosphine **72** to regenerate catalyst **73**. The reader may have noticed the difference in reactivity of Pt complex **80** compared to the similar Pd complex **50** (Section 6.2.3.1, Scheme 6.24). Formation of analogous Pd and Pt phosphido (**54** and **73**) complexes is summarised in Scheme 6.39.

Phosphine **72** does not react with NaOTMS. Complex **50** does not react with NaOTMS either but it does with phosphine **72**. The inverse situation holds for platinum complex **83**.

Analogously to the Pd-catalysed enantioselective phosphination, thermodynamics is responsible for the selectivity, although the relative rates of alkylation are also important. The major diastereomer in **73** (R_P) gives the major product (*R*)-**74** (Scheme 6.34 and Scheme 6.38).

Tandem enantioselective alkylation/arylation of primary monophosphines and diphosphines has also been reported (Scheme 6.40).[57]

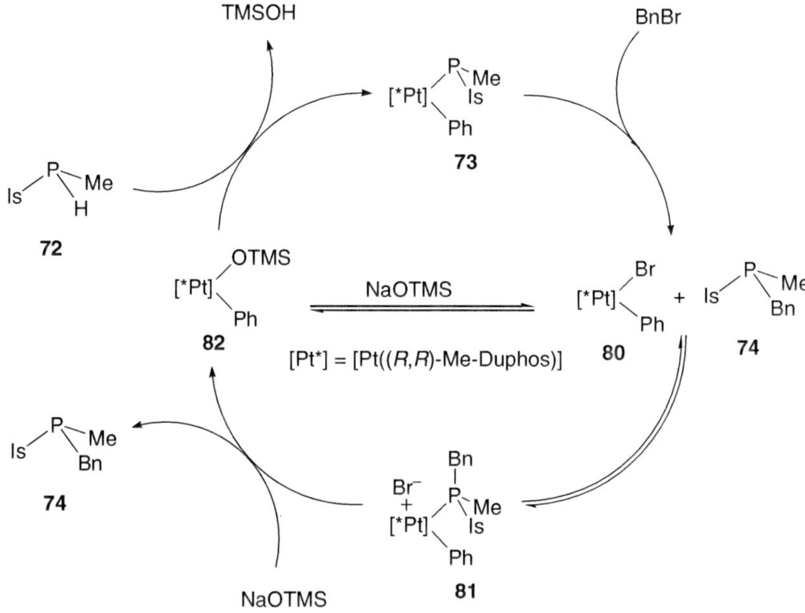

Scheme 6.38 Proposed mechanism of Pt-catalysed alkylation of secondary phosphines.

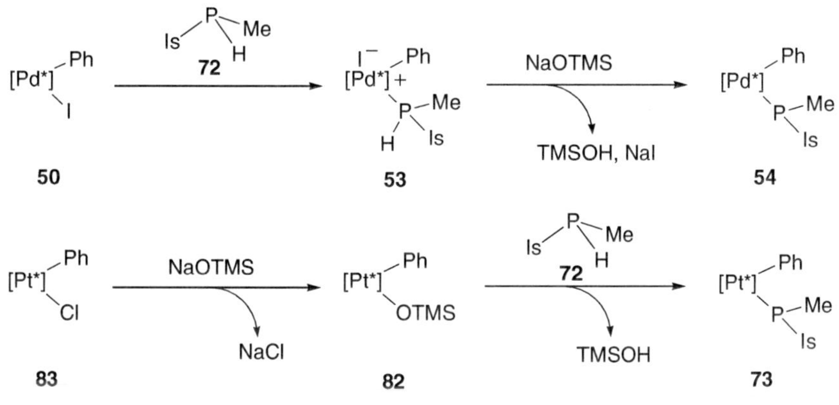

Scheme 6.39 Different mechanisms produce analogous Pd and Pt phosphido complexes.

This reaction constitutes the first synthesis of 1-phosphaacenaphthenes and proceeds with the formation of two (**85**) or four (**86**) P–C bonds. The names AcePhos and DuAcePhos were coined for **85** and **86** respectively. Again, complexes [Pt(DuPhos)(Cl)(Ph)] were found to be suitable catalysts for the reaction. The results are summarised in Table 6.13.

Good yields are obtained for all substrates, while stereoselectivities range from low to good in the case of monophosphines (entries 1–3). For

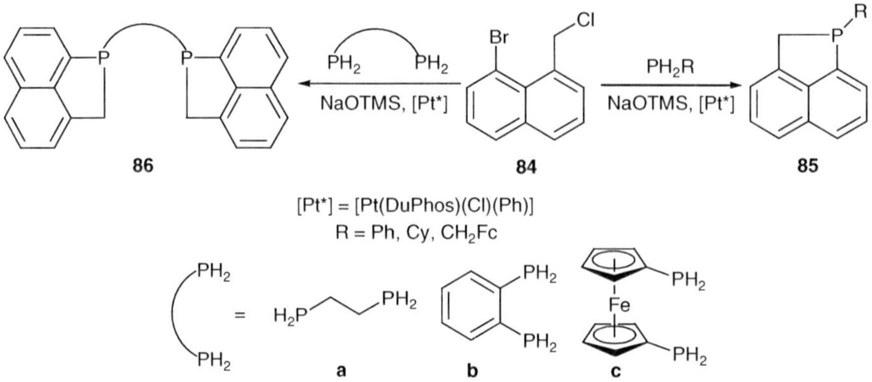

Scheme 6.40   Pt-catalysed one-pot synthesis of phosphaacenaphthenes.

Table 6.13   Results of one-pot Pt-catalysed synthesis of 1-phosphaace-naphtenes.

Entry[a]	Substrate	Yield (%)	ee (%)
1	PH_2Ph	80	74
2	PH_2Cy	85	26
3	PH_2CH_2Fc	85	50
4	**a**	78	20% de 60% ee
5	**b**	85	64% de >98% ee
6	**c**	69	31% de 39% ee

[a]Reaction conditions: 5% [Pt((R,R)-i-Pr-DuPhos)(Ph)(Cl)] for entries 1 and 2, 5–10% (R,R)-Me-DuPhos analogue for other entries, NaOTMS, THF or toluene, room temperature.

diphosphines, diastereoselectivities are low (entries 4 and 6) to moderate (entry 5). Startlingly, for *o-bis*(diphosphino)benzene (**b**) the chiral C_2 was formed in optically pure form.

The proposed mechanism (partly depicted in Scheme 6.41) is a variation of Scheme 6.38.

[Pt(DuPhos)(Cl)(Ph)] produces the secondary phosphido complexes **87**, which after alkylation by **84** and subsequent deprotonation form phosphido complexes **89**. At this point, a proposed intramolecular nucleophilic aromatic substitution gives **90**, bearing a coordinated phosphacycle. Displacement of the product by an incoming phosphine substrate and NaOTMS completes the catalytic cycle.

6.2.4.2   *Ru-Catalysed Alkylation of Secondary Phosphines*

A closely related alkylation of racemic secondary phosphines with Ru catalysis has been reported by Bergman, Toste and co-workers (Scheme 6.42).[58,59]

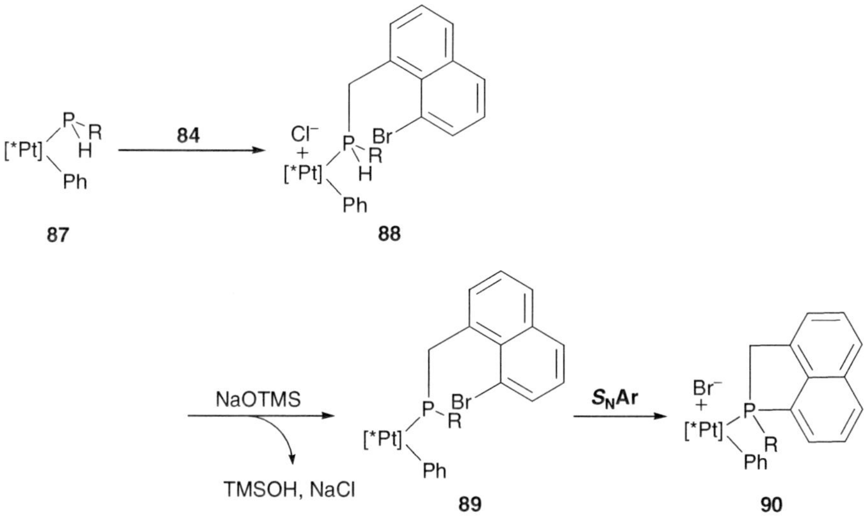

Scheme 6.41 Part of the proposed mechanism for Pt-catalysed alkylation/arylation of primary phosphines.

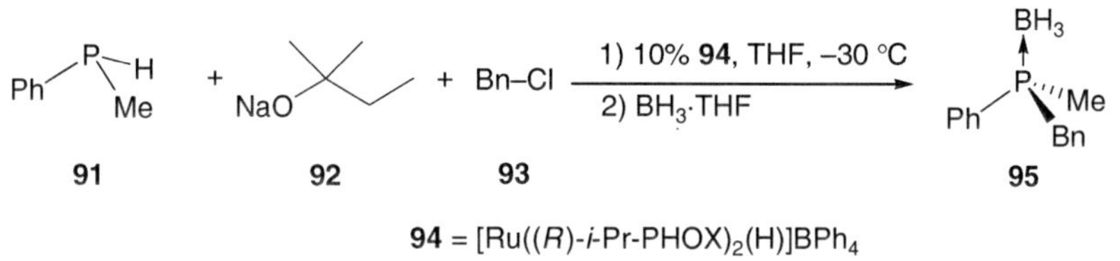

94 = [Ru((*R*)-*i*-Pr-PHOX)$_2$(H)]BPh$_4$

Scheme 6.42 Ru-catalysed enantioselective alkylation of **91**.

Ruthenium complex **94** was chosen after confirming the high nucleophilicity of the phosphorus atom in the related phosphido complex [Ru(dmpe)$_2$(H) (PMePh)]BPh$_4$, dmpe = 1,2-*bis*(dimethylphosphino)ethane. The base (sodium *tert*-amyloxide, **92**) was selected because it regenerates catalyst precursor **94** efficiently but at the same time does not produce the uncatalysed background reaction between deprotonated **91** and benzyl chlorides **93**. The chiral ligand *i*-Pr-PHOX was chosen after screening several typical bidentate ligands (C_2 diphosphines like DIOP and DuPhos gave no enantioselectivity whereas DiPAMP gave 43% ee)[59] because it provided the highest enantioselectivity (79% ee). It was found that low temperatures are required to obtain good enantioselectivities. Under these conditions, the reaction was examined for a variety of benzyl chlorides (Table 6.14).

The fact that the reactions proceed with benzyl chlorides at $-30\,^{\circ}$C with good yields highlight the remarkable nucleophilic character of the ruthenium catalyst. *Para*-substituted electron-donating benzyl chlorides are well tolerated

Table 6.14 Results of Ru-catalysed alkylation of **91** with benzylic substrates.

Entry	Product	Yield^a (%)	ee (%)
1	BH₃, P, Ph, Me, X = H, Cl, Me OMe	91 (H) 96 (Cl) 80 (Me) 85 (OMe)	75 (H) 41 (Cl) 83 (Me) 85 (OMe)
2	BH₃, P, Ph, Me, Me	92	57
3	BH₃, P, Ph, Me	96	59
4	BH₃, P, Ph, Me, N	94	48
5	BH₃, P, Ph, Me, O	80	68
6	H₃B, P, Ph, Me, BH₃, P, Me, Ph	87	74^b
7	BH₃, Me, P, Ph, BH₃, P, Ph, Me	86	95^c
8	BH₃, Me, P, Ph, N, BH₃, P, Ph, Me	89	84^d

^{a}Isolated yields.
^{b}33:67 ratio C_2:*meso* diastereomers.
^{c}74:26 ratio C_2:*meso* diastereomers.
^{d}58:42 ratio C_2:*meso* diastereomers.

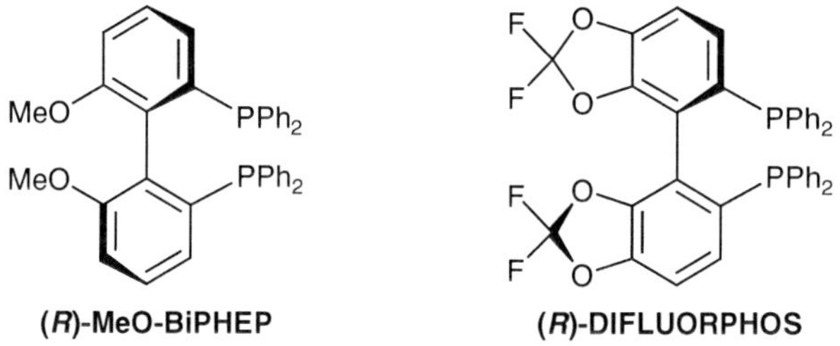

Figure 6.4 Structures of (*R*)-MeO-BiPHEP and (*R*)-DIFLUORPHOS.

(entry 1). Substitution in the *ortho* position did not affect the yield but significantly lowered the enantioselectivity (entries 2 and 3). Heteroaryl substrates (entries 4 and 5) and *bis*(benzylic) substrates (entries 6–8) are also efficiently phosphinated, albeit with lower enantio- or diastereoselectivities. With other type of substrates inferior results were obtained. With ethyl bromide 70% yield and 57% ee were obtained[58,59] whereas bromo(trimethylsilyl)methane furnished the prouct in 76% yield but almost as a racemate.[59]

Further studies were undertaken to develop a second generation of ligands with a broader scope.[59] After extensive experimentation, it was found that mixed-ligand complexes of the type [Ru(dmpe)(H)(P*)]BPh₄, containing a dmpe and one chiral diphosphine, gave the best results. The structures of the best two phosphines, MeO-BiPHEP and DIFLUORPHOS are shown in Figure 6.4.

With the complex containing (*R*)-MeO-BiPHEP the results with mono- and dibenzylic substrates were similar to those obtained with complex **94** (Table 6.14) but at room temperature instead of at −30 °C.[59] Interestingly, in contrast to **94** it was found that the complex with the phosphine (*R*)-DIFLUORPHOS is active in the phosphination of aliphatic substrates (Table 6.15).[59]

In general, good yields and moderate enantioselectivities were found for monophosphinations (entries 1–4) and diphosphinations (entries 5 and 6). The latter reactions are interesting because they provide an easy entry to a wide variety of *P*-stereogenic diphosphines. Compared to the related reactions with platinum complexes seen in Section 6.2.4.1 (Scheme 6.35) the ruthenium systems have the advantage that aliphatic dibromides are much easier to obtain than *bis*(secondary) phosphines. The success of the mixed-ligand catalysts can be attributed to the ability of the biaryl ligand to force a chiral conformation to the dmpe ligand, enhancing the chiral environment of the ruthenium atom.[59]

Although a thorough mechanistic study is lacking, the proposed mechanism for the reaction with benzyl bromide is illustrated in Scheme 6.43.[58,59]

Catalyst precursor **94** forms the Ru phosphido complex **96**, which is alkylated by **93** to give the cationic complex **97**. This complex is thought to dissociate to give the phosphine along with the unsaturated ruthenium hydrido complex **98**, which readily adds **91** to yield complex **99**. Finally, deprotonation of **99** by sodium *tert*-amyloxide regenerates the catalyst. In analogy to the platinum analogue, complex **96** is present as an unequal mixture of diastereomers interconverting by inversion at the phosphorus atom.

Table 6.15   Results of Ru-catalysed alkylation of **91** with aliphatic substrates.

Entry	Product	Yield^a (%)	ee (%)
1	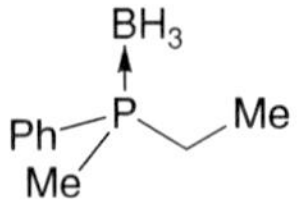 Ph,Me-P(→BH$_3$)–CH$_2$CH$_2$–Me	97	66
2	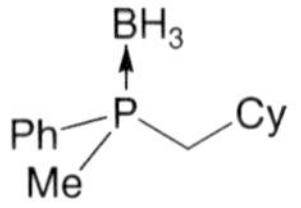 Ph,Me-P(→BH$_3$)–CH$_2$CH$_2$–Cy	77	45
3	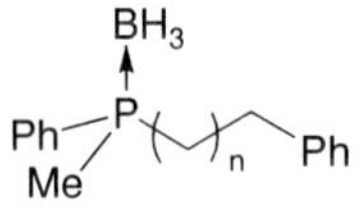 Ph,Me-P(→BH$_3$)–(CH$_2$)$_n$–CH$_2$CH$_2$–Ph	35 ($n=1$) 91 ($n=2$)	68 ($n=1$) 72 ($n=2$)
4	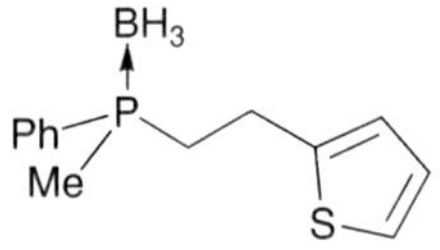 Ph,Me-P(→BH$_3$)–CH$_2$CH$_2$–(2-thienyl)	84	69
5	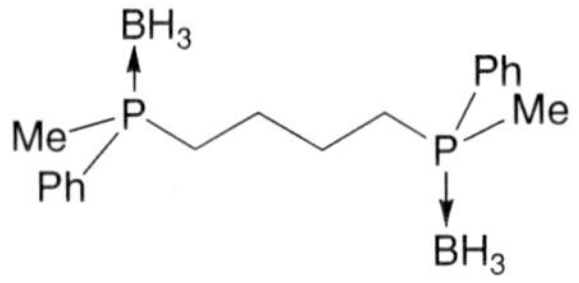 Me,Ph-P(→BH$_3$)–(CH$_2$)$_4$–P(→BH$_3$)(Ph)(Me)	95	73^b
6	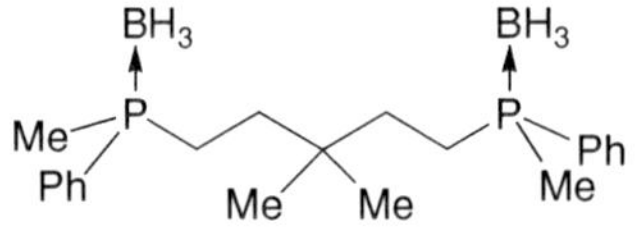 Me,Ph-P(→BH$_3$)–CH$_2$CH$_2$–C(Me)$_2$–CH$_2$CH$_2$–P(→BH$_3$)(Ph)(Me)	54	97^c

^{a}Isolated yields.
^{b}74:26 ratio C_2:*meso* diastereomers.
^{c}71:29 ratio C_2:*meso* diastereomers.

6.2.5   Rh-Catalysed Desymmetrisation of Dialkynylphosphine Oxides

The reactions discussed in the last sections produce enantioenriched *P*-stereogenic compounds by metal-catalysed enantioselective resolutions of chiral but racemic secondary phosphines and derivatives. Another set of methods is based on desymmetrisations of achiral (but prochiral) tertiary phosphine derivatives bearing two enantiotopic groups amenable to functionalisation by chiral transition metal catalysts.

Tanaka and co-workers[60] reported the desymmetrisation of symmetrical dialkynylphosphine oxides mediated by chiral rhodium catalysts (Scheme 6.44).

In this reaction, both triple bonds in 1,6-diynes **100** and one of the prochiral alkynes in the phosphine oxides **101** react enantioselectively in a Rh-catalysed [2 + 2 + 2] cycloaddition, generating an aromatic ring attached to the newly

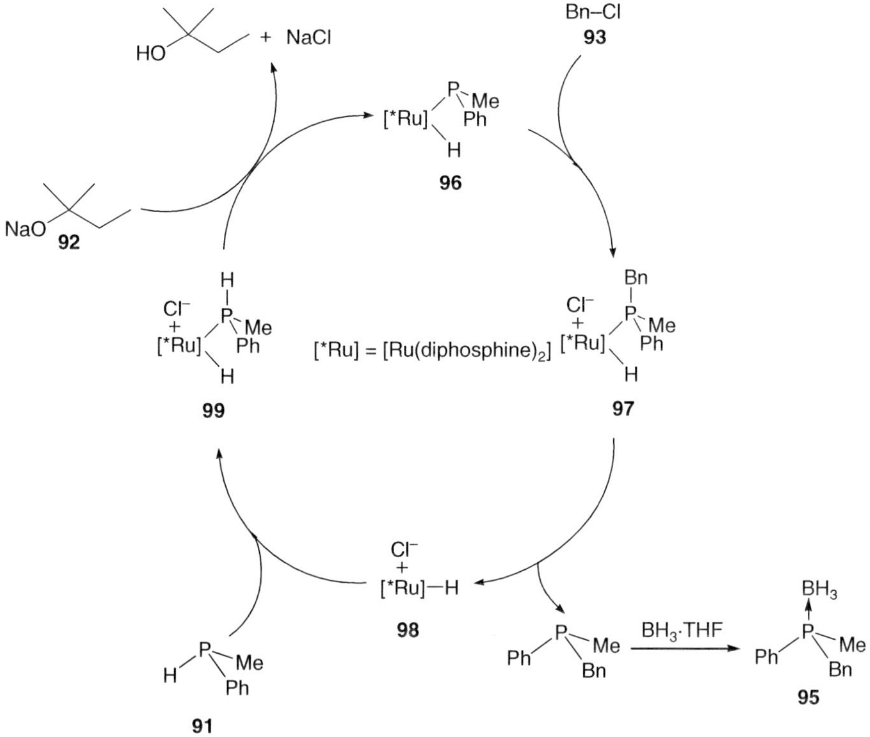

Scheme 6.43　Proposed mechanism for Ru-catalysed enantioselective phosphination of benzyl chlorides.

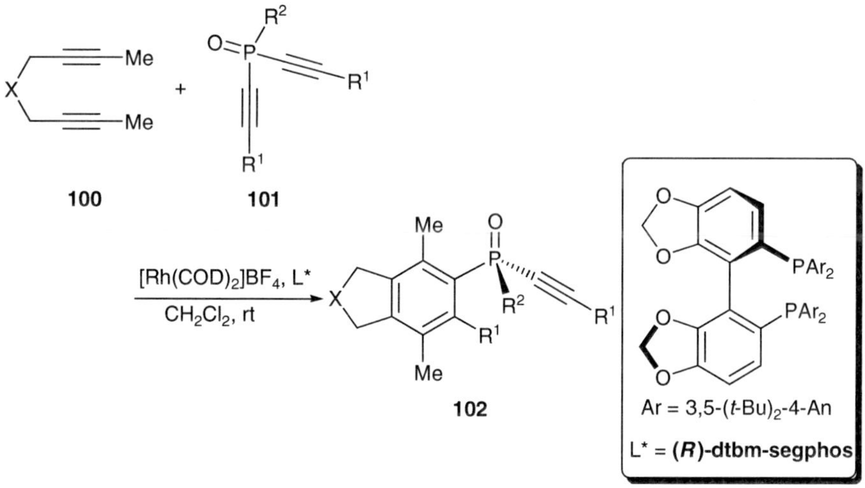

Scheme 6.44　Rh-catalysed enantioselective desymmetrisation of dialkynylphosphine oxides.

Table 6.16   Results of Rh-catalysed desymmetrisation of prochiral dialkynylphosphine oxides.

Entry[a]	X	R[1]	R[2]	Yield[b] (%)	ee[c] (%)
1	O	Ph	Me	>99	93
2[d]	O	Ph	Me	>99	−92
3[e]	O	Ph	Me	91	93
4	CH_2	Ph	Me	98	93
5	NTs	Ph	Me	>99	85
6	$NSO_2(4\text{-}BrC_6H_4)$	Ph	Me	86	87 (S)
7	O	$p\text{-}MeOC_6H_4$	Me	96	95
8	O	$p\text{-}F_3CC_6H_4$	Me	83	91
9	O	$n\text{-}Bu$	Me	71	−35
10[f]	O	Ph	Ph	>99	−50
11[g]	O	Ph	$t\text{-}Bu$	82	41

[a]Reaction conditions: 5% $[Rh(COD)_2]BF_4$, 5% of (R)-dtbm-segphos CH_2Cl_2, room temperature, 1 h.
[b]Isolated yields after chromatography.
[c]Absolute configurations were not determined (except for entry 6), only the sign of the optical rotation.
[d]With (S)-dtbm-segphos.
[e]With 1% of $[Rh(COD)_2]BF_4$.
[f]With (S)-Tol-BINAP.
[g]With (S)-Xyl-BINAP.

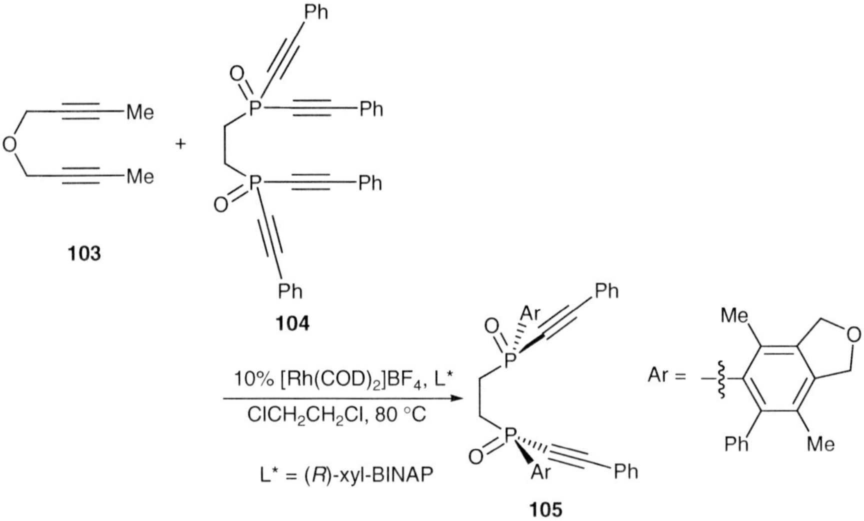

Scheme 6.45   Rh-catalysed double desymmetrisation of a *bis*(dialkynylphosphineoxide).

formed *P*-stereogenic centre. The scope of this enantioselective cycloaddition was screened with several substrates and number of ligands (Table 6.16).

Previous studies[61–63] showed that biaryl phosphines are suitable ligands for this reaction; dtbm-segphos provided the best results under the reaction

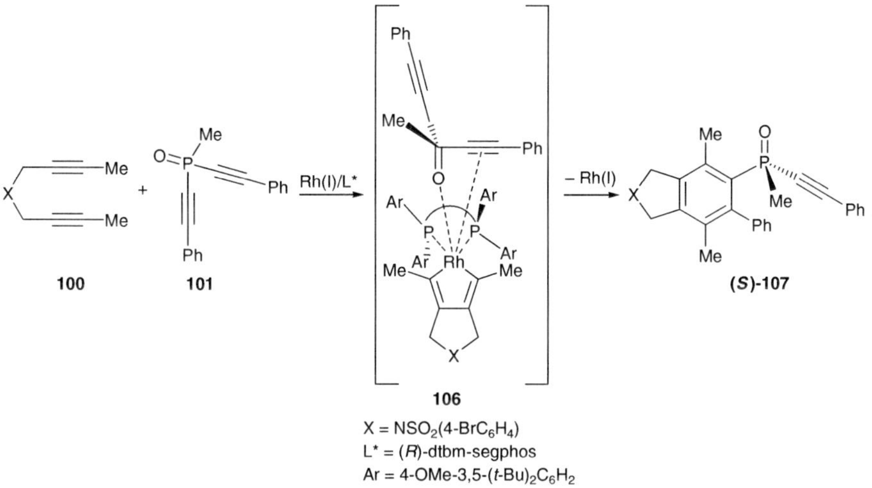

X = NSO$_2$(4-BrC$_6$H$_4$)
L* = (*R*)-dtbm-segphos
Ar = 4-OMe-3,5-(*t*-Bu)$_2$C$_6$H$_2$

Scheme 6.46   Proposed mechanism for the formation of (*S*)-**107**.

conditions specified in Scheme 6.44. The high activity of the catalyst enabled the reaction to proceed with excellent results in the presence of only 1% of catalyst (entry 3). Both electron-donating and electron-withdrawing groups are well tolerated in the alkyne terminus of **101** (entries 7 and 8), although the *n*-butyl group is detrimental to the yield and selectivity (entry 9). Finally, changing the R^2 group in **101** to phenyl or *tert*-butyl furnishes product **103** with a significantly reduced enantioselectivity (entries 10 and 11).

The reaction was extended to the synthesis of a C_2 *P*-stereogenic diphosphine dioxide **105** (Scheme 6.45).

Diyne **103** and tetrayne **104** reacted as expected, giving diphosphine dioxide **105** as a mixture of diastereomers (C_2:*meso* = 60:40; 71% ee) in quantitative yield. A single recrystallisation furnished the C_2 compound in 33% yield with 97% ee.

For entry 6 of Table 6.16, the crystal structure of the product allowed its absolute configuration to be determined as *S*. With this information, formation of annulated *P*-stereogenic phosphine oxides was proposed to occur through rhodium intermediate **106** (Scheme 6.46).

The spatial arrangement in **106** results from the steric interaction between the alkynyl group and the bulky PAr$_2$ groups of (*R*)-dtbm-segphos. Insertion of this alkynyl into the rhodium metallacycle and subsequent reductive elimination gives (*S*)-**107**. Although mentioned as a possibility, no reduction of the phosphine oxides to the parent phosphines was reported.

6.2.6   Ru- and Mo-Catalysed Metathesis

Alkene metathesis has been successfully employed to prepare *P*-stereogenic phosphine oxides and phosphine boranes.[64] Although most of the work has yielded racemates, the use of already resolved precursors or diastereoselective reactions have produced some enantiopure compounds. Moreover, the first enantioselective version of the reaction in the context of *P*-stereogenic compounds has already appeared.[65]

Pietrusiewicz and co-workers[66] used optically pure vinylphosphine oxides in the synthesis of substituted derivatives by alkene cross-metathesis (CM), by (achiral) Grubbs and Hoveyda-type ruthenium systems (Scheme 6.47).

It was demonstrated that the reaction proceeds without racemisation of the stereogenic phosphorus atom and with total selectivity towards the (*E*)-alkenylphosphine oxide. Unexpectedly, it was also possible to dimerise phosphine oxide **109** (with R^1 = Me) with 5% of **112b** to obtain the corresponding optically pure (*E*)-diphosphine oxide in 85% yield. The crystal structure of this compound has been determined by X-ray diffraction[67] and has been used as dipolarophile in 1,3-dipolar cycloadditions with nitrones, yielding several optically pure diphosphine oxides.[68] Similar homometathesis reactions have been investigated in more detail by Grela, Pietrusiewicz, Butenschön and co-workers[69,70] with other (racemic) substrates such as **109** and different catalysts. Gouverneur and co-workers[71] studied a similar dimerisation of

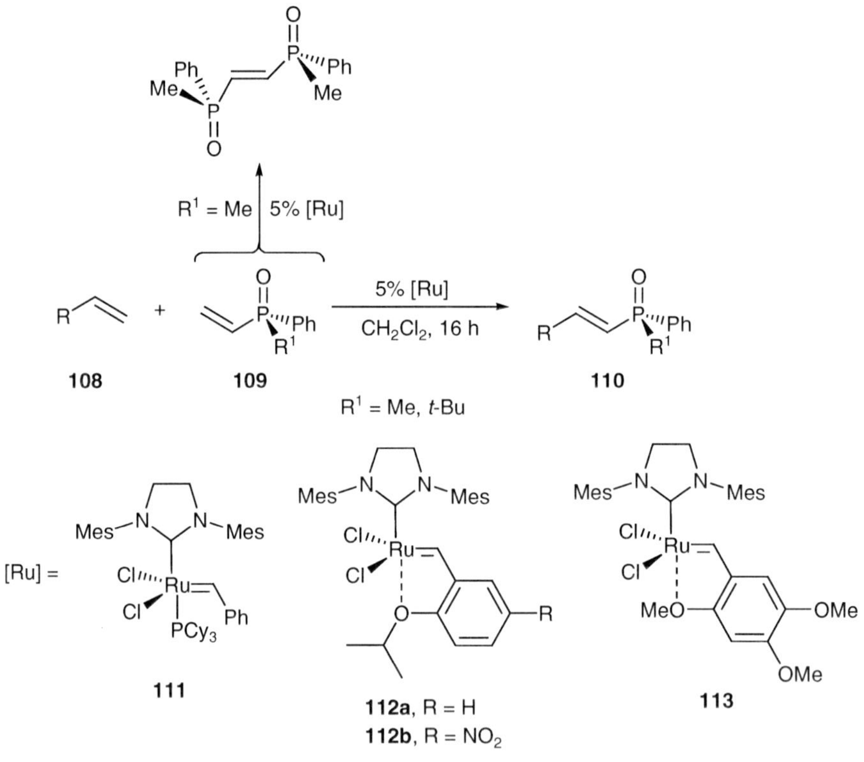

Scheme 6.47   Cross metathesis between alkenes and enantiomerically pure phosphine oxides.

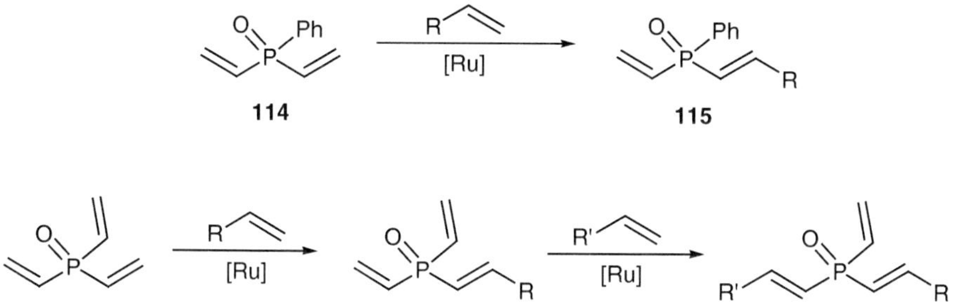

Scheme 6.48 Ru-catalysed desymmetrisation of prochiral (**114**) and achiral (**116**) phosphine oxides *via* CM.

vinylphosphine boranes with catalyst **111** but only starting material was recovered. In contrast, allyphosphine boranes yield the expected homo-dimerisation product. In the same account, they report more success with the CM of a few phosphine boranes and several alkenes, to afford racemic *P*-stereogenic products. In spite of excellent conversions, considerably lower yields of isolated products were obtained due to decomposition upon purification.

Ru-catalysed CM has also been used successfully in desymmetrisation several substrates to generate *P*-stereogenic phosphine oxides by Gouverneur and Bisaro (Scheme 6.48).[72]

These desymmetrisations by CM are challenging because selectivity has to be controlled at several levels. There are many alkenes in the reaction mixture that can undergo undesired self-metathesis reactions. In addition, double metathesis yielding achiral products has to be minimised. Finally, there is the issue of *E*/*Z* selectivity. Despite these hurdles, several examples of *P*-stereogenic (but racemic) phosphine oxides **115** were obtained from prochiral phenyl divinylphosphine oxide (**114**) in 47–86% yield. A three-fold excess of **114** was used to minimise double metathesis. In all but one case none of the *E* isomer of **115** is formed. The vinyl group in **115** was further functionalised by CM with styrene.

Remarkably, two successive CM reactions (R = $(CH_2)_9CH_3$, R′ = Ph and vice versa) allowed the preparation of **118**, whose vinyl group was subjected again to CM with 3-phenylpropene yielding a phosphine oxide with three different *E*-alkenyl groups.

Despite these examples of cross-metathesis, ring-closing metathesis (RCM) has found so far much wider applicability in the preparation of *P*-stereogenic compounds. Gouverneur and co-workers[73,74] reported the preparation of a few phosphine boranes but most of the work has been carried out with P(V) compounds. For instance, Ru-catalysed metathesis was used to prepare a few phosphine oxides[75] and several five-, six- and seven-membered *P*-stereogenic heterocycles that include a phosphorus atom connected to oxygen or nitrogen atoms.[76,77] Two examples are shown in Scheme 6.49.

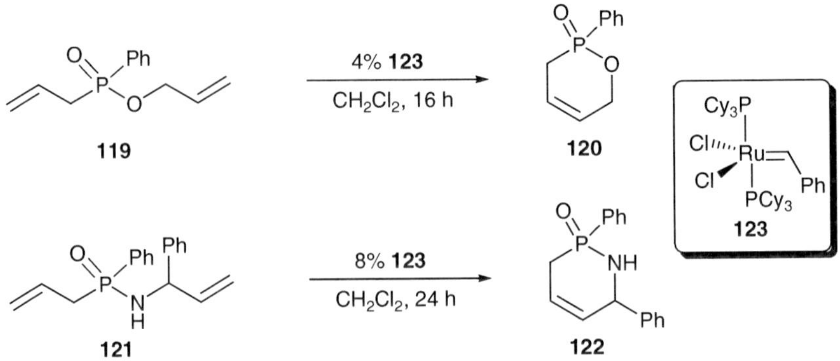

Scheme 6.49 Preparation of *P*-stereogenic cyclic compounds by Ru-catalysed RCM.

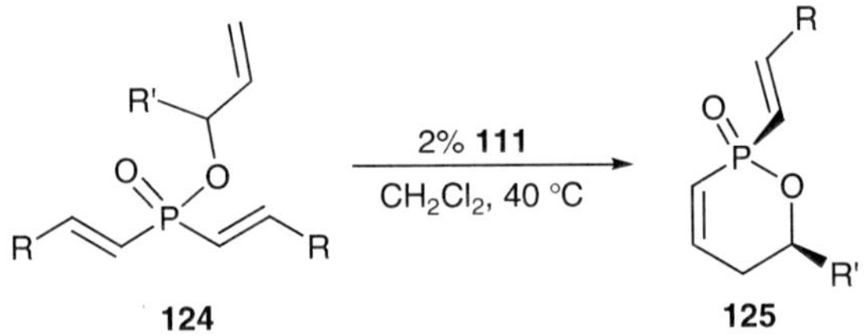

Scheme 6.50 Diastereoselective preparation of *P*-stereogenic phosphinates *via* DSRCM.

Compounds **120** and **122** were obtained in a yield of 92% and 63% respectively, the former in racemic form and the latter as a mixture of diastereomers. The reader will have noticed that **121** has stereogenic carbon and phosphorus atoms, and therefore some diastereoselectivity in **122** can be expected. This aspect was further investigated by the same group,[78] in the preparation of *P*-stereogenic cyclic phosphinates by diastereoselective ring closing metathesis (DSRCM, Scheme 6.50) of trienes lacking a *P*-stereogenic atom.

Starting trienes **124** fulfil two essential conditions: the two diastereotopic alkenes at the chirotopic phosphorus atom do not react with each other and the metathesis catalyst attacks the double bond closest to the stereogenic carbon atom first, since the two identical alkenes are conjugated with the electron-withdrawing phosphinate. Yields of cyclic phosphinates **125** were typically above 60% and with de up to 86%. As expected, both the geometry and substitution at the double bonds are prominent factors in the selectivity of the reaction.

In a related study, Hanson and Stoianova[79] described the application of DSRCM for the desymmetrisation of enantiomerically pure phosphonamides (Scheme 6.51) and phosphonates following previous studies.[80,81]

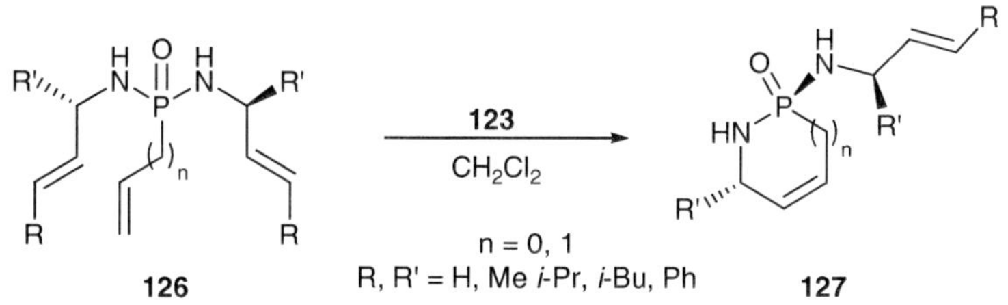

Scheme 6.51 DSRCM of chiral phosphonamides.

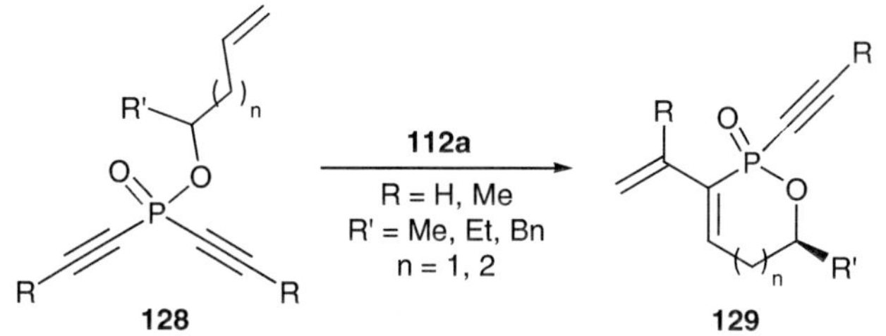

Scheme 6.52 Desymmetrisation of ene-diynes **128** by RCM.

Cyclic products **127** were obtained in moderate to good yields (37–80%) with diastereomeric excesses ranging from 4% to 86%, with the best values corresponding to the five-membered product ($n = 0$). It was also found that the alkeneic substituent R has a much more pronounced effect on the selectivity than R′.

Recently Gouverneur and co-workers[82] reported the desymmetrisation of *bis*(alkynyl)phosphinates by diastereoselective enyne RCM (Scheme 6.52).

With catalyst **112a** good yields and diastereoselectivities (up to 90% de) were generally observed favouring the diastereomer of **129** bearing the alkynyl and R′ substituents in relative *cis* disposition. Racemic substrates were used except in one case in which optically pure **(R)-128** (R = H, R′ = Me, $n = 1$) was employed, yielding **129** in very high optical purity (90% de, >98% ee). The diene present in **129** was found reactive in Diels–Alder reactions whereas the exocyclic alkynyl group in **129** was further derivatised by Cu-catalysed Huisgen cycloaddtions with azides.

DSRCM has found application in the synthesis of many other *P*-stereogenic cyclic compounds such as phosphates,[83,84] phosphonates[85] and phosphonosugars.[86,87] Some of these compounds have been further derivatised by CM.[83,84]

Finally, there is a report on Mo-catalysed asymmetric ring closing metathesis (ARCM) for the enantioselective desymmetrisation of prochiral phosphinates and phosphine oxides (Scheme 6.53).[65] This constitutes the first example where ARCM is used to prepare chiral compounds with a heteroatom as stereogenic centre.

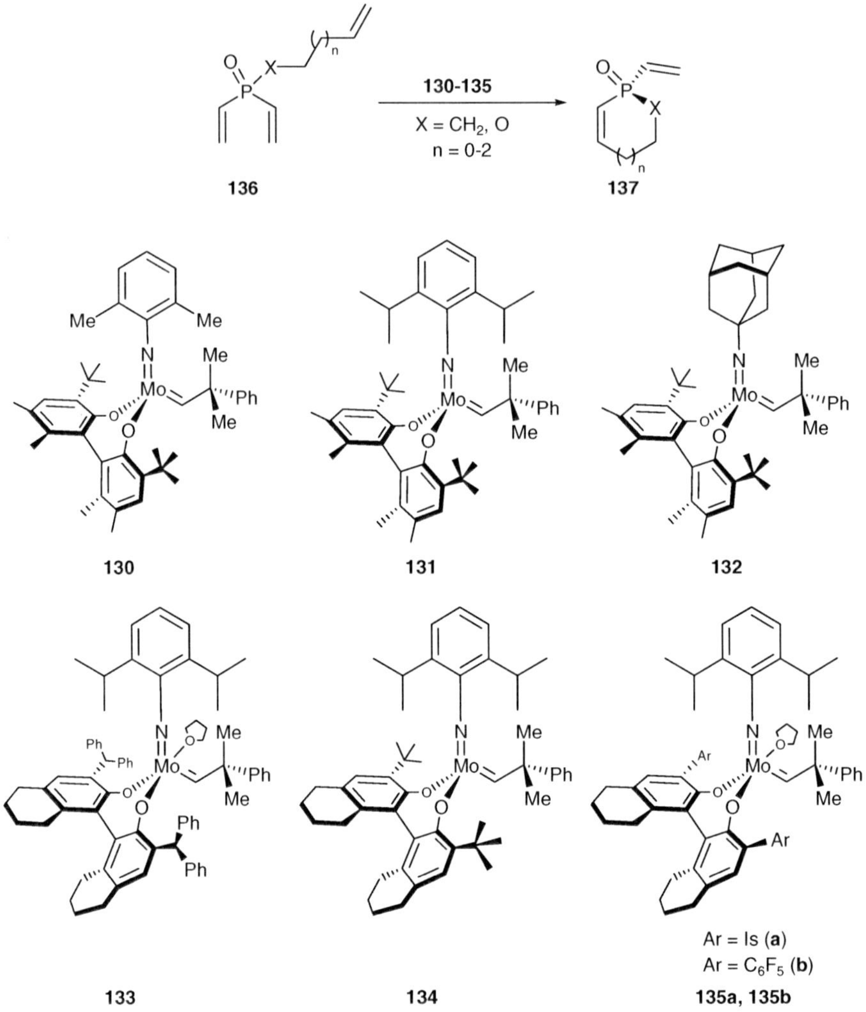

Scheme 6.53 Mo-catalysed ARCM of divinylphosphinates or divinylphosphine oxides.

Chiral molybdenum complexes **130–135** were selected based on their documented ability to promote ARCM in kinetic resolutions or desymmetrisations.[88] They were tested in the catalytic ARCM of several phosphinates and phosphine oxides which were strategically synthesised to provide *P*-stereogenic compounds. The results are listed in Table 6.17.

Most of the phosphinate substrates (entries 1–7) are efficiently cyclised with good to excellent enantioselectivities. Homodimers, formed by intermolecular metathesis, were also detected in the crude reaction mixtures. In entries 1 and 6, with a substrate containing an allyloxy arm, no reaction was observed with any of the catalysts studied. This can be due to the formation of an unreactive alkylidyne *via* chelation of the P(O) group and Lewis-base-promoted

Table 6.17   Results of Mo-catalysed ARCM of phosphinates and phosphine oxides.

Entry[a]	136	137	[Mo*]	Solvent	T (°C)	Conversion[b] (%)	Yield (%)	ee[c] (%)
1	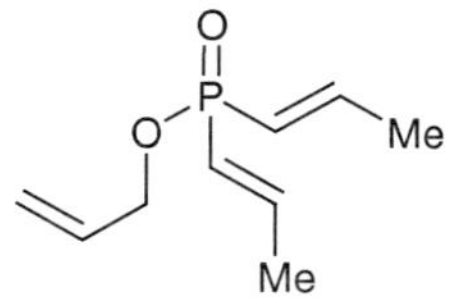	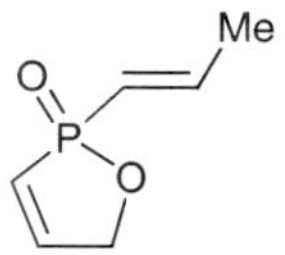	**130–135**	CH_2Cl_2	22–60	<2[d]	–	–
2	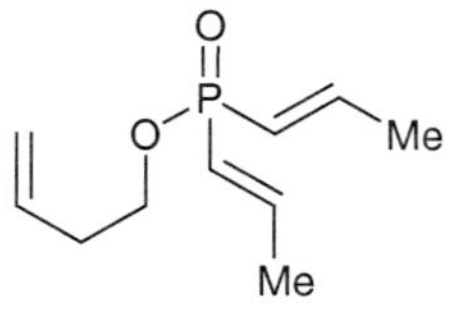	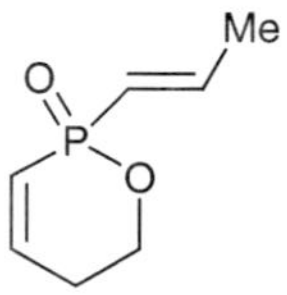	**131**	CH_2Cl_2	22	63	43	60
3	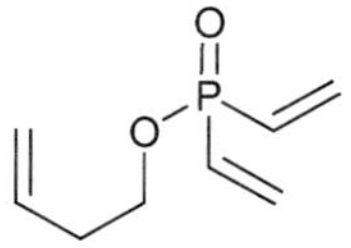	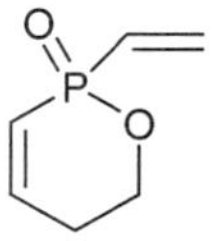	**130**	C_6H_6	22	60	54	86
4	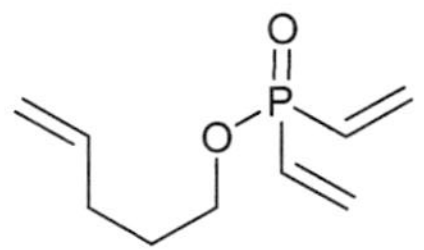	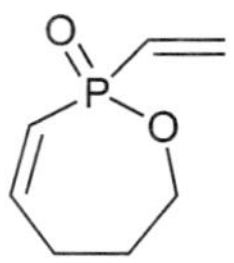	**135a**	C_6H_6	22	48	42	73
5	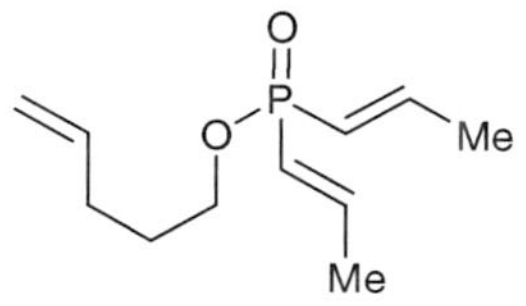	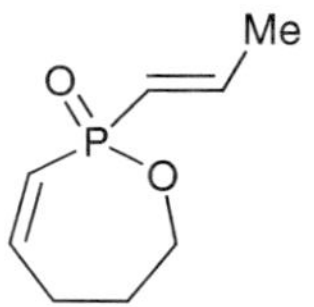	**131**	CH_2Cl_2	22	32	30	27

Table 6.17   (*Continued*).

Entry[a]	136	137	[Mo*]	Solvent	T (°C)	Conversion[b] (%)	Yield (%)	ee[c] (%)
6	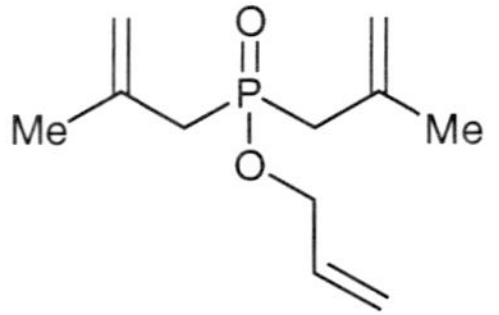	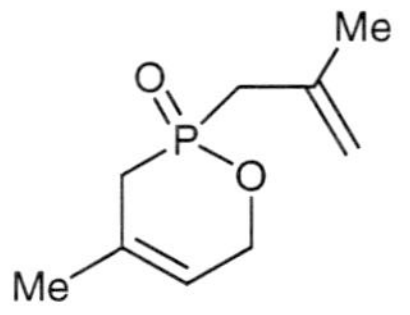	**130–135**	C_6H_6	22	$<2^d$	–	–
7	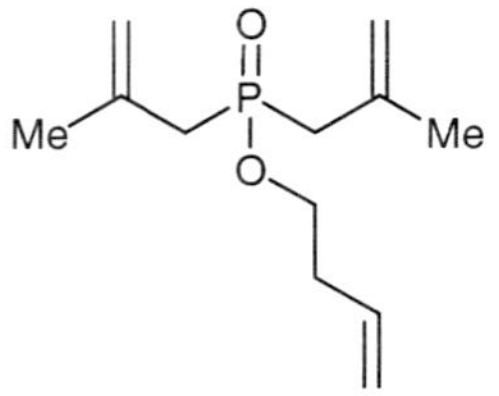	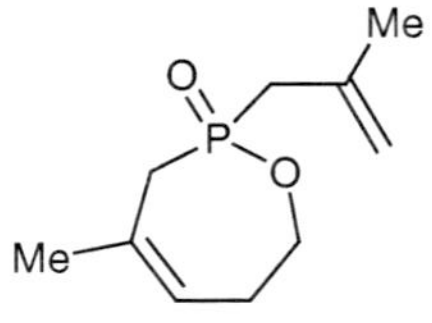	**135a**	C_6D_6	22	81	79	96
8	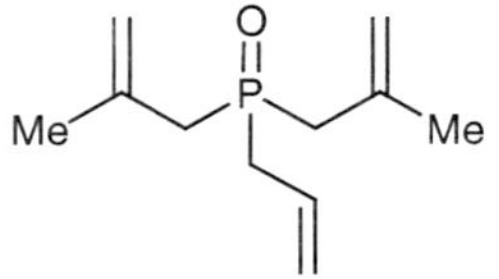	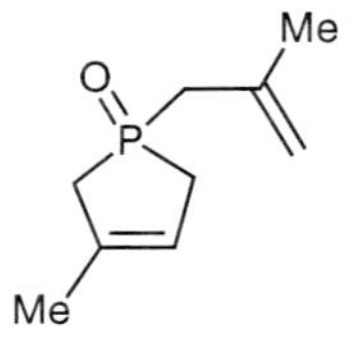	**135a**	CH_2Cl_2	60	96	88	71

| 9 | | | **134** | CH$_2$Cl$_2$ | 22 | 84 | 80 | 74 |
| 10 | | | **134** | C$_6$H$_6$ | 60 | 81 | 79 | 91 |

[a]10% [Mo*], 5% of (*R*)-dtbm-segphos CH$_2$Cl$_2$, room temperature, 1 h.
[b]Measured by ^{1}H NMR, homodimer was present.
[c]Measured by GLC of the purfied material.
[d]Starting material recovered.

tautomerisation. Phosphine oxides are also good substrates for the reaction, including the formation of a five-membered ring in entry 8.

Absolute configurations for the products were not determined, except for the product in entry 10, which was found to be *R* by X-ray diffraction of a derivative. However, additional experiments allowed the study of the relative stereoinduction for each catalyst and substrate, by analysing of the sense of optical rotations. It was found that this sense is strongly dependent on the exact structure of the catalyst. For example, catalysts **134** and **135a** (bearing a different diol) cyclise the substrate of entry 7 with 96% and 93% ee respectively, but giving products with opposite absolute configurations. More surprising is that a similar trend was observed when the substrate in entry 10 was reacted in the presence of catalyst **130** or **132** (differing only in the achiral imido group). This reversal of stereoinduction can be attributed to different reacting alkylidene isomers during the catalytic cycle in each case.

6.3   Biocatalytic Synthesis of *P*-Stereogenic Compounds

6.3.1   Kinetic Resolutions

As early as 1973, Zerner and Dudman[89] reported the kinetic resolution of a *P*-stereogenic phosphate triester (butylmethyl(*p*-nitrophenyl)phosphate) by hydrolysis with beef and horse serums, albeit with very low efficiency. The first efficient resolutions of phosphorus stereocentres with enzymes were independently reported in 1994 by two groups. Serreqi and Kazlauskas[90] studied the enantioselective hydrolysis of pendant acetate groups in chiral racemic phosphines and phosphine oxides mediated by several commercially available lipases and esterases, under kinetic resolution conditions (*i.e.* the reaction was stopped at 50% conversion). In general, enantioselectivities were low to moderate, but for one substrate promising results were obtained (Scheme 6.54).

Cholesterol esterase (CE) and lipase from *Candida rugosa* (CRL) were very efficient in this kinetic resolution, even on a preparative scale. The best result was obtained with CRL: 1 g of racemic **138** yielded 0.48 g of unreacted substrate (*S*)-**138** in 90% ee and 0.39 g of deacetylated product (*R*)-**139** in 88% ee. (*S*)-**138** was hydrolysed with NaOH, producing (*S*)-**139** in 75% yield and in almost optically pure form after recrystallisation. Both enantiomers of **139** were subjected to further chemical transformations. For example, the alcohol in (*S*)-**139** was methylated giving 75% yield of enantiomerically pure (*S*)-**140**, which was finally reduced by trichlorosilane/triethylamine to give phosphine (*S*)-**141** in 96% ee and 72% yield. This last step occurs with inversion of configuration but some erosion of enantiomeric purity was observed.

In the same year, Mikolajczyk and co-workers[91] described the kinetic resolution of racemic phosphinoylacetates **142** by hydrolysis with pig liver esterase (PLE), to give the corresponding phosphinoacetic acids **143** and unreacted esters (Scheme 6.55).

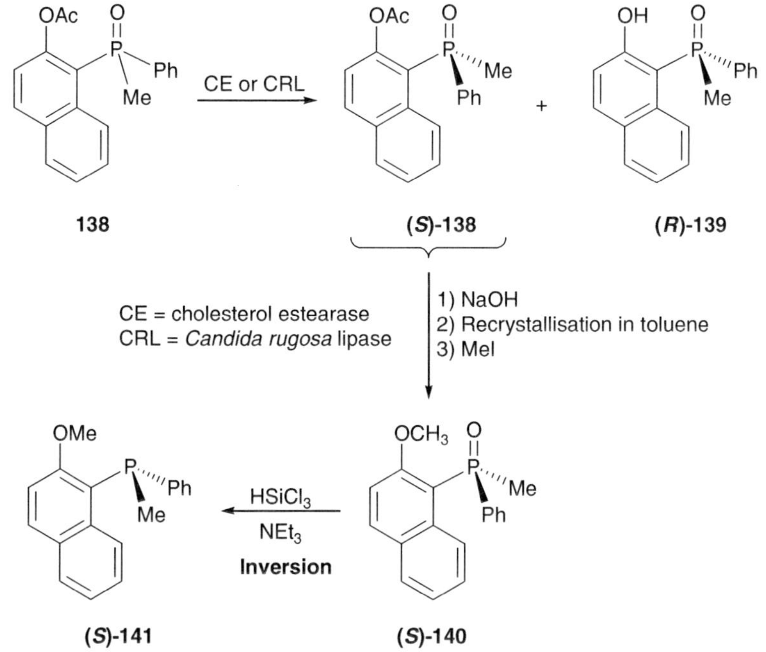

Scheme 6.54 Enzyme-catalysed kinetic resolution of a *P*-stereogenic phosphine oxide.

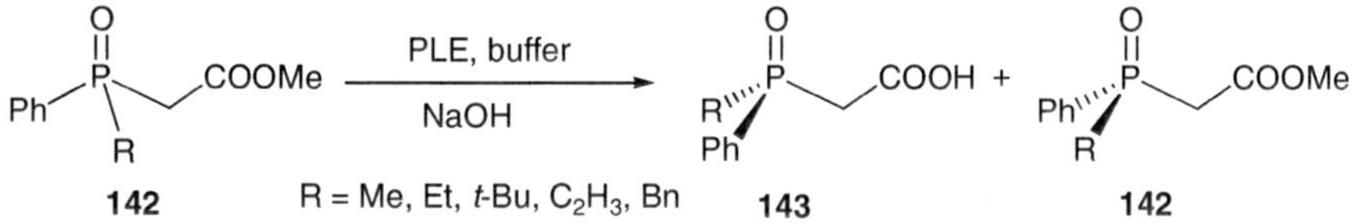

Scheme 6.55 PLE-catalysed kinetic resolution of phosphinoylacetates.

Yields of hydrolysis products **143** were in the range of 18–42% and around 80% ee. For the recovered esters **142**, yields of 36–51% and enantiomeric excesses of 72–100% were found. Interestingly, for R = Me and Bn, major enantiomers of **143** have absolute configurations *S* (and obviously, *R* for recovered **142**) but the contrary was found for R = Et and C_2H_3. The substrate with R = *t*-Bu did not react. The scope was further expanded[92] to other substrates, with up to 95% ee. Previous knowledge of the PLE active site,[93] originally developed for substrates with a stereogenic carbon, allow one to successfully predict which of the enantiomers is preferentially hydrolysed.

In other reports,[94,95] several lipases were used in the kinetic resolution of racemic *P*-stereogenic α-hydroxymethylphosphonates and phosphinates (Scheme 6.56).

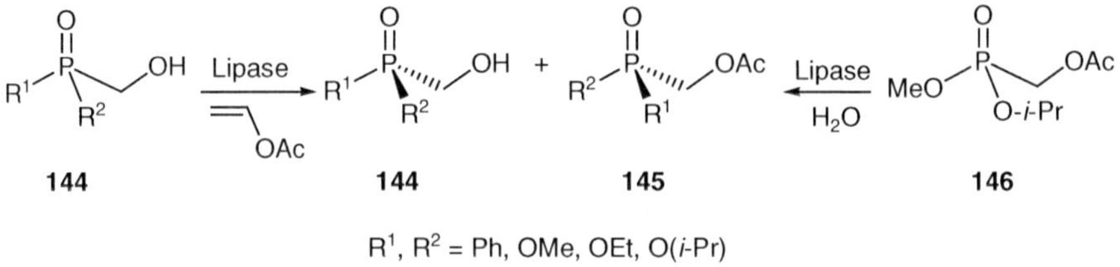

R¹, R² = Ph, OMe, OEt, O(*i*-Pr)

Scheme 6.56 Lipase-catalysed kinetic resolution of α-hydroxyphosphonates and phosphinates.

Pseudomonas fluorescens lipase and *Burkholderia cepacia* lipase were the enzymes tested. In the case of phosphonate **146**, the reverse procedure, *i.e.* the hydrolysis of the acetyl group, provides better results. Good yields (usually limited to 50% due to the kinetic conditions used) and ee values up to 92% were obtained for **144** and **145**.

The same reactions also take place in ionic liquids (BMIM · X, BIMIM = 1-butyl-3-methylimidazolium; X = PF$_6$, BF$_4$).[96] It was found that the anion X has a huge influence on the stereoselectivity of the resolution: reactions in BMIM · PF$_6$ are up to six time more enantioselective than in common organic solvents, whereas in BMIM · BF$_4$ racemic products are obtained. They tentatively attributed this lack of selectivity to the miscibility of BMIM · BF$_4$ with water. An extension of the same chemistry to supercritical carbon dioxide has also been reported by Kielbasinski and co-workers.[97] They found that *Candida antarctica* lipase was reactive for the transformation. The enantioselectivity of the reaction (up to 88% ee for unreacted alcohol **144**) was strongly dependent on the pressure and the substrate.

The same authors[98] explored the kinetic resolutions of *P*-stereogenic phosphinite boranes bearing an hydroxymehyl substituent by lipase-catalysed acetylations but the reactions proceeded slowly and with low stereoselectivity.

A report of Shioji and co-workers[99] studied the kinetic resolution of hydroxymethylphosphine boranes **147** by lipase-catalysed acylation reactions (Scheme 6.57).

With lipase AK from *Pseudomonas fluorescens*, **147** (R = *i*-Pr) could be obtained enantiomerically pure whereas the ee of the acetylated product **148** reached 84% under the best conditions. In contrast, lipase from *Candida antarctica* gave much inferior results for this substrate although high enantioselectivities were found for R = Et and *n*-Bu when vinyl butyrate was employed as acyl donor.

Hydroxymethylphosphine oxides (**149**) proved to be excellent substrates for kinetic resolution, as reported by Shioji and co-workers (Scheme 6.58).[100]

Pseudomonas fluorescens lipase (lipase AK), *Burkholderia cepacia* lipase (lipase PS) and *Candida antarctica* lipase (CAL) were employed, with enol esters such as vinyl acetate, as acyl donors. The results with AK and PS lipases are summarised in Table 6.18.

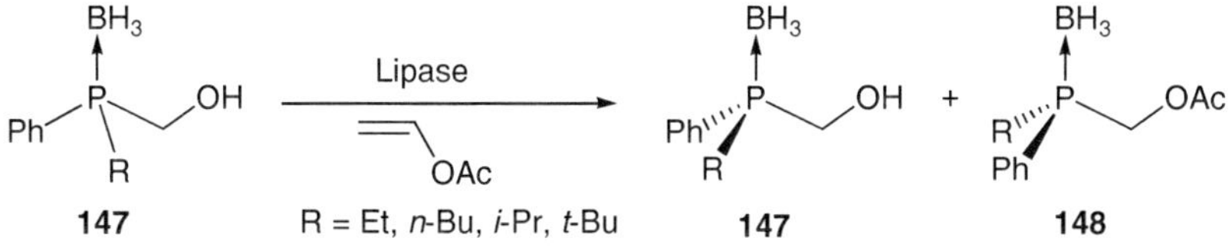

Scheme 6.57 Lipase-catalysed enantioselective acylation of hydroxymethylphosphine boranes.

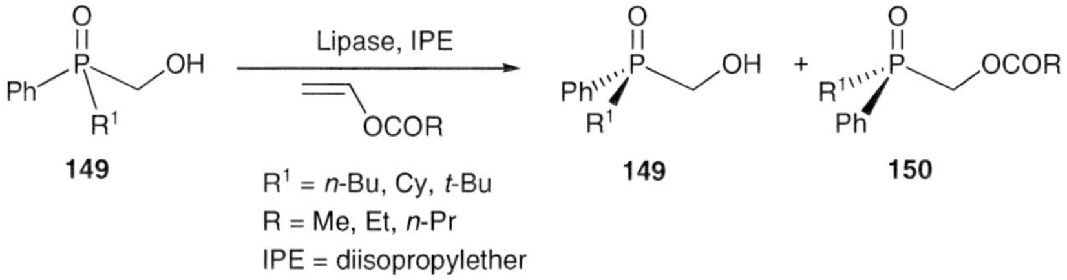

Scheme 6.58 Lipase-catalysed enantioselective acylation of hydroxymethylphosphine oxides.

Table 6.18 Results of *Pseudomonas* lipase-catalysed kinetic resolution of **149**.

Entry[a]	R^1	Lipase	R	Time (h)	Conversion (%)	ee 149 (%)	ee 150 (%)
1	n-Bu	AK	Me	0.75	50	83	28
2	n-Bu	AK	Et	1	56	>98	34
3	n-Bu	AK	n-Pr	2	53	>98	65
4	n-Bu	PS	Me	0.5	52	81	17
5	n-Bu	PS	Et	0.7	58	84	10
6	n-Bu	PS	n-Pr	1	55	76	10
7	Cy	AK	Me	0.75	52	94	50
8	Cy	AK	Et	0.75	48	85	66
9	Cy	AK	n-Pr	1.5	49	90	95
10	Cy	PS	Me	1	37	51	23
11	Cy	PS	Et	1	45	67	20
12	Cy	PS	n-Pr	1	37	51	24
13	t-Bu	AK	Me	3	42	32	4
14	t-Bu	AK	Et	4	58	69	6
15	t-Bu	AK	n-Pr	7	59	>98	27
16	t-Bu	PS	Me	3	50	23	2
17	t-Bu	PS	Et	3	47	34	3
18	t-Bu	PS	n-Pr	6	70	78	4

[a]Lipase (20 mg), **149** (0.023 mmol), acyl donor (0.165 mmol), molecular sieves 3 Å (20 mg), IPE (2 ml), 36 °C.

The enantioselectivity with lipase AK increases with the bulkiness of the acyl moiety (entries 1–3, 7–9 and 13–15) and when the acyl donor is vinyl butyrate the ee of the recovered **147** is improved (entries 3, 9 and 15). Such effects were not observed with the lipase PS-catalysed acylations. The CAL-catalysed

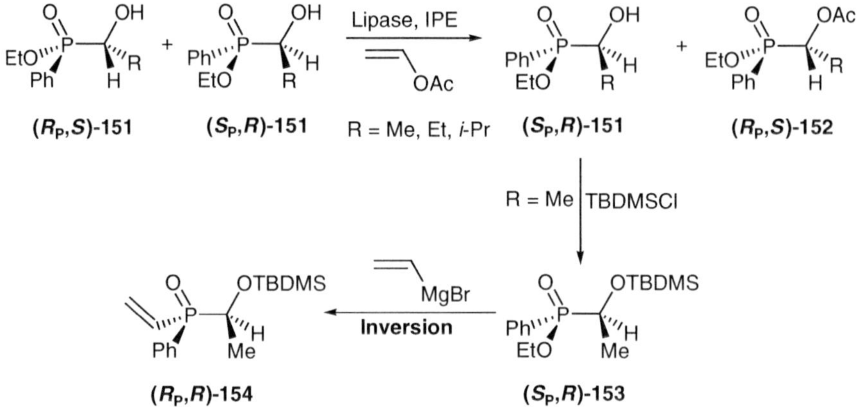

Scheme 6.59 Lipase-catalysed resolution of hydroxyphosphinates.

Table 6.19 Results of lipase-catalysed kinetic resolution of **151**.

Entry[a]	R	Lipase	Time (h)	Conversion (%)	ee 151 (%)	ee 152[b] (%)
1	Me	CAL	1	50	>98	>98
2	Et	CAL	32	17	28	>98
3[b]	i-Pr	CAL	—	—	—	—
4	Me	AK	8	50	>98	>98
5	Me	AK	48	50	>98	>98
6	Et	AK	29	50	>98	>98
7	i-Pr	AK	75	9	2	>98

[a]Lipase (20 mg), **151** (0.023 mmol), vinyl acetate (0.165 mmol), molecular sieves 3 Å (20 mg), IPE (2 ml), 36 °C.
[b]No reaction.

reactions (not listed in the table) proceed with poor stereoselectivities, not being influenced by the R group in the vinyl esters. Interestingly, it was found that lipase AK favours the *R* enantiomer of **150** ($R^1 = t$-Bu), whereas the CAL-catalysed reaction gives preferentially the other enantiomer.

Further work[101] extended this reaction to racemic mixtures of hydroxyphosphinates containing two stereogenic centres: the phosphorus and the carbon at the α position (Scheme 6.59).

CAL and AK lipase were used in these reactions and the results are listed in Table 6.19. In one case, the acetylated product **151** was converted (after protection of the alcohol with *tert*-butyldimethylsilyl chloride) to a phosphine oxide with vinylmagnesium bromide, with clean inversion at the phosphorus atom.

Perfect resolution is achieved for methyl and ethyl derivatives of **151** (entries 1 and 4–6). The *i*-Pr derivative proves more problematic and was not acetylated by CAL (entry 3). Furthermore, with AK the ester **152** can be obtained in enantiopure form but the alcohol **151** was found to be racemic (entry 7).

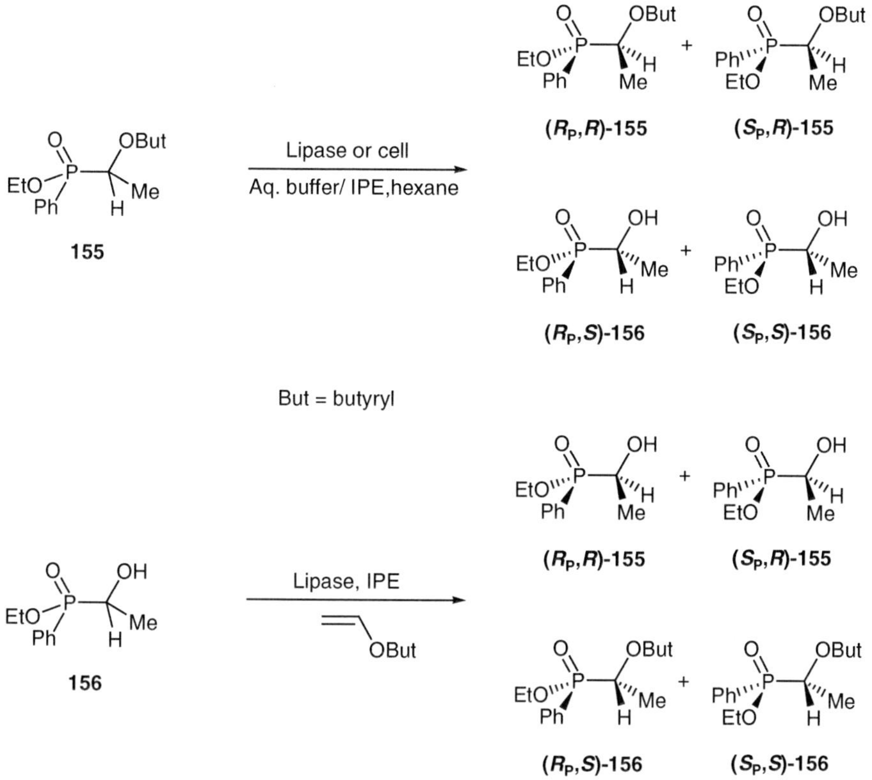

Scheme 6.60 Biocatalytic kinetic resolutions of **155** by hydrolysis and of **156** by acylation.

Majewska and co-workers[102,103] studied kinetic resolutions by esterification or hydrolysis of similar systems (Scheme 6.60).

The biocatalysts were several lipases (different from CAL or AK) and even whole cells from bacteria and fungi. In all cases, isomers bearing an α carbon with *S* as absolute configuration are preferentially hydrolysed. However, since no stereodifferentiation on the phosphinate moiety is observed, the combination of the two biotransformations provides diastereomers of hydroxyphosphinate **156** with enantiomeric excesses above 98%.

6.3.2 Desymmetrisations of Phosphine Oxides and Phosphine Boranes

In 2003 Kielbasinski, Mikolajczyk and co-workers[104] reported the first enzymatic desymmetrisations of prochiral phosphine oxides. They desymmetrised prochiral phosphine oxide **157** by PLE-catalysed hydrolysis (Scheme 6.61).

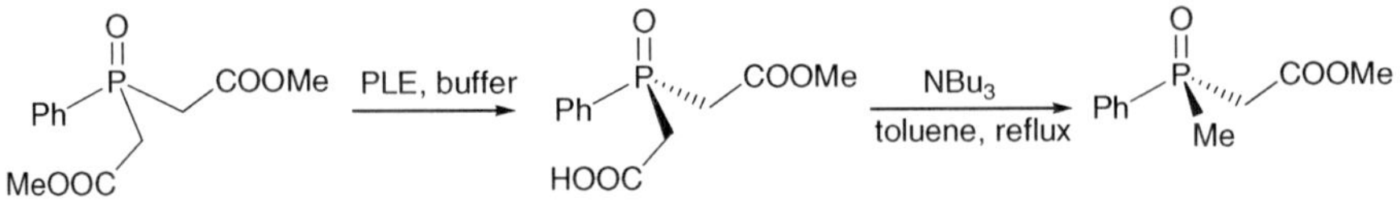

Scheme 6.61   Desymmetrisation of prochiral **157** by hydrolysis.

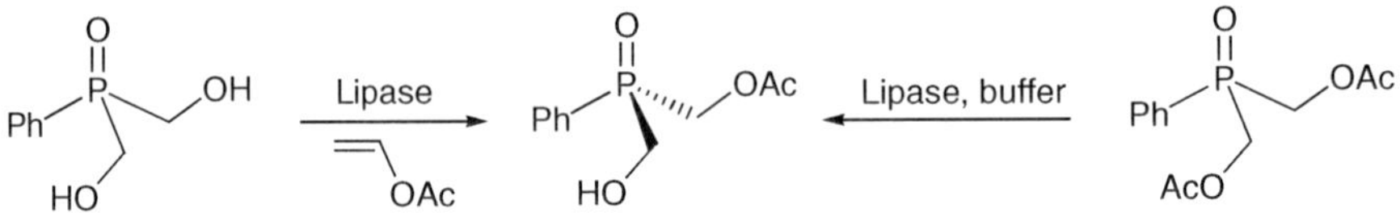

Scheme 6.62   Lipase-catalysed desymmetrisations to give **161**.

The monoacetate phosphine oxide **158** was isolated in 92% yield and 72% ee. The absolute configuration was shown to be *R* by chemical correlation after preparing the known phosphine oxide **159**. As in the previously described kinetic resolutions of phosphoryl derivatives,[91,92] the sense of the chiral induction can be explained by the Jones model of the PLE active site.[93] In the same report the enzyme-catalysed preparation of *P*-stereogenic phosphine oxide **156** from prochiral precursors was described (Scheme 6.62).

Several lipases were employed in the acetylation of **160** and hydrolysis of **162** with reasonable yields (close to the theoretical 50% or surpassing it with erosion of enantioselectivity) and enantioselectivities (up to 79%). The application of reverse procedures made the preparation of both enantiomers of **161** possible.

The same group expanded the desymmetrisation protocol to *bis*(cyanomethyl)phenylphosphine oxides (Scheme 6.63).[105]

The hydrolysis of the nitrile can lead to acetamides **164** or directly to the acids **166** depending on the enzyme. Moreover, other enzymes (amidases) can convert the amide to the acids.

A variety of nitrilases and even whole cells were used in this transformation, with the aim of obtaining different proportions of starting material and compounds **164** and **166** but not **165** or *bis*(hydrolised) products. Products **164** and **166** were obtained in up to 99 and 70% ee respectively, both with *S* absolute configuration.

Wiktelius, Johansson and co-workers[106] developed a similar route to enantioenriched monoacetylated phosphine borane **168** (Scheme 6.64).

The prochiral substrate, *bis*(hydroxymethyl)phenylphosphine borane (**167**), was chosen because it is similar in structure to 1,3-propanediols, which are

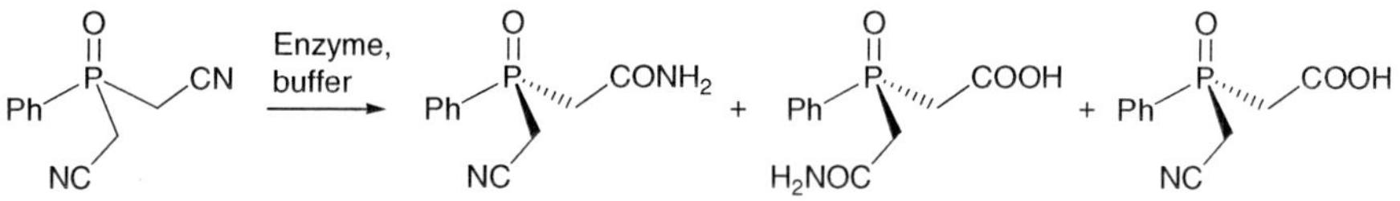

Scheme 6.63   Enzymatic desymmetrisation of **163**. Only one enantiomer of the chiral products are shown.

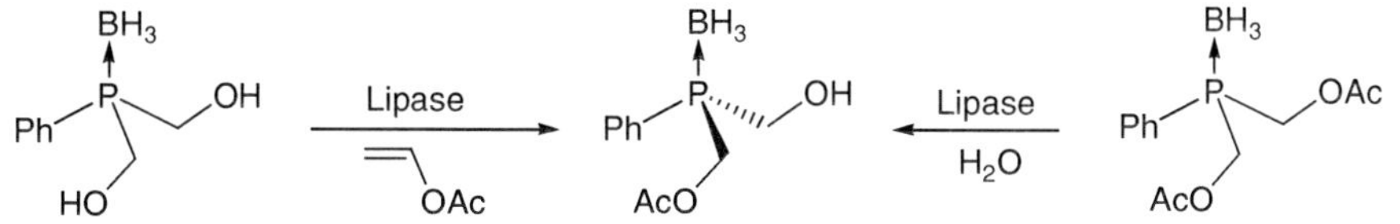

Scheme 6.64   Lipase-catalysed desymmetrisation of **167** or **169** by acetylation or hydrolysis respectively.

known to be adequate substrates for lipase-catalysed desymmetrisations.[107] After screening different lipases, *Candida antarctica* lipase B was found to give the best results. For the acetylation reaction, after 6 h, a mixture of **167**:**168**:**169** in 0:48:52 ratio was found. Compound **168** was highly enantioenriched: 92% ee for the *R* enantiomer. It was found that although the first acetylation is only moderately enantioselective (60% ee for **168** when **169** starts to form), in the second one (to afford achiral **169**) the minor *S* enantiomer of **168** reacts much faster than the *R* and therefore 'corrects' the optical purity of **168**. For the hydrolysis reaction, when the mixture consisted of **167**:**168**:**169** in 48:49:3 ratio, (*S*)-**168** was present in 90% ee. Again, the second hydrolysis improves the low stereoselectivity of the first one. Using this method, both enantiomers of **168** are obtained in high enantioselectivities but at the expense of yield. However, since **167**, **168** and **169** can be interconverted, in theory either enantiomer of **168** can be obtained with minimal loss of material.

Kielbasinski, Mikolajczyk and co-workers[98] reported the desymmetrisation of *bis*(2-hydroxyethyl)phenylphosphine borane with a similar procedure. Several lipases were screened of which only *Pseudomonas fluorescens* was efficient, giving the monoacetylated product in 43% yield and 90% ee.

6.4   Conclusions

This chapter has shown that catalytic enantioselective synthesis is a perfectly viable method for the preparation of *P*-stereogenic compounds in enantioenriched or even enantiopure form.

It is a paradox that whereas the search for *P*-stereogenic compounds is spurred by their application in enantioselective catalysis, little effort has been devoted so far to the preparation of these ligands *via* catalytic methods. However, this situation has started to change in the last decade. In particular, catalytic reactions where a chiral metal complex controls the stereochemistry of the P–C bond formation are extremely promising. Despite this, the field has not reached yet enough maturity to provide practical methods for the synthesis of a wide variety of phosphines. The good news is that modern techniques of characterisation, especially the availability of high field NMR spectrometers, easily provide a wealth of mechanistic information about catalytic reactions, and should assist the development of better processes.[1,108]

References

1. D. S. Glueck, *Chem. Eur. J.*, 2008, **14**, 7108.
2. A. L. Schwan, *Chem. Soc. Rev*, 2004, **33**, 218.
3. D. S. Glueck, *Synlett*, 2007, 2627.
4. D. S. Glueck, *Dalton Trans.*, 2008, 5276.
5. C. Genet, S. J. Canipa, D. F. O'Brien and S. Taylor, *J. Am. Chem. Soc.*, 2006, **128**, 9336.
6. J. J. Gammon, S. J. Canipa, P. O'Brien, B. Kelly and S. Taylor, *Chem. Commun.*, 2008, 3750.
7. H. Itoh, E. Yamamoto, S. Masaoka, K. Sakai and M. Tokunaga, *Adv. Synth. Catal.*, 2009, **351**, 1796.
8. W. E. Buhro, B. D. Zwick, S. Georgiou, J. P. Hutchinson and J. A. Gladysz, *J. Am. Chem. Soc.*, 1988, **110**, 2427.
9. G. T. Crisp, G. Salem and S. B. Wild, *Organometallics*, 1989, **8**, 2360.
10. D. S. Bohle, G. R. Clark, C. E. F. Rickard and W. R. Roper, *J. Organomet. Chem.*, 1990, **393**, 243.
11. M. A. Zhuravel, D. S. Glueck, L. V. Zakharov and A. L. Rheingold, *Organometallics*, 2002, **21**, 3208.
12. *Catalytic Heterofunctionalization. From Hydroamination to Hydrozirconation*, ed. A. Togni and H. Grutzmacher, Wiley-VCH, Weinheim, 2001.
13. K. Bourumeau, A. Gaumont and J. Denis, *Tetrahedron Lett.*, 1997, **38**, 1923.
14. D. Mimeau, O. Delacroix and A. Gaumont, *Chem. Commun.*, 2003, 2928.
15. D. Mimeau, O. Delacroix, B. Join and A. Gaumont, *C. R. Chimie*, 2004, 7, 845.
16. B. Join, O. Delacroix and A. Gaumont, *Synlett*, 2005, 1881.
17. *Aqueous Organometallic Chemistry and Catalysis*, ed. I. T. Horvath and F. Joo, Kluwer, Dordrecht, The Netherlands, 1995.
18. D. K. Wicht, I. V. Kourkine, I. Kovacik, D. S. Glueck, T. H. Concolino, G. P. A. Yap, C. Incarvito and A. L. Rheingold, *Organometallics*, 1999, **18**, 5381.

19. D. K. Wicht, I. V. Kourkine, B. M. Lew, J. M. Nthenge and D. S. Glueck, *J. Am. Chem. Soc.*, 1997, **119**, 5039.
20. D. K. Wicht, I. Kovacik, D. S. Glueck, L. M. Liable-Sands, C. D. Incarvito and A. L. Rheingold, *Organometallics*, 1999, **18**, 5141.
21. I. Kovacik, D. K. Wicht, N. S. Grewal, D. S. Glueck, C. D. Incarvito, I. A. Guzei and A. L. Rheingold, *Organometallics*, 2000, **19**, 950.
22. D. K. Wicht, M. A. Zhuravel, R. V. Gregush and D. S. Glueck, *Organometallics*, 1998, **17**, 1412.
23. D. K. Wicht, D. S. Glueck, L. M. Liable-Sands and A. L. Rheingold, *Organometallics*, 1999, **18**, 5130.
24. C. Scriban, I. Kovacik and D. S. Glueck, *Organometallics*, 2005, **24**, 4871.
25. C. Scriban, D. S. Glueck, L. V. Zakharov, W. S. Kassel, A. G. DiPasquale, J. A. Golen and A. L. Rheingold, *Organometallics*, 2006, **25**, 5757.
26. D. Basavaiah, A. J. Rao and T. Satyanarayana, *Chem. Rev.*, 2003, **103**, 811.
27. M. R. Douglass and T. J. Marks, *J. Am. Chem. Soc.*, 2000, **122**, 1824.
28. M. R. Douglass, C. L. Stern and T. J. Marks, *J. Am. Chem. Soc.*, 2001, **123**, 10221.
29. K. Mislow and J. Siegel, *J. Am. Chem. Soc.*, 1984, **106**, 3319.
30. M. R. Douglass, M. Ogasawara, S. Hong, M. V. Metz and T. J. Marks, *Organometallics*, 2002, **21**, 283.
31. A. Motta, I. L. Fragala and T. J. Marks, *Organometallics*, 2005, **24**, 4995.
32. L. Han, C. Zhao, S. Onozawa, M. Goto and M. Tanaka, *J. Am. Chem. Soc.*, 2002, **124**, 3842.
33. D. Mimeau and A. Gaumont, *J. Org. Chem.*, 2003, **68**, 7016.
34. B. Join, D. Mimeau, O. Delacroix and A. Gaumont, *Chem. Commun.*, 2006, 3249.
35. Y. Xu and J. Zhang, *J. Chem. Soc. Chem. Commun.*, 1986, 1606.
36. J. Zhang, Y. Xu, G. Huang and H. Guo, *Tetrahedron Lett.*, 1988, **29**, 1955.
37. Y. Xu, H. Wei, J. Zhang and G. Huang, *Tetrahedron Lett.*, 1989, **30**, 949.
38. T. Oshiki and T. Imamoto, *J. Am. Chem. Soc.*, 1992, **114**, 3975.
39. T. Imamoto, T. Oshiki, T. Onozawa, M. Matsuo, T. Hikosaka and M. Yanagawa, *Heteroatom. Chem.*, 1992, **3**, 563.
40. M. Al-Masum and T. Livinghouse, *Tetrahedron Lett.*, 1999, **40**, 7731.
41. J. R. Moncarz, N. F. Laritcheva and D. S. Glueck, *J. Am. Chem. Soc.*, 2002, **124**, 13356.
42. N. F. Blank, J. R. Moncarz, T. J. Brunker, C. Scriban, B. J. Anderson, O. Amir, D. S. Glueck, L. V. Zakharov, J. A. Golen, C. D. Incarvito and A. L. Rheingold, *J. Am. Chem. Soc.*, 2007, **129**, 6847.
43. A. Gaumont, M. B. Hursthouse, S. J. Coles and J. M. Brown, *Chem. Commun.*, 1999, 63.
44. T. J. Brunker, B. J. Anderson, N. F. Blank, D. S. Glueck and A. L. Rheingold, *Org. Lett.*, 2007, **9**, 1109.

45. M. Peer, J. C. de Jong, M. Kiefer, T. Langer, H. Rieck, H. Shell, P. Sennhenn, J. Sprinz, H. Steinhagen, B. Wiese and G. Helmchen, *Tetrahedron*, 1996, **52**, 7547.

46. C. Korff and G. Helmchen, *Chem. Commun.*, 2004, 530.

47. N. F. Blank, K. C. McBroom, D. S. Glueck, W. S. Kassel and A. L. Rheingold, *Organometallics*, 2006, **25**, 1742.

48. S. E. Tunney and J. K. Stille, *J. Org. Chem.*, 1987, **52**, 748.

49. V. S. Chan, R. G. Bergman and F. D. Toste, *J. Am. Chem. Soc.*, 2007, **129**, 15122.

50. J. R. Moncarz, T. J. Brunker, D. S. Glueck, R. D. Sommer and A. L. Rheingold, *J. Am. Chem. Soc.*, 2003, **125**, 1180.

51. J. R. Moncarz, T. J. Brunker, J. C. Jewett, M. Orchowski, D. S. Glueck, R. D. Sommer, K. Lam, C. D. Incarvito, T. H. Concolino, C. Ceccarelli, L. V. Zakharov and A. L. Rheingold, *Organometallics*, 2003, **22**, 3205.

52. S. Pican and A. Gaumont, *Chem. Commun.*, 2005, 2393.

53. C. Scriban and D. S. Glueck, *J. Am. Chem. Soc.*, 2006, **128**, 2788.

54. B. J. Anderson, D. S. Glueck, A. G. DiPasquale and A. L. Rheingold, *Organometallics*, 2008, **27**, 4992.

55. T. W. Chapp, D. S. Glueck, J. A. Golen, C. E. Moore and A. L. Rheingold, *Organometallics*, 2010, **29**, 378.

56. C. Scriban, D. S. Glueck, J. A. Golen and A. L. Rheingold, *Organometallics*, 2007, **26**, 1788.

57. B. J. Anderson, M. A. Guino-o, D. S. Glueck, J. A. Golen, A. G. DiPasquale, L. M. Liable-Sands and A. L. Rheingold, *Org. Lett.*, 2008, **10**, 4425.

58. V. S. Chan, I. C. Stewart, R. G. Bergman and F. D. Toste, *J. Am. Chem. Soc.*, 2006, **128**, 2786.

59. V. S. Chan, M. Chiu, R. G. Bergman and F. D. Toste, *J. Am. Chem. Soc.*, 2009, **131**, 6021.

60. G. Nishida, K. Noguchi, M. Hirano and K. Tanaka, *Angew. Chem. Int. Ed.*, 2008, **47**, 3410.

61. A. Kondoh, H. Yorimitsu and K. Oshima, *J. Am. Chem. Soc.*, 2007, **129**, 6996.

62. K. Tanaka, *Synlett*, 2007, **13**, 1977.

63. G. Nishida, K. Noguchi, M. Hirano and K. Tanaka, *Angew. Chem. Int. Ed.*, 2007, **46**, 3951.

64. M. D. McReynolds, J. M. Dougherty and P. R. Hanson, *Chem. Rev.*, 2004, **104**, 2239.

65. J. S. Harvey, S. J. Malcolmson, K. S. Dunne, S. J. Meek, A. L. Thompson, R. R. Schrock, A. H. Hoveyda and V. Gouverneur, *Angew. Chem. Int. Ed.*, 2009, **48**, 762.

66. O. M. Demchuk, K. M. Pietrusiewicz, A. Michrowska and K. Grela, *Org. Lett.*, 2003, **5**, 3217.

67. H. Butenschön, N. Vinokurov, I. Baumgardt and K. M. Pietrusiewicz, *Acta Cryst. Sect. E*, 2009, **E65**, o517.

68. N. Vinokurov, K. M. Pietrusiewicz, S. Frynas, M. Wiebcke and H. Butenschön, *Chem. Commun.*, 2008, 5408.
69. N. Vinokurov, A. Michrowska, A. Szmigielska, Z. Drzazga, G. Wojciuk, O. M. Demchuk, K. Grela, K. M. Pietrusiewicz and H. Butenschön, *Adv. Synth. Catal.*, 2006, **348**, 931.
70. N. Vinokurov, J. R. Garabatos-Perera, Z. Zhao-Karger, M. Wiebcke and H. Butenschön, *Organometallics*, 2008, **27**, 1878.
71. K. S. Dunne, S. E. Lee and V. Gouverneur, *J. Organomet. Chem.*, 2006, **691**, 5246.
72. F. Bisaro and V. Gouverneur, *Tetrahedron*, 2005, **61**, 2395.
73. C. A. Slinn, A. J. Redgrave, S. L. Hind, C. Edlin, S. P. Nolan and V. Gouverneur, *Org. Biomol. Chem.*, 2003, **1**, 3820.
74. M. Schuman, M. Trevitt, A. Redd and V. Gouverneur, *Angew. Chem. Int. Ed.*, 2000, **39**, 2491.
75. M. Trevitt and V. Gouverneur, *Tetrahedron Lett.*, 1999, **40**, 7333.
76. M. Bujard, V. Gouverneur and C. Mioskowski, *J. Org. Chem.*, 1999, **64**, 2119.
77. L. Hetherington, B. Greedy and V. Gouverneur, *Tetrahedron*, 2000, **56**, 2053.
78. K. S. Dunne, F. Bisaro, B. Odell, J. Paris and V. Gouverneur, *J. Org. Chem.*, 2005, **70**, 10803.
79. D. S. Stoianova and P. R. Hanson, *Org. Lett.*, 2000, **2**, 1769.
80. P. R. Hanson and D. S. Stoianova, *Tetrahedron Lett.*, 1998, **39**, 3939.
81. P. R. Hanson and D. S. Stoianova, *Tetrahedron Lett.*, 1999, **40**, 3297.
82. J. S. Harvey, G. T. Guiffredi and V. Gouverneur, *Org. Lett.*, 2010, **12**, 1236.
83. J. D. Waetzig and P. R. Hanson, *Org. Lett.*, 2006, **8**, 1673.
84. J. D. Waetzig and P. R. Hanson, *Org. Lett.*, 2008, **10**, 109.
85. H. Zhang, R. Tsukuhara, G. Tigyi and G. D. Prestwich, *J. Org. Chem.*, 2006, **71**, 6061.
86. D. S. Stoianova and P. R. Hanson, *Org. Lett.*, 2001, **3**, 3285.
87. D. S. Stoianova, A. Whitehead and P. R. Hanson, *J. Org. Chem.*, 2005, **70**, 5880.
88. R. R. Schrock and A. H. Hoveyda, *Angew. Chem. Int. Ed.*, 2003, **42**, 4592.
89. N. P. B. Dudman and B. Zerner, *J. Am. Chem. Soc.*, 1973, **95**, 3019.
90. A. N. Serreqi and R. J. Kazlauskas, *J. Org. Chem.*, 1994, **59**, 7609.
91. P. Kielbasinski, R. Zurawinski, K. M. Pietrusiewicz, M. Zablocka and M. Mikolajczyk, *Tetrahedron Lett.*, 1994, **35**, 7081.
92. P. Kielbasinski, P. Goralczyk, M. Mikolajczyk, M. W. Wieczorek and W. R. Majzner, *Tetrahedron: Asymmetry*, 1998, **9**, 2641.
93. E. J. Toone, M. J. Werth and J. B. Jones, *J. Am. Chem. Soc.*, 1990, **112**, 4946.
94. P. Kielbasinski, J. Omelanczuk and M. Mikolajczyk, *Phosphorus, Sulfur, and Silicon*, 1999, **147**, 179.
95. P. Kielbasinski, J. Omelanczuk and M. Mikolajczyk, *Tetrahedron: Asymmetry*, 1998, **9**, 3283.

96. P. Kielbasinski, M. Albrycht, J. Luczak and M. Mikolajczyk, *Tetrahedron: Asymmetry*, 2002, **13**, 735.

97. M. Albrycht, P. Kielbasinski, J. Drabowicz, M. Mikolajczyk, T. Matsuda, T. Harada and K. Nakamura, *Tetrahedron: Asymmetry*, 2005, **16**, 2015.

98. P. Kielbasinski, M. Albrycht, R. Zurawinski and M. Mikolajczyk, *J. Mol. Catal. B*, 2006, **39**, 45.

99. K. Shioji, Y. Karauchi and K. Okuma, *Bull. Chem. Soc. Jpn.*, 2003, **76**, 833.

100. K. Shioji, Y. Ueno, Y. Kurauchi and K. Okuma, *Tetrahedron Lett.*, 2001, **42**, 6569.

101. K. Shioji, A. Tashiro, S. Shibata and K. Okuma, *Tetrahedron Lett.*, 2003, **44**, 1103.

102. P. Majewska, P. Kafarski, B. Lejczak, I. Bryndal and T. Lis, *Tetrahedron: Asymmetry*, 2006, **17**, 2697.

103. P. Majewska, P. Kafarski and B. Lejczak, *Tetrahedron: Asymmetry*, 2006, **17**, 2870.

104. P. Kielbasinski, R. Zurawinski, M. Albrycht and M. Mikolajczyk, *Tetrahedron: Asymmetry*, 2003, **14**, 3379.

105. P. Kielbasinski, M. Rachwalski, M. Kwiatkowska, M. Mikolajczyk, W. M. Wieczorek, M. Szyrej, L. Sieron and F. P. J. T. Rutjes, *Tetrahedron: Asymmetry*, 2007, **18**, 2108.

106. D. Wiktelius, M. J. Johansson, K. Luthman and N. Kann, *Org. Lett.*, 2005, **7**, 4991.

107. E. Garcia-Urdiales, I. Alfonso and V. Gotor, *Chem. Rev.*, 2005, **105**, 313.

108. D. S. Glueck, *Coord. Chem. Rev.*, 2008, **252**, 2171.

CHAPTER 7

Hydrogenation and Related Reactions

7.1 Introduction

Heterogeneous catalytic hydrogenation is performed on an enormous scale in industry. In spite of that, the importance of homogeneously catalysed hydrogenation is by no means any lower.[1] In fact, transition metal-catalysed hydrogenation of unsaturated substrates is by far the most intensively studied transformation in homogeneous catalysis. Furthermore, it is no exaggeration to say that most of the mechanistic knowledge gained over the years in homogeneous catalysis can be ultimately traced back to the hydrogenation studies of functionalised alkenes by the soluble Wilkinson catalyst, $[RhCl(PPh_3)_3]$.

Asymmetric hydrogenation[2] constitutes the most important reaction in enantioselective homogeneous catalysis and is responsible for the development of the whole field. As discussed in Chapter 1, the earliest studies in this area were performed with chiral versions of the Wilkinson catalyst prepared from *P*-stereogenic ligands, quickly leading to the outstanding ligand DiPAMP. After that, interest in ligands with stereogenic phosphorus atoms rapidly faded due to the discovery that other chiral motifs were more stable, easier to synthesise and led to equally impressive results.

However, as it is often stated in catalysis papers and textbooks, there is no such a thing as a 'universal' catalyst so the number of chiral ligands continues to grow. This search rediscovered, around 20 years after DiPAMP, that *P*-stereogenic phosphines and derivatives are indeed a superb but under-developed class of ligands for homogeneous hydrogenation.

This chapter describes the results on hydrogenation of several families of substrates including alkenes, ketones and imines with Rh, Ru, Ir and Pd complexes bearing *P*-stereogenic ligands. In addition, a section with a brief

RSC Catalysis Series No. 7
P-Stereogenic Ligands in Enantioselective Catalysis
By Arnald Grabulosa
© Arnald Grabulosa 2011
Published by the Royal Society of Chemistry, www.rsc.org

discussion on the mechanism of hydrogenation, focused on the contribution of *P*-stereogenic ligands to unravel it, is also presented.

The chapter ends with a description of catalytic hydrosilylation and hydrogen transfer, which are reductions of ketones groups to the parent alcohols but avoiding the use of dihydrogen.

7.2 Rh-Catalysed Hydrogenation

Since the discovery of the Wilkinson catalyst, most of the work on hydrogenation has been carried out with functionalised alkenes as substrates and Rh(I) complexes as catalytic precursors. These hydrogenations are discussed in the next sections. There are also a few results on hydrogenation of ketones and imines, described in Section 7.2.3.

7.2.1 Hydrogenation of Alkenes

Due to the mechanism of the reaction (Section 7.2.2), the most successful substrates in alkene hydrogenation are functionalised alkenes, which are able to chelate to the rhodium atom during the catalytic cycle through the alkene and an additional donor atom, usually a carbonyl oxygen of an amide or an ester. Although many different alkenes have been employed, the most important substrates can be classified into several families (Figure 7.1).

Figure 7.1 also lists some of the most typically used substrates of each family and their abbreviations.[3,4] Since the early times of hydrogenation, the most important kind of substrates have been α-dehydroamino acids (**MAC** in particular) but nowadays other types of substrates, often more challenging, are also regularly found in the literature. The next sections give an overview of the performance of *P*-stereogenic ligands in the hydrogenation of each type of substrate.

7.2.2.1 α-Dehydroamino Acid Derivatives. Rhodium-catalysed hydrogenation of α-dehydroamino acid derivatives (**1**) has been routinely used to test the efficiency of new chiral ligands in homogeneous catalysis for decades. In addition, it is one of the principal routes to obtain optically pure α-amino acids **2** (Scheme 7.1).

The most commonly employed substrates are (*Z*)-α-acetamidocinnamic acid (**ACA**), its methyl ester (**MAC**), acetamidoacrylic acid (**AAA**) and its methyl ester (**MAA**). Quantitative yields at mild hydrogen pressure (1–3 atm) are usually obtained with these substrates.[2] Values of ee very close to 100% have been reached for dozens ligands, including many *P*-stereogenic phosphines. The most successful examples of the hydrogenation of α-dehydroamino acid derivatives are listed in Table 7.1.

Results on hydrogenation of **AAA** are collected in entries 1–6. Although DiPAMP is a reasonably stereoselective ligand for this substrate (entry 1) the structure of the phosphine has been optimised leading to significantly improved

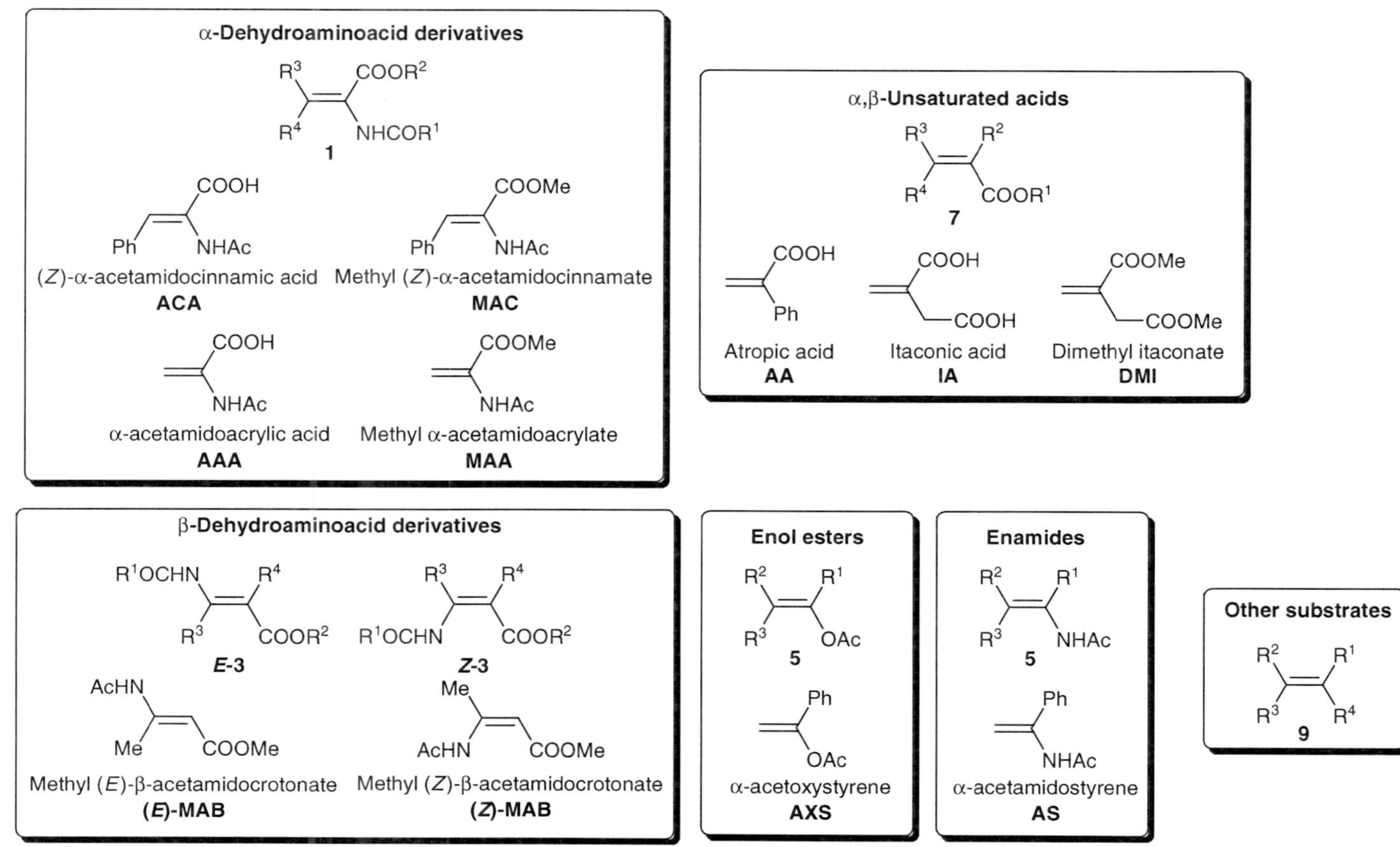

Figure 7.1 Alkene substrates employed in Rh-catalysed hydrogenation.

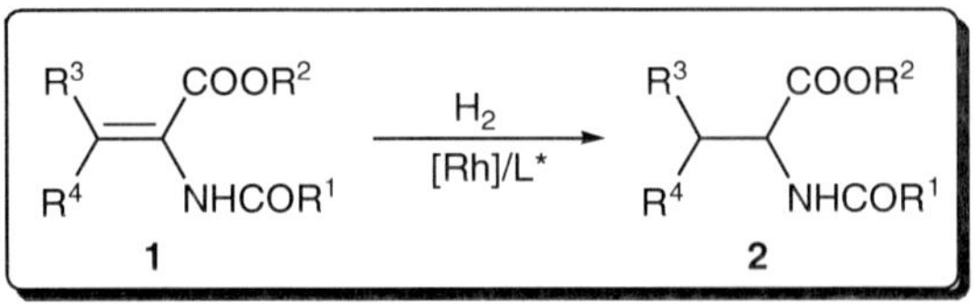

Scheme 7.1   Rh-catalysed hydrogenation of α-dehydroamino acid derivatives.

ligands (entry 2). The best ligands to date are electron-rich phosphines with a methylene bridge between the phosphorus atoms (entries 3 and 4) and the *bis*(phospholane) ligand DuanPhos (entry 6). With some of these ligands almost perfect enantioselectivities have been reached.

Entries 7–22 show that many more results are available on the hydrogenation of **MAA**. With this substrate, excellent results have been obtained with monophosphorus ligands (entries 7 and 8). DiPAMP (entry 9) and some of its modified versions (entries 10 and 11) are also very good ligands for this system. Electron-richer phosphines BisP* (entry 12), MiniPHOS (entry 13) Trichickenfootphos (entry 14) and the diphosphine of entry 15 have lead to outstanding stereoselectivities. Other phosphines bearing phosphetane (entry 16) and phospholane (entries 17–19) moieties are also exceptional stereoselectors. Some C_1 diphosphines (entries 20–22) have also been employed with very good results.

Similar trends have been observed in the hydrogenation of **ACA** (entries 23–36). With this substrate the best enantioselectivities have been obtained with modified DiPAMP ligands (entry 24), Trichickenfootphos (entry 28), TangPhos (entry 30), BIPNOR (entry 34) and MeO-POP (entry 36).

Methyl (*Z*)-α-acetamidocinnamate (**MAC**, entries 37–66) is the most typically used hydrogenation substrate. Table 7.1 contains only the most successful results, but many other *P*-stereogenic ligands have been reported to be active in the hydrogenation of this substrate. All the families of *P*-stereogenic ligands have examples leading to almost flawless enantioselectivities with this substrate: DiPAMP and its modified versions (entries 38–46), more rigid diphosphines (entries 47–49), ferrocenylphosphines (entries 50 and 51) and heterocyclic diphosphines (entries 52–60). Good results have also been obtained with C_1 diphosphines (entries 61–66) and even with the simple oxazaphospholidine of entry 37.

β,β-disubstituted α-dehydroamino acids (entries 67–75) are more challenging substrates because they react sluggishly, require higher hydrogen pressures[16] and often lead to products with lower stereoselectivity,[20] as shown in entry 67 with DiPAMP. In spite of that, very high enantioselectivities have been obtained with a modified DiPAMP ligand (entry 68), with unsymmetrical BisP* phosphines (entries 70–72) and with Trichickenfootphos (entry 74).

Many other α-dehydroamino acids have been successfully hydrogenated with rhodium catalysts containing *P*-stereogenic ligands. A small selection is given in entries 76–82. Entry 78 shows once more that differences in the

Table 7.1 Results of Rh-catalysed hydrogenation of α-dehydroamino acid derivatives (Scheme 7.1).

Entry	R^1	R^2	R^3	R^4	Ligand	ee (%)	References
1	Me	H	H	H	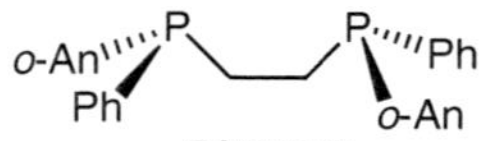 **DiPAMP**	90 (R)	5
2	Me	H	H	H	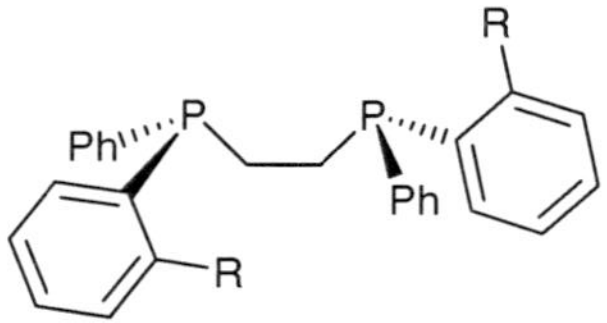	90 (R, Me) 93 (R, Et) 96 (R, i-Pr) 90 (R, OMe) 98.6 (S, O-i-Pr)[a]	3, 5
3	Me	H	H	H	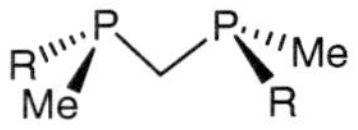 **R-MiniPHOS**	>99.9 (R, t-Bu) 99.1 (R, Cy) 98 (R, i-Pr) 26 (R, Ph)	6, 7
4	Me	H	H	H	 **Trichickenfootphos**	>99 (R)	8, 9
5	Me	H	H	H	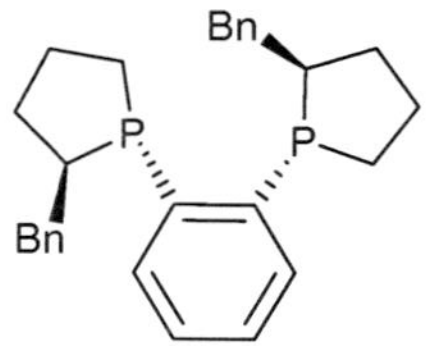	97 (S)	10

Table 7.1 (*Continued*).

Entry	R^1	R^2	R^3	R^4	Ligand	ee (%)	References
6	Me	H	H	H	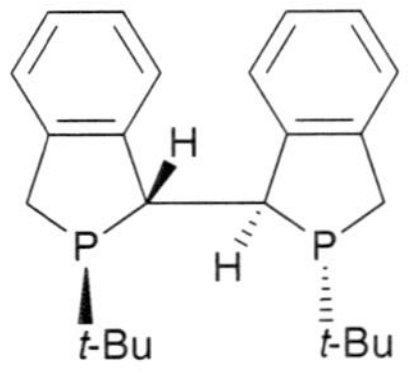 DuanPhos	>99 (*R*)	11
7	Me	Me	H	H	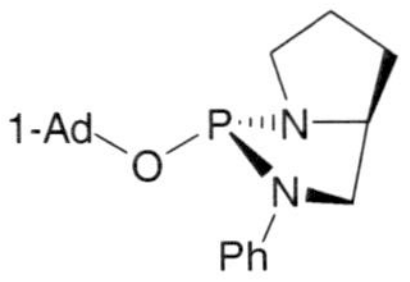	98 (*R*)	12
8	Me	Me	H	H	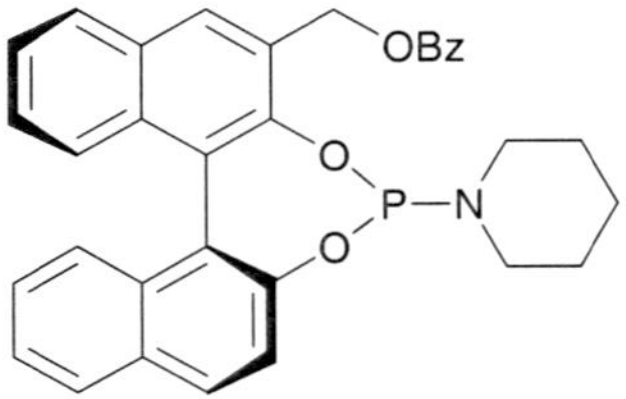	>99 (*R*)	13
9	Me	Me	H	H	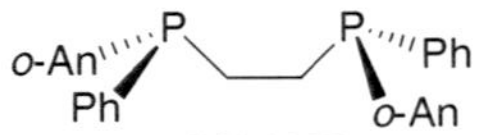 DiPAMP	95 (*S*)	14, 15

					Ligand	ee (%)	Ref.
10	Me	Me	H	H	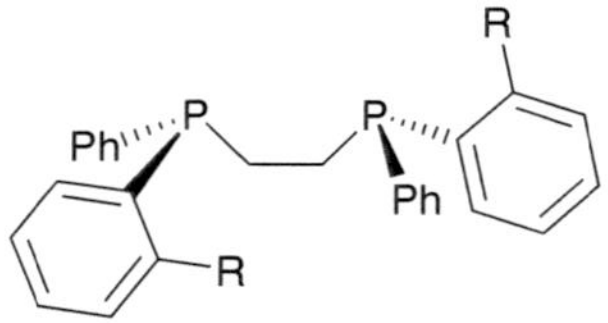	84 (R, Me) 91 (R, Et) 78 (R, i-Pr) 95 (R, OMe) 99.4 (S, O-i-Pr)[a] 99.4 (S, O-i-Bu)[a] 99.9 (S, O-t-Bu)[a]	3–5, 15
11	Me	Me	H	H	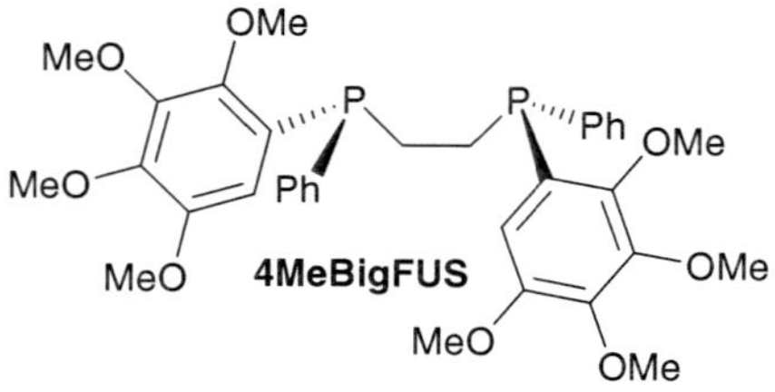 **4MeBigFUS**	99 (S)	15
12	Me	Me	H	H	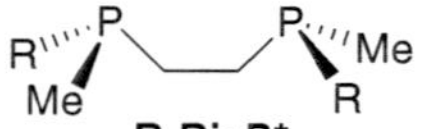 **R-BisP***	98.1 (R, t-Bu) >99.9 (R, 1-Ad) 94.5 (R, t-C_8H_{17})[b] 45 (R, Fc)	16–20
13	Me	Me	H	H	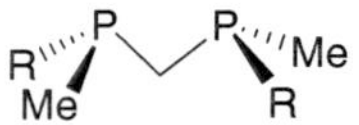 **R-MiniPHOS**	>99.9 (R, t-Bu) 98.9 (R, Cy) 98 (R, i-Pr)	6
14	Me	Me	H	H	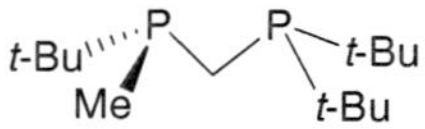 **Trichickenfootphos**	>99 (R)	8, 9
15	Me	Me	H	H	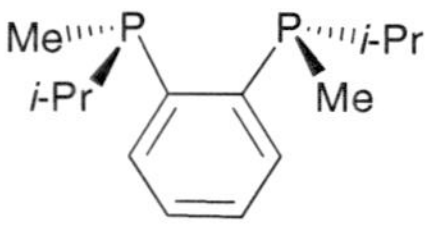	97 (S)	21

Table 7.1 (*Continued*).

Entry	R^1	R^2	R^3	R^4	Ligand	ee (%)	References
16	Me	Me	H	H	**DiSquareP***	99 (*R*)	22
17	Me	Me	H	H		95 (*R*)	10, 23
18	Me	Me	H	H		98 (*S*)	10
19	Me	Me	H	H	**BIBOP**	99 (*R*)	24

20	Me	Me	H	H	**MeO-POP**	98 (S)	25
21	Me	Me	H	H		97	26
22	Me	Me	H	H		97.5 (R)	27
23	Me	H	H	Ph	**DiPAMP**	94	28, 29
24	Me	H	H	Ph		89 (R, Me) 90 (R, Et) 92 (R, i-Pr) 96 (R, OMe) 99.6 (S, O-i-Pr)[a]	3, 5

Table 7.1 (*Continued*).

Entry	R^1	R^2	R^3	R^4	Ligand	ee (%)	References
25	Me	H	H	Ph	 R-BisP*	98 (R, t-Bu) 98.6 (R, 1-Ad) 93.5 (R, t-C$_8$H$_{17}$)^b	7, 16, 17
26	Me	H	H	Ph	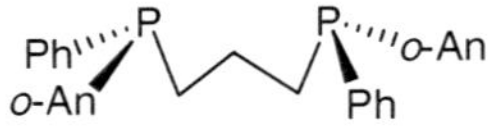	94 (R)	30
27	Me	H	H	Ph	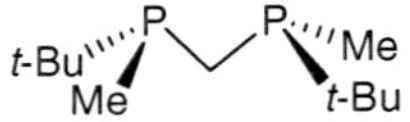 t-Bu-MiniPHOS	98 (R)	6, 7
28	Me	H	H	Ph	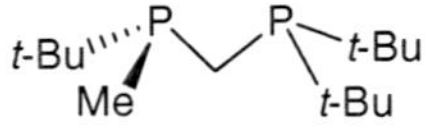 Trichickenfootphos	>99 (R)	8, 9
29	Me	H	H	Ph	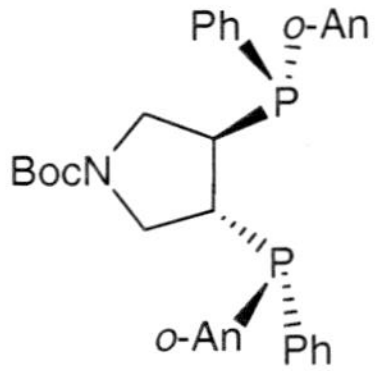	99 (S)	31
30	Me	H	H	Ph	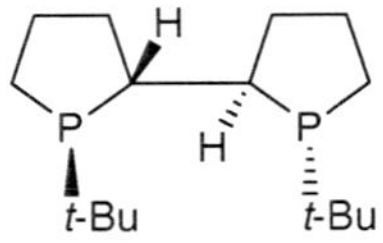 TangPhos	>99	32

				Ligand	ee (%)	Ref.
31	Me	H	H	Ph	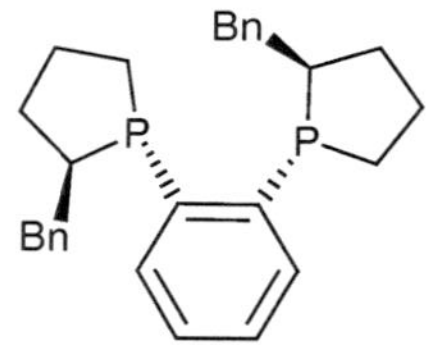	96 (*S*)	10
32	Me	H	H	Ph	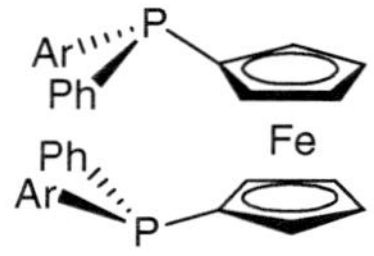 Ar = 9-Phen	98.5 (*S*)	33
33	Me	H	H	Ph	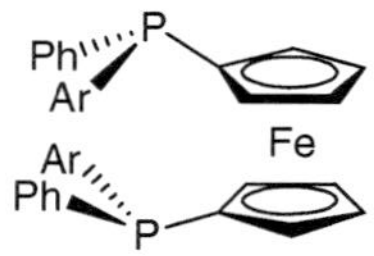 Ar = *o*-An	88 (*R*)	34
34	Me	H	H	Ph	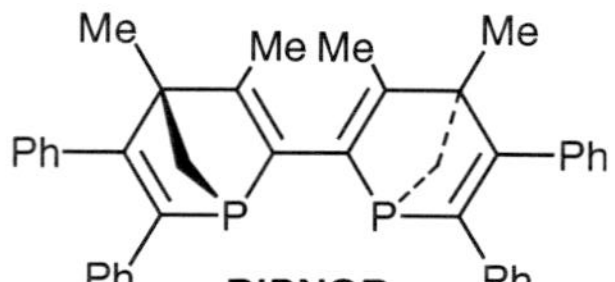 **BIPNOR**	>98 (*S*)	35
35	Me	H	H	Ph	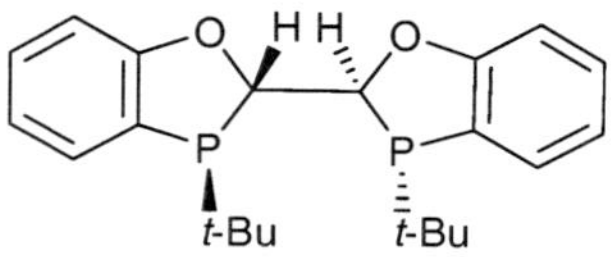 **BIBOP**	98 (*R*)	24

Table 7.1 (*Continued*).

Entry	R^1	R^2	R^3	R^4	Ligand	ee (%)	References
36	Me	H	H	Ph	MeO-POP	99 (S)	25
37	Me	Me	H	Ph		95.2 (S)	36
38	Me	Me	H	Ph	DiPAMP	96 (S)	15, 29
39	Me	Me	H	Ph		95 (R, Me) 97 (R, Et) >99 (R, i-Pr) 97 (R, OMe) 99.7 (S, O-i-Pr)[a] 99.7 (S, O-i-Bu)[a] 99.8 (S, O-t-Bu)[a]	3–5, 15, 37, 38

Entry					Ligand	ee (%)	Ref
40	Me	Me	H	Ph	4MeBigFUS	99 (S)	15
41	Me	Me	H	Ph		98.6 (S)	39
42	Me	Me	H	Ph	R-BisP*	99.9 (R, t-Bu) 99.9 (R, 1-Ad)	7, 16, 18, 20
43	Me	Me	H	Ph		99.6 (R)	40
44	Me	Me	H	Ph	R-MiniPHOS	>99 (R, t-Bu) 99 (R, Cy)	6, 7
45	Me	Me	H	Ph	Trichickenfootphos	>99 (R)	8, 9

Table 7.1 (*Continued*).

Entry	R^1	R^2	R^3	R^4	Ligand	ee (%)	References
46	Me	Me	H	Ph	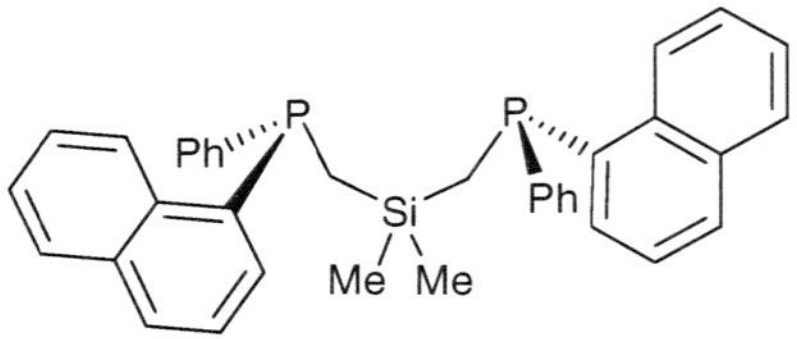	97.7 (S)	41
47	Me	Me	H	Ph	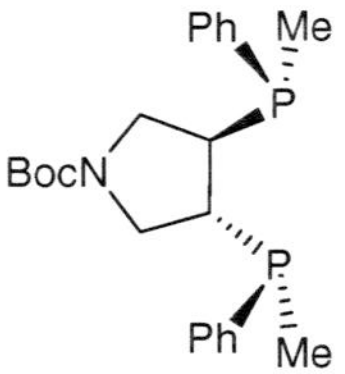	91	42
48	Me	Me	H	Ph	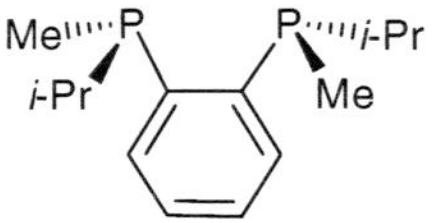	97 (S)	21
49	Me	Me	H	Ph	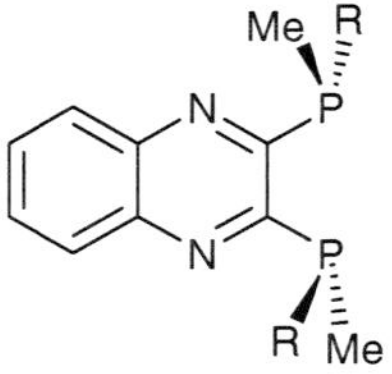 R-QuinoxP*	99.9 (R, t-Bu) 99.9 (R, 1-Ad)	43, 44

Entry				Ligand	% ee	Ref
50	Me	Me	H	Ph	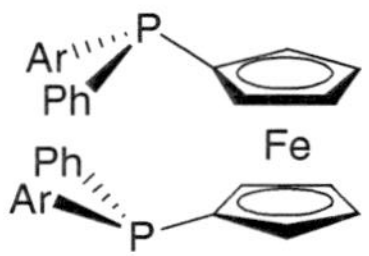 (ferrocene, Ar = 9-Phen)	98.7 (S)	33
51	Me	Me	H	Ph	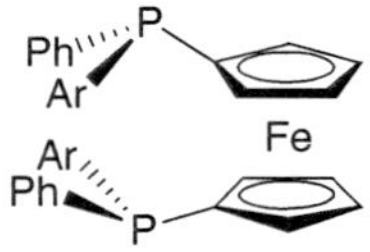 (ferrocene, Ar = o-An)	96 (R)	34
52	Me	Me	H	Ph	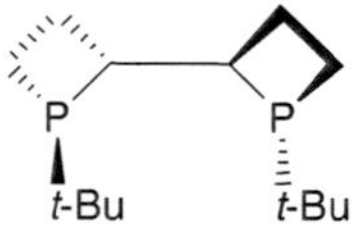 DiSquareP*	>99	22
53	Me	Me	H	Ph	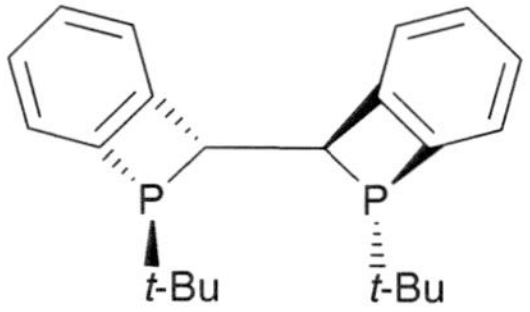 (bisphospholane, t-Bu)	96	45
54	Me	Me	H	Ph	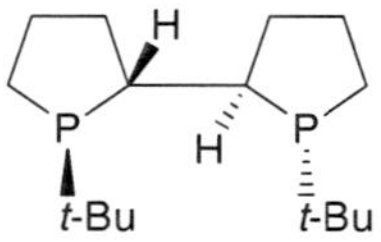 TangPhos	99	32

Table 7.1 (*Continued*).

Entry	R^1	R^2	R^3	R^4	Ligand	ee (%)	References
55	Me	Me	H	Ph	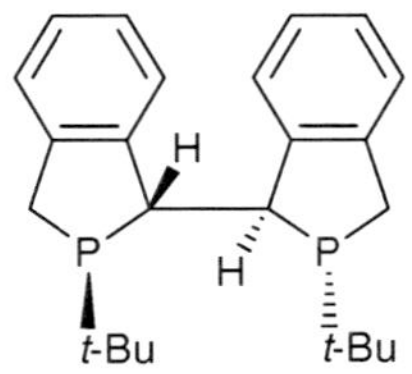 **DuanPhos**	>99 (*R*)	11
56	Me	Me	H	Ph	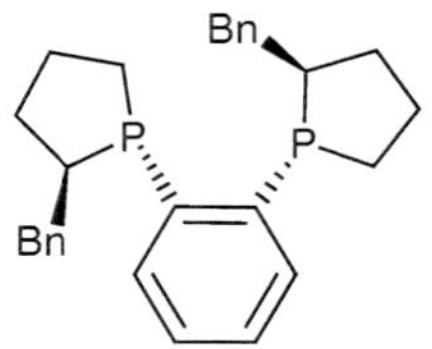	95 (*S*)	10
57	Me	Me	H	Ph	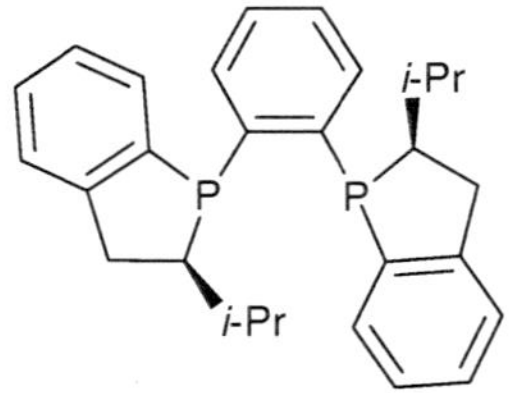 *i*-Pr-BeePHOS	98 (*R*)	46
58	Me	Me	H	Ar	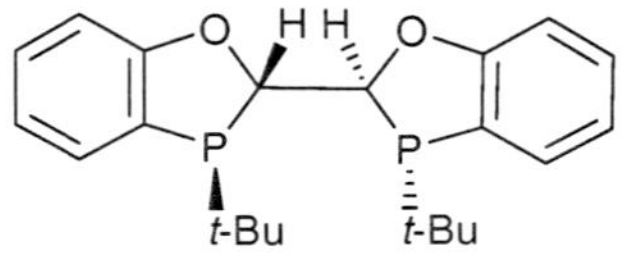 **BIBOP**	97 (*R*, Ph) 96 (*R*, *p*-An)	24

Entry					Ligand	ee (%)	Ref.
59	*t*-Bu	Me	H	Ph	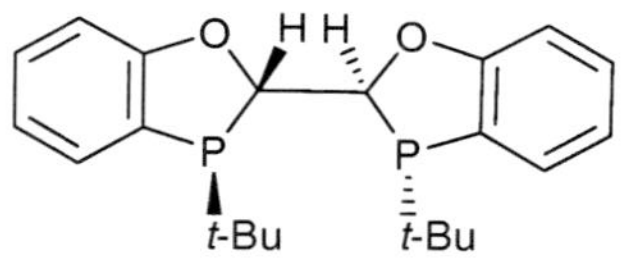 **BIBOP**	99 (*R*)	24
60	Me	Me	H	Ph	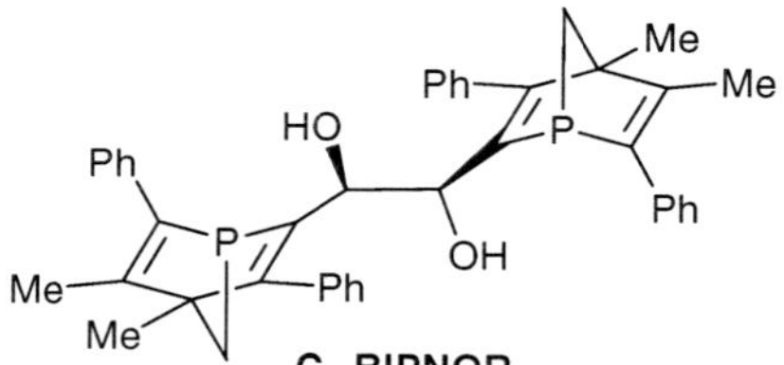 **C₂-BIPNOR**	>98 (*R*)	47
61	Me	Me	H	Ph	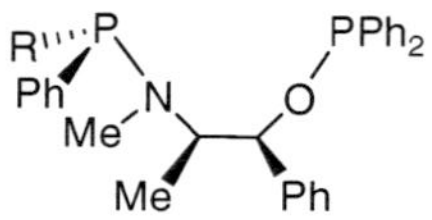	46 (*S*, Ph)[c] / 99 (*S*, *o*-An)	48, 49
62	Me	Me	H	Ph	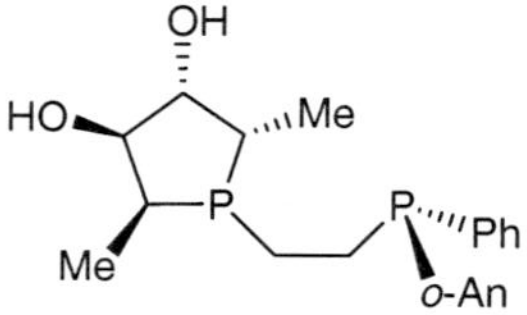	92 (*S*)	50
63	Me	Me	H	Ph	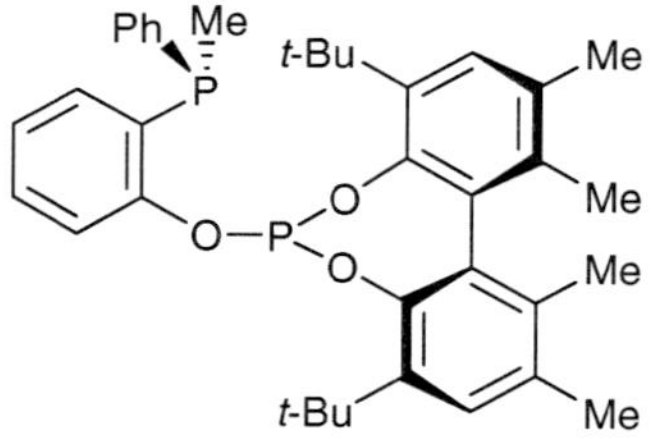	99 (*S*)	51

Table 7.1 (*Continued*).

Entry	R^1	R^2	R^3	R^4	Ligand	ee (%)	References
64	Me	Me	H	Ph	PingFer	99.6 (*S*)	52
65	Me	Me	H	Ph		>99	26
66	Me	Me	H	Ph	MeO-POP	99 (*S*)	25
67	Me	Me	Me	Me	DiPAMP	55 (*S*)	14
68	Me	Me	Me	Me		89 (*R*, *i*-Pr) 98.7 (*S*, O-*i*-Pr)[a]	3, 5

Entry					Ligand	ee (%) (config)	Ref.
69	Me	Me	Me	Me	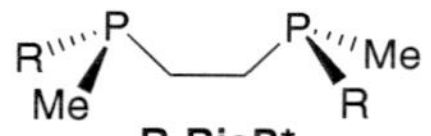 R-BisP*	36 (R, t-Bu) / 66 (R, Fc) / 91 (R, Cy)	7, 16, 18, 19
70	Me	Me	Me	Me	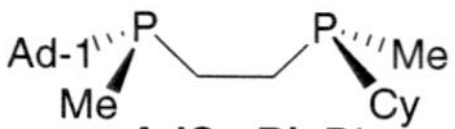 AdCy-BisP*	94 (R)	18, 20
71	Me	Me	Me	Me	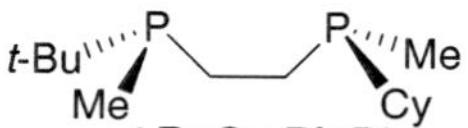 t-BuCy-BisP*	94 (R)	18, 20
72	Me	Me	Me	Me	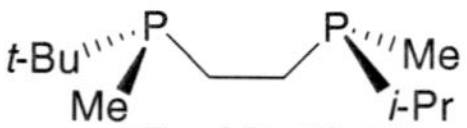 t-Bu-i-Pr-BisP*	95.8 (R)	20
73	Me	Me	Me	Me	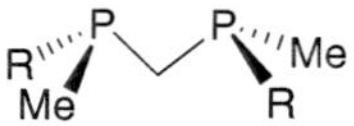 R-MiniPHOS	87 (R, t-Bu) / 85 (R, Cy)	6, 7
74	Me	Me	Me	Me	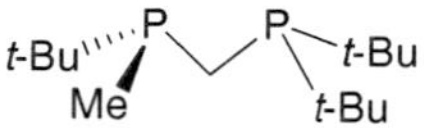 Trichickenfootphos	98 (R)	9
75	Me	Me	Me	Me	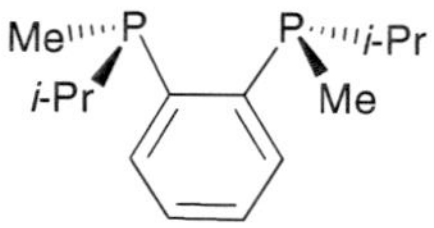	87 (S)	21

Table 7.1 (*Continued*).

Entry	R^1	R^2	R^3	R^4	Ligand	ee (%)	References
76	Ph	H	H	Ph	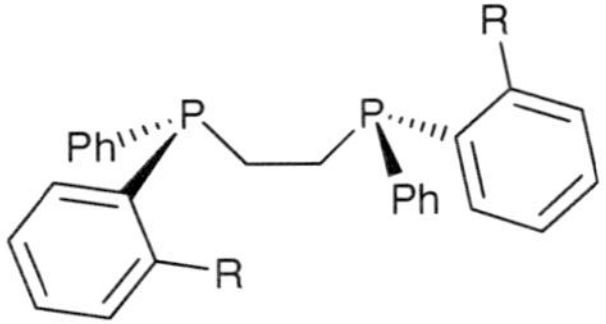	91 (R, Me) 86 (R, Et) 93 (R, i-Pr) 93 (R, OMe)	5
77	Ph	H	H	Ph	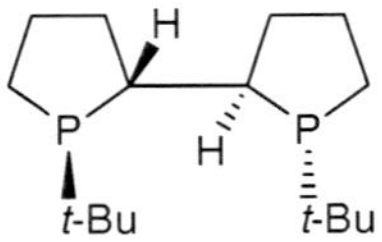 **TangPhos**	>99	32
78	Me	Me	Ph	H	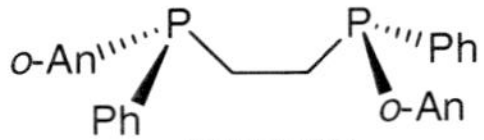 **DiPAMP**	23 (S)	29
79	Me	Me	H	n-Pr	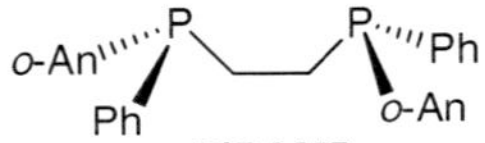 **DiPAMP**	96 (S)	14

80	Me	Me	*n*-Pr	H	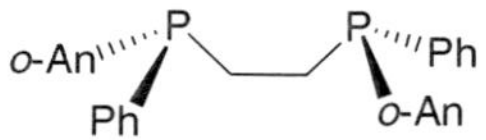 DiPAMP	95 (*S*)	14
81	Me	Me	H	Ph	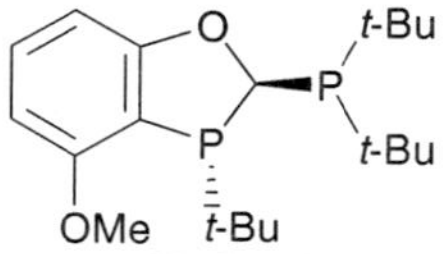 MeO-POP	98 (*S*)	25
82	Ph	Me	H	Ph	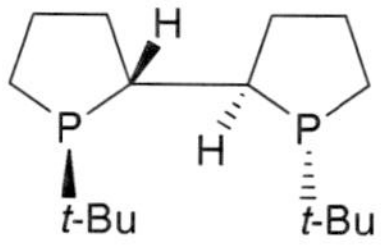 TangPhos	>99	32

[a] The (*R,R*)-isomer of phosphine was employed.
[b] *t*-C$_8$H$_{17}$ = *tert*-octyl = 1,1,3,3-tetramethylbutyl.
[c] The phosphorus atoms are not stereogenic.

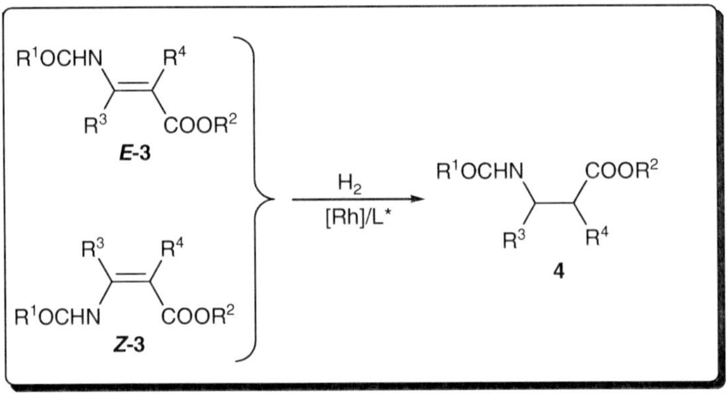

Scheme 7.2   Rh-catalysed hydrogenation of β-dehydroamino acid derivatives.

substrate can lead to dramatic differences in the outcome of the reaction. Whereas **MAC** (*Z* configuration) is efficiently hydrogenated with Rh/DiPAMP (entry 38), the *E* isomer is a very poor substrate under the same conditions (entry 78). In contrast, entries 79 and 80 show that with the *n*-propyl-substituted acrylate no differences in the enantioselectivity are observed employing the same ligand.

7.2.1.2   β-Dehydroamino Acid Derivatives.

The hydrogenation of β-dehydroamino acid derivatives is very interesting since the obtained β-amino acids (**4**, Scheme 7.2) are key precursors of many physiologically active compounds[2,53–55] including β-lactams, the most important class of antibiotics.[55]

The most frequently applied substrates are (*Z*)- and (*E*)-β-acetamidocrotonates (**MAB**). The best results obtained with those substrates and some derivatives using Rh(I)/*P*-stereogenic phosphines are listed in Table 7.2.

The results with ***E*-MAB** are listed in entries 1–10. With this substrate excellent enantioselectivities have been obtained with several C_2-symmetric diphosphines including bulkier versions of DiPAMP (entry 1), *t*-Bu-BisP* (entry 2), methylene-bridged diphosphines (entries 3 and 4), *t*-Bu-QuinoxP* (entry 5) and several phosphines based on the phospholane backbone (entries 6–8 and 10). Other related substrates (entries 11–16) have also been efficiently hydrogenated with similar ligands. Hydrogenation of ***Z*-MAB** (entries 17–24) and other *Z*-substituted β-dehydroamino acids (entries 25–29) is more challenging because lower enantioselectivities are frequently obtained.[55,57,58] This fact is clearly demonstrated by comparison of entry 1 with 18 and entry 7 with 23. In spite of that, exceedingly high enantioselectivities have been found for both substrates with some *P*-stereogenic ligands. For example, Trichickenfootphos, BIBOP and *t*-Bu-QuinoxP* give equally impressive results for both geometrical isomers of **MAB** (compare entries 4 with 20, 5 with 21 and 10 with 24 respectively). Interestingly, BINAPINE gives outstanding results with *Z*-configured substrates including ***Z*-MAB** (entry 27) whereas ***E*-MAB** is hydrogenated with

Table 7.2 Results of Rh-catalysed hydrogenation of β-dehydroamino acid derivatives (Scheme 7.2).

Entry	E/Z	R^1	R^2	R^3	R^4	Ligand	ee (%)	References
1	E	Me	Me	Me	H		93.0 (S, O-i-Pr) 93.4 (S, O-i-Bu) 97.3 (S, O-t-Bu)	3, 4
2	E	Me	Me	Me	H	t-Bu-BisP*	98.7 (R)	56
3	E	Me	Me	Me	H	t-Bu-MiniPHOS	96.4 (R)	56
4	E	Me	Me	Me	H	Trichickenfootphos	99 (R)	9, 57
5	E	Me	Me	Me	H	t-Bu-QuinoxP*	99.7 (R)	43

Table 7.2 (*Continued*).

Entry	E/Z	R^1	R^2	R^3	R^4	Ligand	ee (%)	References
6	E	Me	Me	Me	H	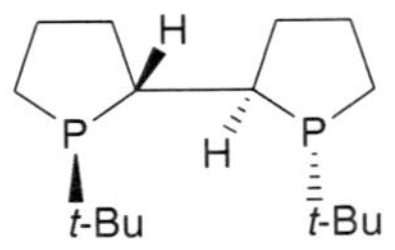 TangPhos	99.6 (R)	58
7	E	Me	Me	Me	H	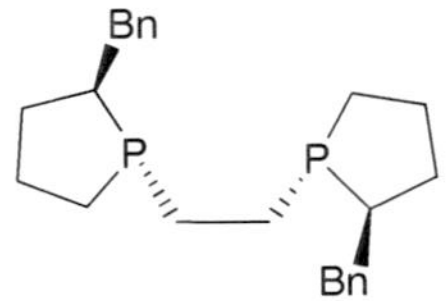	96 (R)	10
8	E	Me	Me	Me	H	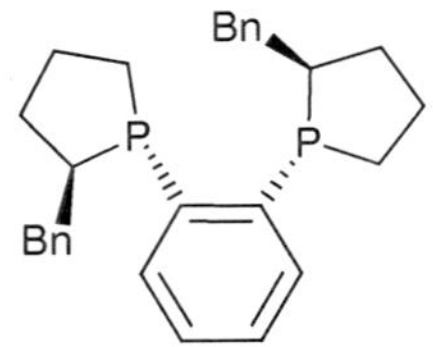	96 (S)	10
9	E	Me	Me	Me	H	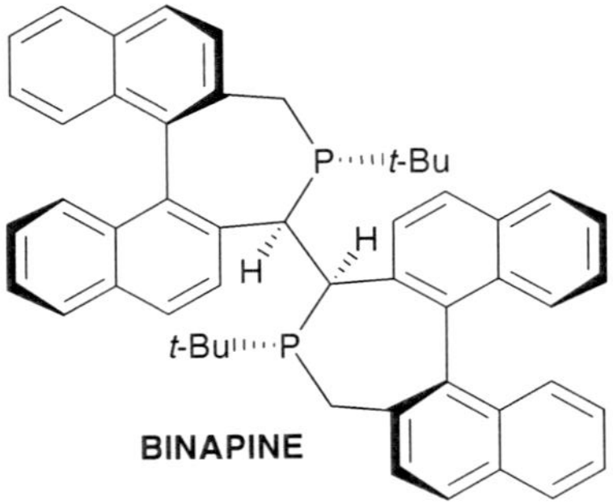 BINAPINE	32.7 (S)	59

10	*E*	Me	Me	Me	H	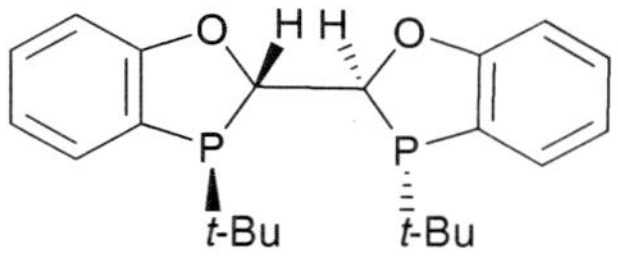 BIBOP	99 (*R*)	24
11	*E*	Me	Me	Et	H	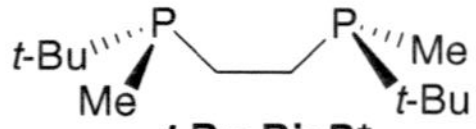 *t*-Bu-BisP*	99.7 (*R*)	56
12	*E*	Me	Me	Et	H	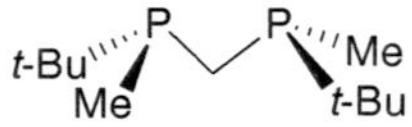 *t*-Bu-MiniPHOS	99.3 (*R*)	56
13	*E*	Me	Me	Et	H	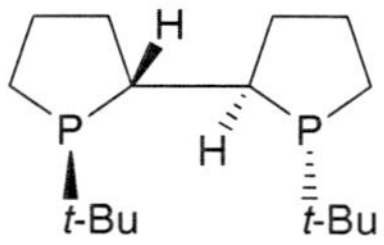 TangPhos	99.6 (*R*)	58
14	*E*	Me	Et	Me	H	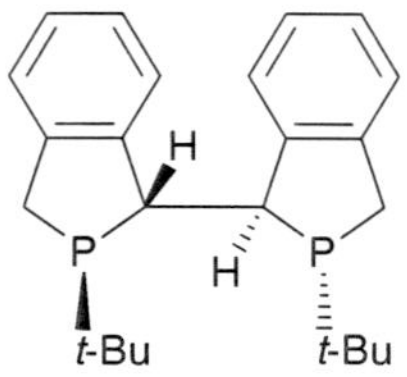 DuanPhos	>99 (*R*)	11

Table 7.2 (*Continued*).

Entry	E/Z	R^1	R^2	R^3	R^4	Ligand	ee (%)	References
15	*E*	Me	Et	Me	H	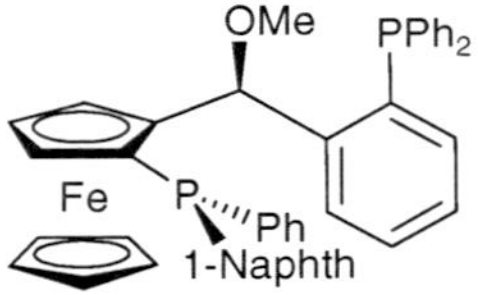 **PingFer**	96.1 (*S*)	52
16	*E*	Me	R	Me	H	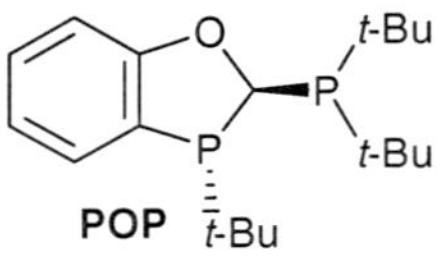 **POP**	95 (*S*, Me) 98 (*S*, *i*-Pr)	25
17	*Z*	Me	Me	Me	H	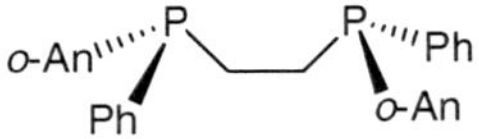 **DiPAMP**	66 (*S*)	15
18	*Z*	Me	Me	Me	H	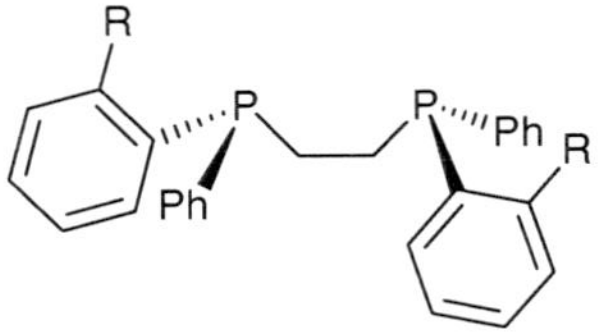	66 (*S*, OMe) 82.1 (*S*, O-*i*-Pr) 70.2 (*S*, O-*i*-Bu) 80.1 (*S*, O-*t*-Bu)	3, 4, 15

19	Z	Me	Me	Me	H	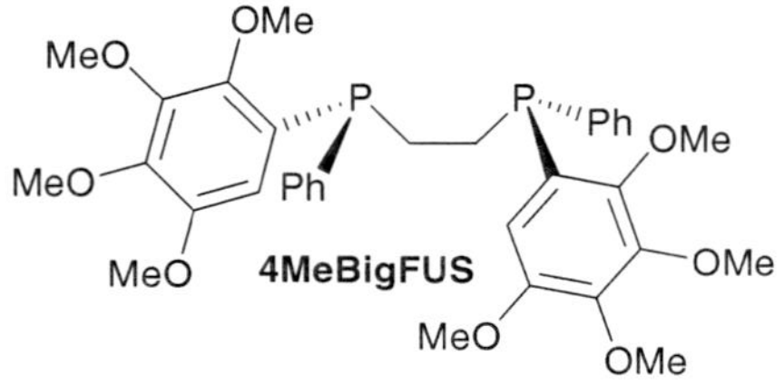 4MeBigFUS	80 (S)	15
20	Z	Me	Me	Me	H	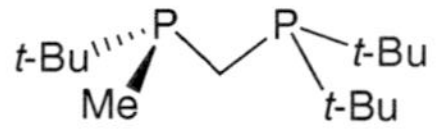 Trichickenfootphos	99 (R)	9, 57
21	Z	Me	Me	Me	H	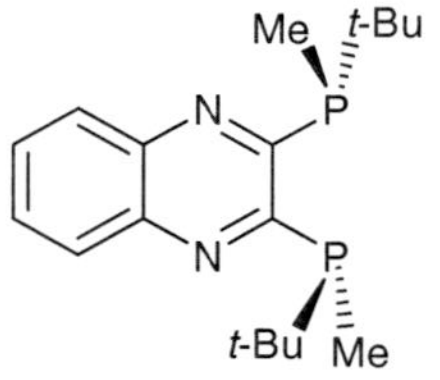 t-Bu-QuinoxP*	99.2 (R)	43
22	Z	Me	Me	Me	H	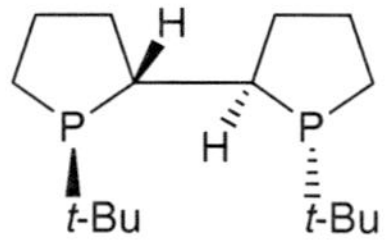 TangPhos	98.5 (R)	58
23	Z	Me	Me	Me	H	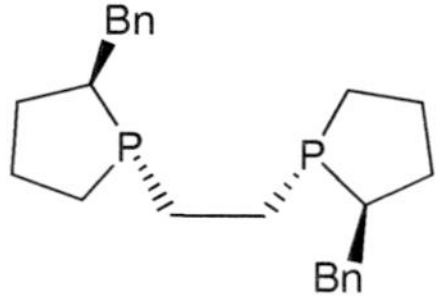	89 (R)	10

Table 7.2 (*Continued*).

Entry	E/Z	R^1	R^2	R^3	R^4	Ligand	ee (%)	References
24	Z	Me	Me	Me	H	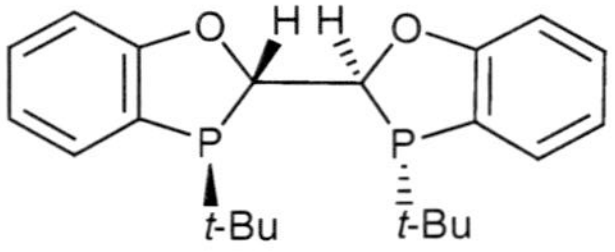 BIBOP	99 (R)	24
25	Z	Me	Et	R	H	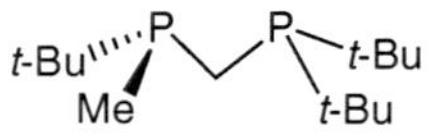 Trichickenfootphos	99 (R, n-Pr) 92 (S, i-Pr) 98 (R, i-Bu) 96 (S, Ph)	57
26	Z	Me	Et	Me	H	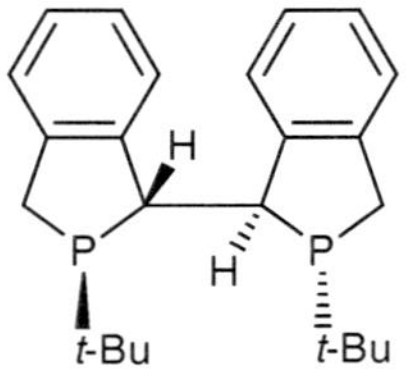 DuanPhos	97 (R)	11

27	Z	Me	Me	R	H	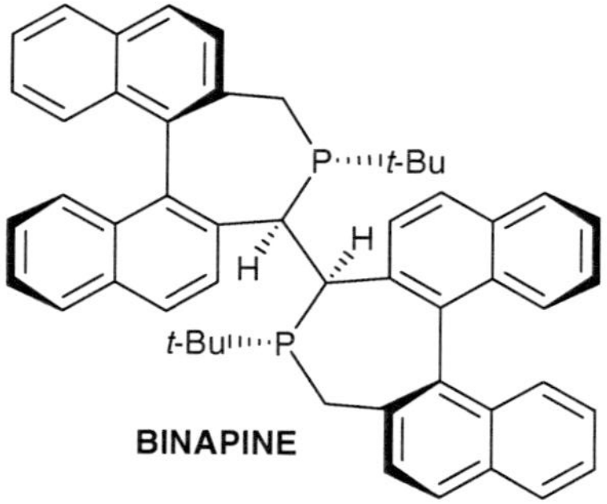 **BINAPINE**	99.2 (S, Me) >99 (S, Ph) 99 (S, o-An) >99 (S, o-Tol)	59
28	Z	Me	Me	R	H	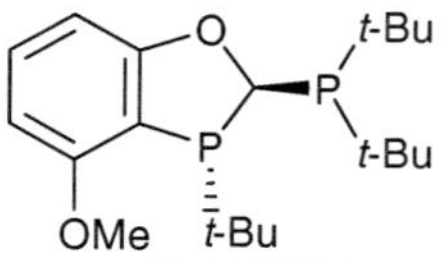 **MeO-POP**	99 (S, Me) 98 (S, Et) 97 (R, Ph)	25
29	Z	Me	Et	Me	H	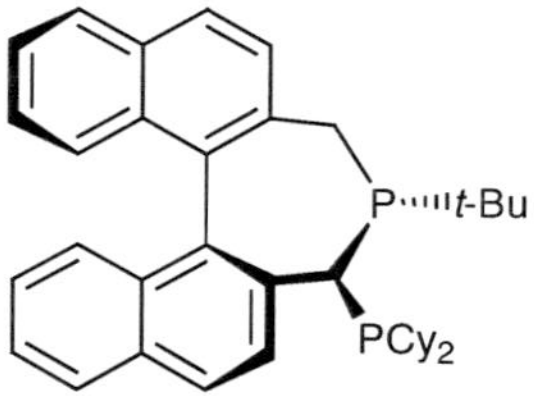	92.6 (R)	27

$$5 \quad X = NHCOR, OCOR \quad 6$$

Scheme 7.3   Rh-catalysed hydrogenation of enol esters and enamides.

very poor enantioselectivity (entry 9). From a practical point of view it is important to design efficient catalysts for both geometrical isomers since many synthetic protocols afford β-dehydroamino acids as E/Z mixtures,[58] with preference for Z isomers.[53,57]

7.2.1.3   Enamides and Enol Esters.   Enol esters and enamides (**5**) are also typical substrates in enantioselective hydrogenation (Scheme 7.3).[2]

Frequently used substrates include α-acetoxystyrene (**AXS**) and α-acetamidostyrene (**AS**). Hydrogenation of enol esters leads to chiral acetates, which can be easily hydrolysed to the parent alcohols.[60] Hydrogenation of enamides is equally interesting because the obtained amines are extremely important synthetic precursors.[61] Some results with those and related substrates are listed in Table 7.3.

There is a relatively small amount of results about the hydrogenation of enol esters (entries 1–6). For **AXS** and related substrates, very high enantioselectivities have been obtained with bulkier DiPAMP analogues (entry 1) and with phosphines based on the phospholane backbone (entries 2–5).

Many more results on hydrogenation of simple enamides have been published (entries 7–29). With the model substrate **AS** and other α-aryl substrates (entries 7–16), DiPAMP leads to relatively low enantioselectivity (entry 7) but its heavier analogues provide much better results (entry 8). Superb enantioselectivities have also been obtained with other C_2-diphosphines such as *t*-Bu-QuinoxP* (entry 12) and phospholane-based phosphines (entries 13–15). With aliphatic enamides (entries 17–21), extremely high enantioselectivities have also been achieved with similar ligands. Interestingly, the absolute configurations of the hydrogenation products of α-aryl enamides are opposite to those of α-alkyl enamides when the same ligands are employed (compare entries 11 and 18, 10 and 19 and entry 21). The origin of this difference has been clarified by mechanistic studies of Gridnev, Imamoto and co-workers,[63,64] who showed that aryl and alkyl enamides follow different coordination pathways. Finally, a few excellent results with β-substituted (entries 22–26) and β,β-disubstituted substrates (entries 27–29) are also on record.

7.2.1.4   α,β-Unsaturated Acids.   Several α,β-unsaturated carboxylic acids (**7**), not included in the preceding sections, have been employed in Rh-catalysed hydrogenation (Scheme 7.4).

Table 7.3 Results of Rh-catalysed hydrogenation of enol esters and enamides (Scheme 7.3).

Entry	X	R^1	R^2	R^3	Ligand	ee (%)	References
1	OAc	Ph	H	H	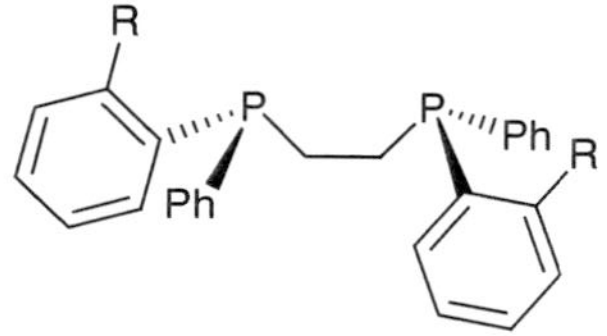	92.8 (*S*, O-*i*-Pr) 98.7 (*S*, O-*t*-Bu)	3, 4
2	OAc	Ph	H	H	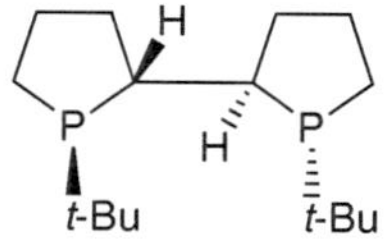 **TangPhos**	96 (*R*)	60
3	OAc	Ph	H	H	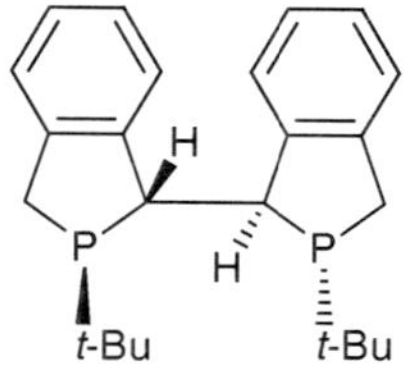 **DuanPhos**	97 (*R*)	11
4	OAc	2-Naphth	H	H	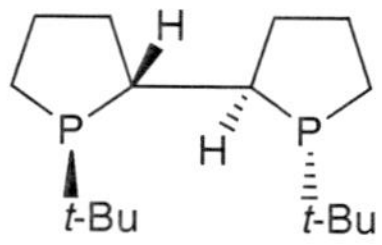 **TangPhos**	97 (*R*)	60

Table 7.3 (*Continued*).

Entry	X	R^1	R^2	R^3	Ligand	ee (%)	References
5	OAc	2-Naphth	H	H	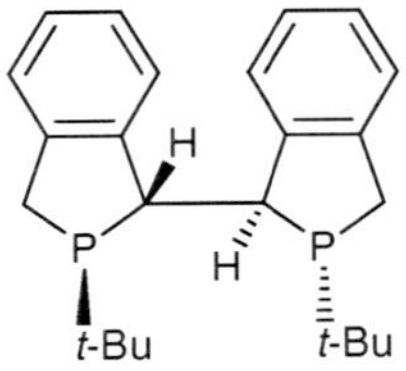 **DuanPhos**	98 (R)	11
6	OAc	COOEt	H	Ph	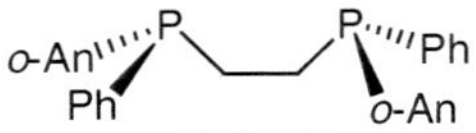 **DiPAMP**	90	29
7	NHAc	Ph	H	H	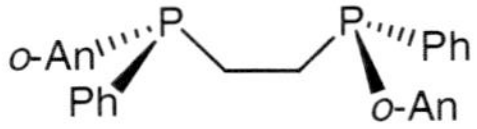 **DiPAMP**	83 (S)	62
8	NHAc	Ph	H	H	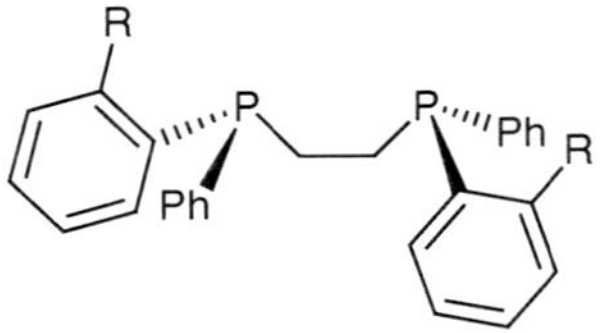	83 (S, OMe) 97.8 (S, O-i-Pr) 96.9 (S, O-i-Bu) 99.4 (S, O-t-Bu)	3, 4, 62
9	NHAc	Ar	H	H	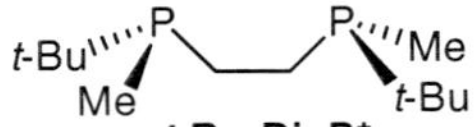 **t-Bu-BisP***	99 (R, Ph) 50 (R, o-An)	63, 64

10	NHAc	Ph	H	H	*t*-Bu-MiniPHOS	66 (*R*)	64
11	NHAc	Ph	H	H	Trichickenfootphos	98 (*R*)	9
12	NHAc	Ph	H	H	*t*-Bu-QuinoxP*	99.9 (*R*)	43
13	NHAc	Ph	H	H	TangPhos	99 (*R*)	32
14	NHAc	Ph	H	H	DuanPhos	>99 (*R*)	11

Table 7.3 (*Continued*).

Entry	X	R^1	R^2	R^3	Ligand	ee (%)	References
15	NHAc	Ar	H	H	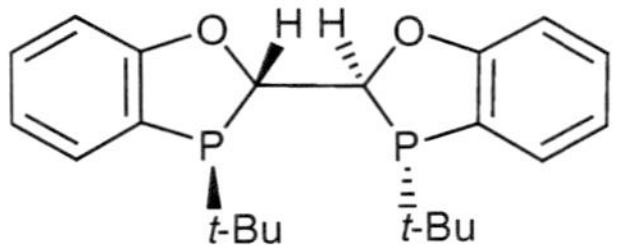 BIBOP	99 (R, Ph) 99 (R, p-An) >99 (R, m-Tol)	24
16	NHAc	Ph	H	H	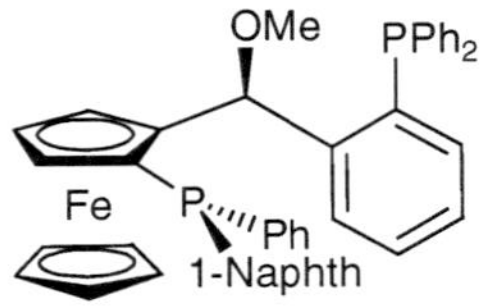 PingFer	96.3 (S)	52
17	NHAc	R	H	H	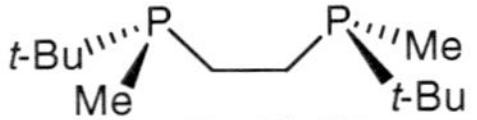 t-Bu-BisP*	99 (S, t-Bu) 99 (S, 1-Ad)	63, 64
18	NHAc	t-Bu	H	H	 Trichickenfootphos	99 (S)	9
19	NHAc	1-Ad	H	H	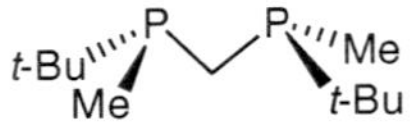 t-Bu-MiniPHOS	99 (S)	64

Entry					Ligand	ee (%)	Ref.
20	NHAc	1-Ad	H	H	*t*-Bu-QuinoxP*	96.3 (*S*)	43
21	NHAc	R	H	H	DiSquareP*	93 (*S*, *t*-Bu) >99 (*R*, Ph) >99 (*R*, *p*-An)	22
22	NHAc	Me	H	Ar	TangPhos	99.3 (*S*, Ph) 99.0 (*S*, *o*-An) 99.1 (*S*, *m*-An) 96.6 (*S*, *p*-An)	61
23	NHAc	Me	Ph	H	TangPhos	31.5 (*S*)	61
24	NHAc	Ph	H	Ph	TangPhos	99.3 (*S*)	61

Table 7.3 (*Continued*).

Entry	X	R^1	R^2	R^3	Ligand	ee (%)	References
25	NHAc	Ph	H	Me	DiSquareP*	>99 (*R*)	22
26[a]	NHAc	Ph	H, *i*-Pr		DuanPhos	97 (*R*)	11
27	NHAc	Ph	Me	Me	*t*-Bu-BisP*	99 (*S*)	64
28	NHAc	Ph	Me	Me	*t*-Bu-MiniPHOS	98 (*S*)	64
29	NHAc	Ph	Me	Me	1-Ad-QuinoxP*	99.9 (*R*)	44

[a]The substrate was a *E/Z* mixture of isomers.

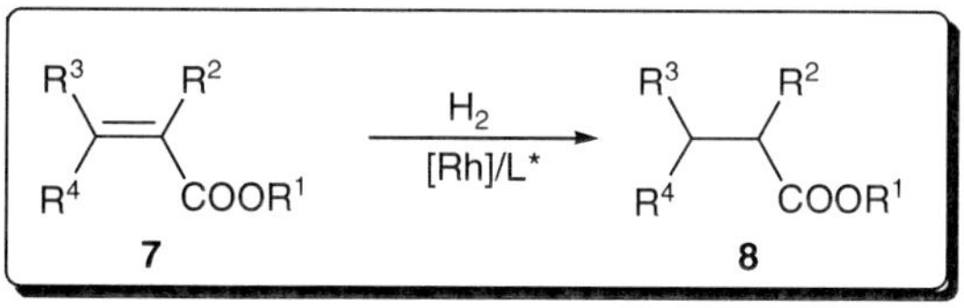

Scheme 7.4   Rh-catalysed hydrogenation of α,β-unsaturated acids.

Typical examples of this class of substrates include atropic acid (**AA**), itaconic acid (**IA**) and their methyl esters. With those substrates enantioselectivities tend to be somewhat lower compared to α-dehydroamino acids,[7] but at present many *P*-stereogenic ligands have provided excellent results (Table 7.4).

Only a few results on the enantioselective hydrogenation of **AA** are available (entries 1–4). Whereas DiPAMP is a very poor ligand for this transformation (entry 1), some of its heavier analogues are very good ligands (entries 2 and 3). BIPNOR gave also promising results with a modified **AA**, as entry 4 demonstrates. Extremely high enantioselectivities have been reported for the hydrogenation of **IA** (entries 5–8) and its monomethyl esters (entries 9–11) employing MiniPHOS (entries 6 and 9) and TangPhos (entry 7) ligands. The most common substrate is **DMI** (entries 12–21), which has been hydrogenated with almost perfect enantioselectivity by a bulkier DiPAMP analogue (entry 14), TangPhos (entry 18) and DuanPhos (entry 19). β-Substituted itaconic acid derivatives are much more challenging substrates[2,60] but entries 22–24 show that among *P*-stereogenic ligands TangPhos and DuanPhos are efficient ligands for these substrates. Finally, entry 25 shows that the system Rh/TriFer efficiently hydrogenates 3-aryl-2-ethoxyacrylic acids, considered to be very challenging substrates. The success in this case is thought to be a result of a secondary electrostatic interaction of the carboxylic group of the substrate with one of the dimethylamino groups of the ligand.[67]

7.2.1.5   Other Substrates.

This section deals with some other functionalised alkenes (**9**) not covered by the preceding sections (Scheme 7.5).

Examples of these substrates include α,β-unsaturated phosphonates, a precursor to the drug pregabalin, and a few other alkenes. A selection of results with these substrates are listed in Table 7.5.

Entries 1–9 show that several Rh/*P*-stereogenic diphosphine ligands are able to hydrogenate α,β-unsaturated phosphonates in very high enantioselectivities. The *t*-butoxy analogue of DiPAMP and Trichickenfootphos (entries 1, 4 and 7) lead to the best results. α,β-Unsaturated phosphonates are interesting because they are precursors of α-hydroxy- and α-aminophosphonic acids, compounds with a broad spectrum of biological activity.[69]

Some results on hydrogenation of a precursor of the drug pregabalin[8] are listed in entries 10–12. Phospholane-based ligands (entries 11 and 12) and Trichickenfootphos (entry 10) lead to very high enantioselectivities with this substrate.

Table 7.4 Results on Rh-catalysed hydrogenation of α,β-unsaturated acid derivatives (Scheme 7.4).

Entry	R^1	R^2	R^3	R^4	Ligand	ee (%)	References
1	H	Ph	H	H	DiPAMP	7 (S)	15
2	H	Ph	H	H		7 (S, OMe) 88.0 (S, O-i-Pr) 86.0 (S, O-i-Bu) 95.6 (S, O-t-Bu)	3, 4, 15
3	H	Ph	H	H	4MeBigFUS	88 (S)	15
4	H	Ar[a]	H	H	BIPNOR	98	35

					Ligand	ee (%)	Ref.
5	H	CH_2CO_2H	H	H	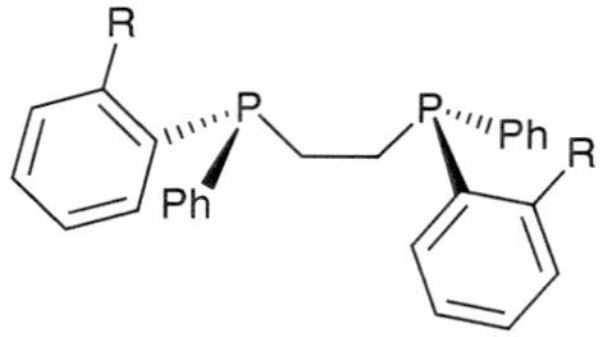	77 (R, OMe) 98.7 (R, O-i-Pr) 99.7 (R, O-t-Bu)	3, 4, 65
6	H	CH_2CO_2H	H	H	 R-MiniPHOS	>99.9 (t-Bu) >99.9 (Cy) 98 (i-Pr)	6, 7
7	H	CH_2CO_2H	H	H	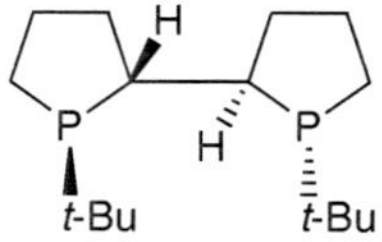 TangPhos	99 (S)	60
8	H	CH_2CO_2H	H	H	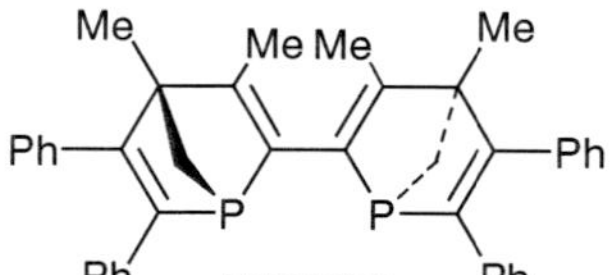 BIPNOR	93 (R)	35, 66
9	Me	CH_2CO_2H	H	H	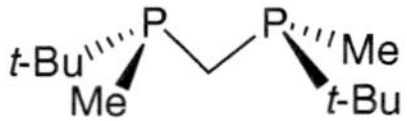 t-Bu-MiniPHOS	99.9	7
10	Me	CH_2CO_2H	H	H	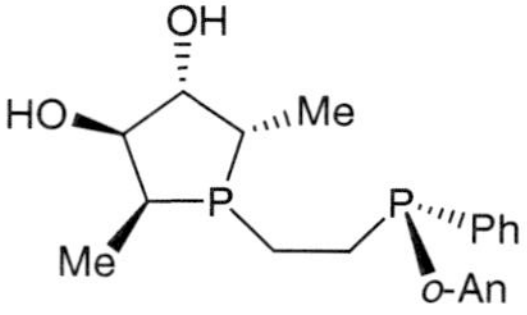	95 (R)	50

Table 7.4 (*Continued*).

Entry	R^1	R^2	R^3	R^4	Ligand	ee (%)	References
11	H	CH_2CO_2Me	H	H	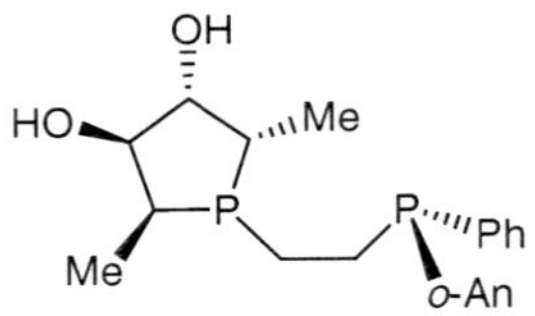	93 (*R*)	50
12	Me	CH_2CO_2Me	H	H	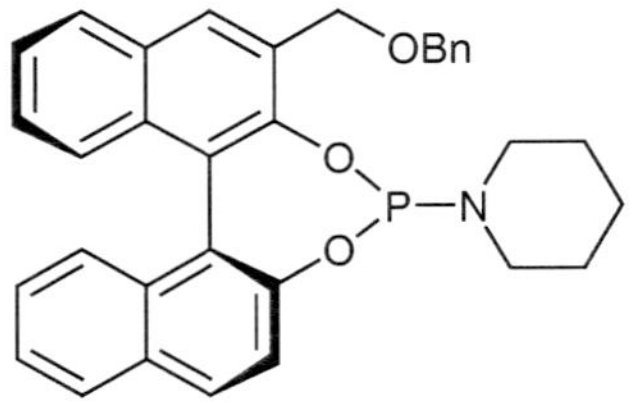	96.8 (*S*)	13
13	Me	CH_2CO_2Me	H	H	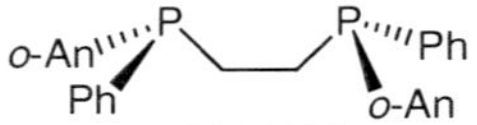 **DiPAMP**	85-88 (*R*)	15, 65
14	Me	CH_2CO_2Me	H	H	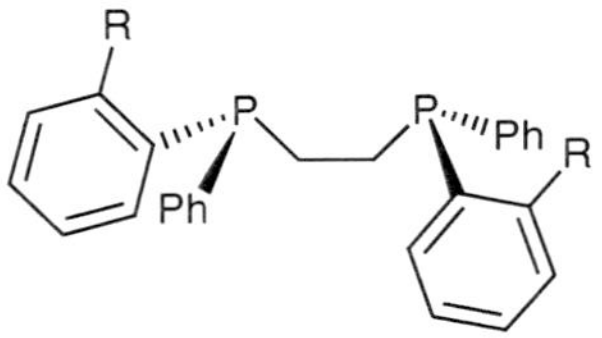	85 (*R*, OMe) 98.1 (*R*, O-*i*-Pr) 98.5 (*R*, O-*i*-Bu) 99.8 (*R*, O-*t*-Bu)	3, 4, 15

15	Me	CH$_2$CO$_2$Me	H	H	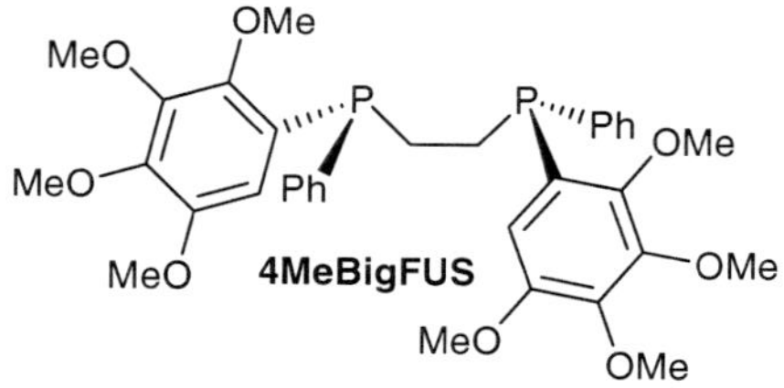 4MeBigFUS	97 (R)	15
16	Me	CH$_2$CO$_2$Me	H	H	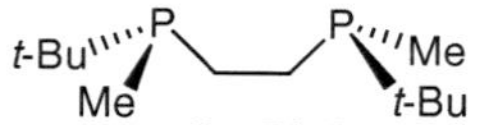 t-Bu-BisP*	98.6	7
17	Me	CH$_2$CO$_2$Me	H	H	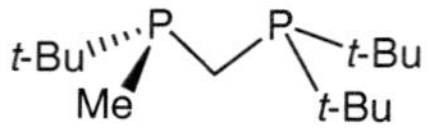 Trichickenfootphos	98 (S)	9
18	Me	CH$_2$CO$_2$Me	H	H	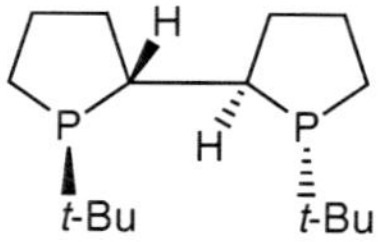 TangPhos	99 (S)	60
19	Me	CH$_2$CO$_2$Me	H	H	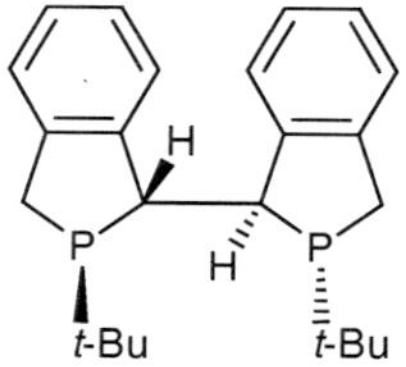 DuanPhos	>99 (S)	11

Table 7.4 (*Continued*).

Entry	R^1	R^2	R^3	R^4	Ligand	ee (%)	References
20	Me	CH_2CO_2Me	H	H	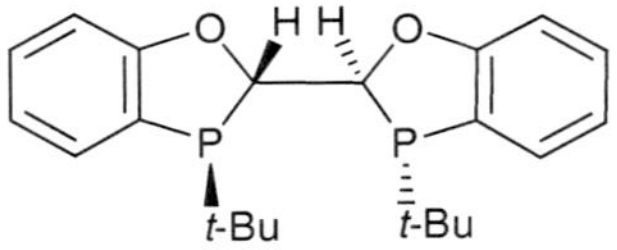 **BIBOP**	94 (*S*)	24
21	Me	CH_2CO_2Me	H	H	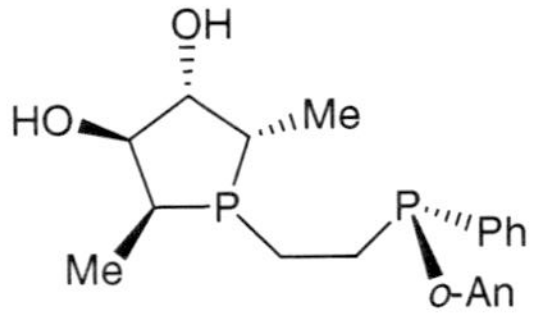	85 (*R*)	50
22[b]	Me	CH_2CO_2H	H, *i*-Pr		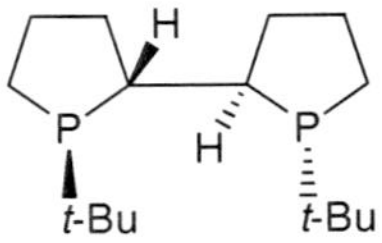 **TangPhos**	96 (*S*)	60
23[b]	Me	CH_2CO_2H	H, *i*-Pr		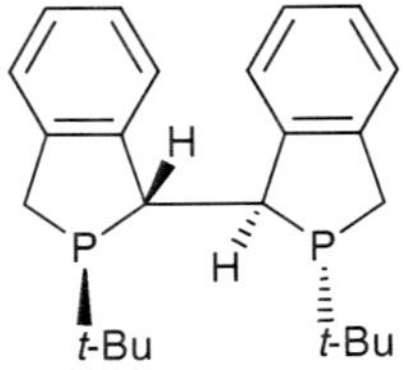 **DuanPhos**	>99 (*S*)	11

24[b]	Me	CH$_2$CO$_2$H	H, Ar		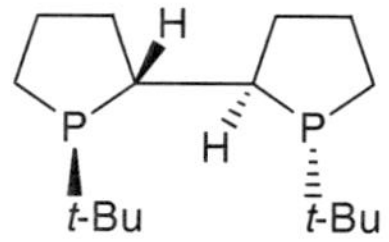 95 (*R*, Ph) 97 (*R*, *p*-An) 97 (*R*, *p*-Tol) 99 (*R*, 1-Naphth)	60
				TangPhos		
25	H	OEt	Ar	H	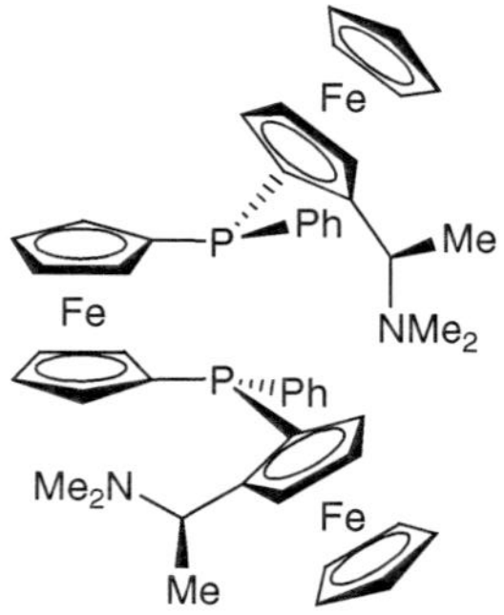 95.2 (*o*-An) 98.0 (*p*-CNC$_6$H$_4$)	67
				TriFer		

[a]Ar = 6-methoxy-2-naphthyl.
[b]The substrate was a mixture of the *E*/*Z* isomers.

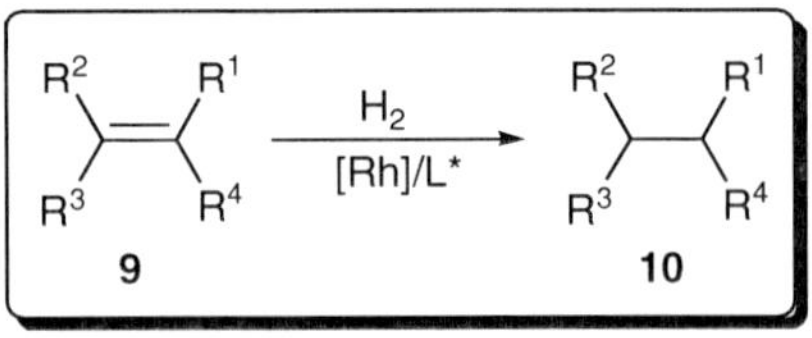

Scheme 7.5   Rh-catalysed hydrogenation of miscellaneous substrates.

Finally, a few examples of successful enantioselective hydrogenation of miscellaneous substrates are given in entries 13–15.

7.2.2   Mechanism

The literature regarding the mechanism of Rh-catalysed hydrogenation of functionalised alkenes is so vast that it would be impossible to give a detailed account in a few paragraphs.[1] Despite being the most extensively studied mechanism in homogeneous catalysis, new details and subtleties are continuously being reported in the primary literature,[75–80] to the point that, although it is a mature field, we are still far from a complete understanding.[81] In this section only an overview is given, completely focused on enantioselective hydrogenation[82] of functionalised alkenes by Rh(I)/diphosphine systems.

For these systems there are basically two general mechanisms, commonly known as *unsaturated* and *dihydride* mechanisms (Scheme 7.6).[69,79,82,83]

As shown in Scheme 7.6, both mechanisms take place through the same elementary steps involving similar Rh(I)/Rh(III) species, the main difference being the order of substrate coordination/oxidative addition of dihydrogen at the catalyst **12**. In the study of both mechanistic pathways, *P*-stereogenic diphosphines have played a vital role.[69,79,80,84–86]

The 'unsaturated' mechanism of hydrogenation of **MAC** with Rh(I)/DiPAMP precursors is depicted in Scheme 7.7 in more detail.[85] Although for the sake of simplicity this particular case is presented in the scheme, the mechanism has a wider scope, being applicable to hydrogenation of α-dehydroamino acids and related substrates. Therefore, each step is discussed in a general context.

The usual precursors **11** are cationic Rh(I)/diphosphine complexes with dialkenes, typically COD or NBD, with weakly coordinating anions such as tetrafluoroborate or BARF. It was found that under hydrogen atmosphere the alkene is easily hydrogenated to form the *bis*(solvato) complexes **12**.[87–90] Often the reactions are carried out at low hydrogen pressure (< 10 atm) in methanol, which stabilises compounds **12**. Under these conditions, complexes **12** show high affinity for dehydroamino acids, prompting reversible bidentate complexation of the substrate to form *bis*(chelated) catalyst–substrate complexes **13**, which have been characterised by NMR spectroscopy and some by X-ray crystallography.[54,88,90–96] When using C_2-symmetric diphosphines, two possible diastereomers can be formed with prochiral substrates, depending on which

Table 7.5 Results of Rh-catalysed hydrogenation of miscellaneous substrates (Scheme 7.5).

Entry	R^1	R^2	R^3	R^4	Ligand	ee (%)	References
1	P(O)(OMe)$_2$	H	H	OBz		99.6 (R)	4
2	P(O)(OMe)$_2$	H	H	OBz	*t*-Bu-BisP*	88 (R)	68, 69
3	P(O)(OMe)$_2$	H	H	OBz	*t*-Bu-MiniPHOS	96 (R)	69
4	P(O)(OMe)$_2$	H	H	OBz	Trichickenfootphos	99 (R)	9
5	P(O)(OMe)$_2$	H	*i*-Pr	OBz	*t*-Bu-BisP*	98 (R)	69
6	P(O)(OMe)$_2$	H	*i*-Pr	OBz	*t*-Bu-MiniPHOS	98 (R)	69

Table 7.5 (*Continued*).

Entry	R^1	R^2	R^3	R^4	Ligand	ee (%)	References
7	$P(O)(OEt)_2$	H	H	NHAc		99.9 (*R*)	4
8	$P(O)(OMe)_2$	H	H	NHAc	**t-Bu-BisP***	90 (*R*)	69
9	$P(O)(OEt)_2$	H	H	Ph	Ar = p-CF$_3$C$_6$H$_4$	95	70
10	CN	*i*-Pr	H	R^a	**Trichickenfootphos**	99 (*S*)	8
11	CN	*i*-Pr	H	R^a		97 (*S*)	23

12	CN	*i*-Pr	H	R^a		96 (*S*)	71
13	OC(O)NMe$_2$	H	H	Ph		84 (*S*)	72
14	NHPh	COOMe	H	Me	**TangPhos**	94.6	73
15	CH$_2$CO$_2$H	H	H	R	**DuanPhos**	97 (*S*, Ph) 74 (*R*, Bn)	74

^{a}R $=$ CH$_2$COO$^-$[NH$_3$*t*-Bu]$^+$.

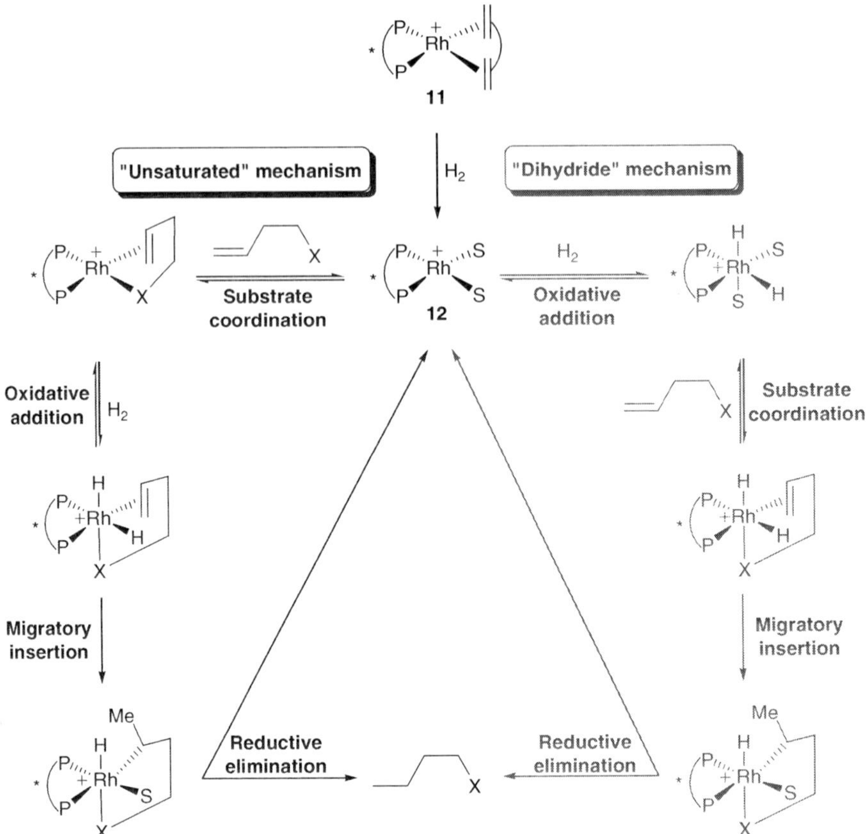

Scheme 7.6   The two accepted mechanisms of hydrogenation of functionalised alkenes with Rh(I)/diphosphine systems.

face of the alkene coordinates to the Rh centre.[79] For example, with DiPAMP precursors **MAC** forms an unequal mixture highly enriched with *si*-**13** in equilibrium with the other diastereomer.[85,88] This equilibrium, which is fast compared to catalytic turnover, was shown to proceed intra- and inter-molecularly, the latter through complex **12**.[84–86] Oxidative addition of hydrogen is thought to render (unobserved) Rh(III) dihydride complexes **14**, which suffer migratory insertion of the alkene into the Rh–H bond to afford alkyl-hydride complexes **15**,[87,97,98] which finally collapse by reductive elimination to afford the product **16** and regenerate the catalyst.

Formation of **14** is thought to be the rate-limiting step.[85] The rate of dihydrogen addition to each isomer of **13** can be very different, with the minor diastereomer, thermodynamically disfavoured, reacting much faster than the major diastereomer. For the exact system of Scheme 7.7 the minor isomer *re*-**13** reacts about 600 times faster than *si*-**13**.[85] This different reactivity and not the isomeric ratio of **13** is the parameter that dictates, to a large extent,[76] the

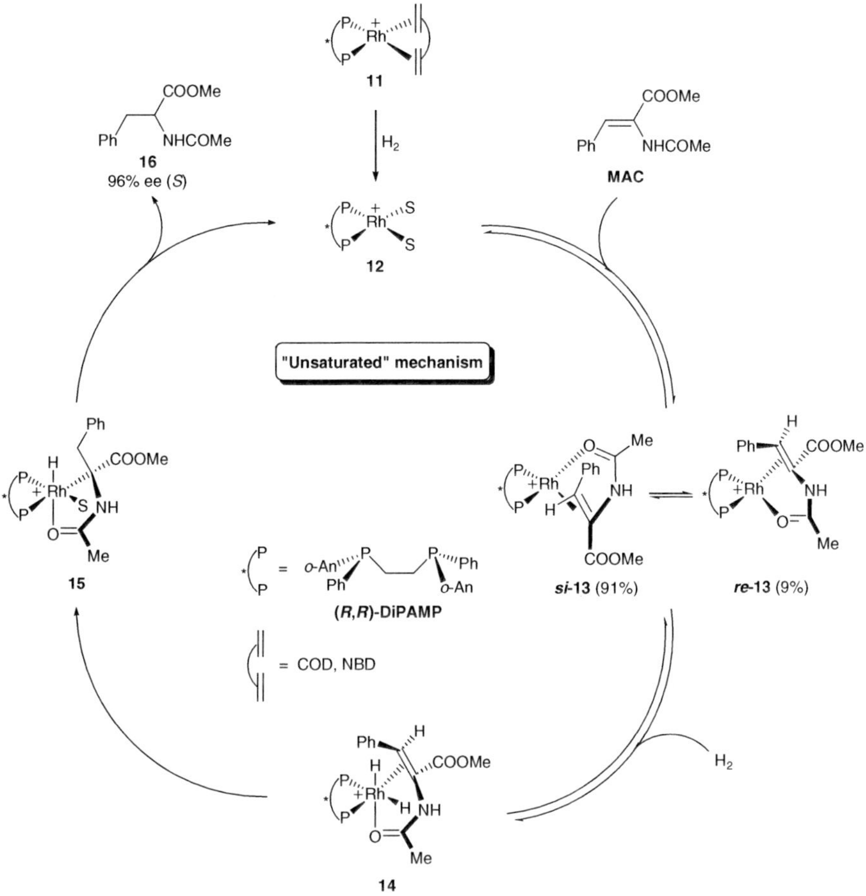

Scheme 7.7 'Unsaturated' mechanism for the hydrogenation of **MAC** with Rh/ DiPAMP precatalysts. Only the reaction of *re*-**13** with H_2 is represented.

enantioselectivity of the reaction and the absolute configuration of the major product.[79,85,92,97] This concept is sometimes referred in literature as *major/ minor* or *anti-lock-and-key* principle.[54,76]

This is the classical asymmetric hydrogenation mechanism, taught in every course on enantioselective homogeneous catalysis. By the end of 1990s, each step of this mechanism had been studied in detail, including the preparation and characterisation of all the intermediates, with the exception of dihydride complexes **14**, whose detection remained elusive.[85]

Several detailed NMR studies with new, electron-richer ligands were undertaken around the year 2000. With these new ligands, Rh(III) solvate dihydrides, $[Rh(P–P)(H)_2(S)_2]^+$, could be easily prepared and characterised for several diphosphines.[7,99,100] Bargon, Brown and co-workers[101] used chiral (but not *P*-stereogenic) diphosphine **PHANEPHOS** and special NMR techniques to detect an agostic intermediate, $[Rh(P–P)(H)_2(\mathbf{MAC})]^+$, with one of the hydrogen atoms being transferred from the rhodium to the **MAC** double bond. In other words, they took a 'photograph' of the migratory insertion step with the hydrogen atom simultaneously bound to rhodium and carbon atoms.

These studies, although they did not deliver the much desired direct characterisation of a dihydride intermediate with the coordinated alkene, opened the door to the alternative mechanism, known as *dihydride* mechanism.[79] An example of this mechanism, the hydrogenation of **MAC** with Rh/*t*-Bu-BisP* complex, is depicted in Scheme 7.8.[78,99] As with the 'unsaturated' mechanism, this particular case is presented in the scheme, the mechanism is discussed in a broader context.

For electron-rich phosphines *bis*(solvato) complexes **12** have moderately (or very)[102] high affinity for dihydrogen, rendering (reversibly) dihydride Rh(III) complexes **17** as mixtures of diastereomers. For *t*-Bu-BisP*, a 1:10 mixture of interconverting diastereomers was detected by NMR at –90 °C.[7,99] Coordination of the substrate is thought to lead to the very reactive (unobserved) dihydride complexes **18**, which suffer migratory insertion very rapidly, even at – 100 °C, leading to monohydride complexes **19**.[7,56,63,64,68,69] These complexes rearrange to afford the more stable complexes **19′**, bearing both carbonylic groups coordinated to the Rh centre. Reductive elimination on **19′** transfers the second hydrogen atom to the substrate to afford weakly bound η^6-arene complexes **20**. A last step of decoordination of the product closes the cycle.

The enantioselectivity of the reaction could be correlated to the isomeric ratio of monohydride intermediate **19′** suggesting that the migratory insertion is the stereodetermining step,[69,78] a conclusion supported by labelling experiments.[7]

To study the possibility of the 'unsaturated' mechanism with electron-rich diphosphines, *bis*(chelated) catalyst–substrate complexes analogous to **13** (Scheme 7.7) were prepared by addition of substrates to solutions of **12**.[7,99] As expected, the catalyst–substrate complexes were present as mixtures of two diastereomers, interconverting intra- and intermolecularly. Stoichiometric hydrogenation experiments of those mixtures were found to yield **19′**, but at lower rate compared to the reaction of substrate with dihydrides **17**. The enantioselectivities obtained from these stoichiometric experiments depended on the exact system. Weakly bound enamides[64] and β-amino acids[56] were hydrogenated at low temperatures (–100 °C) with the same enantioselectivities than using dihydride precursors whereas tightly bound phosphonates could only be hydrogenated at higher temperatures (–30 °C) leading to much reduced enantioselectivities.[102] The conclusion is that Rh(I) complexes with electron-rich diphosphines with strong steric dissymmetry on the phosphorus atoms hydrogenate functionalised alkenes following the 'dihydride' mechanism under catalytic conditions.[79]

Whichever the mechanism, all the steps before the irreversible formation of the monohydride intermediates seem to be reversible, at least for some systems.[79,80] This is an important observation because both mechanisms converge at this intermediate, whose formation by migratory insertion is considered to be

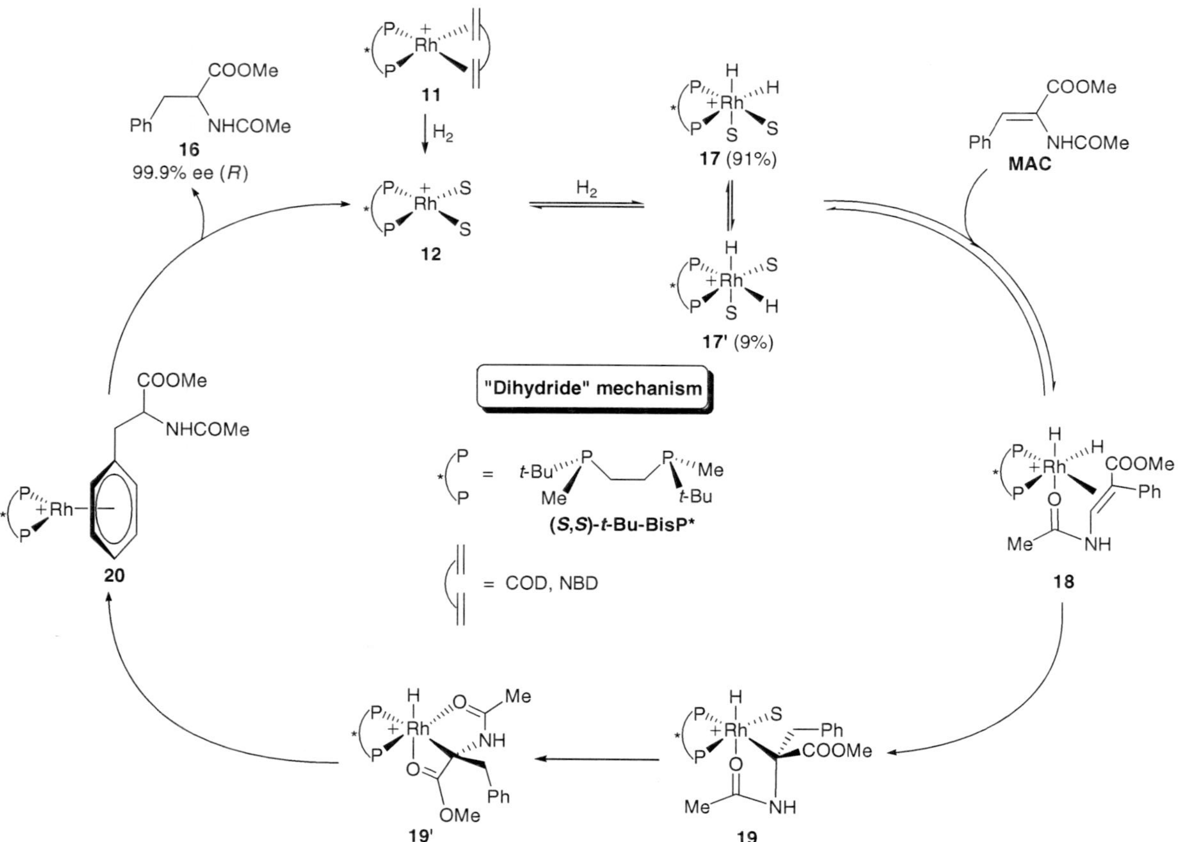

Scheme 7.8 'Dihydride' mechanism for the hydrogenation of **MAC** with Rh/*t*-Bu-BisP* precatalysts.

the enantiodetermining step. This means that the distinction between the 'unsaturated' and 'dihydride' mechanisms is irrelevant for stereoselection.[79]

Nowadays all the possible types of intermediates of the 'unsaturated' and 'dihydride' mechanisms except the much sought dihydride–alkene Rh complex have been detected, although not for the same catalytic system. It is becoming clear that although formal distinction between both mechanisms has enormous historical importance, they are interconnected at many points, forming an overall mechanism of interconverting intermediates.[80] The ligand, the substrate and the catalytic conditions determine in each case which exact pathway is followed.

7.2.3 Hydrogenation of Ketones and Imines

There are a few examples of hydrogenation of functionalised ketones catalysed by rhodium/*P*-stereogenic ligands. Zhang an co-workers[103,104] reported the hydrogenation of protonated α- and β-amino ketones (**21**) in the presence of potassium carbonate (Scheme 7.9). The nitrogen atom of these amino ketones is able to coordinate to the Rh centre during the catalysis forming metal–substrate chelates.

This is an expedient synthetic route to 1,2- and 1,3-amino alcohols **22**, which are important building blocks for the preparation of many natural and pharmaceutical products. Good results have been obtained with the electron-rich diphosphines DuanPhos and BINAPINE (Table 7.6).

Good yields and enantioselectivities could be obtained for both types of substrates. DuanPhos led to better enantioselectivities with primary α-amino ketones (entries 2–4) and secondary β-amino ketones (entries 7–10) whereas BINAPINE was better suited to hydrogenate secondary α-amino ketones (entries 1, 5 and 6).

Another route to 1,3-amino alcohols **24**, in this case bearing two stereogenic centres, was disclosed by Wu, Zhang and co-workers[105] by hydrogenation of β-ketoenamides **23** (Scheme 7.10).

In this transformation, the two stereogenic centres are generated simultaneously with excellent *anti* diastereoselectivity and very good enantioselectivity with the DuanPhos ligand. Inferior results were obtained with the related ligands TangPhos and BINAPINE. Hydrogen pressure had to be carefully adjusted to achieve good yields and minimise the amount of mono-reduced product still containing the carbonyl group.

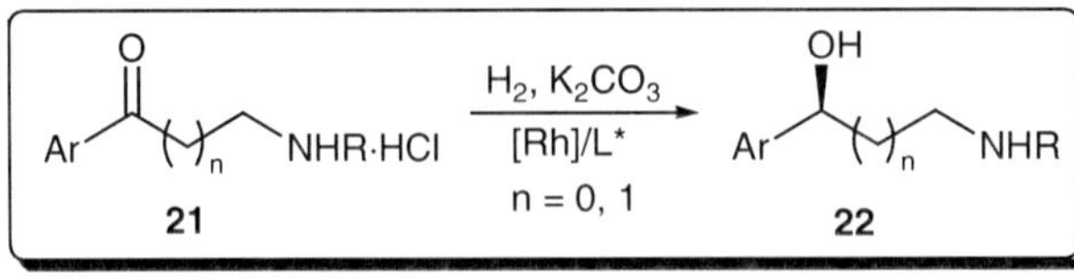

Scheme 7.9 Rh-catalysed hydrogenation of α- and β-amino ketones.

Table 7.6 Results of Rh-catalysed hydrogenation of α- and β-amino ketones (Scheme 7.9).

Entry	Substrate	Ligand	Yield (%)	ee (%)	References
1	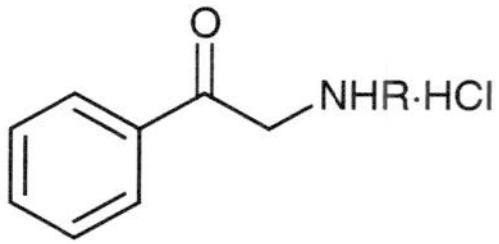	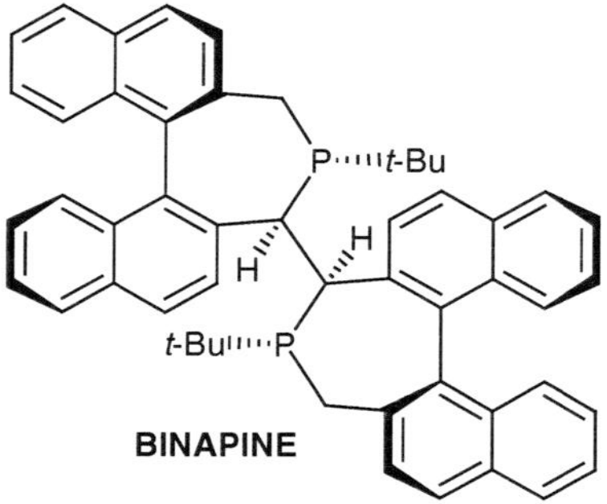 BINAPINE	>95 (H) >95 (Me)	82 (*S*, H) 95 (*S*, Me)	104
2	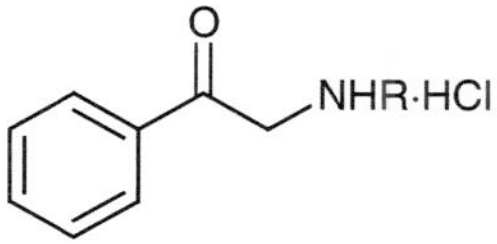	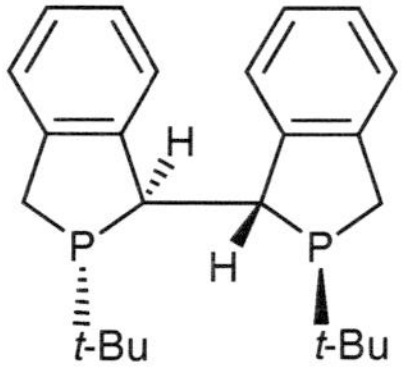 DuanPhos	>95 (H) >95 (Me)	88 (*S*, H) 40 (*S*, Me)	104
3	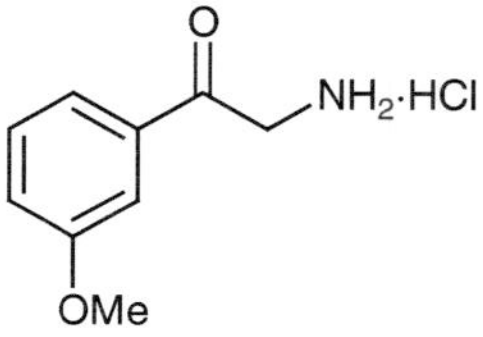	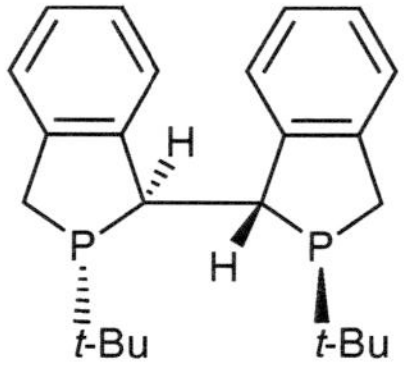 DuanPhos	>95	83	104

Table 7.6 (Continued).

Entry	Substrate	Ligand	Yield (%)	ee (%)	References
4	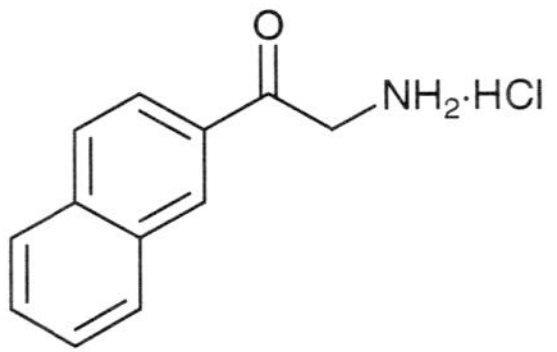 NH₂·HCl (2-acetylnaphthalene aminoketone)	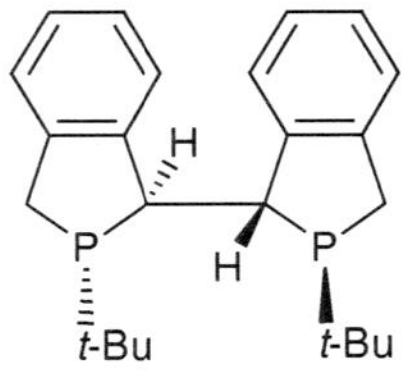 DuanPhos	>95	87	104
5	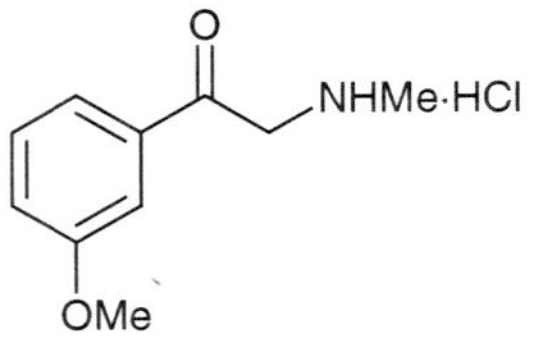 NHMe·HCl (3-methoxyphenyl aminoketone)	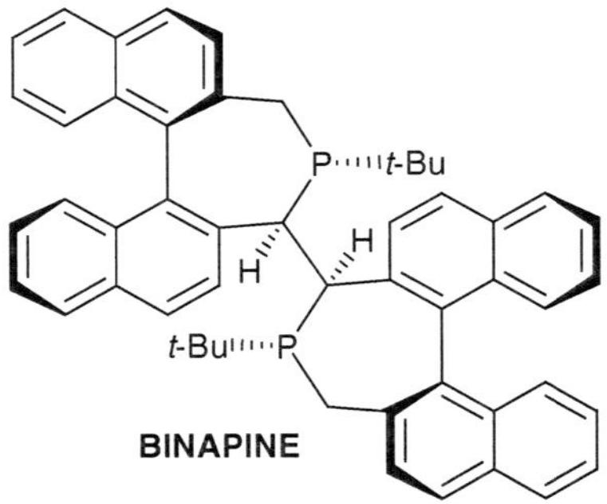 BINAPINE	>95	92	104

	Substrate	Ligand			
6	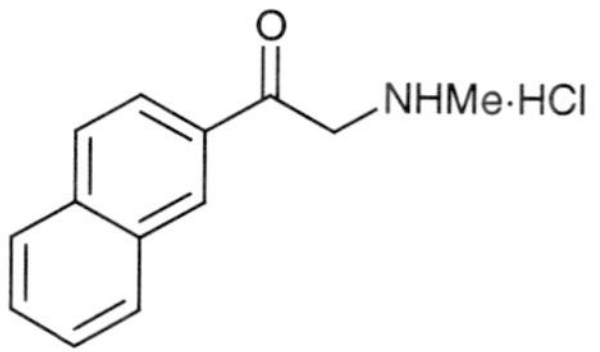	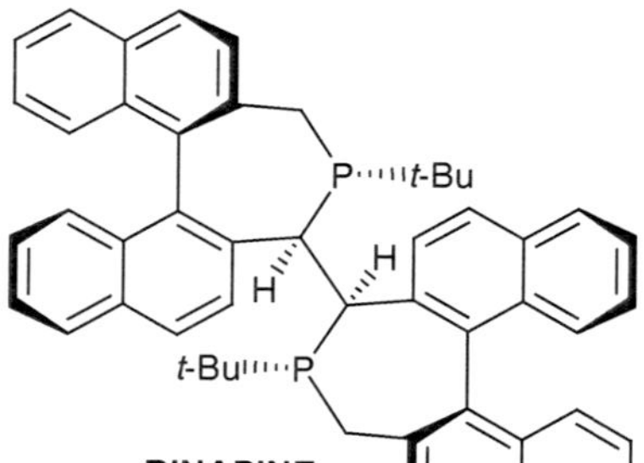 BINAPINE	>95	89	104
7	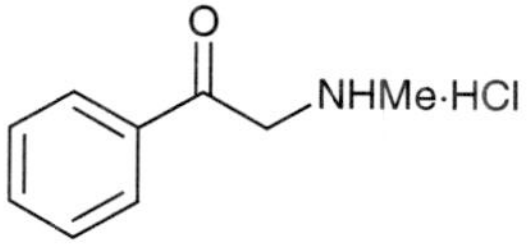	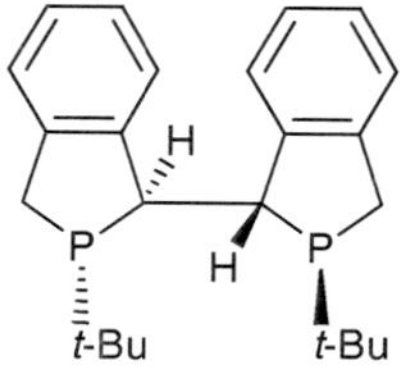 DuanPhos	90	98 (*S*)	103
8	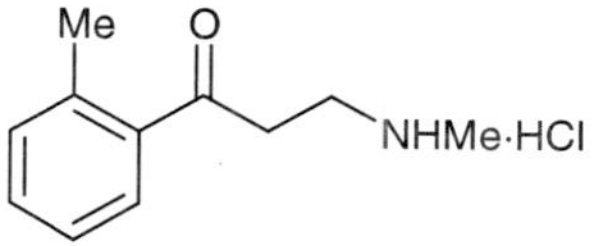	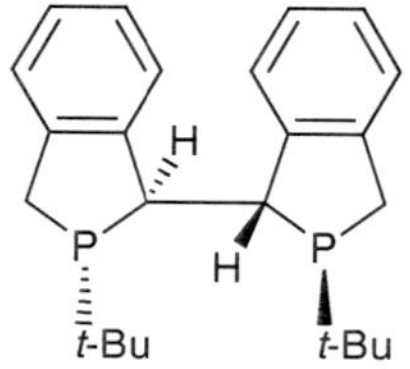 DuanPhos	92	99 (*S*)	103

Table 7.6 (*Continued*).

Entry	Substrate	Ligand	Yield (%)	ee (%)	References
9	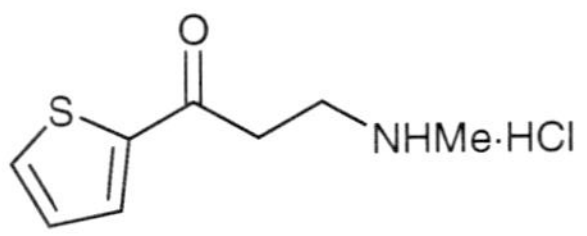 (thiophen-2-yl ketone, NHMe·HCl)	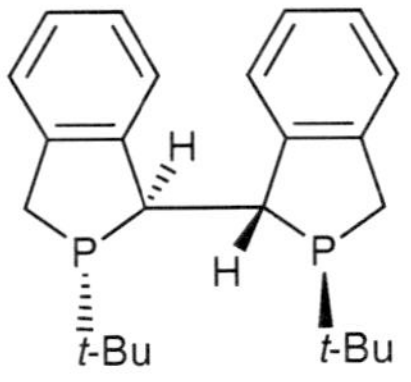 DuanPhos	93	>99 (S)	103
10	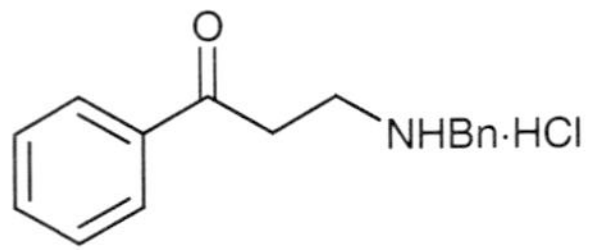 (phenyl ketone, NHBn·HCl)	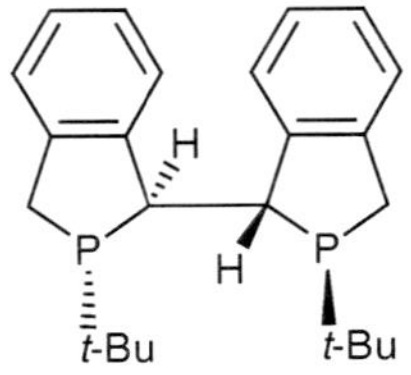 DuanPhos	90	96 (S)	103

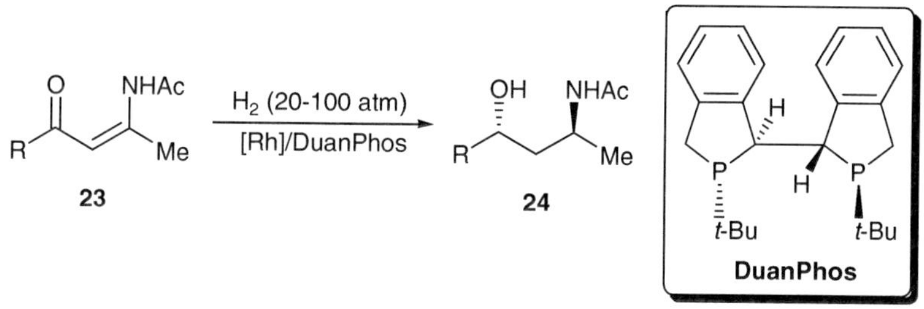

R = Me
96% yield, >98% de, 96% ee
R = Ph
100% yield, 90% de, 99% ee (1*R*,3*S*)
R = *o*-An
100% yield, 72% de, 94% ee
R = *p*-An
95% yield, 95% de, 84% ee
R = 2-Naphth
100% yield, >98% de, 97% ee

Scheme 7.10 Rh-catalysed hydrogenation of β-ketoenamides.

There is also one report of Rh-catalysed hydrogenation of α-aryl imino esters **25**, yielding chiral aryl glycines (Scheme 7.11).[106]

Although a large number of α-amino acids have been prepared by enantio-selective hydrogenation of the parent α-dehydroamino acids (see Section 7.2.1.1), glycines can not be prepared in this way since they lack the β hydrogen. *o*-Anisyl-protected α-aryl imino esters **25** were chosen as substrates because they are more stable than imines and can be conveniently deprotected by cerium ammonium nitrate. Once again, TangPhos was found to be very well suited for the hydrogenation of the imino bond of these substrates, leading to full conversions and good to excellent enantioselectivities with little influence of the properties of the aryl group. Interestingly, the substrate bearing a cyclohexyl group was also efficiently hydrogenated.

7.3 Ru-, Ir- and Pd-Catalysed Hydrogenation

7.3.1 *Ru-Catalysed Hydrogenation*

A clear advantage of ruthenium compared to rhodium is its price. Unfortunately, compared to rhodium and iridium, ruthenium complexes have lower catalytic activities in hydrogenation of alkenes.[107] This can an advantage for regio- and chemoselective reductions of polyalkenes but for other substrates it implies that higher hydrogen pressures and temperatures are required. In contrast, ruthenium complexes are very good at reducing carbonyl groups of simple ketones, β-keto esters and diketones. With these substrates, several very good results have been obtained reactions with *P*-stereogenic ligands.

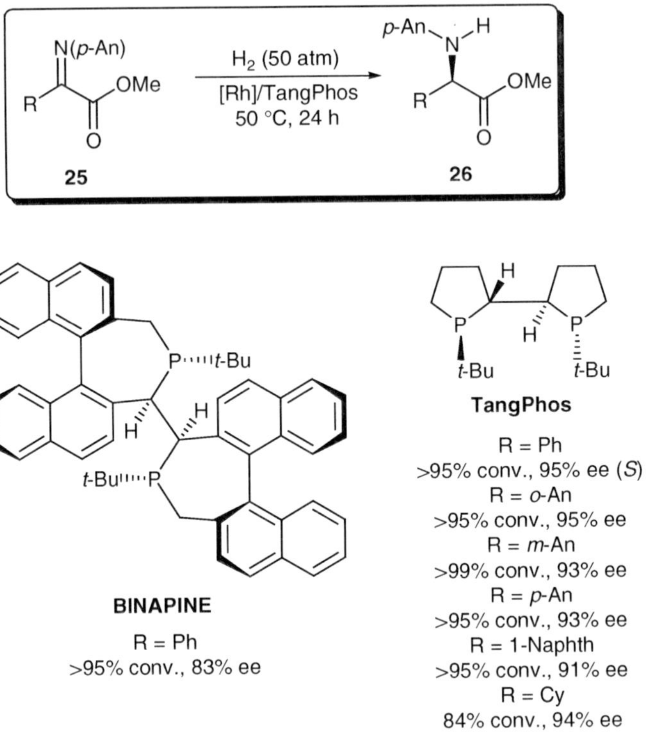

N(p-An)
OMe
R
O
25
H2 (50 atm)
[Rh]/TangPhos
50 °C, 24 h
p-An
N
H
R
OMe
O
26
P
t-Bu
H
t-Bu
P
BINAPINE
R = Ph
>95% conv., 83% ee
H
P
H
P
t-Bu
t-Bu
TangPhos
R = Ph
>95% conv., 95% ee (S)
R = o-An
>95% conv., 95% ee
R = m-An
>99% conv., 93% ee
R = p-An
>95% conv., 93% ee
R = 1-Naphth
>95% conv., 91% ee
R = Cy
84% conv., 94% ee

Scheme 7.11 Rh-catalysed hydrogenation of α-aryl imino esters.

7.3.1.1 Alkenes. There are only a few reports about enantioselective hydrogenation of alkenes with ruthenium complexes bearing *P*-stereogenic ligands. The substrates are the same as those used in rhodium-catalysed hydrogenations. Most of the results are listed in Table 7.7.

Genêt and co-workers[108,111] were the first to obtain ruthenium complexes with a *P*-stereogenic ligand (DiPAMP) and test them in Ru-catalysed hydrogenations (entries 1, 3 and 4). At present, tiglic acid (entries 1 and 2), naproxen and alanine precursors (entries 3 and 4 respectively), (*E*)-2-methylcinnamic acid (entry 5), **MAC** (entry 6) and **DMI** (entries 6 and 7) have been successfully hydrogenated, albeit in low enantioselectivities. The results of these studies show that ruthenium/*P*-stereogenic phosphine are active in alkene hydrogenation but the enantioselectivity is clearly inferior to rhodium, at least with the ligands employed so far.

7.3.1.2 Ketones. The rather disappointing results in Ru-catalysed enantioselective hydrogenation of alkenes with *P*-stereogenic ligands contrast with the much better results on ketone hydrogenation. The most common substrates are β-keto esters (**28**),[112] although simple ketones (**27**) can also be hydrogenated (Scheme 7.12 and Table 7.8).

Table 7.7 Results of enantioselective Ru-catalysed hydrogenation of alkenes.

Entry	Substrate	Ligand	Yield (%)	ee (%)	References
1	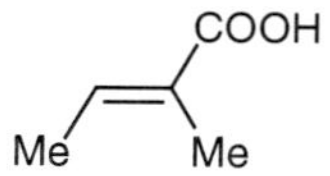	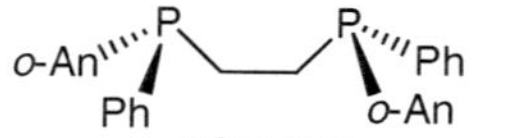 DiPAMP	Quant.	25–40 (S)	108, 109
2	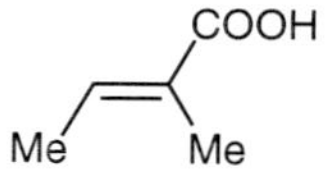	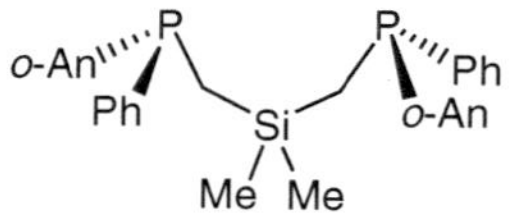 DiPAMPSi	Quant.	15 (S)	109
3	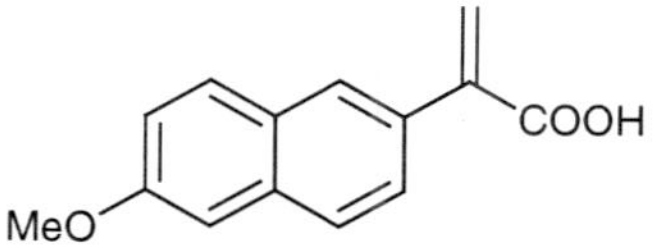	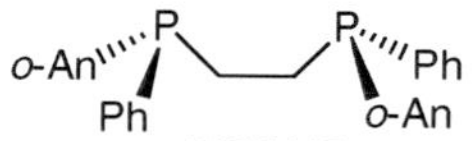 DiPAMP	Quant.	49–55 (R)	108, 109
4	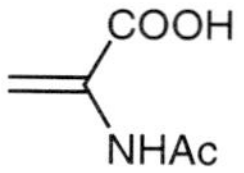 AAA	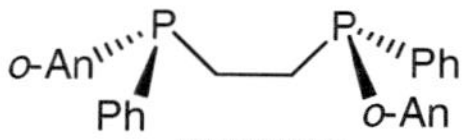 DiPAMP	85	25–38 (R)	108, 109
5	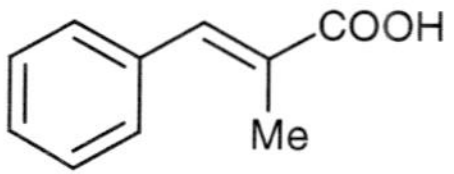	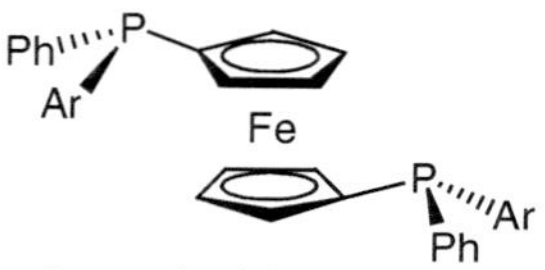 Ar = o-An (a), 1-Naphth (b)	16 (a) Quant. (b)	36 (R, a) 22 (R, b)	110

Table 7.7 (*Continued*).

Entry	Substrate	Ligand	Yield (%)	ee (%)	References
6	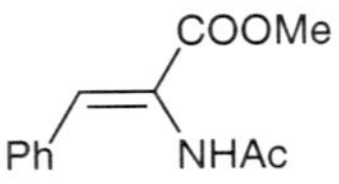 **MAC**	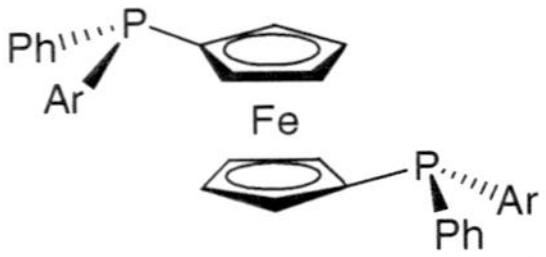 Ar = o-An (**a**), 1-Naphth (**b**)	98 (**a**) Quant. (**b**)	5 (*R*, **a**) 18 (*S*, **b**)	110
7	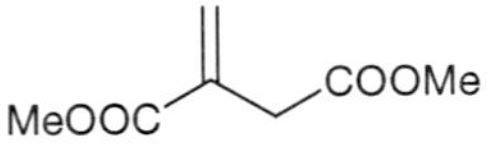 **DMI**	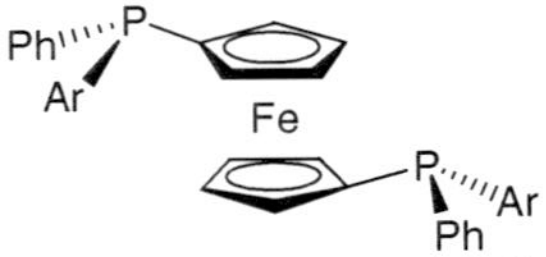 Ar = o-An (**a**), 1-Naphth (**b**)	89 (**a**) Quant. (**b**)	0 (**a**) 17 (*R*, **b**)	110
8	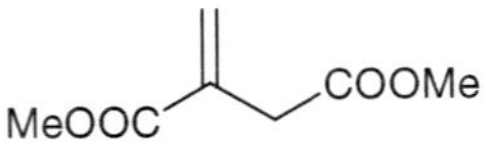 **DMI**	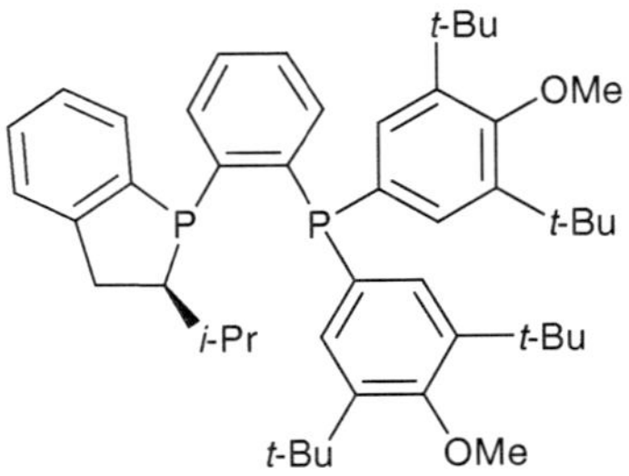 *i*-Pr-mBeePHOS-DTBM	78^a	16 (*R*)	46

^{a}Value for the conversion at 15 h.

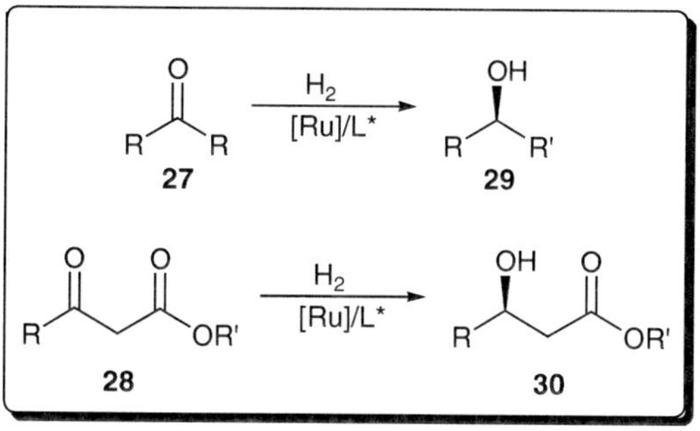

Scheme 7.12 Ru-catalysed enantioselective hydrogenation of ketones (**27**) and β-keto esters (**28**).

The hydrogenation of acetophenone met with little success in terms of enantioselectivity with the systems reported so far (entries 1–3). There are very few reports on hydrogenation of diketones although *t*-Bu-QuinoxP* seems to a very good ligand for these systems (entry 5). The hydrogenation of α-keto esters has also been little studied although encouraging results have been reported (entries 6 and 7). Most of the work with has been carried out with β-keto esters (entries 8–15). Very good to exceptional results in yield and enantioselectivity have been provided by *t*-Bu-BisP* (entries 10 and 13), TangPhos (entries 12 and 14) and *t*-Bu-QuinoxP* (entries 11 and 15). Entries 17 and 18 indicate the same ligands are efficient in reducing β-keto amides. In spite of that, entry 16 shows that even these very stereoselective ligands fail for some substrates.

7.3.2 Ir-Catalysed Hydrogenation

The arena of enantioselective homogeneous hydrogenation is clearly dominated by Rh(I) and Ru(II) complexes, followed at a distance by Ir(I) systems, whose potential is being intensively explored at present. Ir complexes with chiral phosphines have been found to be particularly good ligands for hydrogenation of unfunctionalised alkenes and imines. Very good results for these two kinds of substrates have been obtained with *P*-stereogenic ligands.

7.3.2.1 Alkenes. Trisubstituted (*E*)-alkenes **31** have been hydrogenated in the presence of Ir(I) complexes bearing bidentate *P*-stereogenic ligands (Scheme 7.13 and Table 7.9).

Both unfunctionalised substrates (entries 1–5) and functionalised substrates (entries 6–10) can be quantitatively hydrogenated with high enantioselectivity. With (*E*)-methylstilbene derivatives (entries 1–4), *P*-stereogenic phosphinooxazolines are highly efficient. Entries 1 and 2 show that the absolute configuration of the product is determined by the stereogenic carbon of the oxazoline ring. For (*E*)-2-(*p*-anisyl)-2-butene a good result has been obtained with diazaphospholidine-oxazoline ligand of entry 5. Methyl- and ethyl (*E*)-β-methylcinnamates

Table 7.8 Results in enantioselective Ru-catalysed hydrogenation of ketones (Scheme 7.12).

Entry	Substrate	Ligand	Yield (%)	ee (%)	References
1	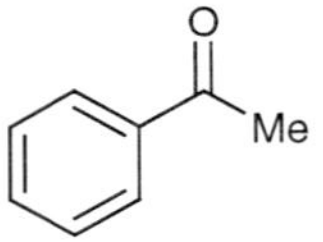	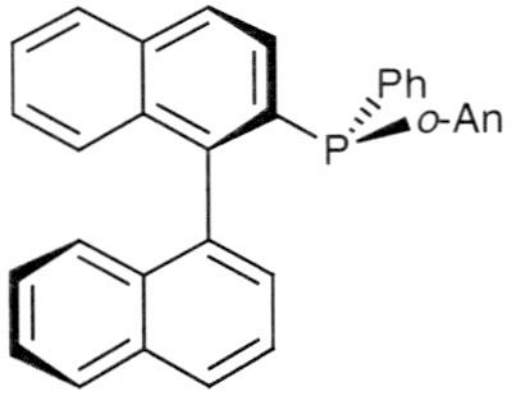	Quant.[a]	13 (S)	113
2	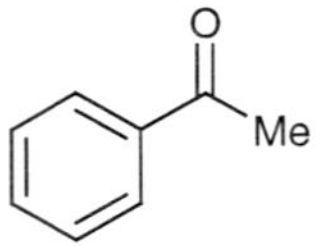	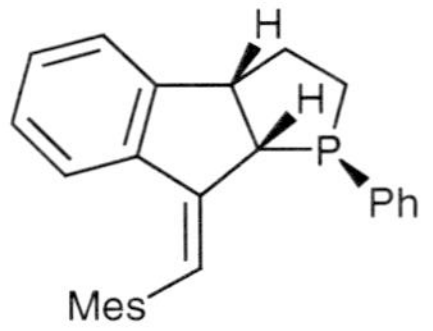	56[a]	33 (R)	114
3	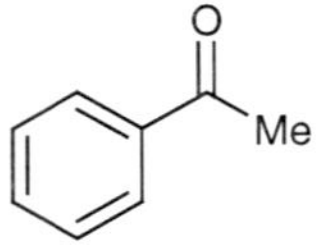	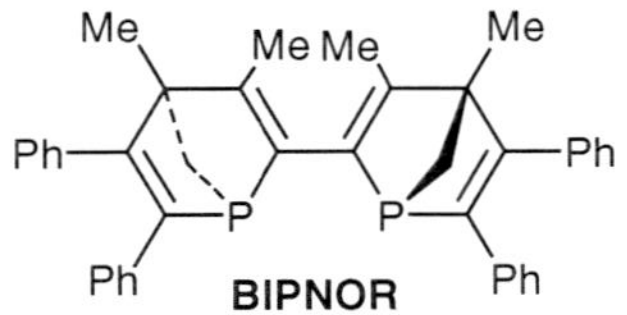 BIPNOR	30[a]	60 (R)	35, 66
4[b]	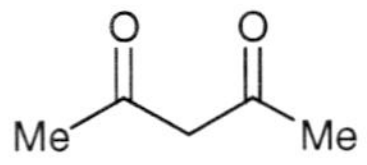	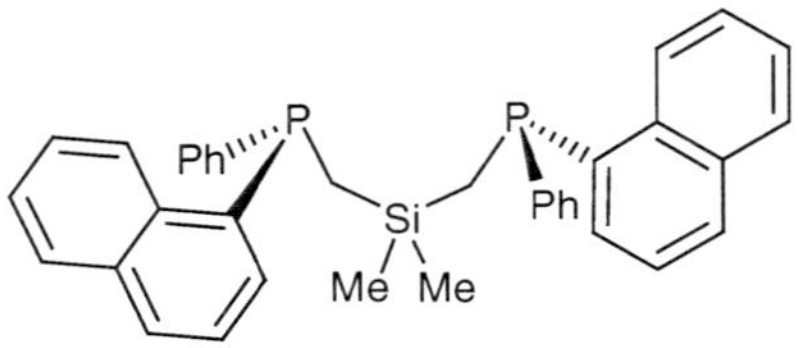	93	56 (S,S)	41

Entry		Ligand	Yield (%)	ee	Ref.
5[c]	Me–CO–CH₂–CO–t-Bu	**t-Bu-QuinoxP***	93	99.0	115
6	Ph–CO–CO–OMe	**t-Bu-BisP***	90	70 (R)	116
7	Me₂-substituted dioxolanone	**DiPAMP**	35	80 (R)	108, 109
8	Me–CO–CH₂–CO–OEt	Ar = 1-Naphth	Quant.	53 (R)	110
9	Me–CO–CH₂–CO–OMe	**C₂-BIPNOR**	Quant.[a]	70 (R)	47

Table 7.8 (*Continued*).

Entry	Substrate	Ligand	Yield (%)	ee (%)	References
10	R—C(O)—CH₂—C(O)—OMe, R = Me, Et	*t*-Bu-BisP*	86 (Me) 96 (Et)	97 (Me, *R*) 98 (Et, *R*)	116
11	Me—C(O)—CH₂—C(O)—OMe	*t*-Bu-QuinoxP*	Quant.	99.7 (*R*)	115
12	R—C(O)—CH₂—C(O)—OMe, R = Me, Et, *i*-Pr	TangPhos	Quant.[a]	99.8 (Me, *R*) 98.8 (Me, *R*) 99.0 (Me, *S*)	117
13	Ph—C(O)—CH₂—C(O)—OEt	*t*-Bu-BisP*	Quant.	89 (*R*)	116
14	Ar—C(O)—CH₂—C(O)—OEt, Ar = Ph, *o*-An, *p*-An	TangPhos	Quant.[a]	96.2 (Ph, *S*) 94.9 (*o*-An, *S*) 97.5 (*p*-An, *S*)	117

15	Ar = Ph, *p*-An, *p*-Tol	*t*-**Bu-QuinoxP***	89 (Ph) 73 (*p*-An) 94 (*p*-Tol)	99.3 (Ph, *S*) 99.3 (*p*-An, *S*) 98.9 (*p*-Tol, *S*)	115
16		**TangPhos**	Quant.[a]	17 (*R*)	118
17		*t*-**Bu-BisP***	Quant.	89 (*R*)	116
18		*t*-**Bu-QuinoxP***	94	97	115

[a]Value for the conversion.
[b]The data refers to the *bis*(hydrogenated) product (2,4-penatanediol).
[c]The data refers to one of the monohydrogenated products (5-hydroxy-2,2-dimethyl-3-hexanone).

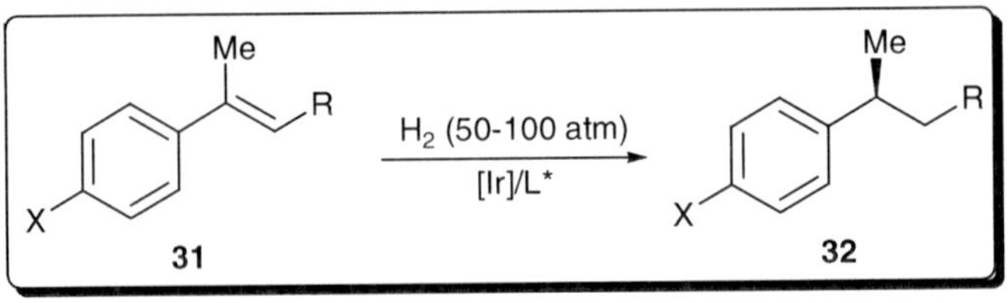

Scheme 7.13   Ir-catalysed hydrogenation of (E)-trisubstituted alkenes.

(entries 6–10) are also competent substrates for the reaction. Once again, the sense of enantioselectivity was determined by the stereogenic carbon atom (compare entries 6 and 7) and the best result has been obtained with the phosphinooxazolines of entries 8 and 9.

Rh- and Ru-based systems are the catalysts of choice to hydrogenate dehydroamino acids. When those systems fail, switching to iridium can lead to improved results. An example is the hydrogenation of the sterically hindered dehydroamino ester **33** with a secondary phosphine oxide as ligand, depicted in Scheme 7.14.

Minnaard, Feringa, de Vries and co-workers[72] found that the Ir-based catalyst was clearly superior to rhodium with using *t*-butylphenyphosphine oxide as ligand. In contrast, with other substrates (*N*-acyl dehydroamino acids, itaconic acid and enol carbamates) rhodium catalysts lead to higher enantioselectivities.

7.3.2.2   Imines. The hydrogenation of imines is challenging because is difficult to find catalysts providing high enantioselectivity, broad applicability and acceptable rates at moderate pressures of hydrogen.[2] Additional complications are the interconversions between the E and Z isomers in case of the acyclic substrates, the tendency to racemisation of the secondary amines obtained after the reaction and the inhibition of the catalyst by amine coordination. Iridium-based catalysts are finding increased application in this reaction. The most common substrates are (E)-imines derived from acetophenone (**35**, Scheme 7.15).

With these substrates, very interesting results have been obtained with several *P*-stereogenic ligands. Some of them are listed in Table 7.10.

Good results have been obtained with *t*-butylphenyphosphine oxide as ligand despite its simplicity (entries 1 and 2) although other secondary phosphine oxides did not lead to any improvement.[121] *N*-phenyl imine of entry 3 gave a racemic product with *t*-butylphenyphosphine oxide (entry 3) and other secondary phosphine oxides. In contrast, C_2-BIPNOR (entry 4) and TangPhos (entry 5) are relatively good ligands for this substrate although they are outperformed by the phosphine-phosphite ligand of entry 6 and by *t*-BuBisP* (entry 7). With this diphosphine, the substrate of entry 8 was hydrogenated in excellent yield and outstanding enantioselectivity with BARF as counterion and only at 1 atm of H_2.[124] The best result for the unsubstituted *N*-phenyl substrate has been obtained with DuanPhos (entry 9) in the presence of BARF.

Table 7.9 Results of Ir-catalysed enantioselective hydrogenation of **31** (Scheme 7.13).

Entry	X	R	Ligand	ee (%)	References
1	H	Ph	(t-Bu, i-Pr oxazolinyl phosphine)	91 (R)	119
2	H	Ph	(t-Bu, i-Pr oxazolinyl phosphine)	77 (S)	119
3	H	Ph	(t-Bu, Ph oxazolinyl phosphine)	95 (R)	119
4	OMe	Ph	(t-Bu, Ph oxazolinyl phosphine)	91 (R)	119
5	OMe	Me	(t-Bu oxazoline phosphoramidite, Ph)	91 (R)	120
6	H	COOMe	(t-Bu, i-Pr oxazolinyl phosphine)	94 (R)	119
7	H	COOMe	(t-Bu, i-Pr oxazolinyl phosphine)	93 (S)	119
8	H	COOMe	(t-Bu, Ph oxazolinyl phosphine)	98 (R)	119
9	Cl	COOMe	(t-Bu, Ph oxazolinyl phosphine)	98 (R)	119
10	H	COOEt	(t-Bu oxazoline phosphoramidite, Ph)	92 (R)	120

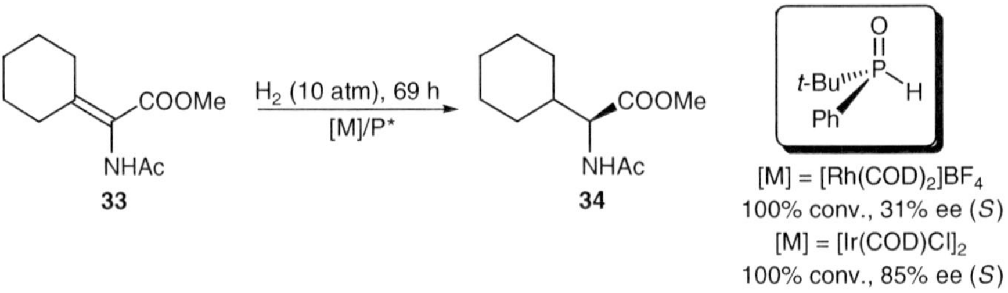

Scheme 7.14 Rh- and Ir-catalysed hydrogenation of **33**.

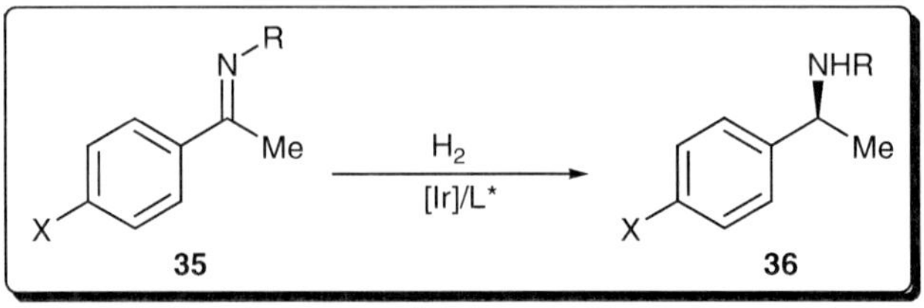

Scheme 7.15 Ir-catalysed hydrogenation of (E)-imines derived from acetophenone.

With this diphosphine many other substrates were hydrogenated quantitatively and in high enantioselectivity.[122]

7.3.3 Pd-Catalysed Hydrogenation

Although heterogeneous palladium catalysts have been extensively used in hydrogenations, there are relatively few reports on homogeneous hydrogenations with palladium complexes. Zhang and co-workers[125] reported the hydrogenation of *N*-tosylimines **37** employing Pd(II) salts and the electron-rich TangPhos ligand, a system leading to exceedingly high enantioselectivities (Scheme 7.16).

N-tosylimines were chosen since they can be easily prepared as pure *E* isomers and because of the electron-withdrawing character of the tosyl group, which reduces the basicity of the formed amines **38** and therefore their inhibition capacity by coordination to the catalytic species.

Although good results were also obtained with Rh catalysts (91% ee and quantitative conversion after 12 h; TangPhos; R = Ph, R′ = Me),[125] the best enantioselectivities were obtained with palladium trifluoroacetate and TangPhos. In all cases, quantitative conversions were found after 24 hours. For arylmethylimines almost perfect enantioselectivities were achieved, decreasing slightly when the methyl was replaced by an ethyl group. An impressive ee was also observed by the *t*-butyl-substituted imine although for the cyclopropyl counterpart a significantly lower enantioselectivity was found, probably due to the smaller steric differentiation between R and R′. A drawback of this transformation is the relatively

Table 7.10 Results in the enantioselective hydrogenation of **35** (Scheme 7.15).

Entry	X	R	Ligand	ee (%)	References
1	H	Bn		82 (S)	121
2	OMe	Bn		83	121
3	H	Ph		0	121
4	H	Ph	**C₂-BIPNOR**	57 (R)	47
5	H	Ph	**TangPhos**	73 (R)	122
6	H	Ph		84 (R)	123
7	H	Ph	**t-Bu-BisP***	86 (R)	124
8	H	p-CF₃C₆H₄	**t-Bu-BisP***	99	124
9	H	Ph	**DuanPhos**	93 (R)	122

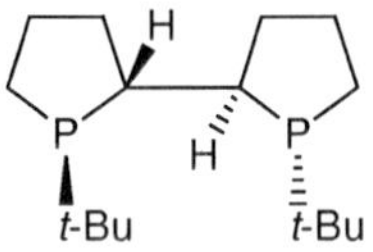

TangPhos

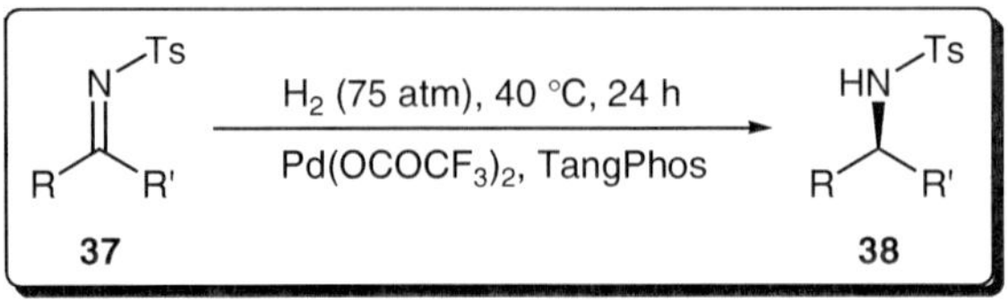

R = Ph, R' = Me
>99% conv., 99% ee (*R*)
R = *p*-An, R' = Me
>99% conv., 99% ee
R = 2-Naphth, R' = Me
>99% conv., >99% ee
R = Ph, R' = Et
>99% conv., 93% ee (*R*)
R = *t*-Bu, R' = Me
>99% conv., 98% ee
R = Cyclopropyl, R' = Me
>99% conv., 75% ee

Scheme 7.16 Pd-catalysed enantioselective hydrogenation of *N*-tosylimines **37**.

high pressure of hydrogen required to achieve good conversions. The BINAPINE ligand was also tested for this reaction (R = Ph, R' = Me), leading only to 15% conversion after 24 hours.

7.4 Hydrosilylation

Hydrosilylation reactions are formal additions of Si–H units across multiple bonds. They are fundamental reactions in organosilicon chemistry. Despite early reports of Marinetti and Ricard[126,127] on Pd-catalysed hydrosilylation of alkenes with phosphetanes (up to 65% ee with styrene) and of Zhang and co-workers[128] on Ru-catalysed hydrosilylation of ketones (up to 54% ee with acetophenone), most of the work on enantioselective hydrosilylation with *P*-stereogenic ligands has been carried out with Rh(I) complexes and prochiral ketones as substrates. Initially, silyl ethers are formed but they are usually cleaved under acidic conditions affording alcohols. As a result, hydrosilylations of ketones are formally identical to hydrogenations but do not involve the manipulation of dihydrogen. The model substrate for enantioselective hydrosilylation is acetophenone (Scheme 7.17).

With this substrate, good results have been obtained with several *P*-stereogenic diphosphine ligands (Table 7.11).

Table 7.11 shows that some *P*-stereogenic ligands are very good ligands for the reaction. Several trends can be observed. Among the electron-rich phosphines, R-BisP* ligands are much faster than R-MiniPHOS (compare entries 1 and 2) showing increased enantioselectivity with increasing bulkiness or R (entries 2 and 3). The ferrocenyl diphosphine of entry 4 and *t*-BuQuinoxP* are also competent ligands for the reaction. The silane has a profound impact on the activity and enantioselectivity of the reaction, with the bulkier

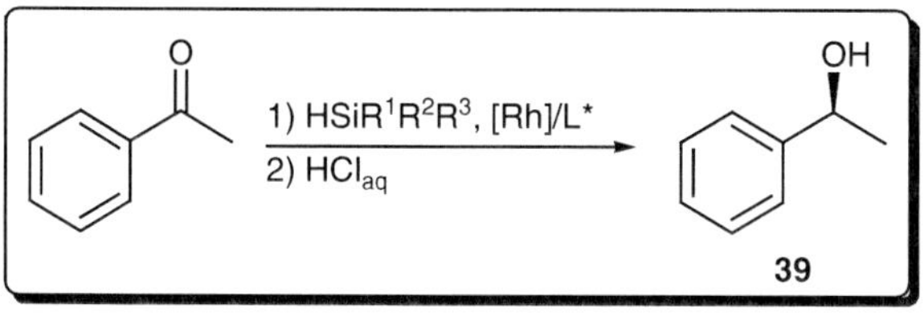

1) HSiR^{1}R^{2}R^3, [Rh]/L*
2) HCl$_{aq}$

39

Scheme 7.17 Model reaction for enantioselective ketone hydrosilylation.

1-naphthylphenylsilane affording better results than diphenylsilane (compare entries 5 and 6).

Imamoto and co-workers[129] studied the mechanism of the reaction, concluding that it is similar to the 'dihydride' mechanism of Rh-catalysed hydrogenations (Section 7.2.2), with the enantioselection occurring at the migratory insertion step.

7.5 Hydrogen Transfer

Hydrogen transfer reactions consist of formal transfers of two hydrogen atoms from a hydrogen donor to a prochiral substrate, catalysed by ruthenium(II) complexes under basic conditions.[131–137] The model substrate for hydrogen transfer of ketones is once again acetophenone with isopropanol as hydrogen source and potassium *tert*-butoxide or KOH as base (Scheme 7.18).

This reaction has the important advantage of not involving hydrogen gas but on the downside the reaction is reversible in some cases, leading to incomplete conversions and racemisation of **39**. This problem is minimised by performing the reaction in isopropanol as a solvent. There are few results with *P*-stereogenic ligands, listed in Table 7.12.

Entry 1 shows that low to moderate enantioselectivities can be obtained with simple monodentate phosphines whereas low enantioselectivities have been obtained with the P,N ligand of entry 2 or with the pincer ligand of entry 3. Much better results have been obtained with β-aminophosphines and their oxides (entries 4–6). Interestingly, it has been found that phosphine oxides provide better activities and similar enantioselectivities than trivalent phosphines.[141,143] The best result has been obtained with the ligand of entry 6, derived from (–)-norephedrine.

7.6 Conclusion

There is no doubt that *P*-stereogenic ligands are ideally suited to hydrogenate a wide variety of substrates in very high enantioselectivity. A quick look at the tables contained in this chapter will convince the reader that in spite of the many intricacies, some favoured ligands have emerged. Although it was designed more than three decades ago, DiPAMP (and its modern 'heavier' analogues) is still a valuable ligand for hydrogenation of many substrates.

Table 7.11 Results in the enantioselective hydrosilylation of acetophenone (Scheme 7.17).

Entry	Ligand	Silane	Time (h)	Yield (%)	ee (%)	References
1	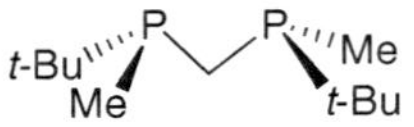 *t*-Bu-MiniPHOS	1-Naphth(Ph)SiH$_2$	3–4[a]	86	91	6, 129
2	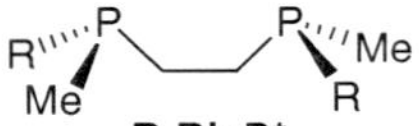 R-BisP*	1-Naphth(Ph)SiH$_2$	30 (*t*-Bu) / 30 (1-Ad)	99 (*t*-Bu) / 86 (1-Ad)	93 (*t*-Bu) / 95 (1-Ad)	129
3	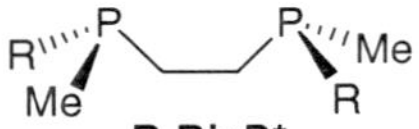 R-BisP*	Ph$_2$SiH$_2$	30 (Cy) / 30 (*t*-Bu)	80 (Cy) / 89 (*t*-Bu)	75 (Cy) / 82 (*t*-Bu)	129
4	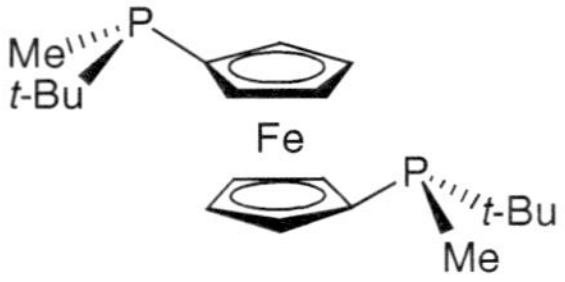	1-Naphth(Ph)SiH$_2$	–	96	92 (*S*)	130
5	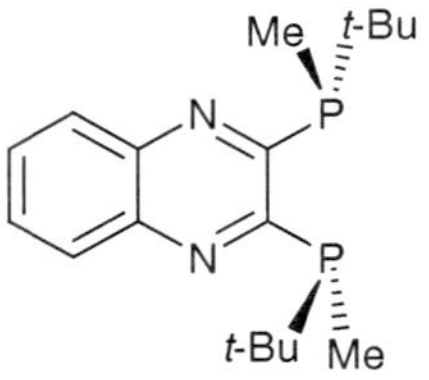 *t*-Bu-QuinoxP*	Ph$_2$SiH$_2$	2	22	71	129
6	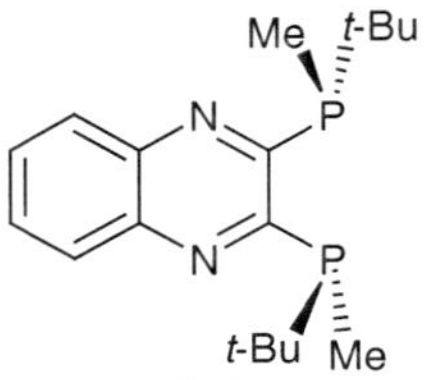 *t*-Bu-QuinoxP*	1-Naphth(Ph)SiH$_2$	2	98	89	129

[a]Days

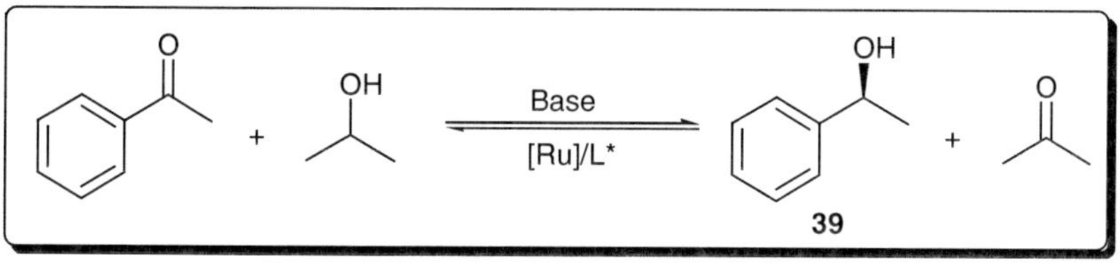

Scheme 7.18 Model reaction for enantioselective hydrogen transfer.

Table 7.12 Results of enantioselective hydrogen transfer to acetophenone (Scheme 7.18).

Entry	Ligand	Base	T (°C)	ee (%)	References
1	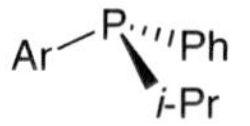 Ar–P(''Ph)(i-Pr); Ar = 1-Naphth; **a**, Ar = 2-Biph; **b**	KO^{t}Bu	82	20 (**a**), 45 (**b**)	138
2	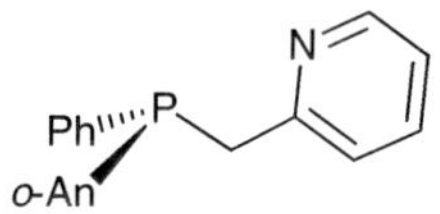 Ph'''P(o-An)CH$_2$-(2-pyridyl)	KO^{i}Pr	45	12 (*R*)	139
3	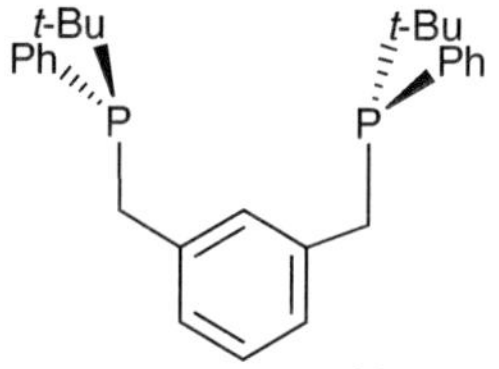 (t-Bu)(Ph)P–CH$_2$–C$_6$H$_4$–CH$_2$–P(Ph)(t-Bu) (1,3-xylylene)	KOH	82	18 (*R*)	140
4	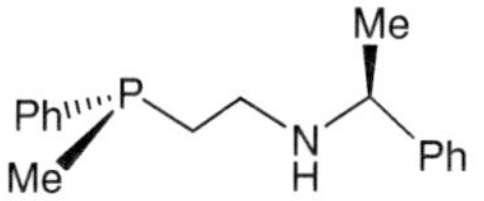 Ph'''P(Me)CH$_2$CH$_2$N(H)CH(Me)Ph	KOH	20	57 (*S*)	141
5	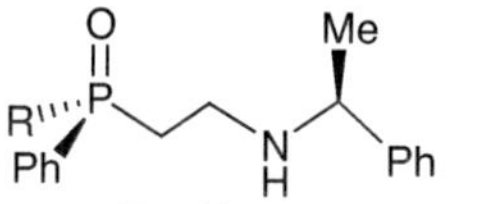 O=P(R)(Ph)CH$_2$CH$_2$N(H)CH(Me)Ph; R = Me; **a**, R = t-Bu; **b**	KOH	20	55 (*R*, **a**), 50 (*R*, **b**)	141
6	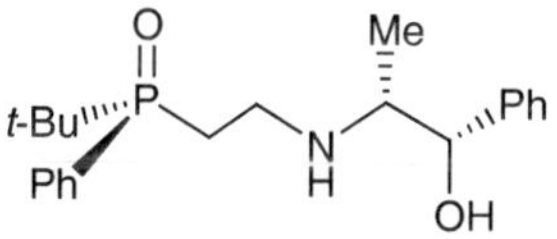 O=P(t-Bu)(Ph)CH$_2$CH$_2$N(H)CH(Me)CH(Ph)OH	KOH	29	93 (*R*)	142

Related C_2-symmetric phosphines BisP* and MiniPHOS, bearing an alkyl bridge like DiPAMP but possessing electron-richer and sterically differentiated phosphorus atoms, afford outstanding results for many substrates. Equal or even superior performance is delivered with the C_1-symmetric diphosphine Trichickenfootphos. This ligand is even more remarkable given its simplicity; its chirality is due to a single *P*-stereogenic atom. Another successful family of ligands is composed of heterocyclic diphosphines. In particular, phospholane-based

TangPhos, DuanPhos and BINAPINE have shown wide applicability and exceptional enantioselectivities.

The extraordinary results in Rh-catalysed enantioselective hydrogenation have run in parallel to tremendous progress into mechanistic understanding since the early days of the reaction. At present, the border between the classical *unsaturated* and *dihydride* reaction mechanisms is more blurred than ever, leading to an overall mechanism that is dependent on (and can be accommodated to) the exact substrate, ligand and reaction conditions.

Apart from intrinsic interest, enantioselective hydrogenation is also the arena where the performance of virtually every new chiral phosphorus ligand is tested before being used in other reactions. Comparison between the present chapter and Chapter 8 shows that *P*-stereogenic ligands are no exception. Indeed, some of the best ligands for C–C bond forming reactions also form exceptional hydrogenation catalysts. Based on that, with the improvement of synthetic methodologies, much progress in the use of *P*-stereogenic ligands in many reactions can be confidently expected.

References

1. *Handbook of Homogeneous Hydrogenation*, ed. J. G. de Vries and C. J. Elsevier, Wiley-VCH, Weinheim, 2007.
2. W. Tang and X. Zhang, *Chem. Rev.*, 2003, **103**, 3029.
3. M. Stephan, D. Sterk and B. Mohar, *Adv. Synth. Catal.*, 2009, **351**, 2779.
4. B. Zupancic, B. Mohar and M. Stephan, *Org. Lett.*, 2010, **12**, 1296.
5. Y. Wada, T. Imamoto, H. Tsuruta, K. Yamaguchi and I. D. Gridnev, *Adv. Synth. Catal.*, 2004, **346**, 777.
6. Y. Yamanoi and T. Imamoto, *J. Org. Chem.*, 1999, **64**, 2988.
7. I. D. Gridnev, Y. Yamamoi, N. Higashi, H. Tsuruta, M. Yasutake and T. Imamoto, *Adv. Synth. Catal.*, 2001, **343**, 118.
8. G. Hoge, H. Wu, W. S. Kissel, D. A. Pflum, D. J. Greene and J. Bao, *J. Am. Chem. Soc.*, 2004, **126**, 5966.
9. I. D. Gridnev, T. Imamoto, G. Hoge, M. Kouchi and H. Takahashi, *J. Am. Chem. Soc.*, 2008, **130**, 2560.
10. G. Hoge and B. Samas, *Tetrahedron: Asymmetry*, 2004, **15**, 2155.
11. D. Liu and X. Zhang, *Eur. J. Org. Chem.*, 2005, **646**.
12. K. N. Gavrilov, E. B. Benetskiy, T. B. Grishina, E. A. Rastorguev, M. G. Maksimova, S. V. Zheglov, V. A. Davankov, B. Schaffner, A. Börner, J. M. Rosset, G. Bailat and A. Alexakis, *Eur. J. Org. Chem.*, 2009, 3923.
13. M. T. Reetz, J. Ma and R. Goddard, *Angew. Chem. Int. Ed.*, 2005, **44**, 412.
14. J. W. Scott, D. D. Keith, G. Nix, D. R. Parrish, S. Remington, G. P. Roth, J. M. Townsend, D. Valentine and R. Yang, *J. Org. Chem.*, 1981, **46**, 5086.
15. B. Zupancic, B. Mohar and M. Stephan, *Adv. Synth. Catal.*, 2008, **350**, 2024.

16. T. Imamoto, J. Watanabe, Y. Wada, H. Masuda, H. Yamada, H. Tsuruta, S. Matsukawa and K. Yamaguchi, *J. Am. Chem. Soc.*, 1998, **120**, 1635.
17. M. Sugiya and H. Nohira, *Bull. Chem. Soc. Jpn.*, 2000, **73**, 705.
18. A. Ohashi and T. Imamoto, *Org. Lett.*, 2001, **3**, 373.
19. N. Oohara, K. Katagiri and T. Imamoto, *Tetrahedron: Asymmetry*, 2003, **14**, 2171.
20. A. Ohashi, S. Kikuchi, M. Yasutake and T. Imamoto, *Eur. J. Org. Chem.*, 2002, 2535.
21. T. Miura and T. Imamoto, *Tetrahedron Lett.*, 1999, **40**, 4833.
22. T. Imamoto, N. Oohara and H. Takahashi, *Synthesis*, 2004, 1353.
23. G. Hoge, *J. Am. Chem. Soc.*, 2003, **125**, 10219.
24. W. Tang, B. Qu, A. G. Capacci, S. Rodriguez, X. Wei, N. Haddad, B. Narayanan, S. Ma, N. Grinberg, N. K. Yee, D. Krishnamurthy and C. H. Senanayake, *Org. Lett.*, 2010, **12**, 176.
25. W. Tang, A. G. Capacci, A. White, S. Ma, S. Rodriguez, B. Qu, J. Savoie, N. D. Patel, X. Wei, N. Haddad, N. Grinberg, N. K. Yee, D. Krishnamurthy and C. H. Senanayake, *Org. Lett.*, 2010, **12**, 1104.
26. W. Chen, W. Mbafor, S. M. Roberts and J. Whittall, *J. Am. Chem. Soc.*, 2006, **128**, 3922.
27. Q. Dai, W. Li and X. Zhang, *Tetrahedron*, 2008, **64**, 6943.
28. W. S. Knowles, M. J. Sabacky, B. D. Vineyard and D. J. Weinkauff, *J. Am. Chem. Soc.*, 1975, **97**, 2567.
29. B. D. Vineyard, W. S. Knowles, M. J. Sabacky, G. L. Bachman and D. J. Weinkauff, *J. Am. Chem. Soc.*, 1977, **99**, 5946.
30. C. R. Johnson and T. Imamoto, *J. Org. Chem.*, 1987, **52**, 2170.
31. U. Nagel and T. Krink, *Angew. Chem. Int. Ed. Engl.*, 1993, **32**, 1052.
32. W. Tang and X. Zhang, *Angew. Chem. Int. Ed.*, 2002, **41**, 1612.
33. U. Nettekoven, P. C. J. Kamer, P. W. N. M. van Leeuwen, M. Widhalm, A. L. Spek and M. Lutz, *J. Org. Chem.*, 1999, **64**, 3996.
34. F. Maienza, M. Wörle, P. Steffanut, A. Mezzetti and F. Spindler, *Organometallics*, 1999, **18**, 1041.
35. F. Mathey, F. Mercier, F. Robin and L. Ricard, *J. Organomet. Chem.*, 1998, **577**, 117.
36. O. G. Bondarev and R. Goddard, *Tetrahedron Lett.*, 2006, **47**, 9013.
37. T. Imamoto, H. Tsuruta, Y. Wada, H. Masuda and K. Yamaguchi, *Tetrahedron Lett.*, 1995, **36**, 8271.
38. H. Tsuruta, T. Imamoto, K. Yamaguchi and I. D. Gridnev, *Tetrahedron Lett.*, 2005, **46**, 2879.
39. F. Maienza, F. Spindler, M. Thommen, B. Pugin, C. Malan and A. Mezzetti, *J. Org. Chem.*, 2002, **67**, 5239.
40. T. Imamoto, Y. Saitoh, A. Koide, T. Ogura and K. Yoshida, *Angew. Chem. Int. Ed.*, 2007, **46**, 8636.
41. R. M. Stoop, A. Mezzetti and F. Spindler, *Organometallics*, 1998, **17**, 668.
42. U. Nagel and B. Rieger, *Organometallics*, 1989, **8**, 1534.
43. T. Imamoto, K. Sugito and K. Yoshida, *J. Am. Chem. Soc.*, 2005, **127**, 11934.

44. T. Imamoto, A. Kumada and K. Yoshida, *Chem. Lett.*, 2007, **36**, 500.

45. T. Imamoto, K. V. L. Crépy and K. Katagiri, *Tetrahedron: Asymmetry*, 2004, **15**, 2213.

46. H. Shimizu, T. Saito and H. Kumobayashi, *Adv. Synth. Catal.*, 2003, **345**, 185.

47. M. Siutkowski, F. Mercier, L. Ricard and F. Mathey, *Organometallics*, 2006, **25**, 2585.

48. D. Moulin, C. Darcel and S. Jugé, *Tetrahedron: Asymmetry*, 1999, **10**, 4729.

49. C. Darcel, D. Moulin, J. Henry, M. Lagrelette, P. Richard, P. D. Harvey and S. Jugé, *Eur. J. Org. Chem.*, 2007, 2078.

50. D. Carmichael, H. Doucet and J. M. Brown, *Chem. Commun.*, 1999, 261.

51. A. Suárez, M. A. Méndez Rojas and A. Pizzano, *Organometallics*, 2002, **21**, 4611.

52. W. Chen, S. M. Roberts, J. Whittall and A. Steiner, *Chem. Commun.*, 2006, 2916.

53. H. Drexler, J. You, S. Zhang, C. Fischer, W. Baumann, A. Spannenberg and D. Heller, *Org. Process Res. Dev.*, 2003, **7**, 355.

54. H. Drexler, W. Baumann, T. Schmidt, S. Zhang, A. Sun, A. Spannenberg, C. Fischer, H. Bushcmann and D. Heller, *Angew. Chem. Int. Ed.*, 2005, **44**, 1184.

55. B. Weiner, W. Szymanski, D. B. Janssen, A. J. Minnaard and B. L. Feringa, *Chem. Soc. Rev*, 2010, **39**, 1656.

56. M. Yasutake, I. D. Gridnev, N. Higashi and T. Imamoto, *Org. Lett.*, 2001, **3**, 1701.

57. H. Wu and G. Hoge, *Org. Lett.*, 2004, **6**, 3645.

58. W. Tang and X. Zhang, *Org. Lett.*, 2002, **4**, 4159.

59. W. Tang, W. Wang, Y. Chi and X. Zhang, *Angew. Chem. Int. Ed.*, 2003, **42**, 3509.

60. W. Tang, D. Liu and X. Zhang, *Org. Lett.*, 2003, **5**, 205.

61. J. Chen, W. Zhang, H. Geng, W. Li, G. Hou, A. Lei and X. Zhang, *Angew. Chem. Int. Ed.*, 2009, **48**, 800.

62. B. Zupancic, B. Mohar and M. Stephan, *Tetrahedron Lett.*, 2009, **50**, 7382.

63. I. D. Gridnev, N. Higashi and T. Imamoto, *J. Am. Chem. Soc.*, 2000, **122**, 10486.

64. I. D. Gridnev, M. Yasutake, N. Higashi and T. Imamoto, *J. Am. Chem. Soc.*, 2001, **123**, 5268.

65. W. C. Christopfel and B. D. Vineyard, *J. Am. Chem. Soc.*, 1979, **101**, 4406.

66. F. Robin, F. Mercier, L. Ricard, F. Mathey and M. Spagnol, *Chem. Eur. J.*, 1997, **3**, 1365.

67. W. Chen, P. J. McCormack, K. Mohammed, W. Mbafor, S. M. Roberts and J. Whittall, *Angew. Chem. Int. Ed.*, 2007, **46**, 4141.

68. I. D. Gridnev, N. Higashi and T. Imamoto, *J. Am. Chem. Soc.*, 2001, **123**, 4631.

69. I. D. Gridnev, M. Yasutake, T. Imamoto and I. P. Beletskaya, *Proc. Natl. Acad. Sci.*, 2004, **101**, 5385.

70. D. Wang, X. Hu, J. Deng, S. Yu, Z. Duan and Z. Zheng, *J. Org. Chem.*, 2009, **74**, 4408.

71. G. Hoge, *J. Am. Chem. Soc.*, 2004, **126**, 9920.

72. X. Jiang, M. van den Berg, A. J. Minnaard, B. L. Feringa and J. G. de Vries, *Tetrahedron: Asymmetry*, 2004, **15**, 2223.

73. Q. Dai, W. Yang and X. Zhang, *Org. Lett.*, 2005, **7**, 5343.

74. X. Sun, L. Zhou, C. Wang and X. Zhang, *Angew. Chem. Int. Ed.*, 2007, **46**, 2623.

75. A. Preetz, W. Baumann, H. Drexler, C. Fischer, J. Sun, A. Spannenberg, O. Zimmer, W. Hell and D. Heller, *Chem. Asian J.*, 2008, **3**, 1979.

76. T. Schmidt, Z. Dai, H. Drexler, M. Hapke, A. Preetz and D. Heller, *Chem. Asian J.*, 2008, **3**, 1170.

77. T. Schmidt, H. Drexler, J. Sun, Z. Dai, W. Baumann, A. Preetz and D. Heller, *Adv. Synth. Catal.*, 2009, **351**, 750.

78. K. V. L. Crépy and T. Imamoto, *Adv. Synth. Catal.*, 2003, **345**, 79.

79. I. D. Gridnev and T. Imamoto, *Acc. Chem. Res.*, 2004, **37**, 633.

80. I. D. Gridnev and T. Imamoto, *Chem. Commun.*, 2009, 7447.

81. K. Rossen, *Angew. Chem. Int. Ed.*, 2001, **40**, 4611.

82. J. M. Brown, in *Handbook of Homogeneous Hydrogenation*, ed. J. G. de Vries and C. J. Elsevier, Wiley-VCH, Weinheim, 2007, p. 1073.

83. L. A. Oro and D. Carmona, in *Handbook of Homogeneous Hydrogenation*, ed. J. G. de Vries, and C. J. Elsevier, Wiley-VCH, Weineheim, 2007, p. 3.

84. J. M. Brown, P. A. Chaloner and G. A. Morris, *J. Chem. Soc., Chem. Commun.*, 1983, 664.

85. C. R. Landis and J. Halpern, *J. Am. Chem. Soc.*, 1987, **109**, 1746.

86. J. A. Ramsden, T. D. W. Claridge and J. M. Brown, *J. Chem. Soc., Chem. Commun.*, 1995, 2469.

87. J. Halpern, D. P. Riley, A. S. C. Chan and J. J. Pluth, *J. Am. Chem. Soc.*, 1977, **99**, 8055.

88. J. M. Brown and P. A. Chaloner, *J. Am. Chem. Soc.*, 1980, **102**, 3040.

89. A. Preetz, H. Drexler, C. Fischer, Z. Dai, A. Börner, W. Baumann, A. Spannenberg, R. Thede and D. Heller, *Chem. Eur. J.*, 2008, **14**, 1445.

90. A. Preetz, W. Baumann, C. Fischer, H. Drexler, T. Schmidt, R. Thede and D. Heller, *Organometallics*, 2009, **28**, 3673.

91. J. M. Brown and P. A. Chaloner, *J. Chem. Soc., Chem. Commun.*, 1978, 321.

92. A. S. C. Chan, J. J. Pluth and J. Halpern, *J. Am. Chem. Soc.*, 1980, **102**, 5952.

93. J. M. Brown and D. Parker, *J. Org. Chem.*, 1982, **47**, 2722.

94. B. McCulloch, J. Halpern, M. R. Thompson and C. R. Landis, *Organometallics*, 1990, **9**, 1392.

95. T. Schmidt, W. Baumann, H. Drexler, A. Arrieta, D. Heller and H. Bushcmann, *Organometallics*, 2005, **24**, 3842.

96. T. Schmidt, Z. Dai, H. Drexler, W. Baumann, C. Jager, D. Pfeifer and D. Heller, *Chem. Eur. J.*, 2008, **14**, 4469.

97. J. M. Brown and P. A. Chaloner, *J. Chem. Soc. Chem. Commun.*, 1980, 344.

98. A. S. C. Chan and J. Halpern, *J. Am. Chem. Soc.*, 1980, **102**, 838.

99. I. D. Gridnev, N. Higashi, K. Asakura and T. Imamoto, *J. Am. Chem. Soc.*, 2000, **122**, 7183.

100. H. Heinrich, R. Giernoth, J. Bargon and J. M. Brown, *Chem. Commun.*, 2001, 1296.

101. R. Giernoth, H. Heinrich, N. J. Adams, R. J. Deeth, J. Bargon and J. M. Brown, *J. Am. Chem. Soc.*, 2000, **122**, 12381.

102. I. D. Gridnev, N. Higashi and T. Imamoto, *Organometallics*, 2001, **20**, 4542.

103. D. Liu, W. Gao, C. Wang and X. Zhang, *Angew. Chem. Int. Ed.*, 2005, **44**, 1687.

104. G. Shang, D. Liu, S. E. Allen, Q. Yang and X. Zhang, *Chem. Eur. J.*, 2007, **13**, 7780.

105. H. Geng, W. Zhang, J. Chen, G. Hou, L. Zhou, Y. Zou, W. Wu and X. Zhang, *Angew. Chem. Int. Ed.*, 2009, **48**, 6052.

106. G. Shang, Q. Yang and X. Zhang, *Angew. Chem. Int. Ed.*, 2006, **45**, 6360.

107. T. Naota, H. Takaya and S. Murahashi, *Chem. Rev.*, 1998, **98**, 2599.

108. J. P. Genêt, C. Pinel, S. Mallart, S. Jugé, N. Cailhol and J. A. Laffitte, *Tetrahedron Lett.*, 1992, **33**, 5343.

109. J. P. Genêt, C. Pinel, V. Ratovelomanana-Vidal, S. Mallart, X. Pfister, L. Bischoff, M. C. C. de Andrade, S. Darses, C. Galopin and J. A. Laffitte, *Tetrahedron: Asymmetry*, 1994, **5**, 675.

110. F. Maienza, F. Santoro, F. Spindler, C. Malan and A. Mezzetti, *Tetrahedron: Asymmetry*, 2002, **13**, 1817.

111. J. P. Genêt, C. Pinel, V. Ratovelomanana-Vidal, S. Mallart, X. Pfister, M. C. Caño de Andrade and J. A. Laffitte, *Tetrahedron: Asymmetry*, 1994, **5**, 665.

112. D. J. Ager and S. A. Laneman, *Tetrahedron: Asymmetry*, 1997, **8**, 3327.

113. M. Jahjah, R. Jahjah, S. Pellet-Rosting and M. Lemaire, *Tetrahedron: Asymmetry*, 2007, **18**, 1224.

114. M. L. Christ, M. Zablocka, S. Spencer, R. J. Lavender, M. Lemaire and J. P. Majoral, *J. Mol. Catal. A*, 2006, **245**, 210.

115. T. Imamoto, M. Nishimura, A. Koide and K. Yoshida, *J. Org. Chem.*, 2007, **72**, 7413.

116. T. Yamano, N. Taya, M. Kawada, T. Huang and T. Imamoto, *Tetrahedron Lett.*, 1999, **40**, 2577.

117. C. Wang, H. Tao and X. Zhang, *Tetrahedron Lett.*, 2006, **47**, 1901.

118. Y. Xu, H. Wei, J. Zhang and G. Huang, *Tetrahedron Lett.*, 1989, **30**, 949.

119. W. Tang, W. Wang and X. Zhang, *Angew. Chem. Int. Ed.*, 2003, **42**, 943.

120. R. Hilgraf and A. Pfaltz, *Adv. Synth. Catal.*, 2005, **347**, 61.

121. X. Jiang, A. J. Minnaard, B. Hessen, B. L. Feringa, A. L. L. Duchateau, J. G. O. Andrien, J. A. F. Boogers and J. G. de Vries, *Org. Lett.*, 2003, **5**, 1503.
122. W. Li, G. Hou, M. Chang and X. Zhang, *Adv. Synth. Catal.*, 2009, **351**, 3123.
123. S. Vargas, M. Rubio, A. Suárez, D. del Río, E. Álvarez and A. Pizzano, *Organometallics*, 2006, **25**, 961.
124. T. Imamoto, N. Iwadate and K. Yoshida, *Org. Lett.*, 2006, **8**, 2289.
125. Q. Yang, G. Shang, W. Gao, J. Deng and X. Zhang, *Angew. Chem. Int. Ed.*, 2006, **45**, 3832.
126. A. Marinetti, *Tetrahedron Lett.*, 1994, **35**, 5861.
127. A. Marinetti and L. Ricard, *Organometallics*, 1994, **13**, 3956.
128. G. Zhu, M. Terry and X. Zhang, *J. Organomet. Chem.*, 1997, **547**, 97.
129. T. Imamoto, T. Itoh, Y. Yamanoi, R. Narui and K. Yoshida, *Tetrahedron: Asymmetry*, 2006, **17**, 560.
130. H. Tsuruta and T. Imamoto, *Tetrahedron: Asymmetry*, 1999, **10**, 877.
131. G. Zassinovich, G. Mestroni and S. Gladiali, *Chem. Rev.*, 1992, **92**, 1051.
132. R. Noyori and S. Hashiguchi, *Acc. Chem. Res.*, 1997, **30**, 97.
133. R. Noyori, M. Yamakawa and S. Hashiguchi, *J. Org. Chem.*, 2001, **66**, 7931.
134. J. E. Bäckvall, *J. Organomet. Chem.*, 2002, **652**, 105.
135. S. E. Clapham, A. Hadzovic and R. H. Morris, *Coord. Chem. Rev.*, 2004, **248**, 2201.
136. S. Gladiali and E. Alberico, *Chem. Soc. Rev.*, 2006, **35**, 226.
137. J. S. M. Samec, J. E. Bäckvall, P. G. Andersson and P. Brandt, *Chem. Soc. Rev.*, 2006, **35**, 237.
138. A. Grabulosa, PhD Thesis, Universitat de Barcelona, Barcelona, 2005.
139. H. Yang, M. Alvarez-Gressier, N. Lugan and R. Mathieu, *Organometallics*, 1997, **16**, 1401.
140. S. Medici, G. M. S. B. Williams, P. A. Chase, S. Gladiali, M. Lutz, A. L. Spek, G. P. M. van Klink and G. van Koten, *Helv. Chim. Acta*, 2005, **88**, 694.
141. A. M. Maj, K. M. Pietrusiewicz, I. Suisse, F. Agbossou and A. Mortreux, *J. Organomet. Chem.*, 2001, **626**, 157.
142. X. Cheng, P. N. Horton, M. Hursthouse and K. K. Hii, *Tetrahedron: Asymmetry*, 2004, **15**, 2241.
143. A. M. Maj, K. M. Pietrusiewicz, I. Suisse, F. Agbossou and A. Mortreux, *Tetrahedron: Asymmetry*, 1999, **10**, 831.

CHAPTER 8

Miscellaneous Reactions

8.1 Introduction

The use of *P*-stereogenic ligands in transformations other than hydrogenation, covered in this chapter, is still in its infancy. Most of the chapter (except Sections 8.4.2 and 8.14) is devoted to C–C bond-forming reactions, which is an extremely important transformation in organic chemistry. Ligands with *P*-stereogenic atoms have proved to be superior stereoinductors in several reactions such as alkylative ring-openings (Section 8.7) and hydroacylation (Section 8.11) but at the same time give very poor results in others such as hydroboration (Section 8.14).

A quick glance at the references section shows that many references are from the last decade, convincing the reader that interest in *P*-stereogenic ligands for enantioselective catalysis is by no means limited to hydrogenation.

8.2 Hydrovinylation

The hydrovinylation of alkenes can be defined as the formal addition of ethylene across a double bond or, alternatively, as a codimerisation of ethylene and an activated alkene (Scheme 8.1).[1]

This reaction has a long history, starting in 1965, when it was found that hydrated Rh and Ru chloride catalysed codimerisations between ethylene and a variety of alkenes.[2] As early as 1972, Wilke and co-workers[3] reported the enantioselective hydrovinylation of 1,3-cyclooctadiene with a system containing Ni(II) and P(*i*-Pr)(Men)$_2$, which constitutes the second metal-catalysed enantioselective C–C bond-forming reaction ever reported.

Most of the work has been carried out with vinylarenes as substrates, particularly styrenes, with Ni and Pd systems as catalytic precursors.[4,5] Enantioselective hydrovinylation is very interesting from an industrial point of view because 3-aryl-1-butenes can be easily transformed without racemisation to

RSC Catalysis Series No. 7
P-Stereogenic Ligands in Enantioselective Catalysis
By Arnald Grabulosa
© Arnald Grabulosa 2011
Published by the Royal Society of Chemistry, www.rsc.org

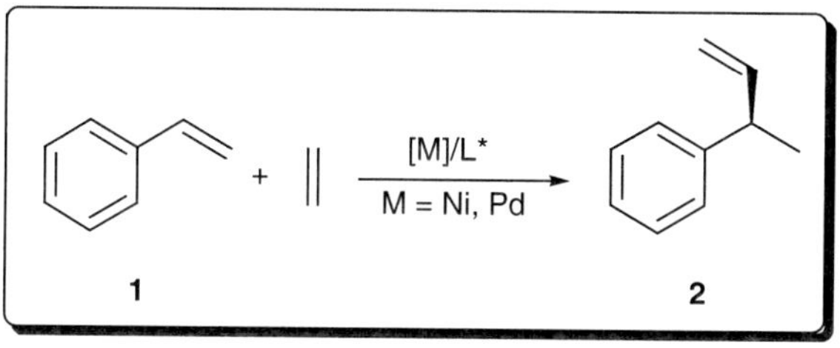

Scheme 8.1 Model reaction for enantioselective hydrovinylation.

M = Ni, Pd
A = BF$_4$, BARF...

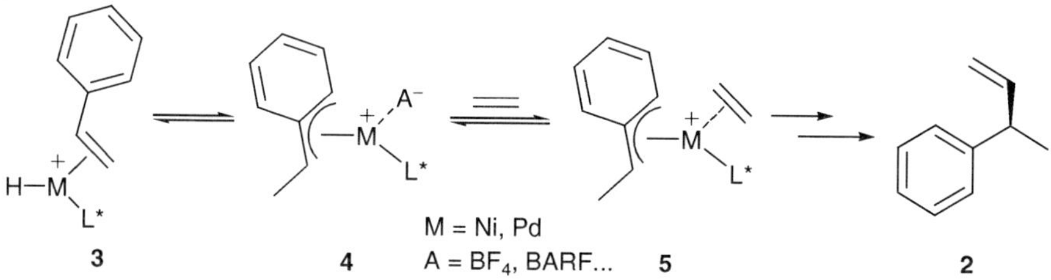

Scheme 8.2 Part of the accepted mechanism for hydrovinylation of styrene.

2-arylpropionic acids, a extensive family of antiinflammatory drugs such as ibuprofen or naproxen.

The true catalyst is thought to be a very reactive cationic metal hydride (**3**), whose addition to styrene forms an unsaturated complex (**4**) stabilised through η^3-coordination (Scheme 8.2). It is here where the chiral phosphine can discriminate between the two prochiral faces of styrene.

This complex coordinates ethylene to form **5**, which after insertion and β-elimination produces the product and regenerates the catalyst. In contrast to most of the catalytic processes, for Ni and Pd systems the reaction is inhibited by chelating phosphines since they would saturate intermediate **4** making impossible the coordination of ethylene. It is therefore a good process to test *P*-stereogenic monophosphines as chiral inductors. One of the main difficulties of the reaction is controlling the selectivity towards formation of **2**. This is a challenging task due to the number of different alkenes present in the reaction medium, which can lead to many undesired di- and oligomerisations. Furthermore, nickel and especially palladium systems isomerise **2** to the more stable conjugated 2-butenes. In general, very good results have been obtained with nickel at very low temperatures while palladium catalysts can be used at room temperature but they tend to isomerise the product, which complicates their use.

A breakthrough in enantioselective hydrovinylation was the discovery that an intricate *P*-stereogenic azaphospholene (**6**, Figure 8.1) gives excellent results in hydrovinylation of styrenes with nickel.

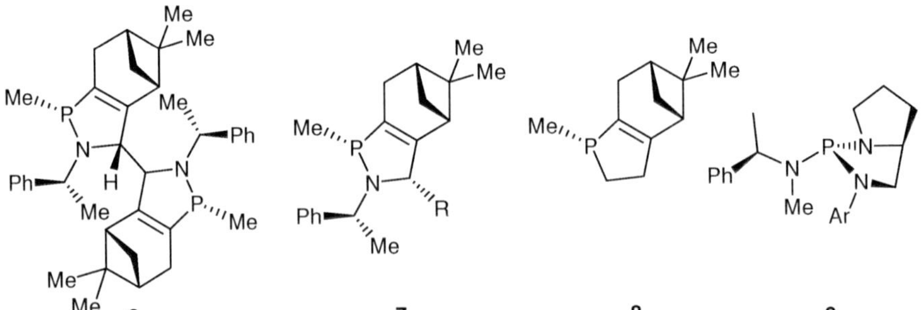

6

97% yield, 93% ee

7

R = Ph: >99% yield, 60% ee
R = Me, Ph: no conversion

8

No conversion

9

77% yield, 63% ee

Figure 8.1 Performance of *P*-stereogenic azaphospholenes **6**, **7**, **8** and **9** in Ni-catalysed hydrovinylation of styrene.

This ligand was developed by Wilke and remained the best ligand for Ni-catalysed hydrovinylation for many years.[6,7] Compound **6** possesses 2 phosphorus and 10 carbons as stereogenic centres and is prepared from (*R*)-myrtenal and (*R*)-1-phenylethylamine. It acts as a *bis*(monodentate) ligand, coordinating to two nickel atoms. The active catalyst was prepared from [Ni(allyl)Cl]$_2$, **6** and [Et$_3$Al$_2$Cl$_3$] as activator and it was necessary to carry out the reactions at temperatures below $-60\,°C$ to obtain good results. Wegner and Leitner[8] reported the use of **6** changing the chloroaluminate (corrosive and pyrophoric) by NaBARF and performing the reaction in supercritical CO_2. At room temperature, full conversion, 71% selectivity to **2** and 86% ee were achieved. Further work from the same group[9] using a continuous flow system $sc(CO_2)$/ionic liquid achieved similar enantioselectivity, but lower yield. In this study the ionic liquid was at the same time solvent and activator. Some simplified versions of **6** containing only one phosphorus donor have been also prepared, such as **7** and **8** and very recently **9**[10] but none of them is comparable to **6** in terms of enantioselectivity.[4]

In spite of the exciting results with **6**, no other accounts on Ni-catalysed hydrovinylation reactions with *P*-stereogenic ligands have been reported.

In contrast, there are a few studies with palladium-based systems. The catalytic precursors are usually prepared by activating neutral [PdX(η^3-allyl)(L*)] with silver salts of weakly coordinating anions or with NaBARF. The reactions are often carried out at room temperature in dichloromethane or THF. A selection of catalytic results is given in Table 8.1.

Entries 1–3 show that good enantioselectivities have been obtained with phosphinites, if an appropriate counterion is chosen (compare entries 1 and 2). Comparison of entries 2 and 3 highlights the importance of the *P*-stereogenic centre in these ligands. Tertiary phosphines have shown good selectivities for hydrovinylation at low conversions although the enantioselectivity is low to moderate depending on the substituents at the phosphine (entries 4–10). The best results in terms of enantioselectivity in Pd-catalysed reactions have been

Table 8.1 Enantioselective hydrovinylation of styrene by [Pd(allyl)(L*)]Y.

Entry	Ligand Counterion	Selectivity[a] (%)	ee of 2 (%)	References
1	(S_P)-PPh(t-Bu)(OMen) SbF_6^-	84	86 (S)	11
2	(S_P)-PPh(t-Bu)(OMen) OTf$^-$	24	0	11
3[b]	PPh(t-Bu)(OMen) SbF_6^-	79	37 (S)	11
4	(R)-PPhBnCy BF_4^-	99.0	61 (S)	12
5	(S)-PPhBnMes BF_4^-	93	40 (S)	13
6	(S)-PPh(o-Biph)Me BF_4^-	97	23 (S)	14
7	(S)-PPh(o-Biph)(i-Pr) BF_4^-	96	40 (S)	14
8	(S)-PPh(1-Naphth)Me BF_4^-	96	0	14
9	(S)-PPh(9-Phen)Me BF_4^-	94	20 (S)	14
10	(S)-PPh(o-Biph)(CH_2TMS) BF_4^-	97	68 (S)	15
11	(S)-PPh(o-Biph)(CH_2TMS) BARF	99.9	83 (S)	16
12	(S)-**94**[c], Ar $=$ o-Biph BF_4^-	97	63 (S)	15
13	(S)-**94**[c], Ar $=$ o-Biph BARF	98.5	75 (S)	15
14	(S)-**100**[c] BARF	99.1	81 (S)	16

[a]Selectivity to **2**.
[b]Epimeric mixture of R_P and S_P.
[c]See Chapter 4, Section 4.4.2.3.

obtained with dendritic systems (entries 12–14). Here the activation of the precursor with NaBARF has a very positive effect on the enantioselectivity (compare entries 12 and 13). Interestingly, the same dendrimers have been tested as precatalysts in supercritical CO_2 medium.[17] Although activities were lower than in dichloromethane, the enantioselectivities remained practically unchanged.

Finally, there is one example of cobalt-catalysed hydrovinylation of styrene with diphosphine ligands in the presence of alkylaluminium activators.[18] With DiPAMP, 30% conversion after 90 minutes and 26% ee were obtained. The most interesting result of these systems is that no isomerisation of **2** to 2-phenyl-2-butenes was observed until full consumption of styrene, even at 60 °C.

8.3 Hydroformylation

The hydroformylation reaction is the formal addition of formaldehyde across a C–C double bond (Scheme 8.3).

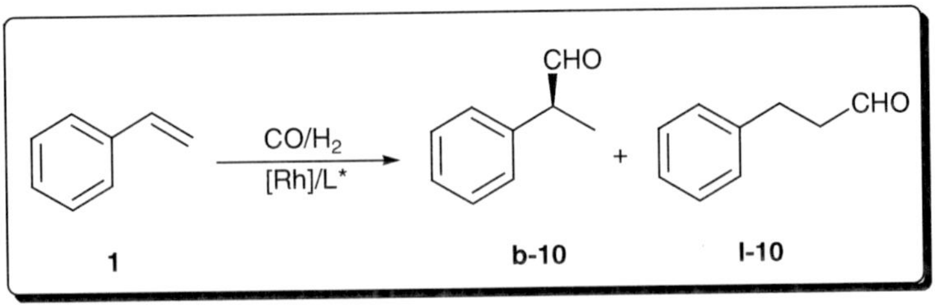

Scheme 8.3 Rh-catalysed enantioselective hydroformylation of styrene.

Discovered more than 70 years ago,[19] hydroformylation is nowadays one of the most important reactions in the chemical industry because aldehydes can be transformed to many other products. In the enantioselective version, rhodium/diphosphorus ligand complexes are the most important catalytic precursors, although cobalt and platinum complexes have also been widely used.[19,20] For these systems, the active species are pentacoordinated trigonal–bipyramidal rhodium hydride complexes, $[HRh(P–P)(CO)_2]$. In those complexes, the coordination mode of the bidentate ligand (equatorial–equatorial or equatorial–apical) is an important parameter to explain the outcome of the process.[21,22]

The most common substrates of enantioselective hydroformylation are styrenes followed by vinyl acetate and allyl cyanide. With these substrates, mixtures of the branched (b, chiral) and linear (l, not chiral) aldehydes are usually obtained. In addition, some hydrogenation of the double bond is often observed. Therefore, chemo- and regioselectivity are prerequisites to enantioselectivity and all of them must be controlled. An additional complication is that chiral aldehydes are prone to racemise in the presence of rhodium species.[23] The first examples of enantioselective hydroformylation were reported nearly 30 years ago but it is still an underdeveloped field[24] in contrast to the achiral version. A survey of the results of enantioselective hydroformylation with *P*-stereogenic ligands, some of them showing very good enantioselectivities, is given in Table 8.2.

Entries 1–8 list the results of the hydroformylation of styrene. Good yield but poor enantioselectivity was obtained with the monophosphine of entry 1, as well as with DiPAMP in the hydroformylation of *N*-vinylimides.[30] AMPP ligand of entry 2 gives much better results for regio- and enantioselectivity, in contrast to the non *P*-stereogenic EPHOS (entry 3). The origin of stereoinduction for this type of ligand has been investigated with a theoretical approach.[31] Ligands of entries 7 and 8 were originally developed for Rh-catalysed hydrogenation reactions, but were identified as good stereoinductors for hydroformylation during a screening of 47 commercially available ligands.[23] The downside of these two ligands is that as they are very electron-rich, they are sluggish in hydroformylation. In the hydroformylation of vinyl acetate (entries 9–13), very good results are obtained with ESPHOS (entry 10, compare with entry 4), in contrast to the poor performance of the monodentate SEMI-ESPHOS (entry 11) and some dendritic derivatives.[32,33] With vinyl acetate

Table 8.2 Alkene enantioselective hydroformylation with Rh/*P*-stereogenic ligands.

Entry	Substrate	Ligand	b/l^a	ee^b (%)	References
1	Styrene	(+)-PBnMePh	9	18	25
2	Styrene	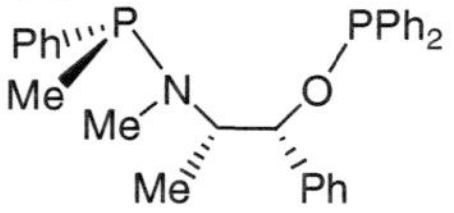	49	75 (*S*)	26
3^c	Styrene	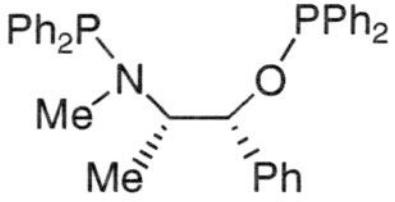 EPHOS	49	10 (*S*)	26
4	Styrene	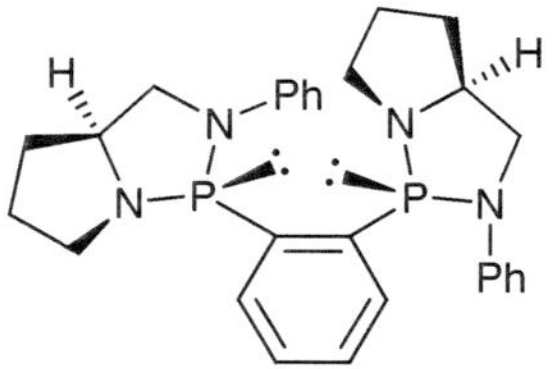 ESPHOS	3.5	0	27
5	Styrene	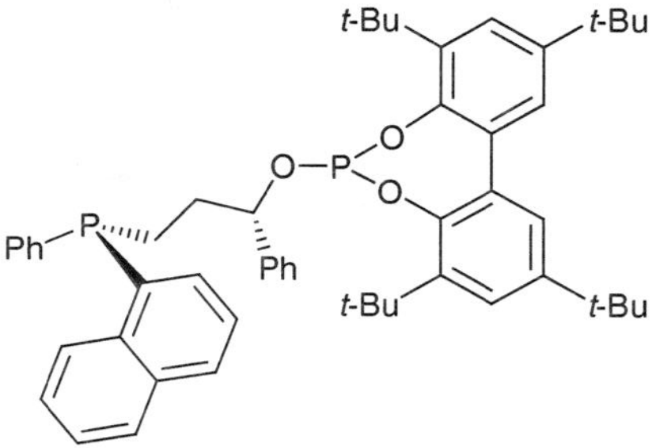	11	63 (*S*)	22

Table 8.2 (*Continued*).

Entry	Substrate	Ligand	b/l^a	ee^b (%)	References
6	Styrene	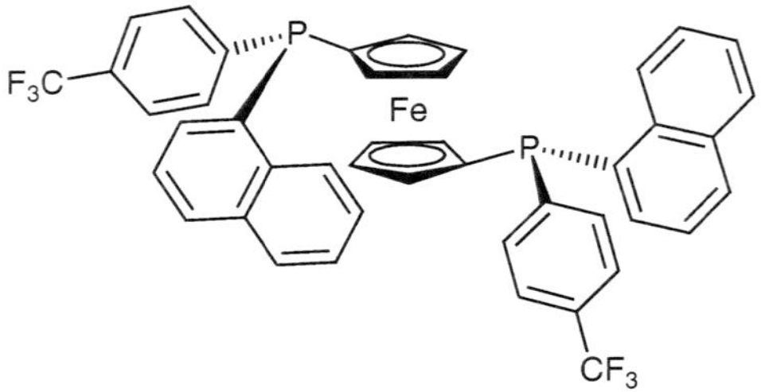	1.5	50 (S)	21
7	Styrene	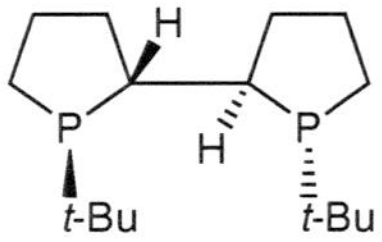 TangPhos	14	63 (S)	23
8	Styrene	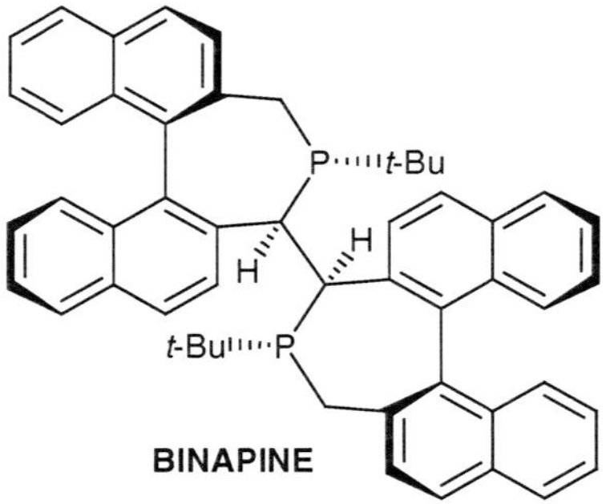 BINAPINE	9.5	94 (S)	23

	Substrate	Ligand			
9	AcO—	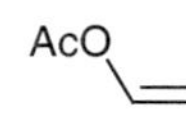	4.6	62	26
10	AcO—	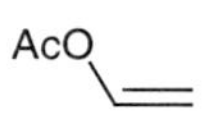ESPHOS	17	89 (S)	27, 28
11	AcO—	SEMI-ESPHOS	2.3	0	27
12	AcO—	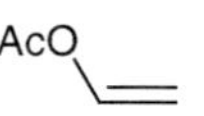TangPhos	59	80 (R)	23

Table 8.2 (*Continued*).

Entry	Substrate	Ligand	b/l^a	ee^b (%)	References
13	AcO	BINAPINE	32	87 (*R*)	23
14	NC		10	40 (*R*)	29
15	NC	TangPhos	7.4	84 (*S*)	23

| 16 | NC⟶ (allyl cyanide) | BINAPINE | 6.7 | 94 (S) | 23 |

[a] Branched to linear ratio.
[b] ee of the branched isomer.
[c] The ligand is not *P*-stereogenic.

as well as with allyl cyanide (entries 14–16) TangPhos and BINAPINE give high enantioselectivities (entries 12, 13, 15 and 16). TangPhos has also been shown to be hydroformylate several [2.2.1]-bicyclic alkenes in high enantio-selectivity.[34]

8.4 Allylic Substitution

The substitution of a leaving group located at an allylic position by a nucleo-phile, catalysed by a transition metal complex, is known as allylic substitution. Unlike most of the transition metal-catalysed reactions, allylic substitution involves reaction at sp^3 rather than sp^2 centres. As an example, the reaction with the most studied substrate, *rac*-3-acetoxy-1,3-diphenyl-1-propene (**11**), is depicted in Scheme 8.4.

Many transition metal complexes catalyse the reaction[35–37] but palladium systems are the most widely used.[38] Allylic substitution can be used to create C–C as well as C–X (X = heteroatom) bonds under very mild conditions, which are compatible with many functional groups. The allylic substitution reaction is unique in the sense that there are many mechanisms that can be responsible for asymmetric induction and because chiral elements can be placed at the nucleophile, the electrophile or both.

The catalytic cycle for stabilised or 'soft' nucleophiles (which are the con-jugate bases of acids whose pK_a is lower than 25)[38] involves the direct attack of the nucleophile on the alkene coordinated to palladium. Part of the mechanism for substrate **11** is depicted in Scheme 8.5.

After alkene coordination to Pd(0) species and ionisation, a Pd(II) π-allylic intermediate is formed. This intermediate undergoes nucleophilic attack at either terminal carbon. In special cases, the nucleophile can attack the central allylic position, leading to substituted cyclopropanes (see Section 8.10).

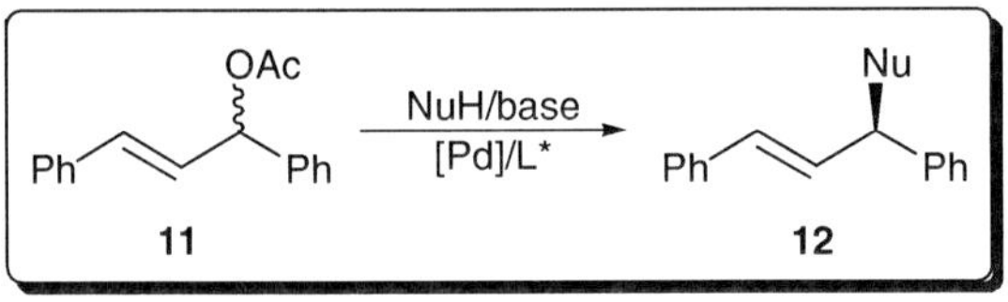

Scheme 8.4 Pd-catalysed enantioselective allylic substitution of **11**.

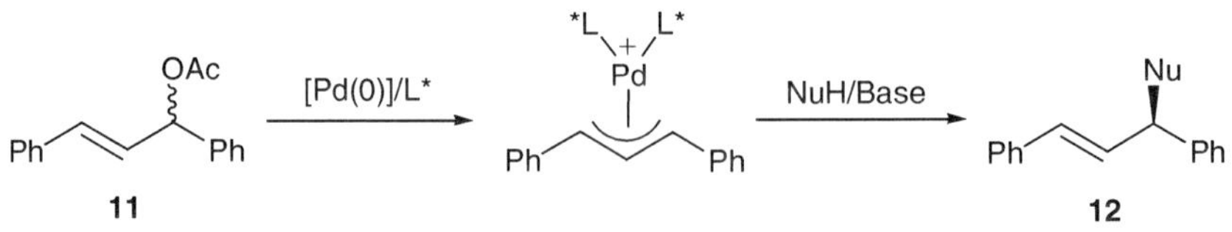

Scheme 8.5 Part of the accepted mechanism for allylic substitution with **11**.

The reaction was discovered more than 40 years ago,[38] and since then a huge number of publications on allylic substitution have appeared. Some of them deal with *P*-stereogenic ligands.

8.4.1 Carbon Nucleophiles

Allylic substitutions with stabilised carbon nucleophiles, known as allylic alkylations, are the most important type of allylic substitutions. The benchmark reaction of enantioselective allylic substitution is the reaction of **11** with the stabilised carbanion derived from dimethylmalonate (DMM) in the presence of *N,O-bis*(trimethylsilyl)acetamide (BSA) and a palladium catalyst, as depicted in Scheme 8.6.

This reaction has been screened with a tremendous quantity of chiral catalysts and indeed is routinely used to test the performance of new ligands in spite of its limited synthetic usefulness and that good or bad results with this system do not necessarily translate to other substrates.[39]

An early report of Bosnich and co-workers[40] describes that DiPAMP yields a racemic alkylated product but since then many mono- and bidentate *P*-stereogenic ligands have been used in the benchmark reaction for allylic substitution with very good results in some cases. Table 8.3 lists some the systems giving rise to good conversions and enantioselectivities.

Although Table 8.3 is not comprehensive, it illustrates the diversity of scaffolds of the ligands applied to enantioselective allylic substitution. Entry 1 shows that some simple monodentate phosphines can lead to very good results, but a good choice is required since small variations in the ligand profoundly affect the outcome, especially the enantioselectivity. Entries 2–7 demonstrate that diazaphospholidines are excellent ligands for this transformation in terms of enantioselectivity although they are less active than phosphines. Similarly, very high enantioselectivities have been obtained with bidentate P,N ligands (entries 8–10). Entries 13–22 show that C_2 ligands can also provide good results. Although DiPAMP is not enantioselective, other *P*-stereogenic diphosphines are very effective (entries 13–19) and the same happens,

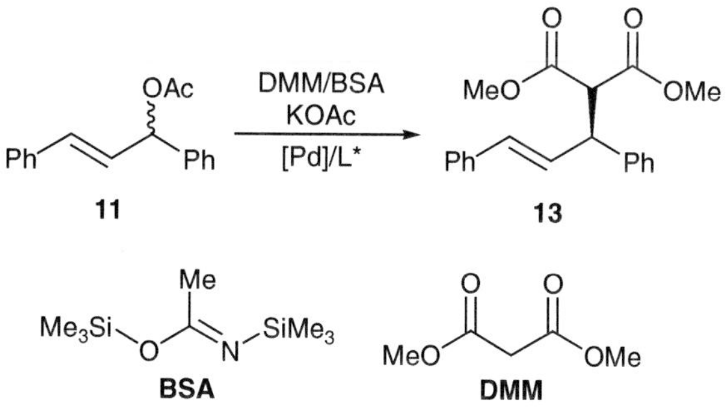

Scheme 8.6 Benchmark system for Pd-catalysed allylic alkylations.

Table 8.3 Pd-catalysed allylic alkylation of **11** with **DMM** (Scheme 8.6).

Entry	Ligand	t (h)	Conv. (%)	ee of 13 (%)	References
1	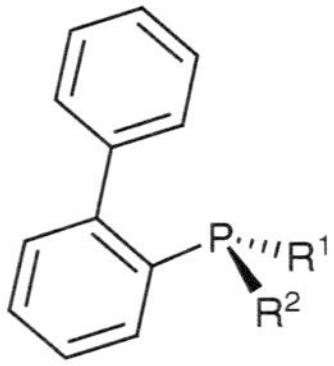 R^1 = Me, R^2 = Cy; **a** R^1 = t-Bu, R^2 = Me; **b** R^1 = Ph, R^2 = Me; **c** R^1 = Ph, R^2 = i-Pr; **d** R^1 = OMe, R^2 = Ph; **e**	17 (**a**) 15 (**b**) 17 (**b**) 1 (**d**) 1 (**e**)	87-quant.[a]	91 (**a**) 80 (**b**) 8 (R) (**c**) 14 (S) (**d**) 74 (S) (**e**)	41, 42
2	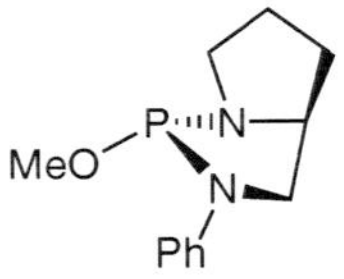	48	98	97 (S)	43
3	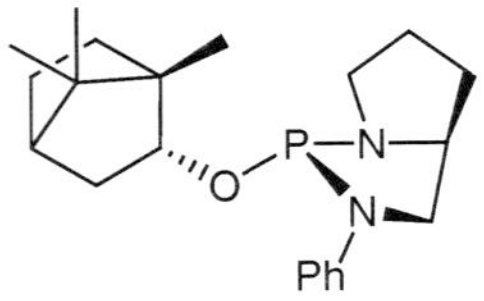	48	79	99 (S)	44
4	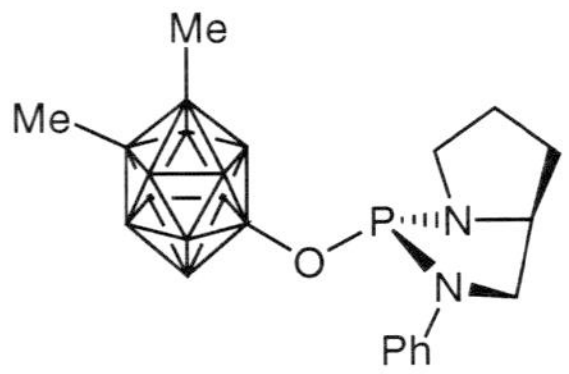	36	100	95 (S)	45

Entry					
5[b]	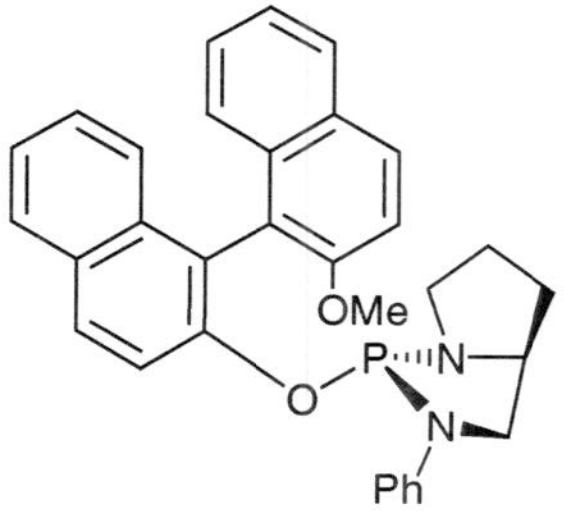	48	71	99 (*R*)	46
6	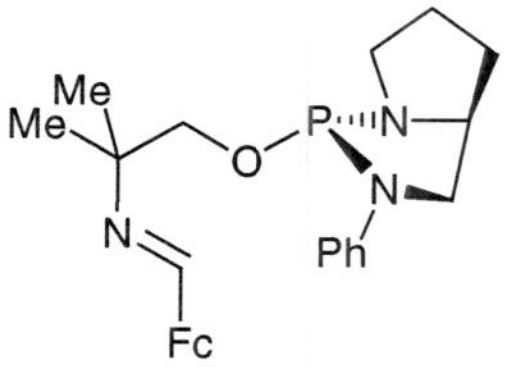	48	99	97 (*S*)	47
7	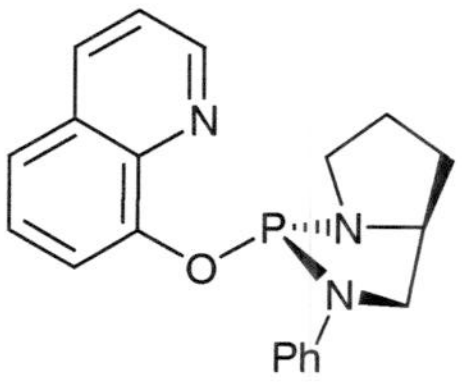 **QUIPHOS**	16	Quant.	85 (*R*)	48–50
8	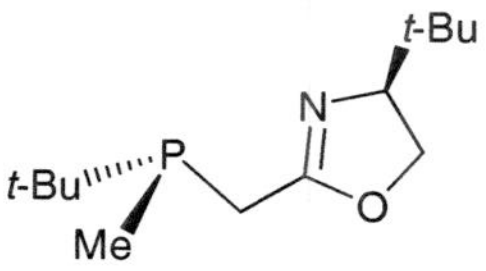	2	99[a]	96 (*S*)	51

Table 8.3 (*Continued*).

Entry	Ligand	*t (h)*	*Conv. (%)*	*ee of 13 (%)*	*References*
9	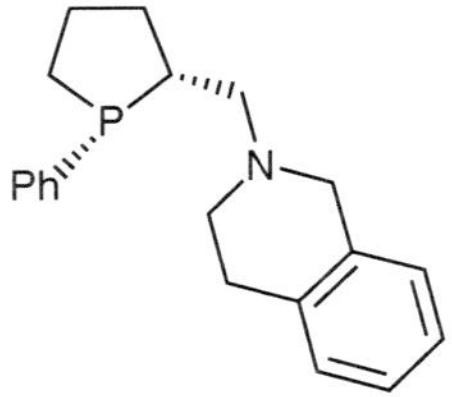	1	99[a]	96 (*S*)	52
10	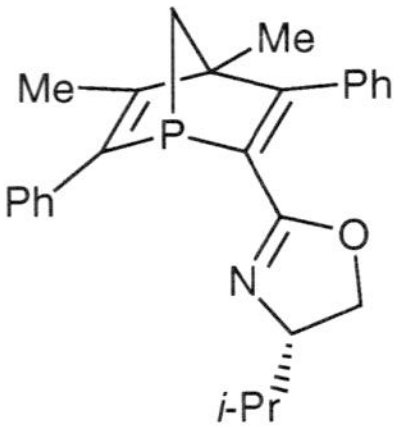	–	Quant.	94 (*S*)	53
11	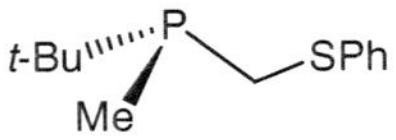	36	94[a]	85 (*R*)	54
12	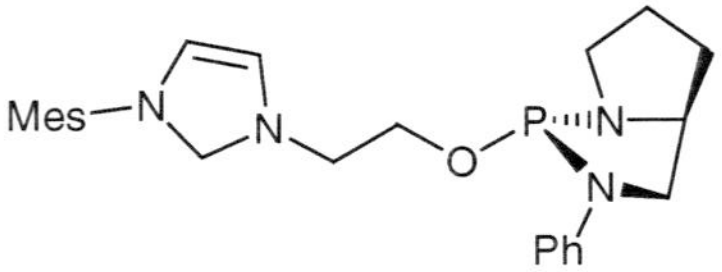	48	47	89 (*S*)	55
13		1 (Fc)	99[a] (Fc)	95 (*S*) (Fc)	56
		16 (*t*-Bu)	96 (*t*-Bu)	88 (*S*) (*t*-Bu)	

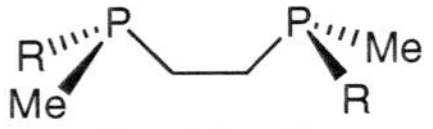

14	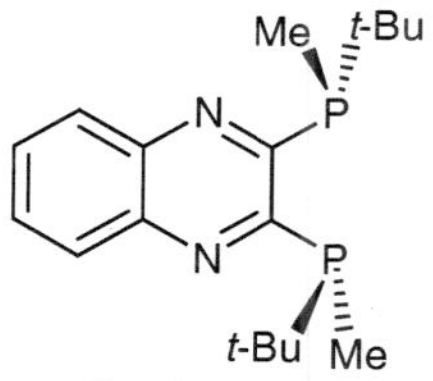 **t-Bu-QuinoxP***	1	85[a]	92 (S)	57
15	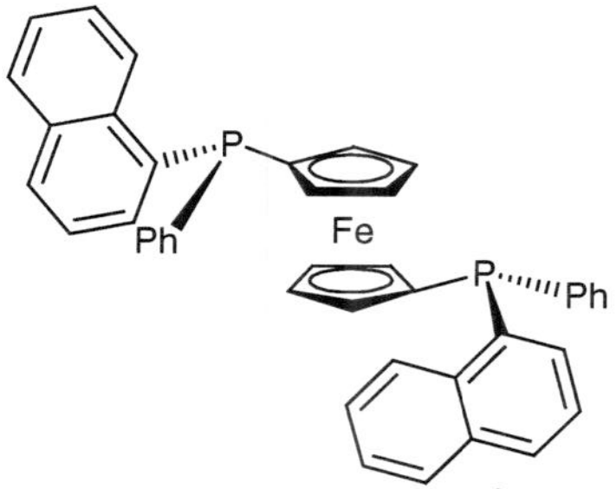	15	84[a]	73 (R)	58
16	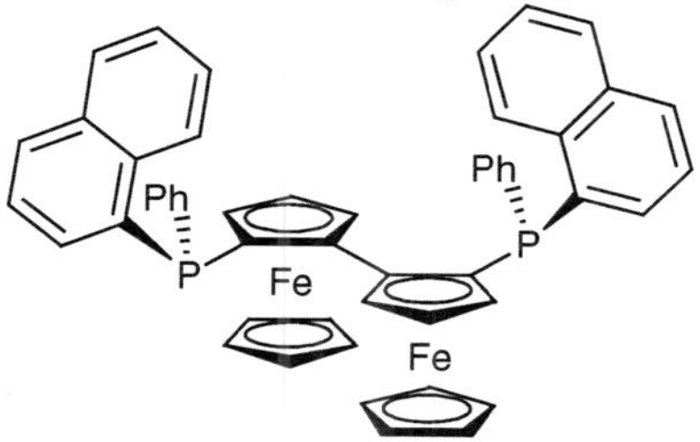	2–48	87[a]	88 (R)	59

Table 8.3 (*Continued*).

Entry	Ligand	t (h)	Conv. (%)	ee of 13 (%)	References
17	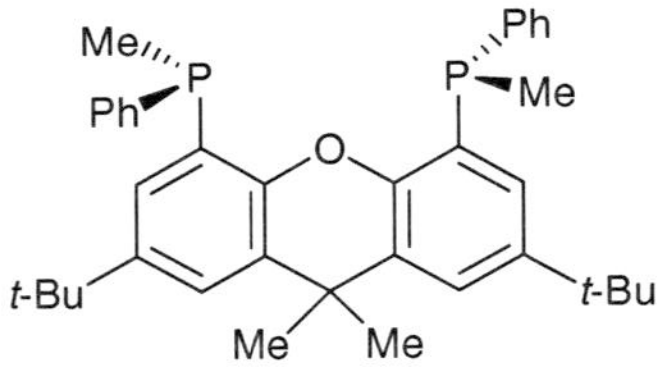	7	96^a	85 (R)	60
18	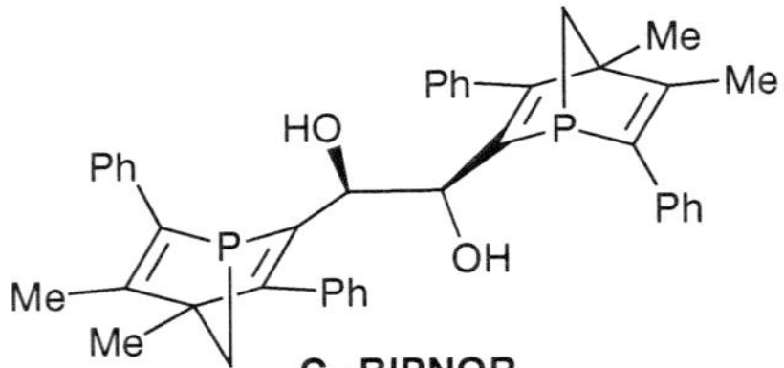 C₂-BIPNOR	24	95^a	86 (S)	61
19	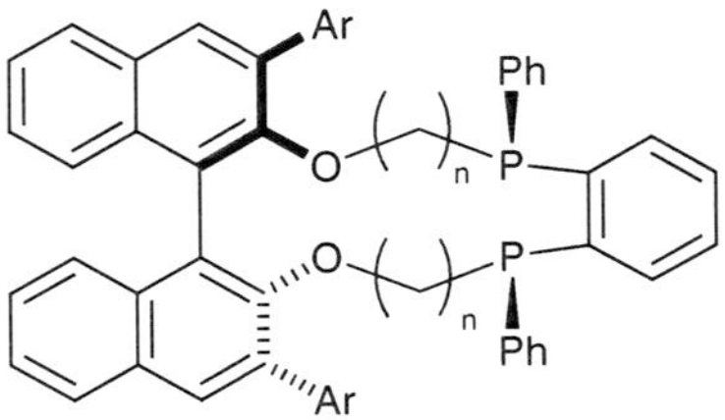 Ar = 2-Naphth, n = 2	—	95^a	98 (S)	62

20	ESPHOS	3	38[a]	96 (R)	63
21	FerriESPHOS	24	87[a]	92 (S)	28
22		48	98	96 (R)	64

Table 8.3 (*Continued*).

Entry	Ligand	t (h)	Conv. (%)	ee of 13 (%)	References
23	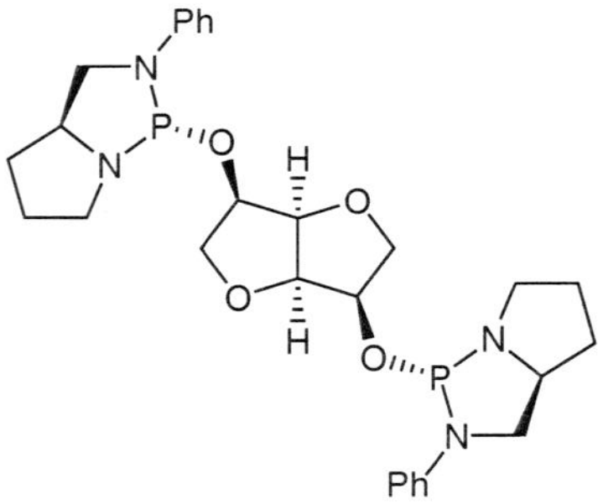	48	72	98 (S)	65
24	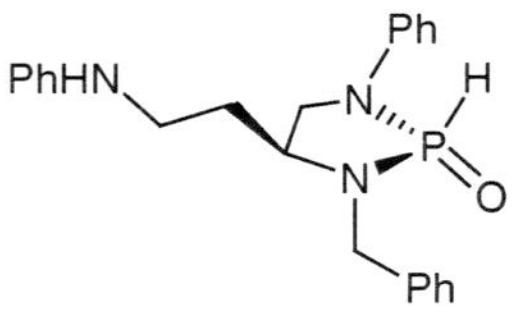 **Ph-DIAPHOX**	34	99[a]	99 (S)	66

[a]Value of the yield.
[b]The binaphthyl group has *R* absolute configuration.

unsurprisingly, with *bis*(diazaphospholidines) of entries 20–23. Finally, the P(V) compound of entry 24 is an outstanding ligand for enantioselective allylic alkylation because of BSA-induced isomerisation to the P(III) tautomer, discussed in Chapter 3, Section 3.2.3.

One of the reasons for the popularity of **11** as a model for allylic alkylations is that many ligands give particularly good enantioselectivity with this substrate.[67] Changing the phenyl groups for smaller alkyl groups leads to much poorer results and constitute a challenge in allylic substitutions. A model for these substrates is (*E*)-pent-3-en-2-yl acetate (**14**). A few results obtained with *P*-stereogenic ligands are shown in Scheme 8.7.

The diminished enantioselectivity of this substrate is due to the much smaller steric hindrance of methyl groups to compared to phenyl groups and to the absence of π-stacking interactions of **14** with the chiral ligands.[58] Ferrocenyl ligands **16**[58,59] and biarylic **17**[62] have been used, leading to good yields but low stereoselectivities.

Small carbocyclic substrates are also challenging, often giving low stereoselectivities. Some selected examples of alkylations of these substrates using *P*-stereogenic ligands are shown in Scheme 8.8.

Substrates **18** give very low asymmetric inductions with C_2 ligands **16**[58,59] and **17**.[62] In contrast, a respectable enantioselectivity was found with P,N-ligands **20**[68] and **21**[69–73] although the best enantioselectivities have been achieved with cymantrenyl phosphinooxazolines **22**,[74] bearing a 2-biphenylyl group at the stereogenic phosphorus atom, developed by Helmchen and co-workers.[71] Further work from the same group[72] has applied Pd-catalysed allylic

Scheme 8.7   Some results of Pd-catalysed allylic alkylation of **14**.

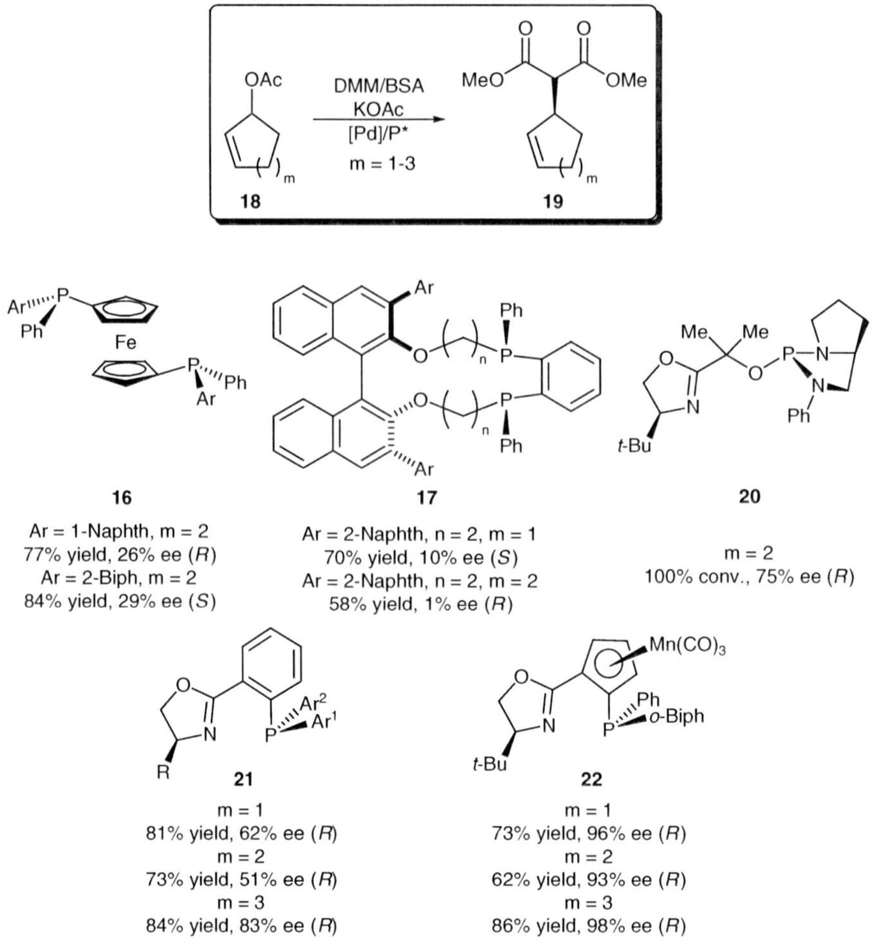

18 → **19**

$$\text{DMM/BSA, KOAc, [Pd]/P*, } m = 1\text{-}3$$

16

Ar = 1-Naphth, m = 2
77% yield, 26% ee (*R*)
Ar = 2-Biph, m = 2
84% yield, 29% ee (*S*)

17

Ar = 2-Naphth, n = 2, m = 1
70% yield, 10% ee (*S*)
Ar = 2-Naphth, n = 2, m = 2
58% yield, 1% ee (*R*)

20

m = 2
100% conv., 75% ee (*R*)

21

m = 1
81% yield, 62% ee (*R*)
m = 2
73% yield, 51% ee (*R*)
m = 3
84% yield, 83% ee (*R*)

22

m = 1
73% yield, 96% ee (*R*)
m = 2
62% yield, 93% ee (*R*)
m = 3
86% yield, 98% ee (*R*)

Scheme 8.8   Some results of Pd-catalysed allylic alkylation of cyclic substrates **18**.

alkylations with **21** in several enantioselective syntheses of chiral piperidines,[75] amino acid derivatives[67] and natural products.[76–78]

An even more challenging type of substrate are those that are unsymmetrically substituted, because regio- and enantioselectivity must be controlled (Scheme 8.9).

With these substrates, achiral **l-24** is strongly favoured with Pd catalysts (for other metals, the situation is different)[79] except in special cases. Diazaphospholidine-oxazoline **20**[68] shows moderate performance in the alkylation of butenyl acetate compared to the exceptionally good results with SiocPhox phosphonamidate ligands for alkyl- and aryl-substituted substrates.[80,81] The same type of ligands have also been applied to the alkylation of other challenging substrates such as polyenyl esters[82] and acyclic ketone enolates.[83]

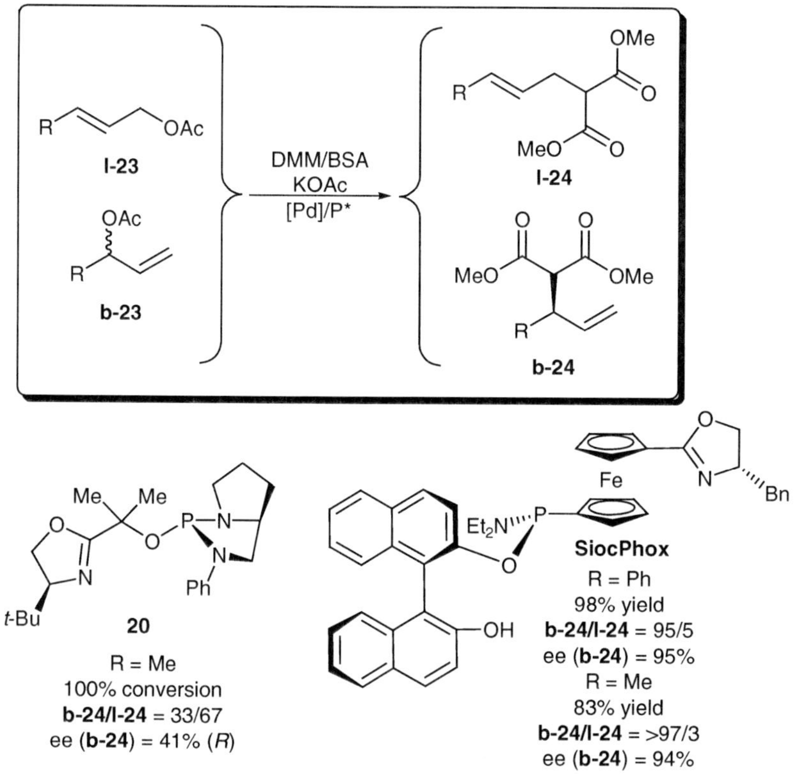

Scheme 8.9 Some results of Pd-catalysed allylic alkylation of unsymmetrically substituted substrates **23**.

They are also useful in the kinetic resolution of 2,3-dihydro-4-quinolones by enantioselective allylic allylation.[84]

Allylic alkylations have also been used to create optically pure all-carbon quaternary stereocentres, which is a daunting task in organic synthesis. Although some encouraging results have been obtained with SiocPhox ligands with acyclic nucleophiles,[85] most of the work involves cyclic nucleophiles. For instance, a few examples of alkylations of β-keto esters leading to quaternary centres are depicted in Scheme 8.10.

Buono, Tenaglia and Brunel[86] obtained good results with QUIPHOS for the allylation of a cyclopentanone but the six-membered counterpart gave very low enantioselectivities. More recently, Hamada and co-workers[66,87,88] have established that DIAPHOX ligands are very well suited for these kinds of reactions in the presence of Zn(II) cations. Further work from the same group[89,90] has extended considerably the scope of DIAPHOX ligands to other enantioselective allylic substitution reactions. For instance, they have successfully employed it in allylic alkylations with nitrometane (up to 98% ee with substrate **11**),[91] synthesis of quaternary α-amino acid derivatives,[92] allylic alkylations of 2-substituted

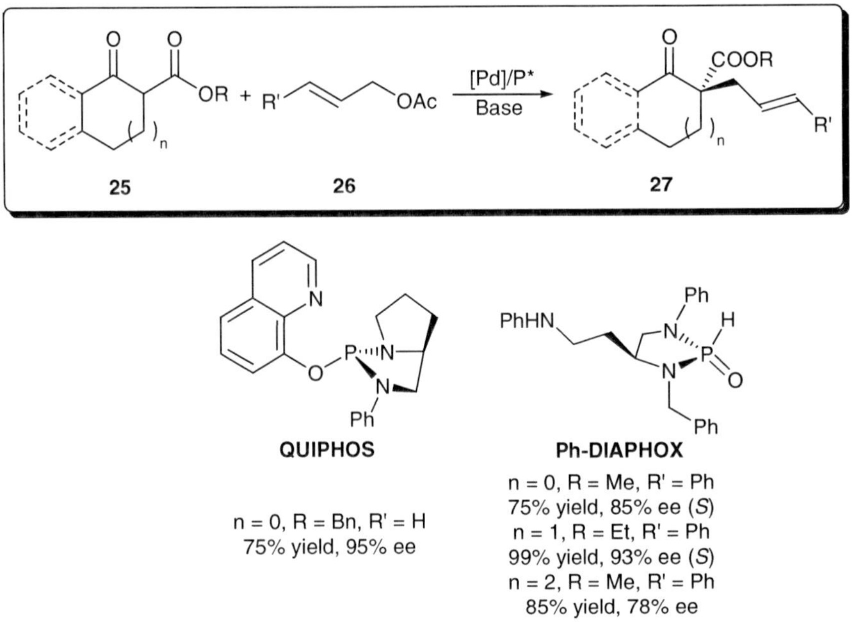

QUIPHOS

n = 0, R = Bn, R' = H
75% yield, 95% ee

Ph-DIAPHOX

n = 0, R = Me, R' = Ph
75% yield, 85% ee (*S*)
n = 1, R = Et, R' = Ph
99% yield, 93% ee (*S*)
n = 2, R = Me, R' = Ph
85% yield, 78% ee

Scheme 8.10   Examples of Pd-catalysed construction of quaternary stereocentres.

cycloalkenyl carbonates (up to 92% ee),[93] allylic alkylations of allenyl acetates (up to 99% ee),[94] iridium-catalysed allylic alkylations of unsymmetrically substituted substrates[79] and very recently in the synthesis of the natural product tangutorine, possessing a pentacyclic skeleton and three stereogenic centres.[95]

8.4.2   Other Nucleophiles

Many non-carbon nucleophiles have been successfully used in Pd-catalysed enantioselective allylic substitutions. With *P*-stereogenic ligands, there are reports mainly of allylic aminations and sulfonylations, with formation of C–N and C–S bonds respectively. Not surprisingly, the benchmark substrate is **11** for both reactions.

Allylic aminations have been performed with a variety of primary and secondary amines (Scheme 8.11 and Table 8.4).

Good conversions and enantioselectivities exceeding 90% ee have been achieved with monodentate ligands (entries 2, 3, 6), monodentate ligands bearing hemilabile oxygen atoms (entries 1, 4, 5) and bidentate ligands (entries 8–12). In analogy to allylic alkylation, diazaphospholidine ligands are found to be excellent stereoinductors (entries 1–8, 12 and 13) as well as DIAPHOX (entry 14).

There are also a few reports on allylic amination of more challenging substrates such as unsymmetrically substituted acetates,[80] dienyl acetates,[82] cyclic substrates,[103–105] and Ir-catalysed allylic aminations.[106]

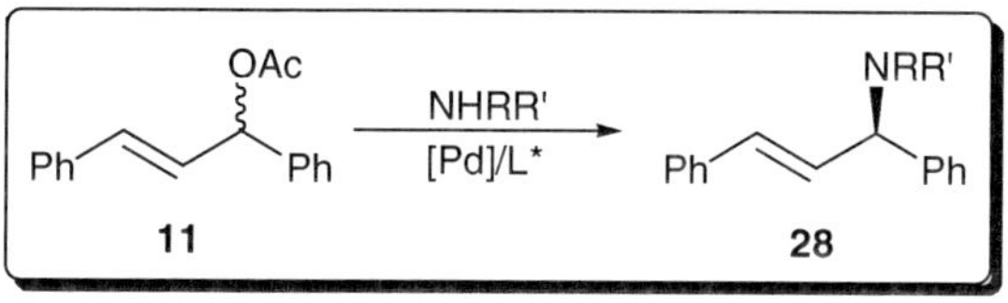

Scheme 8.11 Pd-catalysed allylic amination of **11**.

P-stereogenic diazaphospholidines are good ligands for Pd-catalysed allylic sulfonylation, in which sodium *para*-toluenesulfinate acts as a *S*-nucleophile furnishing allylic sulfones (Scheme 8.12 and Table 8.5).

From Table 8.5 is clear that the diazaphospholidine moiety is an excellent chiral inductor for allylic sulfonylation, although the yield is sometimes unsatisfying (entries 1, 3 and 5). The ligand of entry 1 is remarkable because of its extremely high enantioselectivity given its very simple structure.

8.5   Reductive Coupling of Alkynes and Aldehydes

Ni-catalysed reductive coupling of alkynes and aldehydes in the presence of excess of reductant (triethylborane) forms a new C–C bond and provides allylic alcohols (Scheme 8.13), which are useful building blocks in organic synthesis.

The challenges of these types of reactions are the control of the regio- (**30** *vs.* **30'**), enantio- and *E*/*Z*-selectivity in the formation of the double bond.

Jamison and co-workers[109] found that ferrocenyldiphenylphosphine is a good ligand for this transformation when dialkylacetylenes are used. They subsequently developed the enantioselective version of the reaction using *P*-stereogenic ferrocenyl monophosphines.[109] A few results with these systems are collected in Scheme 8.14.

The best regio- and enantioselectivity was achieved with a ligand bearing one *o*-tolyl group although overall the best results were obtained with the phosphine bearing a methyl group, perhaps because of the high steric dissymmetry of phosphorus atom.[109] Part of the mechanism of the reaction is shown in Scheme 8.15.[109]

It is thought that in complex **32** the aldehyde-BEt$_3$ adduct approaches the coordinated alkyne-Ni(0) complex in a way that the BEt$_3$ group is away from the large ferrocenyl group (and *syn* to the small methyl group), eventually leading to oxametallacyclopentene intermediate **33**.

Reductive nickel-catalysed couplings between alkynes and aldehydes provide (*E*)-trisubstituted allylic alcohols, a moiety found to possess significant biological effects.[110] For this reason, reductive couplings have been used in the synthesis of natural products.[111,112] An example, reported by Jamison and Chan,[111,113] is given in Scheme 8.16.

The coupling between **34** and **35** using tributylphosphine was completely nondiastereoselective. In contrast, using the depicted ferrocenylphosphine, the desired diastereomer (3:1 selectivity) and regiosiomer (2:1 selectivity) of **36** was

Table 8.4 Examples of Pd-catalysed allylic amination of **11** (Scheme 8.11).

Entry	Ligand	Amine	ee of 28 (%)	References
1	SEMI-ESPHOS	Benzylamine	86–95 (R)	63, 96
2		Benzylamine	95 (R)	43
3		HN(n-Pr)$_2$	90	44
4[a]		Pyrrolidine	97 (R)	46

5	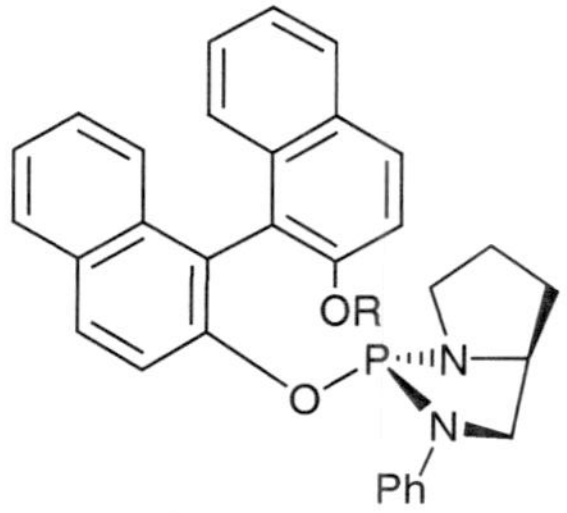 R = *t*-BuCO–	Pyrrolidine	99 (*R*)	97
6	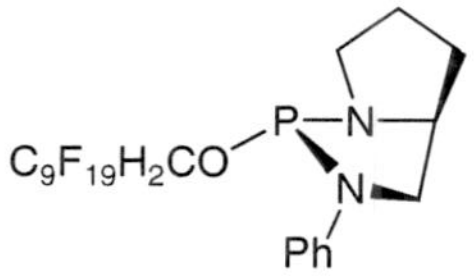 $C_9F_{19}H_2CO$	HN(*n*-Pr)$_2$	91–95	98
7	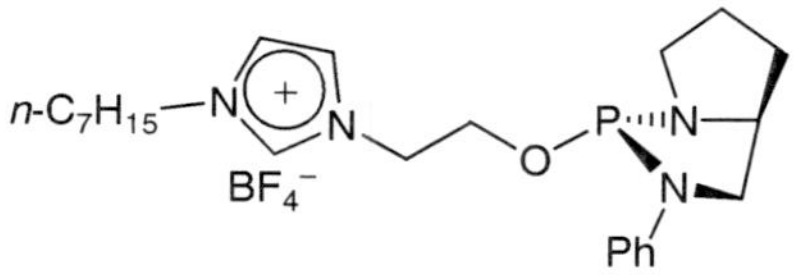 *n*-C_7H_{15} · BF_4^-	HN(*n*-Pr)$_2$	97	99
8	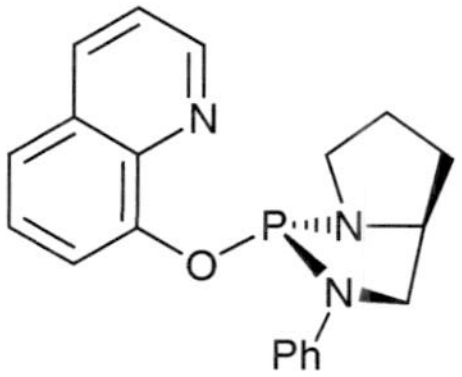 QUIPHOS	Benzylamine	93 (*S*)	100

Table 8.4 (*Continued*).

Entry	Ligand	Amine	ee of 28 (%)	References
9	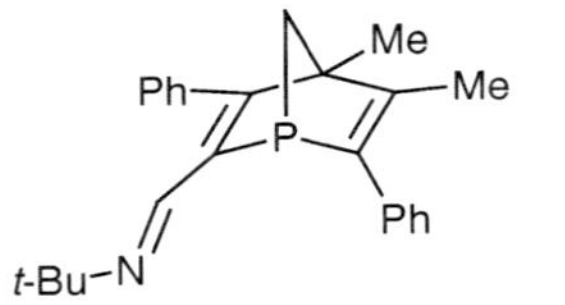	Benzylamine	90 (*S*)	101
10	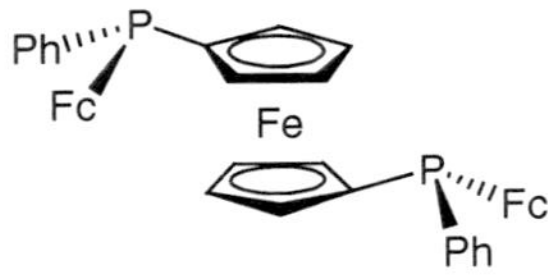	Phthalimide	93 (*S*)	59
11	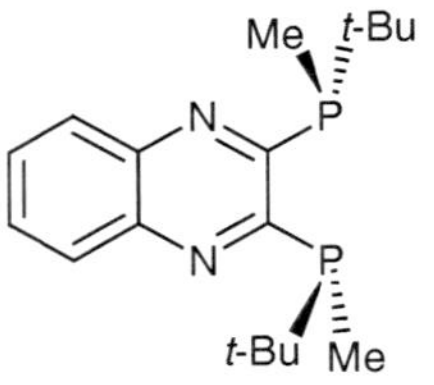 *t*-Bu-QuinoxP*	Morpholine	90 (*R*)	57
12	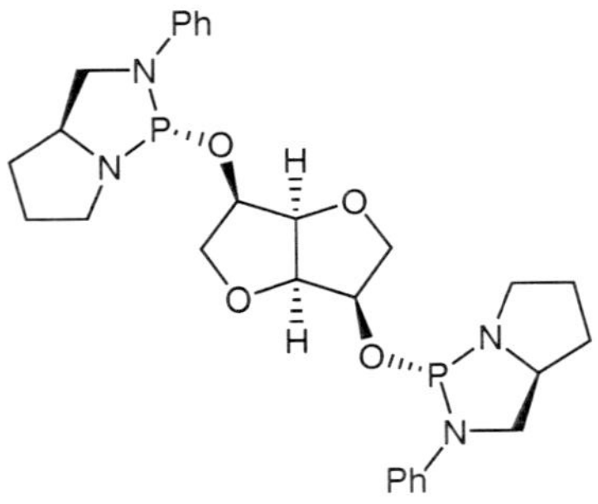	Benzylamine	92 (*R*)	65

| 13 | 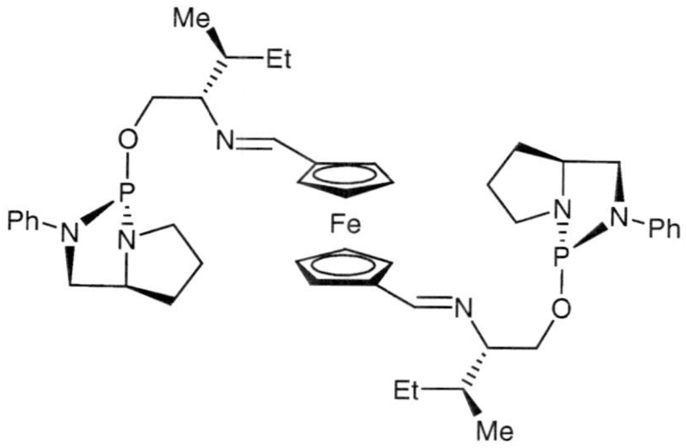 | Pyrrolidine | 85 (*R*) | 102 |
| 14 | 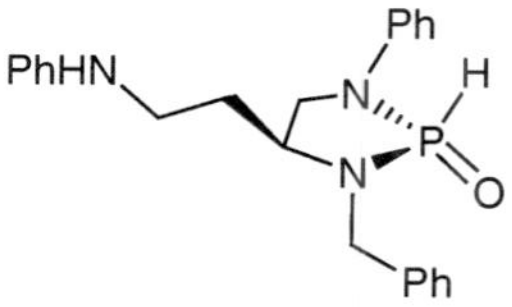 **Ph-DIAPHOX** | Benzylamine | 98 (*R*) | 103 |

[*a*]The binaphthyl group has *R* absolute configuration.

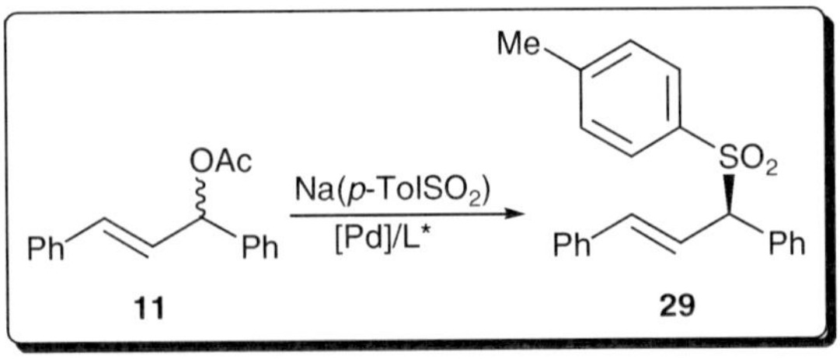

Scheme 8.12   Pd-catalysed allylic sulfonylation of **11**.

obtained. This compound was transformed into the natural product (–)-ter-pestacin after several steps.[111]

Ni-catalysed reductive couplings with *P*-stereogenic ferrocenyl monophosphines have been expanded to couplings between 1,3-enynes with substituted benzaldehydes[110] and to 1,3-enynes with ketones.[114] Both transformations occurred in excellent regioselectivity (>95/5) whereas in the latter case, 1,3-dienes containing a quaternary carbinol stereocentre were prepared in up to 70% ee. There is also one report[115] of enantioselective catalytic coupling of organoboranes, alkynes and aliphatic imines furnishing enantioenriched tetrasubstituted allylic amines in high yields with up to 89% ee.

8.6   Conjugate Additions

8.6.1   Cu-Catalysed Additions

There are many reports of copper-catalysed conjugate additions (1,4-additions)[116] of organometallic reagents to α,β-unsaturated ketones, which constitute examples of Michael reactions.[117–119] The model Michael acceptors for cyclic and acyclic substrates are 2-cyclohexen-1-one (**37**) and *trans*-chalcone (**38**) respectively[119] whereas diethylzinc is the most widely used carbanionic reagent. These model reactions are depicted in Scheme 8.17.

Although Cu(II) salts are often used as the copper source, the mechanism of these reactions[120] appears to involve Cu(I) species as true catalysts, formed by alkyl transfer from dialkylzinc reagent to the copper centre (Scheme 8.18).

It is believed that the 'hard' Zn(II) cation coordinates to the carbonyl while the 'soft' Cu(I) centre forms a π-complex with the enone double bond, forming a bimetallic species (**41**). Oxidative addition to the enone forms the Cu(III) complex **42**, which suffers reductive elimination with concomitant C–C bond formation. After aqueous work-up, **43** delivers the product **39**. The reductive elimination is likely to be the rate-determining step.[120]

There are several reports on the performance of *P*-stereogenic ligands in enantioselective additions of diethylzinc to enones. Selected results are collected in Table 8.6.

For both chalcone and substrate **37** many mono- and bidentate ligands give almost complete conversions but good enantioselectivities are difficult to reach.

Table 8.5 Selected examples of Pd-catalysed allylic sulfonylation of **11** (Scheme 8.12).

Entry	Ligand	t (h)	Conv. (%)	ee of 29 (%)	References
1	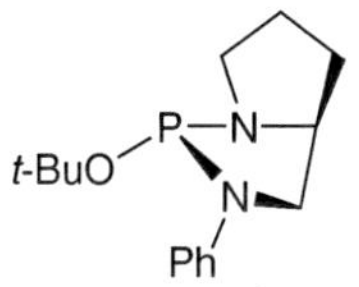	48	44^a	97 (S)	43
2^b	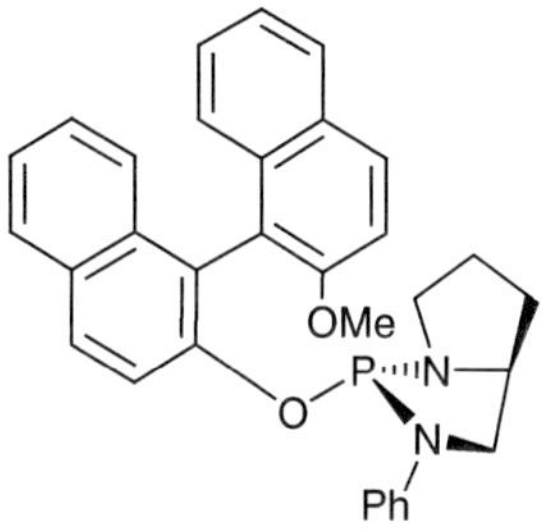	48	95	99 (R)	46
3	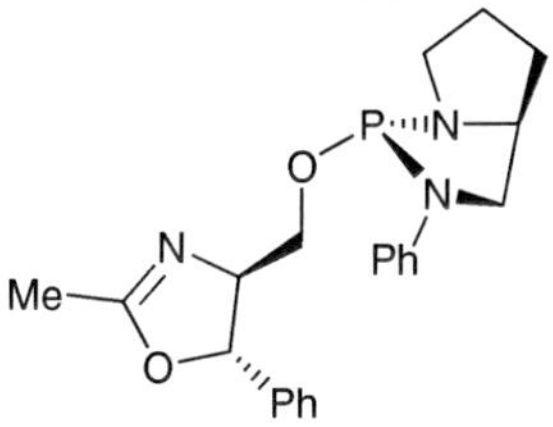	48	33^a	92 (S)	107
4	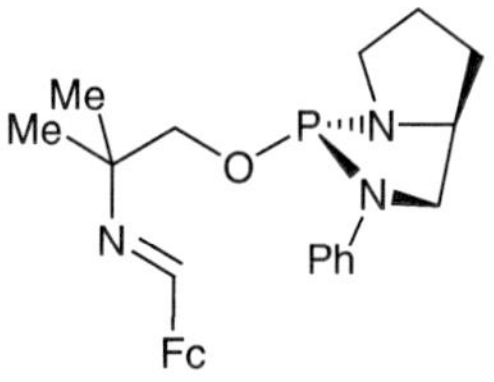	48	85^a	97 (S)	47
5	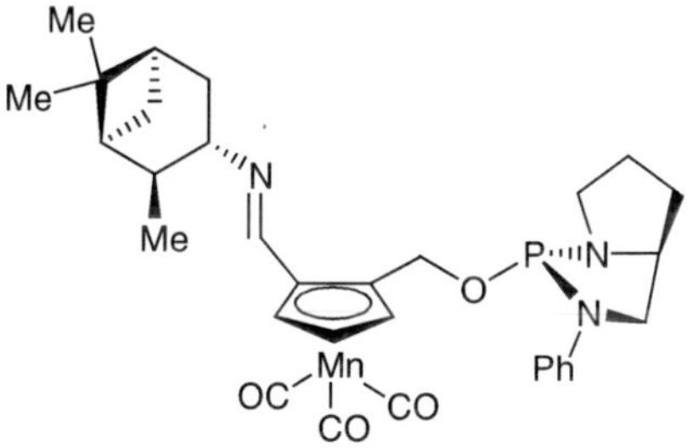	48	42	84 (S)	108
6	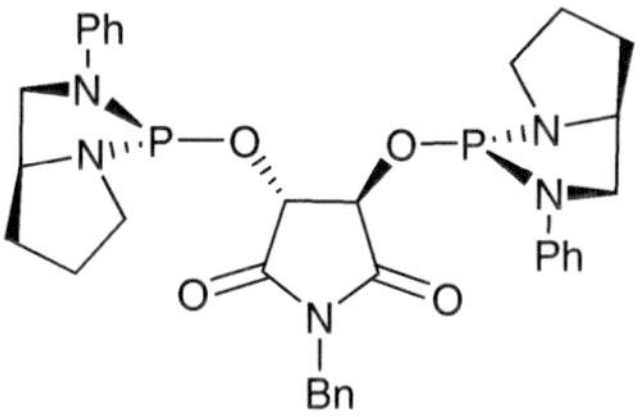	48	90^a	92 (R)	64

^{a}Value of the yield.
^{b}The binaphthyl group has R absolute configuration.

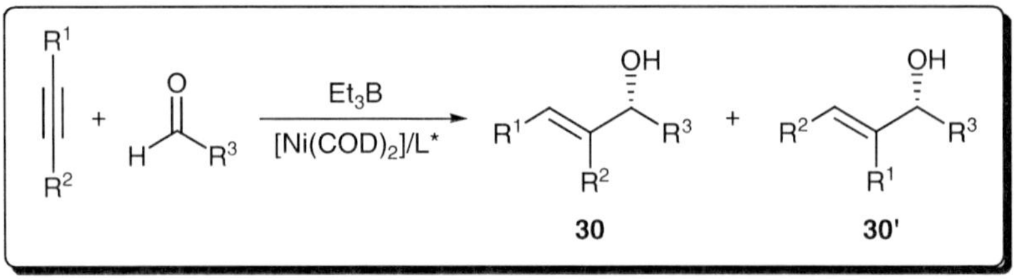

Scheme 8.13   Ni-catalysed reductive coupling of alkynes and aldehydes.

65% yield
31/31' = 69/31
ee (**31**) = 46%
ee (**31'**) = 45%

46% yield
31/31' = 85/15
ee (**31**) = 55%
ee (**31'**) = 19%

60% yield
31/31' = 70/30
ee (**31**) = 2%
ee (**31'**) = 4%

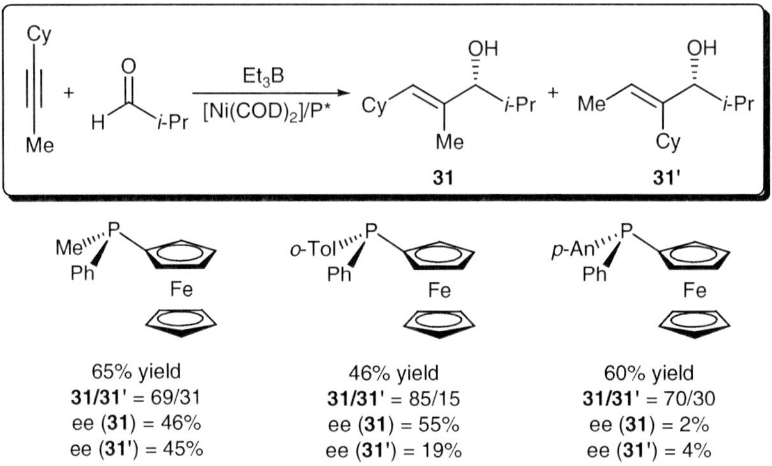

Scheme 8.14   Selected examples of reductive coupling of alkynes and aldehydes using
P-stereogenic ferrocenylphosphines.

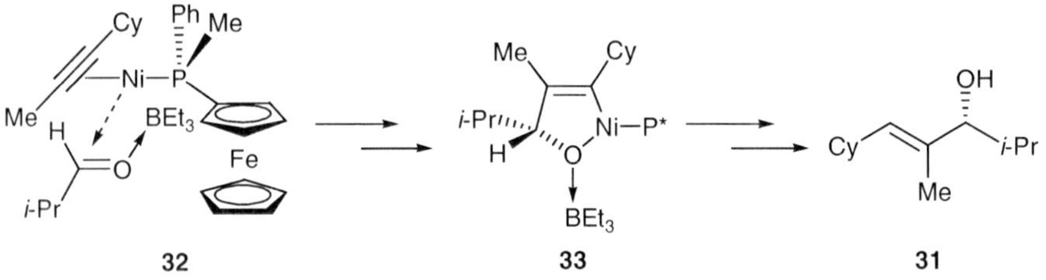

Scheme 8.15   Part of the mechanism of Ni-catalysed reductive coupling of alkynes
and aldehydes.

Entry 3 corresponds to one of the first reports about enantioselective catalytic
Michael additions. For **37** the best results are obtained with MiniPHOS
diphosphines (entries 7 and 8). In contrast, other diphosphines such as BisP*
and DiPAMP provide product **39** in low optical purity (entries 10 and 7

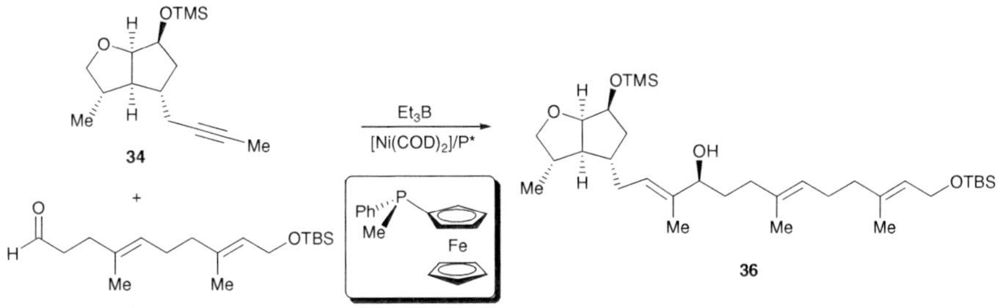

Scheme 8.16 Example of a Ni-catalysed reductive coupling in natural product synthesis.

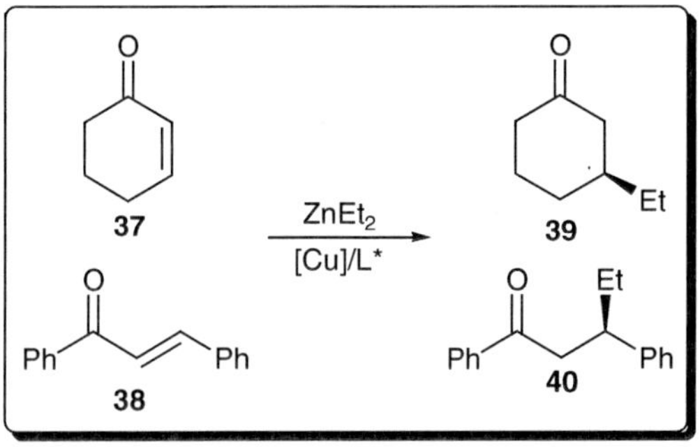

Scheme 8.17 Cu-catalysed enantioselective Michael additions of diethylzinc to α,β-unsaturated ketones.

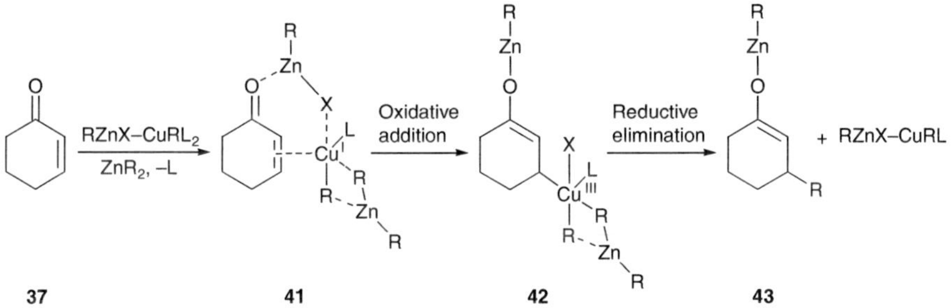

Scheme 8.18 Part of the mechanism of Cu-catalysed conjugate addition reactions. Usually R = Et.

respectively). On the other hand an interesting result is given in entry 1 with a partially fluorinated ligand. Chalcone is usually a more demanding substrate compared to **37** due to *cis/trans* conformational interconversions.[119] In spite of that, good results have been obtained with the phenolic phosphine of entries 11

Table 8.6 Results of Cu-catalysed Michael addition of $ZnEt_2$ to enones (Scheme 8.17).

Entry	Substrate	Ligand	Yield (%)	ee of 39 or 40 (%)	References
1	**37**	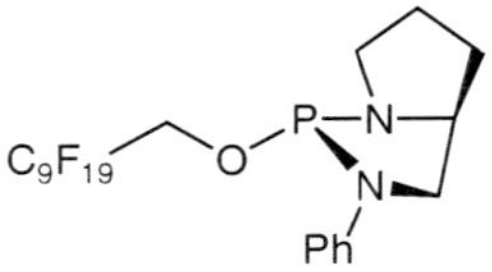	$>99^a$	70 (S)	97, 98
2^b	**37**	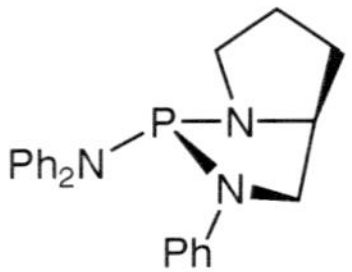	99^a	58 (S)	10
3	**37**	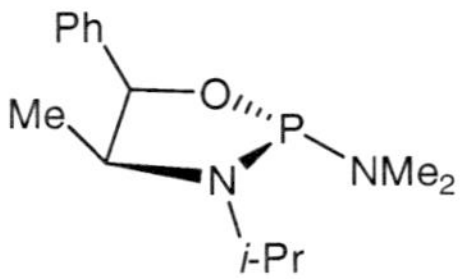	70	32	121
4^c	**37**	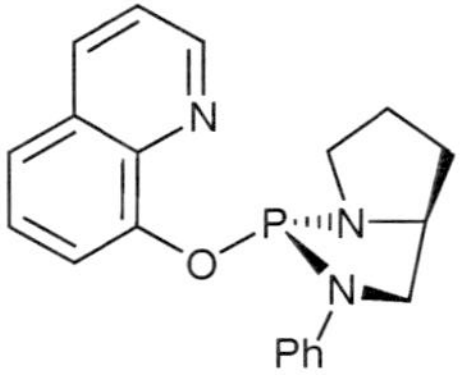 QUIPHOS	100	53–61 (R)	50, 122
5	**37**	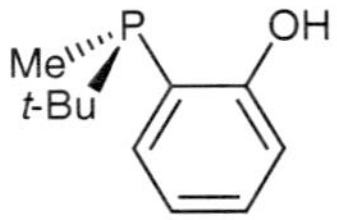	5	8	123

6[d]	**37**	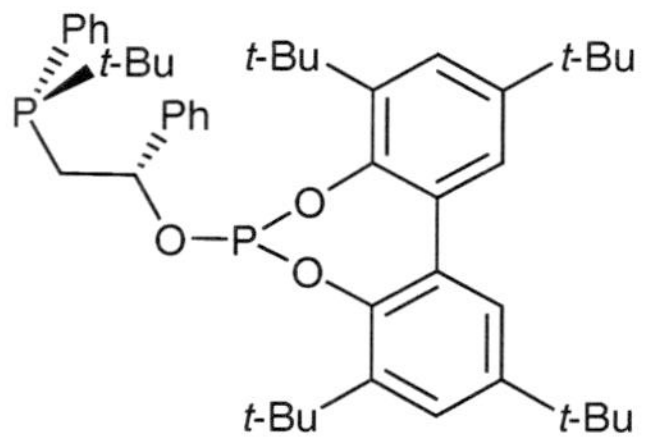	100[a]	62 (S)	124
7	**37**	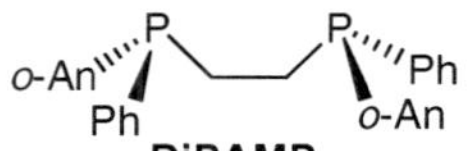 DiPAMP	73	10 (S)	125
8	**37**	 Cy-MiniPHOS	79	83 (S)	126
9	**37**	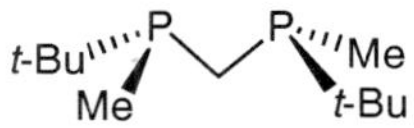 *t*-Bu-MiniPHOS	92	85 (S)	127
10	**37**	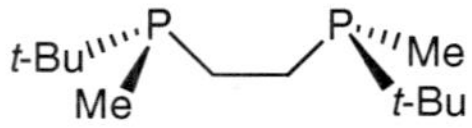 *t*-Bu-BisP*	>99[a]	27 (S)	127
11	**38**	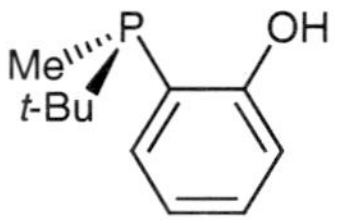	88	85 (R)	123

Table 8.6 (*Continued*).

Entry	Substrate	Ligand	Yield (%)	ee of 39 or 40 (%)	References
12	**Chalcone**[e]		94	82 (*R*)	123
13	**38**	*t*-Bu-MiniPHOS	96	71 (*R*)	126

[a]Value of the conversion.
[b]H_2O was added.
[c]$Zn(OH)_2$ was added.
[d]$AlEt_3$ was used instead of $ZnEt_2$.
[e]*Cis*-chalcone was used.

and 12. Interestingly, the same phosphine gives very poor results with substrate **37** (entry 5).

8.6.2 Rh-Catalysed Additions

A related transformation is the Rh(I)-catalysed addition of arylboronic acids to electron deficient alkenes.[128–130] The model reaction, again with substrate **37**, is depicted in Scheme 8.19.

The catalysts are thought to be hydroxorhodium species (**45**), formed upon hydrolysis of the rhodium precursor. Part of the mechanism for the addition of phenylboronic acid to **37** is shown in Scheme 8.20.

Transmetallation of **45** forms the phenylrhodium complex **46**, which firstly coordinates and then inserts the substrate into the Rh–Ph bond, forming the oxa-π-allylic complex **47**. Hydrolysis of this intermediate furnishes the product **44** and regenerates **45**.

The use of *P*-stereogenic ligands in this reaction is very recent but very good results have been obtained. Some selected examples are listed in Table 8.7.

Monophosphine of entry 1 acts as a bidentate ligand coordinating to Rh atom *via* the phosphorus atom and the double bond in the side arm.[131] It is an excellent ligand for this reaction. The relatively electron-rich phosphines of entries 2 and 3 are even better, providing very high yield of **44** in almost optically pure form. In contrast, ligands of entries 4 and 5 are notoriously inferior in spite of being similar to the ligand of entry 3.

Although Table 8.7 only refers to the model reaction, other substrates and arylboronic acids have also been used.[132,133] An example, reported by Tedrow and co-workers,[134] is the conjugate addition to 4-oxobutenamide **48** (Scheme 8.21).

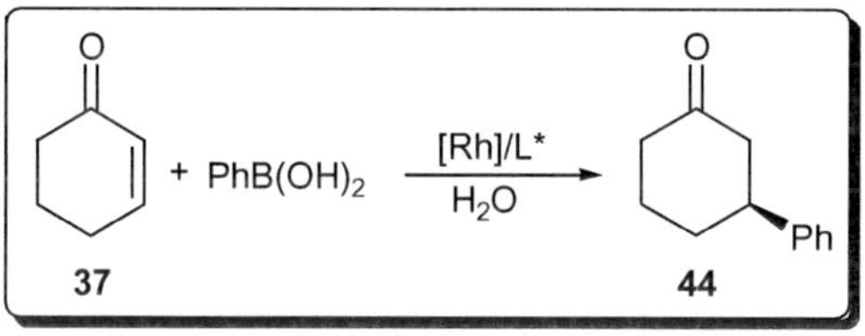

Scheme 8.19   Rh-catalysed conjugate addition of phenylboronic acid to **37**.

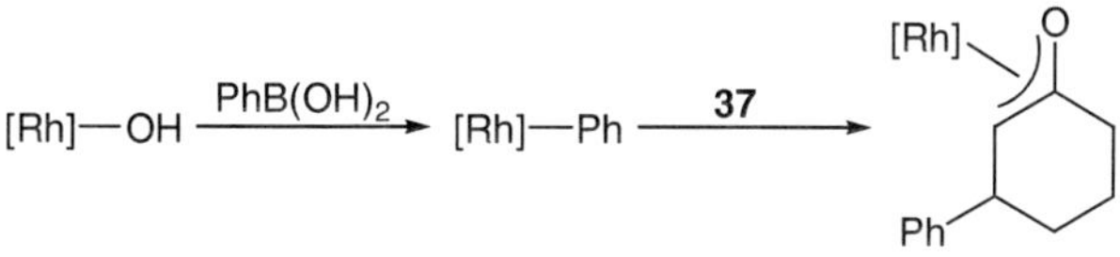

Scheme 8.20   Part of the mechanism for Rh-catalysed addition of phenylboronic acid to **37**. [Rh] = [Rh(diphosphine)].

Table 8.7 Results of Rh-catalysed conjugate addition of phenylboronic acid to **37** (Scheme 8.19).

Entry	Ligand	t (h)	Yield (%)	ee of 44 (%)	References
1		4	88	98.0 (S)	131
2	*t*-Bu-QuinoxP*	1	93	98.2 (R)	132
3		2	93	99.4 (R)	133
4		12	94	46 (R)	133
5	*t*-Bu-BisP*	2	37	20	133

These substrates are particularly challenging because both alkeneic carbons are possible conjugate acceptors. After some screening, it was found that electron-rich and sterically demanding *P*-stereogenic ligands TangPhos and DuanPhos provided high regioselectivity and very high enantioselectivity towards the desired isomer **49**. The Rh/DuanPhos system was used for other substrates and arylboronic acids with very good results.[134]

8.7 Alkylative Ring Opening

Alkylative ring opening is a palladium-catalysed C–C bond-forming reaction in which a dialkylzinc reagent opens a heterocycle, with the formation of two stereogenic centres.[135] The model substrate is oxabenzonorbornadiene (**50**, Scheme 8.22).

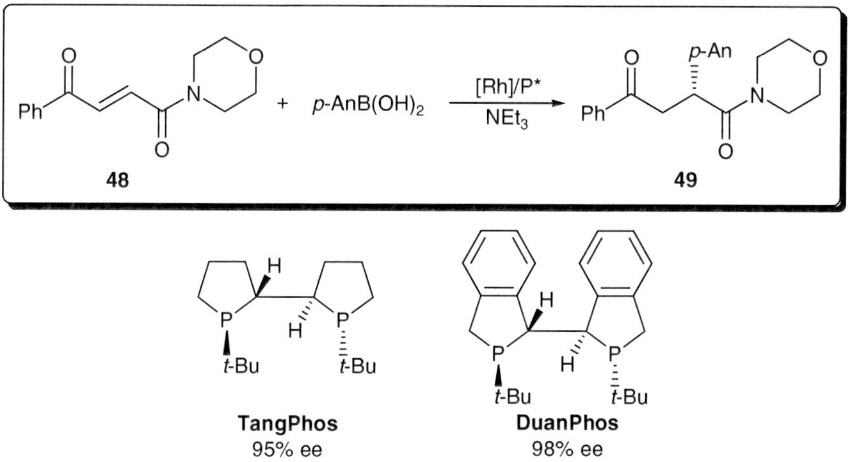

TangPhos
95% ee

DuanPhos
98% ee

Scheme 8.21 Some results in Rh-catalysed conjugate addition of 4-oxobutenamides.

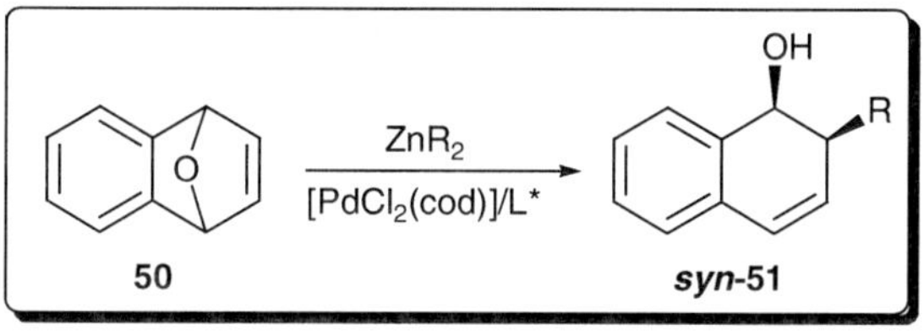

Scheme 8.22 Pd-catalysed alkylative ring opening of **50**.

Certain *P*-stereogenic diphosphines have recently provided outstanding results in this transformation, giving exclusively the *syn* diastereomers of **51** in high enantioselectivities. Some selected results, reported by Imamoto and co-workers,[132,133,136,137] are collected in Table 8.8.

The electron-rich *t*-Bu-MiniPHOS ligand provided good yield but moderate enantioselectivity (entry 1). Much better results have been obtained with the related ligand *t*-Bu-BisP* (entry 2) and with the QuinoxP* ligands (entries 3–5). The diphosphine of entries 6 and 7, containing an unsubstituted ethynyl group, leads to excellent yields and almost perfect enantioselectivities.

Further work from the same group has expanded the scope of the reaction to alkylative ring opening of azabenzonorbornadienes (up to 99% ee with ZnMe₂ and *t*-Bu-MiniPHOS)[137] and has allowed the connection of two catalytic reactions: the ring-opened product (using *t*-Bu-QuinoxP*) has been used, without quenching, as a chiral Lewis acid catalyst for the addition of diethylzinc to aldehydes.[138]

Table 8.8   Results of Pd-catalysed alkylative ring opening of **50** (Scheme 8.22).

Entry	Ligand	R	Time (h)	Yield (%)	ee of 51 (%)	References
1	**t-Bu-MiniPHOS**	Me	4	74	63	137
2	**t-Bu-BisP***	Et	6	93	94 (*S,S*)	133
3	**t-Bu-QuinoxP***	Me	2	90	96 (*S,S*)	132
4	**t-Bu-QuinoxP***	Et	15	88	98 (*S,S*)	132
5	**1-Ad-QuinoxP***	Et	2	67	98.5 (*S,S*)	136
6		Me	2	93	99.9 (*S,S*)	133
7		Et	6	90	99.9 (*S,S*)	133

8.8   Heck Reaction

The Heck reaction is a Pd-catalysed coupling between aryl halides, tosylates or triflates with alkenes and is one of the most important C–C bond-forming reactions. Although in the original version no stereocentres were formed, the

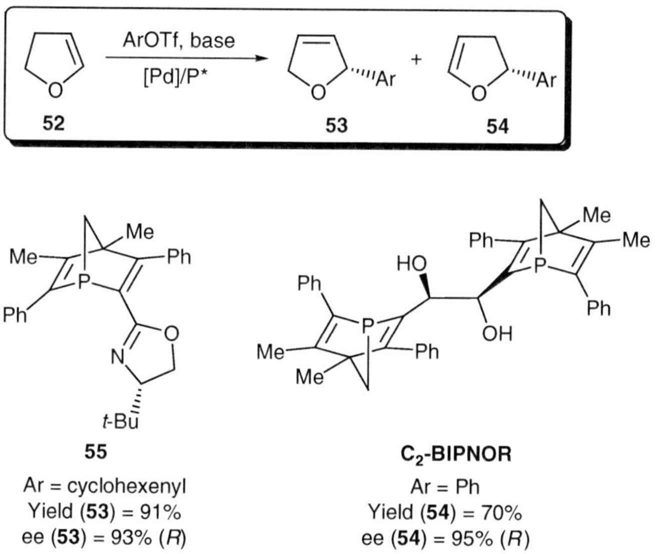

Scheme 8.23 Examples of Pd-catalysed enantioselective Heck reaction of **52** with aryl triflates.

enantioselective variant was developed and has received much attention for the last two decades.[139,140] Nowadays the enantioselective intramolecular Heck reaction constitutes a reliable method to build tertiary and quaternary stereo-centres[139] in complex molecules and is an important tool in natural product synthesis. In contrast, the intermolecular version is much less developed.[140] The most typical substrate for this reaction is 2,3-dihydrofuran (**52**, Scheme 8.23).

The reports with phosphinooxazolines such as **55**[53] and the diphosphine C_2-BIPNOR[61] are the only two accounts dealing with the use of *P*-stereogenic ligands in enantioselective Heck reactions. The good yields and enantioselectivities obtained, however, may indicate that *P*-stereogenic ligands can have great potential in this reaction.

8.9 Cycloaddition Reactions

In cycloaddition reactions two unsaturated molecules are combined to form a cyclic adduct. They constitute an extremely important group of C–C bond-forming reactions in organic chemistry. Although they occur either thermally or photochemically, transition metal-catalysed cycloaddition reactions are interesting since they take place under milder conditions, allowing a better control of enantioselectivity. The most important reaction is the well-known Diels–Alder reaction (a 4 + 2 cycloaddition), which can be catalysed by chiral Lewis acids. A few *P*-stereogenic ligands have been applied to the Diels–Alder reaction between *N*-acrylamide dienophiles (**56**) and cyclopentadiene (**57**). A pair of examples are shown in Scheme 8.24.

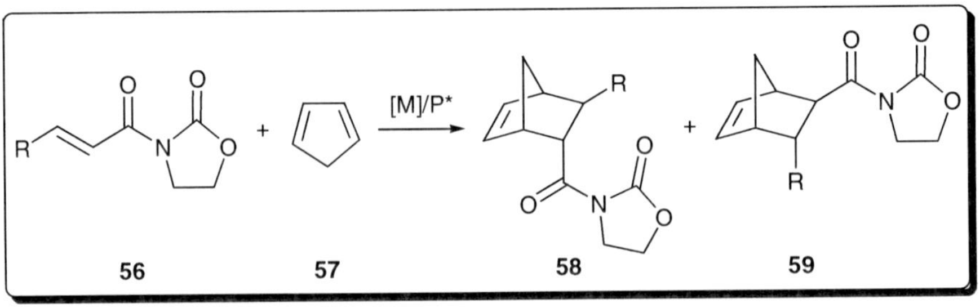

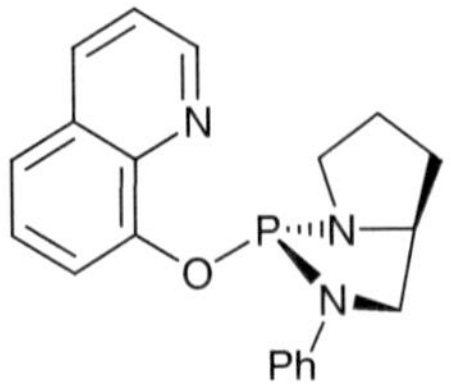

56   57   58   59

QUIPHOS

Ph
N

1-Ad-BisP*(O)₂

[M] = Cu(OTf)₂, R = H
58/59 = >98/2
Conversion = Quant.
ee (58) = >99% (S)

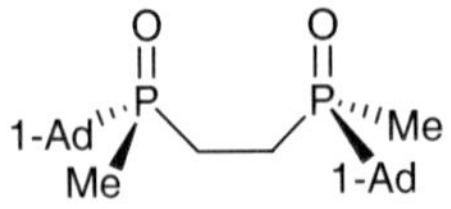

[M] = FeI₃/I₂, R = Me
58/59 = 18/82
41% yield
ee (58) = 70% (S)
ee (59) = 75% (S)

Scheme 8.24   Catalytic enantioselective Diels–Alder reaction.

With this system two isomers can be formed: **58** (*endo*) and **59** (*exo*). With the QUIPHOS ligand and copper catalysis, Buono and co-workers[141] obtained very high *endo/exo* ratios and excellent conversion and enantioselectivity if the reagents were mixed at $-78\,^{\circ}\mathrm{C}$ and slowly warmed to $25\,^{\circ}\mathrm{C}$. Imamoto and co-workers[142] used the oxide, the 1-Ad-BisP* ligand and iron catalysis at $0\,^{\circ}\mathrm{C}$ with modest results of diastereo- and enantioselectivity. Other BisP* and MiniPHOS ligands led to inferior results.

Also on record there is an interesting example of Rh-catalysed intramolecular enantioselective catalytic Diels–Alder reaction (Scheme 8.25).[143]

The cycloisomerisation of the ene-diene **60** produced the bicyclic product **61** with excellent diastereo- and enantioselectivity.

There are a few reports of other cycloaddition reactions other than Diels–Alder. Two examples are given in Scheme 8.26.

Buono and co-workers[144] showed that palladium(II) complexes of secondary phosphine oxide **63** catalyse the $2+1$ cycloaddition between norbornadiene and phenylacetylene, affording benzylidienecyclopropane (**62**). This compound shows what is called *cis–trans* enantiomerism, also called geometrical enantiomorphic isomerism. Using optically pure **63**, moderate enantioselectivity was found.

Hu, Zheng and co-workers[145] employed the ferrocenylphosphine-phosphoramidite **66** in the Ag-catalysed enantioselective $3+2$ cycloaddition of **64** with dimethyl maleate. Very good yields and impressive enantioselectivities were found.

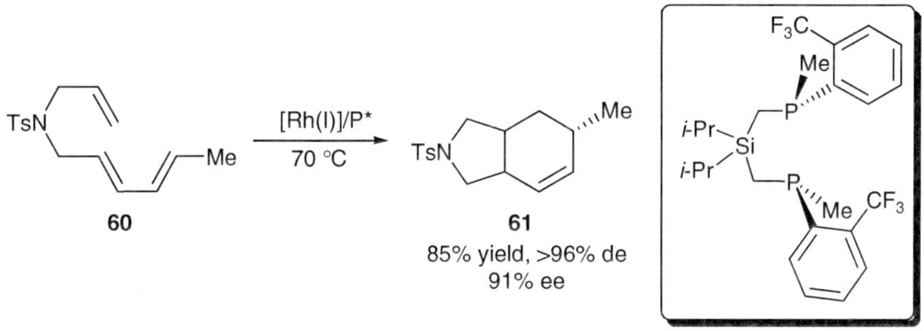

Scheme 8.25 Example of an intramolecular catalytic enantioselective Diels–Alder reaction.

8.10 Cyclopropanation

Several transition metals catalyse the cyclopropanation of alkenes by diazoacetates, through the formation of intermediate metallic carbenes. The model reaction is the cyclopropanation of styrene by ethyl diazoacetate to give the corresponding chiral 2-phenylcyclopropane carboxylates **67** as a Z/E diastereomeric mixture. Very few results with *P*-stereogenic ligands are available; some of them are shown in Scheme 8.27.

In this reaction the thermodynamically favoured enantiomeric pair **Z-67** is preferentially formed. A complication is that ethyl diazoacetate can dimerise to form ethyl fumarate and maleate so the rate of addition of ethyl diazoacetate has to be slow to achieve high conversions.

Very good results have been obtained with iminophospholidine ligand **68** and copper catalysis.[146] In contrast, monophosphine **69**[42] and diphosphine α-bnpe[147] in ruthenium systems led to low activity, diastereo- and enantioselectivity, probably due to catalyst decomposition.

There is one recent example of Pd-catalysed cyclopropanation of allylic carbonates with amides furnishing cyclopropanes with three stereogenic carbon atoms (Scheme 8.28).[148]

This reaction proceeds by nucleophilic attack of the stabilised carbanion of **70** to the central carbon of the π-allylic complex derived from carbonates **71**. It was found that for these substrates SiocPhox ligands favour cyclopropanation over the most common allylic alkylation. After removal of the allylic substitution products and oxidative treatment,[148] cyclopropanes **72** were generally obtained in good yields and excellent diastereo- and enantioselectivities.

8.11 Hydroacylation

Hydroacylation reactions are formal additions of aldehydes across alkeneic or carbonylic bonds. Some *P*-stereogenic ligands have been found to be

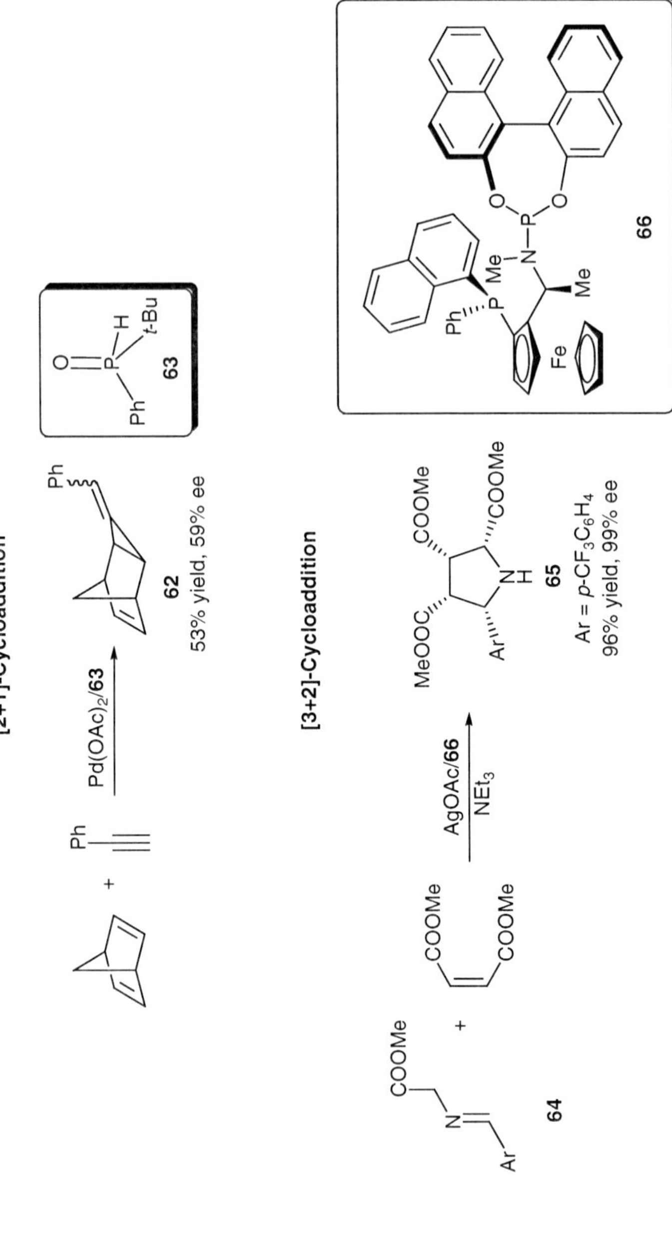

Scheme 8.26 Examples of cycloadditions other than Diels–Alder reactions.

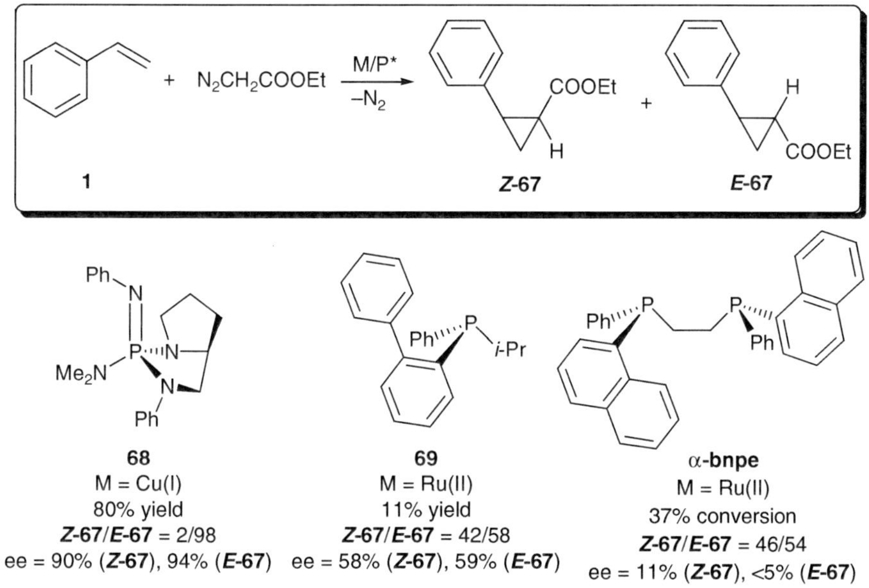

Scheme 8.27 Some results of the model enantioselective cyclopropanation reaction.

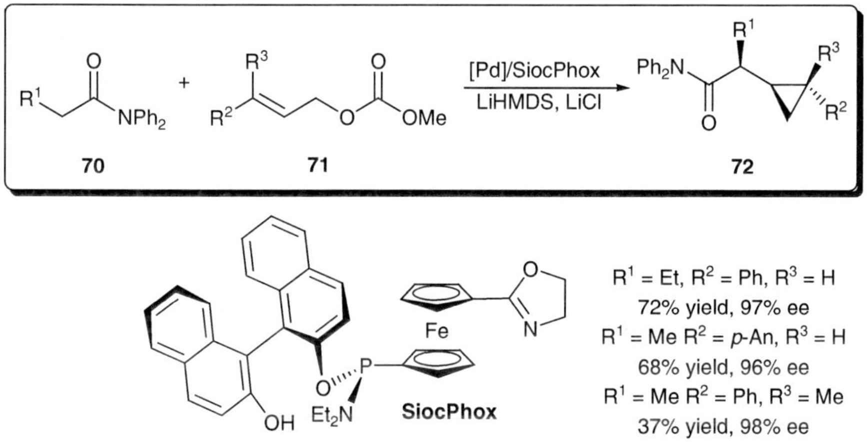

Scheme 8.28 Pd-catalysed cyclopropanations with a SiocPhox ligand.

exceptional ligands for Rh-catalysed hydroacylation of both alkenes and ketones with aldehydes (Scheme 8.29).

Tanaka and Shibata[149] reported the intermolecular hydroacylation of 1,1-disubstituted alkenes (acrylamides) with aldehydes by using a Rh(I)/t-Bu-QuinoxP* complex as catalytic precursor. Good yields and outstanding enantioselectivities in **74** were found for many aliphatic aldehydes. In contrast, benzaldehyde reacted

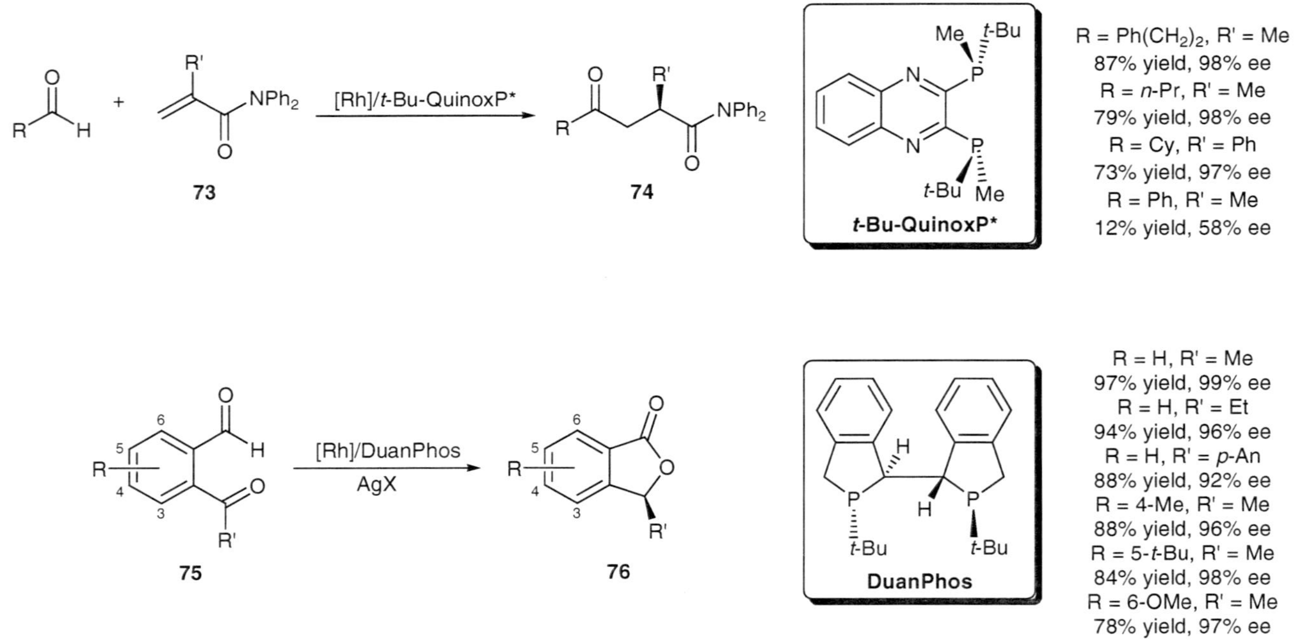

Scheme 8.29 Rh-catalysed enantioselective hydroacylation of acrylamides **73** and ketones **75**.

sluggishly and with low enantioselectivity and pivalaldehyde (R = *t*-Bu) was not reactive at all.

Dong and co-workers[150] reported the synthesis of chiral phthalides (**76**) by intramolecular enantioselective ketone hydroacylation. Excellent yields and exceedingly high optical purities of **76** were found. Good tolerance to substitutions at positions 4, 5 or 6 (but not at 3) of the aromatic ring of **75** was observed. To achieve high enantioselectivities and avoid undesired reactions, an appropriate choice of the silver salt was found to be crucial.

8.12 Arylcyanation

Jacobsen and Watson[151] reported a nickel-catalysed intramolecular addition of arylnitriles across an alkeneic double bond (arylcyanation), giving access to chiral indanes bearing a quaternary stereocentre (**78**, Scheme 8.30).

The catalyst is a Ni(0) species, generated *in situ* by reduction of the Ni(II) salt by metallic zinc. Triphenylborane is thought to coordinate to the cyano group, prompting the oxidative addition of the C–CN bond to the Ni catalyst. Subsequent migratory insertion of the alkene into the Ni–CN bond and reductive elimination complete the cycle. Several ligands, including *P*-stereogenic phosphines (*t*-BuBisP*, *t*-Bu-MiniPHOS and others) were tested but only TangPhos was found to deliver **78** in good yield and excellent enantioselectivity.

8.13 Arylation of Enolates

Buchwald and co-workers[152,153] studied the α-arylation and α-vinylation of enolates derived from ketones with a binaphthyl monophosphine bearing a *P*-stereogenic atom (**80**, Scheme 8.31).

The best results were obtained by coupling the enolate of oxindole **79** with aryl and vinyl bromides.[153] Arylated products **81**, bearing an all-carbon quaternary stereocentre, were formed in good yield and excellent enantiomeric purity for several aryl and vinyl bromides. Interestingly, other binaphthyl-based phosphorus ligands without *P*-stereogenic centres led to inferior results.

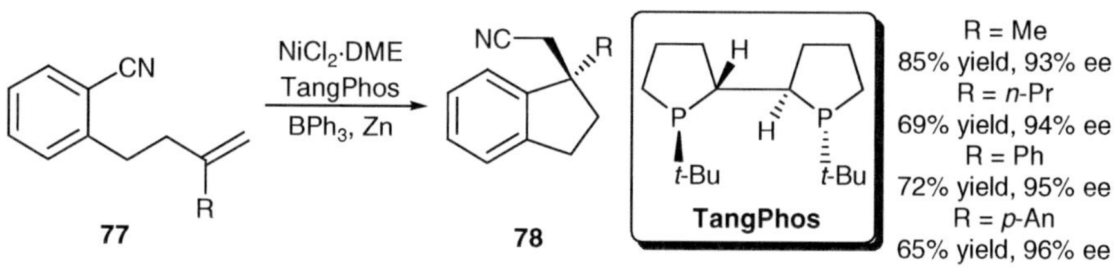

Scheme 8.30 Ni-catalysed intramolecular enantioselective arylcyanation.

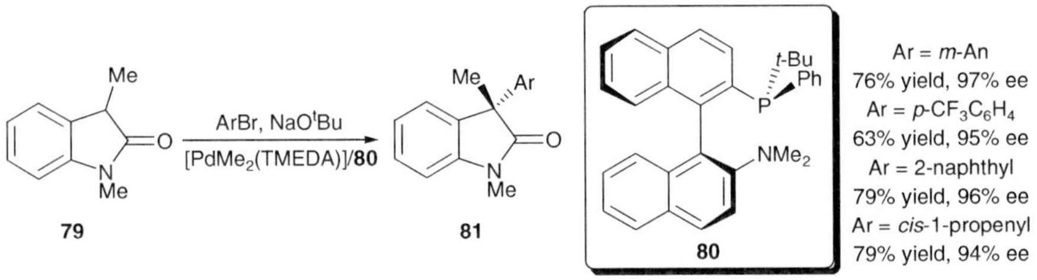

Me
ArBr, NaO^{t}Bu
[PdMe$_2$(TMEDA)]/80
N
Me
79
Me
Ar
81
t-Bu
Ph
P
NMe$_2$
80
Ar = m-An
76% yield, 97% ee
Ar = p-CF$_3$C$_6$H$_4$
63% yield, 95% ee
Ar = 2-naphthyl
79% yield, 96% ee
Ar = cis-1-propenyl
79% yield, 94% ee

Scheme 8.31 Pd-catalysed arylation of enolates derived from **79**.

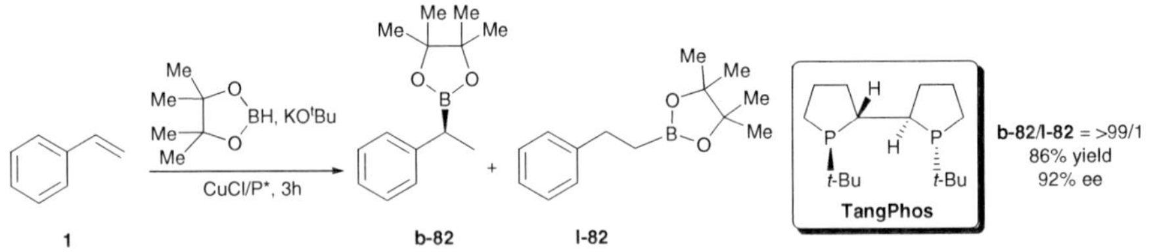

BH, KO^{t}Bu
CuCl/P*, 3h
1
b-82
l-82
TangPhos
t-Bu
b-82/l-82 = >99/1
86% yield
92% ee

Scheme 8.32 Copper-catalysed enantioselective hydroboration of styrene.

LiO^{t}Bu, DMSO
[Cu(I)]/t-BuQuinoxP*
83
84
85
t-Bu-QuinoxP*
n = 1, R = Me
91% yield, 81% ee
n = 1, R = i-Pr
91% yield, 94% ee
n = 1, R = Ph
88% yield, 98% ee (R)
n = 1, R = m-Tol
89% yield, 98% ee
n = 1, R = p-An
84% yield, 93% ee
n = 0, R = Ph
85% yield, 98% ee
n = 2, R = Ph
99% yield, 98% ee

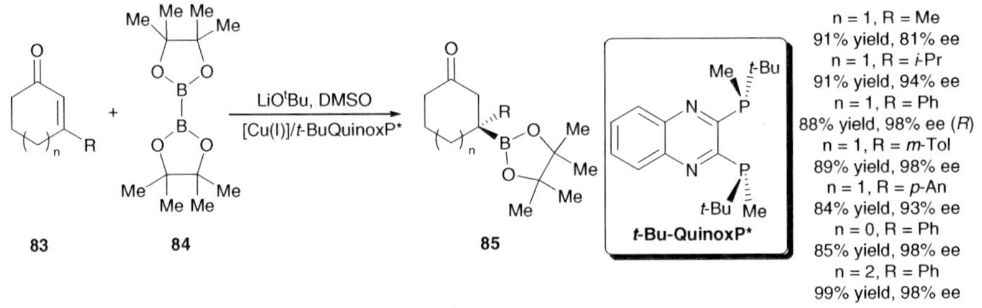

Scheme 8.33 Copper-catalysed enantioselective conjugate boration.

8.14 Hydroboration

The hydroboration of alkenes is the addition of a borane across an alkeneic double bond. It is a very versatile reaction because the C–B bond can be cleaved under a variety of conditions to afford many products.

Yun and co-workers[154] reported the Cu-catalysed enantioselective hydroboration of styrene with pinacolborane using the ligand TangPhos (Scheme 8.32).

The reaction in the presence of a catalytic amount of TangPhos afforded **b-82** in high regio- and enantioselectivity. Many other substrates, including examples of β-substituted (*E*)-styrenes, also provided very good results. In contrast, (*Z*)-alkenes such as indene, were not reactive.

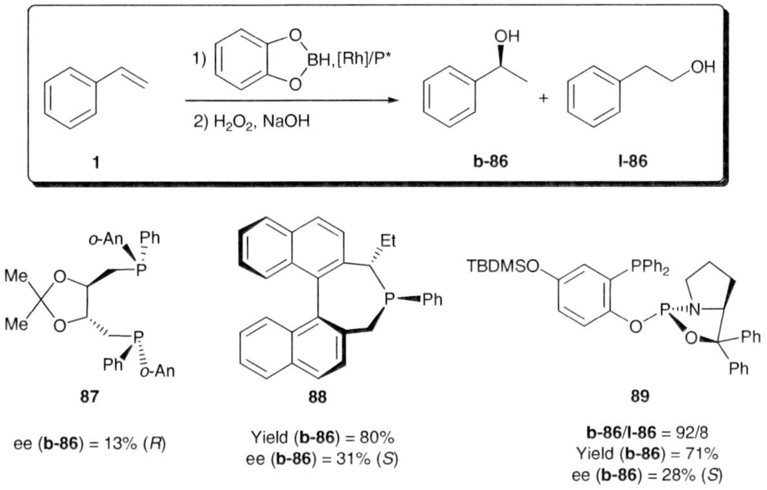

Scheme 8.34   Model reaction for enantioselective hydroboration–oxidation.

Shibasaki and co-workers[155] described a Cu-catalysed conjugate boration of β-substituted α,β-unsaturated cyclic ketones[156] with *bis*(pinacolato)diboron (Scheme 8.33).

Excellent results of **85** for yield and enantioselectivity were found. The reaction is rather insensitive to the nature of the R group since both akyl- and aryl-substituted substrates are well tolerated. Even more interesting is the fact that equally good results were obtained for 5-, 6- or 7-membered substrates.

In spite of these results, the intermediate boranes are not usually isolated, but are transformed into more stable compounds. The most common reaction sequence is hydroboration then oxidation, which formally adds water across a double bond of an alkene, giving an alcohol.[157,158] The model reaction for the enantioselective version is the hydroboration of styrene by catecholborane followed by oxidation of the intermediate boranes by hydrogen peroxide (Scheme 8.34).

Very few reports of this reaction with *P*-stereogenic ligands have been published. An early study with DIOP-DiPAMP hybrids **87** led to very poor enantioselectivities,[159] which have been slightly improved more recently by using ligands **88**[160] and **89**.[161]

References

1. L. J. Goossen, *Angew. Chem. Int. Ed.*, 2002, **41**, 3775.
2. T. Alderson, E. L. Jenner and R. V. Lindsey, *J. Am. Chem. Soc.*, 1965, **87**, 5638.
3. B. Bogdanovic, B. Henc, B. Meister, H. Pauling and G. Wilke, *Angew. Chem. Int. Ed. Engl.*, 1972, **11**, 1023.

4. T. V. RajanBabu, *Chem. Rev.*, 2003, **103**, 2845.
5. T. V. RajanBabu, *Synlett*, 2009, 853.
6. G. Wilke and J. Monkiewicz, *Chem. Abstr.*, 1988, **109**, P6735.
7. G. Wilke, *Angew. Chem. Int. Ed. Engl.*, 1988, **27**, 185.
8. A. Wegner and W. Leitner, *J. Chem. Soc., Chem. Commun.*, 1999, **16**, 1583.
9. A. Bösmann, G. Franciò, E. Janssen, M. Solinas, W. Leitner and P. Wasserscheid, *Angew. Chem. Int. Ed. Engl.*, 2001, **40**, 2697.
10. K. Barta, M. Holscher, G. Franciò and W. Leitner, *Eur. J. Org. Chem.*, 2009, 4102.
11. R. Bayersdörfer, B. Ganter, U. Englert, W. Keim and D. Vogt, *J. Organomet. Chem.*, 1998, **552**, 187.
12. J. Albert, J. M. Cadena, J. R. Granell, G. Muller, J. I. Ordinas, D. Panyella, C. Puerta, C. Sañudo and P. Valerga, *Organometallics*, 1999, **18**, 3511.
13. J. Albert, R. Bosque, J. Magali Cadena, S. Delgado, J. R. Granell, G. Muller, J. I. Ordinas, M. Font Bardia and X. Solans, *Chem. Eur. J.*, 2002, **8**, 2279.
14. A. Grabulosa, G. Muller, J. I. Ordinas, A. Mezzetti, M. A. Maestro, M. Font-Bardia and X. Solans, *Organometallics*, 2005, **24**, 4961.
15. L. Rodríguez, O. Rossell, M. Seco, A. Grabulosa, G. Muller and M. Rocamora, *Organometallics*, 2006, **25**, 1368.
16. L. Rodríguez, O. Rossell, M. Seco and G. Muller, *Organometallics*, 2008, **27**, 1328.
17. L. Rodríguez, O. Rossell, M. Seco, A. Orejon and A. M. Masdéu-Bultó, *J. Organomet. Chem.*, 2008, **693**, 1857.
18. M. M. P. Grutters, C. Müller and D. Vogt, *J. Am. Chem. Soc.*, 2006, **128**, 7414.
19. F. Agbossou, J. Carpentier and A. Montreux, *Chem. Rev.*, 1995, **95**, 2485.
20. S. Gladiali, J. C. Bayón and C. Claver, *Tetrahedron: Asymmetry*, 1995, **6**, 1453.
21. U. Nettekoven, P. C. J. Kamer and P. W. N. M. van Leeuwen, *Organometallics*, 2000, **19**, 4596.
22. S. Deerenberg, P. C. J. Kamer and P. W. N. M. van Leeuwen, *Organometallics*, 2000, **19**, 2065.
23. A. T. Axtell, J. Klosin and K. A. Abboud, *Organometallics*, 2006, **25**, 5003.
24. J. Klosin and C. R. Landis, *Acc. Chem. Res.*, 2007, **40**, 1251.
25. I. Ogata and Y. Ikeda, *Chem. Lett.*, 1972, 487.
26. R. Ewalds, E. B. Eggeling, A. C. Hewat, P. C. J. Kamer, P. W. N. M. van Leeuwen and D. Vogt, *Chem. Eur. J.*, 2000, **6**, 1496.
27. S. Breeden, D. J. Cole-Hamilton, D. F. Foster, G. J. Schwarz and M. Wills, *Angew. Chem. Int. Ed.*, 2000, **39**, 4106.
28. G. J. Clarkson, J. R. Ansell, D. J. Cole-Hamilton, P. J. Pogorzelec, J. Whittell and M. Wills, *Tetrahedron: Asymmetry*, 2004, **15**, 1787.

29. M. Rubio, A. Suarez, E. Alvarez, C. Bianchini, W. Oberhauser, M. Peruzzini and A. Pizzano, *Organometallics*, 2007, **26**, 6428.

30. Y. Becker, A. Eisenstadt and J. K. Stille, *J. Org. Chem.*, 1980, **45**, 2145.

31. J. J. Carbó, A. Lledós, D. Vogt and C. Bo, *Chem. Eur. J.*, 2006, **12**, 1457.

32. N. R. Vautravers and D. J. Cole-Hamilton, *Chem. Commun.*, 2009, 92.

33. N. R. Vautravers, P. Andre and D. J. Cole-Hamilton, *Dalton Trans.*, 2009, 3413.

34. J. Huang, E. Bunel, A. Allgeier, J. Tedow, T. Storz, S. J. Preston, T. Correll, D. Manley, T. Soukup, R. Jensen, R. Syed, G. Moniz, R. Larsen, M. Martinelli and P. J. Reider, *Tetrahedron Lett.*, 2005, **46**, 7831.

35. H. Ito, S. Ito, Y. Sasaki, K. Matsuura and M. Sawamura, *J. Am. Chem. Soc.*, 2007, **129**, 14856.

36. M. Kimura and Y. Uozumi, *J. Org. Chem.*, 2007, **72**, 707.

37. Z. Lu and S. Ma, *Angew. Chem. Int. Ed.*, 2008, **47**, 258.

38. B. M. Trost and D. L. van Vranken, *Chem. Rev.*, 1996, **96**, 395.

39. B. M. Trost and M. L. Crawley, *Chem. Rev.*, 2003, **103**, 2921.

40. P. R. Auburn, P. B. Mackenzie and B. Bosnich, *J. Am. Chem. Soc.*, 1985, **107**, 2033.

41. H. Tsuruta and T. Imamoto, *Synlett*, 2001, 999.

42. A. Grabulosa, PhD Thesis, Universitat de Barcelona, Barcelona, 2005.

43. V. N. Tsarev, S. E. Lyubimov, A. A. Shiryaev, S. V. Zheglov, O. G. Bondarev, V. A. Davankov, A. A. Kabro, S. K. Moiseev, V. N. Kalinin and K. N. Gavrilov, *Eur. J. Org. Chem.*, 2004, 2214.

44. K. N. Gavrilov, E. B. Benetsky, T. B. Grishina, S. V. Zheglov, E. A. Rastorguev, P. V. Petrovskii, F. Z. Macaev and V. A. Davankov, *Tetrahedron: Asymmetry*, 2007, **18**, 2557.

45. S. E. Lyubimov, I. V. Kuchurov, A. A. Vasil'ev, A. A. Tyutyunov, V. N. Kalinin, V. A. Davankov and S. G. Zlotin, *J. Organomet. Chem.*, 2009, **694**, 3047.

46. K. N. Gavrilov, S. E. Lyubimov, S. V. Zheglov, E. B. Benetsky, P. V. Petrovskii, E. A. Rastorguev, T. B. Grishina and V. A. Davankov, *Adv. Synth. Catal.*, 2007, **349**, 1085.

47. V. N. Tsarev, S. E. Lyubimov, O. G. Bondarev, A. A. Korlyukov, M. Y. Antipin, P. V. Petrovskii, V. A. Davankov, A. A. Shiryaev, E. B. Benetsky, P. A. Vologzhanin and K. N. Gavrilov, *Eur. J. Org. Chem.*, 2005, 2097.

48. J. M. Brunel, T. Constantieux, A. Labande, F. Lubatti and G. Buono, *Tetrahedron Lett.*, 1997, **38**, 5971.

49. J. M. Brunel, T. Constantieux and G. Buono, *J. Org. Chem.*, 1999, **64**, 8940.

50. G. Delapierre, J. M. Brunel, T. Constantieux and G. Buono, *Tetrahedron: Asymmetry*, 2001, **12**, 1345.

51. H. Danjo, M. Higuchi, M. Yada and T. Imamoto, *Tetrahedron Lett.*, 2004, **45**, 603.

52. X. Sun, M. Koizumi, K. Manabe and S. Kobayashi, *Adv. Synth. Catal.*, 2005, **347**, 1893.

53. S. R. Gilberston, D. G. Genov and A. L. Rheingold, *Org. Lett.*, 2000, **2**, 2885.

54. H. Sugama, H. Saito, H. Danjo and T. Imamoto, *Synthesis*, 2001, **15**, 2348.

55. N. Toselli, D. Martin and G. Buono, *Org. Lett.*, 2008, **10**, 1453.

56. N. Oohara, K. Katagiri and T. Imamoto, *Tetrahedron: Asymmetry*, 2003, **14**, 2171.

57. T. Imamoto, M. Nishimura, A. Koide and K. Yoshida, *J. Org. Chem.*, 2007, **72**, 7413.

58. U. Nettekoven, M. Widhalm, P. C. J. Kamer and P. W. N. M. van Leeuwen, *Tetrahedron: Asymmetry*, 1997, **8**, 3185.

59. U. Nettekoven, M. Widhalm, H. Kalchhauser, P. C. J. Kamer, P. W. N. M. van Leeuwen, M. Lutz and A. L. Spek, *J. Org. Chem.*, 2001, **66**, 759.

60. Y. Hamada, F. Matsuruta, M. Oku, K. Hatano and T. Shiori, *Tetrahedron Lett.*, 1997, **38**, 8961.

61. M. Siutkowski, F. Mercier, L. Ricard and F. Mathey, *Organometallics*, 2006, **25**, 2585.

62. Y. Yan and M. Widhalm, *Tetrahedron: Asymmetry*, 1998, **9**, 3607.

63. S. Breeden and M. Wills, *J. Org. Chem.*, 1999, **64**, 9735.

64. K. N. Gavrilov, S. V. Zheglov, E. B. Benetsky, A. S. Safronov, E. A. Rastorguev, N. N. Groshkin, V. A. Davankov, B. Schaffner and A. Börner, *Tetrahedron: Asymmetry*, 2009, 20.

65. K. N. Gavrilov, S. V. Zheglov, P. A. Vologzhanin, M. G. Maksimova, A. S. Safronov, S. E. Lyubimov, V. A. Davankov, B. Schaffner and A. Börner, *Tetrahedron Lett.*, 2008, **49**, 3120.

66. T. Nemoto, T. Masuda, T. Matsumoto and Y. Hamada, *J. Org. Chem.*, 2005, **70**, 7172.

67. T. D. Weiss, G. Helmchen and U. Kazmaier, *Chem. Commun.*, 2002, 1270.

68. R. Hilgraf and A. Pfaltz, *Adv. Synth. Catal.*, 2005, **347**, 61.

69. P. Sennhenn, B. Gabler and G. Helmchen, *Tetrahedron Lett.*, 1994, **35**, 8595.

70. H. Rieck and G. Helmchen, *Angew. Chem. Int. Ed. Engl.*, 1995, **34** **2687**.

71. G. Helmchen, S. Kudis, P. Sennhenn and H. Steinhagen, *Pure Appl. Chem.*, 1997, **69**, 513.

72. G. Helmchen, *J. Organomet. Chem.*, 1999, **576**, 203.

73. G. Helmchen and A. Pfaltz, *Acc. Chem. Res.*, 2000, **33**, 336.

74. S. Kudis and G. Helmchen, *Angew. Chem. Int. Ed.*, 1998, **37**, 3047.

75. S. Schleich and G. Helmchen, *Eur. J. Org. Chem.*, 1999, 2515.

76. S. Kudis and G. Helmchen, *Tetrahedron*, 1998, **54**, 10449.

77. M. Ernst and G. Helmchen, *Angew. Chem. Int. Ed.*, 2002, **41**, 4054.

78. E. J. Bergner and G. Helmchen, *Eur. J. Org. Chem.*, 2000, 419.

79. T. Nemoto, T. Sakamoto, T. Fukuyama and Y. Hamada, *Tetrahedron Lett.*, 2007, **48**, 4977.
80. S. You, X. Zhu, Y. Luo, X. Hou and L. Dai, *J. Am. Chem. Soc.*, 2001, **123**, 7471.
81. L. Dai, T. Tu, S. You, W. Deng and X. Hou, *Acc. Chem. Res.*, 2003, **36**, 659.
82. W. Zheng, N. Sun and X. Hou, *Org. Lett.*, 2005, **7**, 5151.
83. W. Zheng, B. Zheng, Y. Zhang and X. Hou, *J. Am. Chem. Soc.*, 2007, **129**, 7718.
84. B. Lei, C. Ding, X. Yang, X. Wan and X. Hou, *J. Am. Chem. Soc.*, 2009, **131**, 18250.
85. X. Hou and N. Sun, *Org. Lett.*, 2004, **6**, 4399.
86. J. M. Brunel, A. Tenaglia and G. Buono, *Tetrahedron: Asymmetry*, 2000, **11**, 3585.
87. T. Nemoto, T. Matsumoto, T. Masuda, T. Hitomi, K. Hatano and Y. Hamada, *J. Am. Chem. Soc.*, 2004, **126**, 3690.
88. T. Nemoto, T. Fukuda, T. Matsumoto, T. Hitomi and Y. Hamada, *Adv. Synth. Catal.*, 2005, **347**, 1504.
89. T. Nemoto and T. Hamada, *Chem. Rec.*, 2007, **7**, 150.
90. T. Nemoto, *Chem. Pharm. Bull.*, 2008, **56**, 1213.
91. T. Nemoto, L. Jin, H. Nakamura and Y. Hamada, *Tetrahedron Lett.*, 2006, **47**, 6577.
92. T. Nemoto, T. Harada, T. Matsumoto and Y. Hamada, *Tetrahedron Lett.*, 2007, **48**, 6304.
93. L. Jin, T. Nemoto, H. Nakamura and Y. Hamada, *Tetrahedron: Asymmetry*, 2008, **19**, 1106.
94. T. Nemoto, M. Kanematsu, S. Tamura and Y. Hamada, *Adv. Synth. Catal.*, 2009, **351**, 1773.
95. T. Nemoto, E. Yamamoto, R. Franzén, T. Fukuyama, R. Wu, T. Fukamachi, H. Kobayashi and Y. Hamada, *Org. Lett.*, 2010, **12**, 872.
96. C. W. Edwards, M. R. Shipton, N. W. Alock, H. Clase and M. Wills, *Tetrahedron*, 2003, **59**, 6473.
97. K. N. Gavrilov, E. B. Benetskiy, T. B. Grishina, E. A. Rastorguev, M. G. Maksimova, S. V. Zheglov, V. A. Davankov, B. Schaffner, A. Börner, J. M. Rosset, G. Bailat and A. Alexakis, *Eur. J. Org. Chem.*, 2009, 3923.
98. E. B. Benetskii, V. A. Davankov, P. V. Petrovskii, E. A. Rastorguev, T. B. Grishina, K. N. Gavrilov, S. Rosset, G. Bailat and A. Alexakis, *Russ. J. Org. Chem.*, 2008, **44**, 1846.
99. K. N. Gavrilov, S. E. Lyubimov, O. G. Bondarev, M. G. Maksimova, S. V. Zheglov, P. V. Petrovskii, V. A. Davankov and M. T. Reetz, *Adv. Synth. Catal.*, 2007, **349**, 609.
100. T. Constantieux, J. Brunel, A. Labande and G. Buono, *Synlett*, 1998, 49.
101. F. Mercier, F. Brebion, R. Dupont and F. Mathey, *Tetrahedron: Asymmetry*, 2003, **14**, 3137.

102. V. N. Tsarev, S. I. Konkin, A. A. Shyryaev, V. A. Davankov and K. N. Gavrilov, *Tetrahedron: Asymmetry*, 2005, **16**, 1737.
103. T. Nemoto, T. Masuda, Y. Akimoto, T. Fukuyama and Y. Hamada, *Org. Lett.*, 2005, **7**, 4447.
104. Z. Pakulski, O. M. Demchuk, J. Frelek, R. Luboradzki and K. M. Pietrusiewicz, *Eur. J. Org. Chem.*, 2004, 3913.
105. T. Nemoto, T. Fukuyama, E. Yamamoto, S. Tamura, T. Fukuda, T. Matsumoto, Y. Akimoto and Y. Hamada, *Org. Lett.*, 2007, **9**, 927.
106. T. Nemoto, T. Sakamoto, T. Matsumoto and Y. Hamada, *Tetrahedron Lett.*, 2006, **47**, 8737.
107. K. N. Gavrilov, O. G. Bondarev, V. N. Tsarev, A. A. Shiryaev, S. E. Lyubimov, A. S. Kucherenko and V. A. Davankov, *Russ. Chem. Bull. Int. Ed.*, 2003, **52**, 122.
108. K. N. Gavrilov, V. N. Tsarev, S. I. Konkin, N. M. Loim, P. V. Petrovskii, E. S. Kelbyscheva, A. A. Korlyukov, M. Y. Antipin and V. A. Davankov, *Tetrahedron: Asymmetry*, 2005, **16**, 3224.
109. E. A. Colby and T. F. Jamison, *J. Org. Chem.*, 2003, **68**, 156.
110. K. M. Miller, C. Molinaro and T. F. Jamison, *Tetrahedron: Asymmetry*, 2003, **14**, 3619.
111. J. Chan and T. F. Jamison, *J. Am. Chem. Soc.*, 2004, **126**, 10682.
112. E. A. Colby, K. C. O'Brien and T. F. Jamison, *J. Am. Chem. Soc.*, 2005, **127**, 4297.
113. J. Chan and T. F. Jamison, *J. Am. Chem. Soc.*, 2003, **125**, 11514.
114. K. M. Miller and T. F. Jamison, *Org. Lett.*, 2005, **7**, 3077.
115. S. J. Patel and T. F. Jamison, *Angew. Chem. Int. Ed.*, 2004, **43**, 3941.
116. J. Christoffers, G. Koripelly, A. Rosiak and M. Rössle, *Synthesis*, 2007, 1279.
117. B. E. Rossiter and N. M. Swingle, *Chem. Rev.*, 1992, **92**, 771.
118. B. L. Feringa, *Acc. Chem. Res.*, 2000, **33**, 346.
119. A. Alexakis and C. Benhaim, *Eur. J. Org. Chem.*, 2002, 3221.
120. T. Pfretzschner, L. Kleemann, B. Janza, K. Harms and T. Schrader, *Chem. Eur. J.*, 2004, **10**, 6048.
121. A. Alexakis, J. Frutos and P. Mangeney, *Tetrahedron: Asymmetry*, 1993, **4**, 2427.
122. G. Delapierre, T. Constantieux, J. M. Brunel and G. Buono, *Eur. J. Org. Chem.*, 2000, 2507.
123. Y. Takahashi, Y. Yamamoto, K. Katagiri, H. Danjo, K. Yamaguchi and T. Imamoto, *J. Org. Chem.*, 2005, **70**, 9009.
124. M. Diéguez, S. Deerenberg, O. Pàmies, C. Claver, P. W. N. M. van Leeuwen and P. C. J. Kamer, *Tetrahedron: Asymmetry*, 2000, **11**, 3161.
125. A. Alexakis, J. Burton, J. Vastra and P. Mangeney, *Tetrahedron: Asymmetry*, 1997, **8**, 3987.
126. Y. Yamanoi and T. Imamoto, *J. Org. Chem.*, 1999, **64**, 2988.
127. S. Taira, K. V. L. Crépy and T. Imamoto, *Chirality*, 2002, **14**, 386.
128. T. Hayashi and K. Yamasaki, *Chem. Rev.*, 2003, **103**, 2829.
129. T. Hayashi, *Pure Appl. Chem.*, 2004, **76**, 465.

130. K. Fagnou and M. Lautens, *Chem. Rev.*, 2003, **103**, 169.

131. P. Kasák, V. B. Arion and M. Widhalm, *Tetrahedron: Asymmetry*, 2006, **17**, 3084.

132. T. Imamoto, K. Sugito and K. Yoshida, *J. Am. Chem. Soc.*, 2005, **127**, 11934.

133. T. Imamoto, Y. Saitoh, A. Koide, T. Ogura and K. Yoshida, *Angew. Chem. Int. Ed.*, 2007, **46**, 8636.

134. J. L. Zigterman, J. C. S. Woo, S. D. Walker, J. S. Tedrow, C. J. Borths, E. E. Bunel and M. M. Faul, *J. Org. Chem.*, 2007, **72**, 8870.

135. M. Lautens, K. Fagnou and S. Hiebert, *Acc. Chem. Res.*, 2003, **36**, 48.

136. T. Imamoto, A. Kumada and K. Yoshida, *Chem. Lett.*, 2007, **36**, 500.

137. T. Ogura, K. Yoshida, A. Yanagisawa and T. Imamoto, *Org. Lett.*, 2009, **11**, 2245.

138. K. Yoshida, T. Toyoshima, N. Akashi, T. Imamoto and A. Yanagisawa, *Chem. Commun.*, 2009, 2923.

139. A. B. Dounay and L. E. Overman, *Chem. Rev.*, 2003, **103**, 2945.

140. M. Shibasaki, E. M. Vogl and T. Ohshima, *Adv. Synth. Catal.*, 2004, **346**, 1533.

141. J. M. Brunel, B. Del Campo and G. Buono, *Tetrahedron Lett.*, 1998, **39**, 9663.

142. S. Matsukawa, H. Sugama and T. Imamoto, *Tetrahedron Lett.*, 2000, **41**, 6461.

143. H. Heath, B. Wolfe, T. Livinghouse and S. K. Bae, *Synthesis*, 2001, 2341.

144. J. Bigeault, L. Giordano and G. Buono, *Angew. Chem. Int. Ed.*, 2005, **44**, 4753.

145. S. Yu, X. Hu, J. Deng, D. Wang, Z. Duan and Z. Zheng, *Tetrahedron: Asymmetry*, 2009, **20**, 621.

146. J. M. Brunel, O. Legrand, S. Reymond and G. Buono, *J. Am. Chem. Soc.*, 1999, **121**, 5807.

147. R. M. Stoop, C. Bauer, P. Setz, M. Wörle, T. Y. H. Wong and A. Mezzetti, *Organometallics*, 1999, **18**, 5691.

148. W. Liu, D. Chen, X. Zhu, X. Wan and X. Hou, *J. Am. Chem. Soc.*, 2009, **131**, 8734.

149. Y. Shibata and K. Tanaka, *J. Am. Chem. Soc.*, 2009, **131**, 12552.

150. D. H. T. Phan, B. Kim and V. M. Dong, *J. Am. Chem. Soc.*, 2009, **131**, 15608.

151. M. P. Watson and E. N. Jacobsen, *J. Am. Chem. Soc.*, 2008, **130**, 12594.

152. T. Hamada and S. L. Buchwald, *Org. Lett.*, 2002, **4**, 999.

153. A. M. Taylor, R. A. Altman and S. L. Buchwald, *J. Am. Chem. Soc.*, 2009, **131**, 9900.

154. D. Noh, H. Chea, J. Ju and J. Yun, *Angew. Chem. Int. Ed.*, 2009, **48**, 6062.

155. I. Chen, L. Yin, W. Itano, M. Kanai and M. Shibasaki, *J. Am. Chem. Soc.*, 2009, **131**, 11664.

156. L. Mantilli and C. Mazet, *ChemCatChem*, 2010, **2**, 501.

157. C. M. Crudden and D. Edwards, *Eur. J. Org. Chem.*, 2003, 4695.
158. A. Carroll, T. P. O'Sullivan and P. J. Guiry, *Adv. Synth. Catal.*, 2005, **347**, 609.
159. K. Burgess, M. J. Ohlmeyer and K. H. Whitmire, *Organometallics*, 1992, **11**, 3588.
160. P. Kasák, K. Mereiter and M. Widhalm, *Tetrahedron: Asymmetry*, 2005, **16**, 3416.
161. F. Blume, S. Zemolka, T. Fey, R. Kranich and H. Schmalz, *Adv. Synth. Catal.*, 2002, **344**, 868.

Subject Index